AF327507

Retinoid Protocols

METHODS IN MOLECULAR BIOLOGY™

John M. Walker, SERIES EDITOR

METHODS IN MOLECULAR BIOLOGY™

Retinoid Protocols

Edited by

Christopher P. F. Redfern

University of Newcastle Upon Tyne, UK

Humana Press ✳ Totowa, New Jersey

© 1998 Humana Press Inc.
999 Riverview Drive, Suite 208
Totowa, New Jersey 07512

This publication is printed on acid-free paper. ∞
ANSI Z39.48-1984 (American Standards Institute) Permanence of Paper for Printed Library Materials.

Cover illustration: Fig. 1 from Chapter 3, "Detection and Measurement of Retinoic Acid Production by Isolated Tissues Using Retinoic Acid-Sensitive Reporter Cell Lines," by Michael Wagner.

Cover design by Patricia F. Cleary.

For additional copies, pricing for bulk purchases, and/or information about other Humana titles, contact Humana at the above address or at any of the following numbers: Tel: 973-256-1699; Fax: 973-256-8341; E-mail: humana@mindspring.com or visit our website at http://www.humanapress.com

Printed in the United States of America. 10 9 8 7 6 5 4 3 2 1

Library of Congress Cataloging in Publication Data

Main entry under title:

Methods in molecular biology™.

Retinoid protocols/edited by Christopher P. F. Redfern.
    p.   cm.—(Methods in molecular biology; vol. 89)
  Includes index.
  ISBN 0-89603-438-0
  1. Retinoids—Laboratory manuals. 2. Retinoids—Analysis—Laboratory manuals. I. Redfern, Christopher P. F. II. Series: Methods in molecular biology (Totowa, NJ); 89.
QP801.R47R46   1998
612.3'99—dc21   97-44455
                                                                        CIP

Preface

Interest in retinoic acid, the main biologically active derivative of vitamin A or retinol, increased dramatically between 1989 and 1993, following the cloning of nuclear receptors or RARs reported in 1987 (**Fig. 1**). Important discoveries since then have shown how RARs work as all-*trans* retinoic acid-dependent heterodimers with related nuclear receptors for 9-*cis* retinoic acid called RXRs. This has stimulated the development of synthetic analogs specific for each type of receptor, and opens the way to develop new methods for regulating pharmacologically the activity of retinoic acid-dependent pathways of gene activation. The potential for the development of new drugs by the pharmaceutical industry is now a major factor driving forward our understanding of vitamin A-regulated pathways in animal development and homeostasis. However, apart from the real potential of retinoid analogs as novel pharmacological agents, there remains the considerable intellectual challenge of understanding the way in which vitamin A and its derivatives function in cell development and differentiation. *Retinoid Protocols* is an attempt to bring together various methodologies that will be vital for rising to this challenge in the future.

Retinoid molecular biology has few methods of its own, but is reliant on standard molecular biology methods applied to this particular research area. Researchers using this book will thus find methods for Northern blotting, reverse transcriptase-polymerase chain reaction (RT-PCR), *in situ* hybridization (ISH), gel-shift analysis, differential display, two-hybrid screening, and gene targeting included in the specific contexts of studying and cloning genes critically involved in retinoid-signaling pathways. I have aimed to bring together methods that are, in my view, fundamental to retinoid research and that will lead to important advances in the future. The scope of the book is deliberately wider than "conventional" molecular biology because molecular biologists studying retinoid receptors and binding proteins need to understand the properties of the ligand they are working with. For example, when cells are incubated with retinoic acid, it is important to know whether or not there is significant metabolism or isomerization, since this could affect the interpretation of experimental results. Thus, the first section of the book, Part I, contains chapters describing the properties and handling of retinoids, and methods

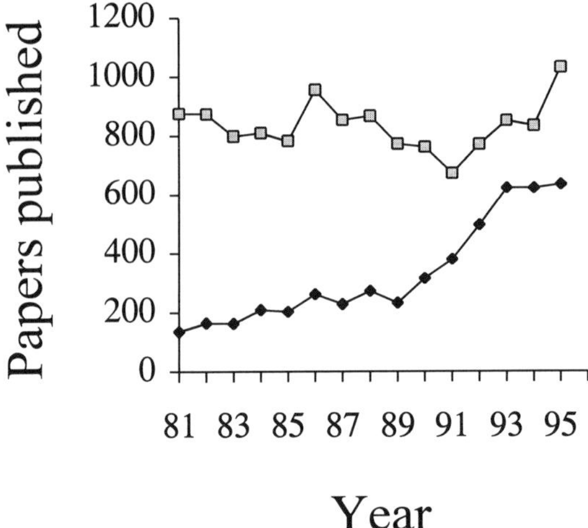

Year

Fig. 1. The number of papers in the Science Citation Index published each year from 1981 to 1995 that included the words "retinoic acid" (♦) or "estrogen" (□) in the title.

for the quantitative analysis of retinoids in biological samples. These two chapters are complemented by a third chapter, describing an elegant bioassay in which stably transfected reporter cells are used to measure biologically active retinoids released from tissues, and which has been adapted for measuring small amounts of biologically active retinoids in tissue and cell extracts.

At least some of the activities of nuclear retinoid receptors are dependent on the presence of bound ligand. The next section of the book, Part II, therefore covers the extracellular and cytosolic retinoid-binding proteins that regulate intracellular concentrations of retinol and its derivatives. The need to purify binding proteins from animal tissues to study the retinoid–protein and protein–protein interactions central to regulating the "flow" of ligands within the cell has now been largely superseded by methods for the efficient purification of recombinant-binding proteins from bacterial hosts. Chapters on cloning via RT-PCR, recombinant protein purification, and characterization in Part II are complemented by others describing the measurement of dissociation rates from interphotoreceptor matrix retinoid-binding protein (IRBP), the use of antisense oligonucleotides to study the role of CRABPs, and methods for studying conformational changes in these proteins. Fluorimetry is an increasingly valuable tool for understanding retinoid-binding protein interactions, and recent substantial improvements in the design and analysis of fluorescent detection experiments are described in this section of the book. Though the binding of retinoid ligands within a hydrophobic pocket formed by antiparallel β-strands is a common feature to many cytosolic and extracellular binding proteins, there is a greater diversity of proteins and protein function within this group than is apparent among nuclear retinoid receptors. This has made it difficult to include specific methods for working with all types of retinoid-binding proteins, but I hope that the methods described may prove to be adequate models for future research and development.

Although the nuclear retinoid receptors are structurally and functionally less diverse than the retinoid-binding proteins, Part III includes a wide variety of techniques for analyzing retinoid receptor expression, structure, and function, ranging from immunological and ISH methods to the analysis of receptor phosphorylation, photoaffinity labeling, gel-shift analysis, differential display, gene targeting, and two-hybrid analysis. ISH methods for mRNA are included in Parts II and III, but each of the three ISH chapters describes a different approach: whole-mount ISH with nonisotopic probes is treated in the context of cellular retinoic acid-binding protein in Part II, whereas ISH using isotopic and nonisotopic probes in tissue sections and cultured cells is described with respect to nuclear receptor mRNAs in Part III. As with the cytosolic-binding proteins, RT-PCR is an important method for cloning receptor isoforms, and for quantitative and qualitative analyses of their expression. Thus, Part III contains a chapter giving detailed methodology and primer combinations for studying N-terminal RAR isoforms. In addition, RT-PCR has become an important tool for the diagnosis and monitoring of acute promyelocytic leukemia; the chapter describing these methods in detail will be invaluable to clinical laboratories setting up a diagnostic service.

I hope this book will be of value to all retinoid researchers by making many advanced techniques more accessible. Many of the chapters in the book represent landmarks in retinoid methodology by giving step-by-step instructions for techniques central to the understanding of cellular control by retinoids. Inevitably, there are omissions, and it was not possible to cover all methods specific for particular genes and proteins. I am particularly pleased that so many colleagues who have made outstanding contributions to retinoid molecular biology have found time to write the chapters in this book. To all authors, I offer my thanks and admiration, and hope that this book proves to be a worthy vehicle for their work.

Christopher P. F. Redfern

Contents

Contributors

ELIZABETH A. ALLEGRETTO • *Department of Retinoid Research, Ligand Pharmaceuticals, San Diego, CA*

MYRIAM I. BAES • *Laboratory of Clinical Chemistry, Faculty of Pharmaceutical Sciences, Catholic University of Leuven, Belgium*

ARUN B. BARUA • *Department of Biochemistry and Biophysics, Iowa State University, Ames, IA*

RODOLFO BERNI • *Science Faculty, Institute of Biochemical Sciences, University of Parma, Italy*

DAVIDE CAVAZZINI • *Science Faculty, Institute of Biochemical Sciences, University of Parma, Italy*

YANG CHEN • *Protein Chemistry, W. Alton Jones, Cell Science Center, Lake Placid, NY*

JOHN W. CRABB • *Protein Chemistry, W. Alton Jones, Cell Science Center, Lake Placid, NY*

ANN K. DALY • *Department of Pharmacological Sciences, Medical School, University of Newcastle Upon Tyne, UK*

PETER E. DECLERCQ • *Laboratory of Clinical Chemistry, Faculty of Pharmaceutical Sciences, Catholic University of Leuven, Belgium*

PASCAL DOLLÉ • *Institut de Génétique et de Biologie Moléculaire et Cellulaire (IGMBC), CNRS/INSERM/ULP/Collége de France, CU de Strasbourg, France*

JACQUELINE A. DYCK • *Cellular and Molecular Medicine, University of California, San Diego, CA*

ULF ERIKSSON • *Ludwig Institute for Cancer Research (Stockholm Branch), Stockholm, Sweden*

JOHN B. C. FINDLAY • *Department of Biochemistry and Molecular Biology, University of Leeds, UK*

PAUL D. FIORELLA • *Department of Biochemistry, School of Medicine and Biomedical Sciences, State University of New York at Buffalo, NY*

CLAUDIA FOLLI • *Science Faculty, Institute of Biochemical Sciences, University of Parma, Italy*

PIERRE FORMSTECHER • *CJF INSERM, Laboratoire de Biochimie Structurale, Faculté de Médecine de Lille, France*

HAROLD C. FURR • *Department of Nutritional Sciences, University of Connecticut, Storrs, CT*

STEVE GOLDFLAM • *Protein Chemistry, W. Alton Jones Cell Science Center, Lake Placid, NY*

ROBERT M. GREENE • *The Daniel Baugh Institute, Department of Pathology, Anatomy, and Cell Biology, Jefferson Medical College, Thomas Jefferson University, Philadelphia, PA*

DAVID GRIMWADE • *Somatic Cell Genetics Laboratory, Imperial Cancer Research Fund, London, UK; Present Address: Division of Medical and Molecular Genetics, Guy's Hospital, London, UK*

ANNE-LEE GUSTAFSSON • *Department of Pharmaceutical Biosciences, Division of Toxicology, Uppsala Biomedical Center Uppsala University, Sweden*

YUICHI HASHIMOTO • *Institute of Molecular and Cellular Biosciences, University of Tokyo, Japan*

RONALD L. HORST • *Metabolic Diseases and Immunobiology Research Unit, Agricultural Research Services, National Animal Disease Center, USDA, Ames, IA*

KATHY HOWE • *Somatic Cell Genetics Laboratory, Imperial Cancer Research Fund, London, UK; Present Address: Division of Medical and Molecular Genetics, Guy's Hospital, London, UK*

ROBERT S. JAMISON • *Department of Biochemistry, School of Medicine, Vanderbilt University, Nashville, TN*

JAMES KAPRON • *Protein Chemistry, W. Alton Jones Cell Science Center, Lake Placid, NY*

VICKIE J. LAMORTE • *Gene Expression Laboratory, The Salk Institute for Biological Studies, La Jolla, CA*

STEPHEN LANGABEER • *Cytogenetics Laboratory, Department of Haematology, University College Hospital, London, UK*

PHILIPPE LEFEBVRE • *INSERM, Faculté de Médecine H. Warembourg, Lille, France*

ELLEN LI • *Department of Medicine, Washington University School of Medicine, St. Louis, MO*

DAVID LOHNES • *Clinical Research Institute of Montreal, Quebec, Canada*

REUBEN LOTAN • *Department of Tumor Biology, MD Anderson Cancer Center, University of Texas, Houston, TX*

ANDREW D. LOUGHNEY • *Medical Molecular Biology Group, Medical School, University of Newcastle upon Tyne, UK*

SALLY D. LYN • *Department of Molecular Immunology, SmithKline Beecham Pharmaceuticals, King of Prussia, PA*

PAUL N. MACDONALD • *Department of Pharmacological and Physiological Sciences, St. Louis University School of Medicine, St. Louis, MO*

GIORGIO MALPELI • *Science Faculty, Institute of Biochemical Sciences, University of Parma, Italy*

JOSEPH L. NAPOLI • *Department of Biochemistry, School of Medicine and Biomedical Sciences, State University of New York at Buffalo, NY*

MARCIA E. NEWCOMER • *Department of Biochemistry, School of Medicine, Vanderbilt University, Nashville, TN*

KAREN NIEDERREITHER • *Institut de Génétique et de Biologie Moléculaire et Cellulaire (IGMBC), CNRS/INSERM/ULP/Collége de France, CU de Strasbourg, France*

ANDREW W. NORRIS • *Department of Medicine, Washington University School of Medicine, St. Louis, MO*

NOA NOY • *Division of Nutritional Sciences, Cornell University, Ithaca, NY*

PAUL NUGENT • *The Daniel Baugh Institute, Department of Pathology, Anatomy, and Cell Biology, Jefferson Medical College, Thomas Jefferson University, Philadelphia, PA*

DAVID E. ONG • *Department of Biochemistry, School of Medicine, Vanderbilt University, Nashville, TN*

MARTIN PETKOVICH • *Cancer Research Laboratories, Botterell Hall, Queen's University, Kingston, Ontario, Canada*

CHRISTOPHE RACHEZ • *CJF INSERM, Laboratoire de Biochimie Structurale, Faculté de Médecine de Lille, France*

CHRISTOPHER P. F. REDFERN • *Medical Molecular Biology Group, Department of Medicine, Medical School, University of Newcastle upon Tyne, UK*

GIOVANNI SARTORI • *Department of Organic and Industrial Chemistry, Science Faculty, University of Parma, Italy*

TORU SASAKI • *Institute of Molecular and Cellular Biosciences, University of Tokyo, Japan*

ASIPU SIVAPRASADARAO • *Department of Pharmacology, University of Leeds, UK*

MAREK SPERKA • *Department of Biochemistry, School of Medicine and Biomedical Sciences, State University of New York at Buffalo, NY*

ZUZANA SPERKOVA • *Department of Biochemistry, School of Medicine and Biomedical Sciences, State University of New York at Buffalo, NY*

ELLEN SOLOMON • *Somatic Cell Genetics Laboratory, Imperial Cancer Research Fund, London, UK; Present Address: Division of Medical and Molecular Genetics, Guy's Hospital, London, UK*

Manickavasagam Sundaram • *Department of Biochemistry and Molecular Biology, University of Leeds, UK*

Ali Tahayato • *CJF INSERM, Laboratoire de Biochimie Structurale, Faculté de Médecine de Lille, France*

Michael Wagner • *Department of Anatomy and Cell Biology, SUNY at Brooklyn, NY*

Karen West • *Protein Chemistry, W. Alton Jones Cell Science Center, Lake Placid, NY*

Jay A. White • *Cancer Research Laboratories, Botterell Hall, Queen's University, Kingston, Ontario, Canada*

Xiaochun Xu • *Department of Epidemiology, MD Anderson Cancer Center, University of Texas, Houston, TX*

Arthur Zelent • *Leukemia Research Fund Centre at the Institute of Cancer Research, Chester Beatty Laboratories, London, UK*

I

Handling and Analysis of Retinoids

1

Properties of Retinoids

Structure, Handling, and Preparation

Arun B. Barua and Harold C. Furr

1. Structure

Retinoids have been defined as a class of compounds consisting of four isoprenoid units ($H_2C=C(CH_3)$-$CH=CH_2$) joined in a head-to-tail manner. The retinoid molecule can be divided into three parts: a trimethylated cyclohexene ring, a conjugated tetraene side chain, and a polar carbon–oxygen functional group. Retinol (I), retinaldehyde (II), and retinoic acid (III), as well as their derivatives whose structures are shown in **Structure 1** *(1)*, are included by this definition.

The conventional numbering of carbon atoms in the retinoid molecule is shown in the structure of retinol (I). On the basis of this numbering scheme, geometric isomers and substituted compounds can be named unambiguously, e.g., 13-*cis* retinoic acid (IV), 3-hydroxyretinoic acid (V), and 9-*cis* retinoic acid (VI).

To name retinoids systematically (IUPAC nomenclature), however, a different numbering scheme must be used: the carbon atom bonded to the functional group is given number 1. The numbering of carbon atoms by this scheme is shown in structure III for all-*trans* retinoic acid. Accordingly, the systemic name for all-*trans* retinoic acid is (all-*trans*) 3,7-dimethyl-9-(2,6,6-trimethylcyclohex-1-en-1-yl)-nona-2,4,6,8-tetraen-1-oic acid, or more simply (all-*trans*) 3,7-dimethyl-9-(2,6,6-trimethylcyclohexene-1-yl)-2,4,6,8-nonatetraenoic acid.

Also note that the terms E and Z are used frequently for *trans* and *cis*, respectively. Thus all-*trans* retinoic acid is also known as all-*E* retinoic acid. Other names for all-*trans* and 13-*cis* (or 13-*Z*) retinoic acid are tretinoin and

From: *Methods in Molecular Biology, Vol. 89: Retinoid Protocols*
Edited by: C. P. F. Redfern © Humana Press Inc., Totowa, NJ

Structure 1.

isotretinoin, respectively. The same numbering system is also used for the aromatic retinoids, such as etretinate (VII) and acetretin (VIII).

Retinoids are essential for several biological processes, including growth and development, reproduction, and cellular differentiation. However, retinoids are toxic when taken in excess, are irritating to the skin, and are highly teratogenic. In an attempt to produce retinoids that are efficacious, yet lack toxicity, many new retinoids have been synthesized and tested biologically. Molecular modifications have been made to all three units of the retinoid molecule: the cyclohexene ring, the polyene chain, and the polar end group. Many such retinoids, such as TTNPB (IX), TTNN (X), Ch-55 (XI), and Am-580 (XII)—which are cyclic, nonpolyisoprenoid compounds—have been shown to be more active than retinol or retinoic acid in several accepted assays for retinoid activity. Hence the above definition of retinoids has become obsolete. Sporn et al. *(2)* recently redefined a retinoid as a substance that can elicit specific biological responses by binding to and activating a specific receptor or set of receptors, with the program for the biological response of the target cell residing in the retinoid receptor rather than in the retinoid ligand itself. However, 13-*cis* retinoic acid, which has no known unique receptor but has biological activity, and anhydroretinol, which has little biological activity, are undisputably con-

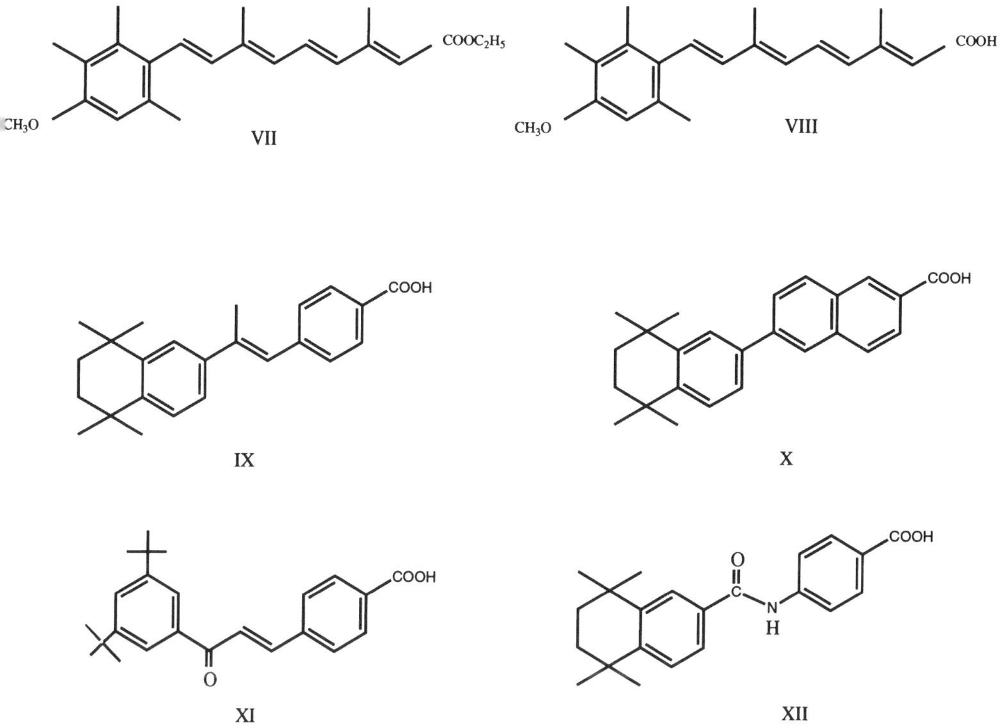

Structure 2.

sidered retinoids. We therefore suggest that retinoids be defined as compounds that are structurally similar to retinol, or that can elicit specific biological responses by binding to and activating a specific receptor or set of receptors.

In this chapter, emphasis will be given to the naturally occurring retinoids in regard to their handling, stability, storage, and other physical and chemical properties. The synthetic-aromatic retinoids, though they have similar biological activity, are different from naturally occurring retinoids in their structure and chemical properties. There is, however, no harm in treating them in the same way as the naturally occurring retinoids. The reader is also referred to previous publications relative to chemistry, physical properties, storage, and handling of retinoids *(3–6)*.

Because of the presence of conjugated double bonds, retinoids in general are unstable compounds. They are readily oxidized and/or isomerized to altered products, especially in the presence of oxidants (including air), light, and excessive heat. They are also labile toward strong acids and solvents that have dissolved oxygen or peroxides.

2. Physical Properties

Commercial preparations of retinoic acid, retinyl acetate, retinyl palmitate, retinaldehyde, and retinol are pale-yellow-to-yellow crystalline or amorphous solids. Retinol and its esters are low-melting (around 60°C) compounds, and if the room temperature is very high, they may turn to oil. Pure preparations of 3,4-didehydroretinol (vitamin A_2 alcohol) can be crystalline or oil.

2.1. Handling and Storage

Some retinoids are very irritating to the skin, and some are highly teratogenic. Hence when large amounts of retinoids are handled, adequate care should be taken so that neither solid particles nor solutions come into contact with the body. Because retinoids are lipid soluble, organic solvents are used to dissolve them. If a solution of retinoid falls on the body, the solvent quickly evaporates, and the retinoid is absorbed by the skin. Therefore, it is recommended that gloves be worn while working with retinoids.

The following three general rules apply towards handling, storage, and analysis of retinoids:

1. Because retinoids are isomerized and degraded by light, exposure of retinoids to bright daylight or bright electric light should be avoided. Black curtains should be used to cover doors and windows in laboratories designated for working with retinoids. Laboratories should be equipped with Gold fluorescent lights. If that is not possible, dim light should be used, and all labware should be amber colored, or if transparent, wrapped with aluminium foil.
2. Because retinoids are isomerized or degraded by exposure to excessive heat, the retinoids and samples containing them should be kept frozen at or below –20°C until time of analysis, and kept on ice (4°C) during handling.
3. Because retinoids are easily oxidized to inactive products, exposure of retinoids or samples containing them to air or oxygen should be avoided at all times. Retinoids, whether solid reference compounds or solutions and samples containing them, should always be stored under nitrogen or argon. Because of their unstable nature, retinoids should be stored in vacuum-sealed, amber-colored ampules, vials, or bottles, protected from light and heat, at –20°C or below. When stored unopened at –20°C or below, the retinoids keep well for a long time. Once containers are opened, the retinoids must be stored under nitrogen or argon, protected well from light at –20°C or below. If not all of the content from a large sample is used immediately, it is recommended that aliquots of solid or solution be transferred to ampules, vials or bottles, the containers flushed with an inert gas (nitrogen or argon) and sealed by torch under partial pressure of the inert gas. Argon may be preferable to nitrogen because of its lower residual oxygen content.

In blood and other tissues, retinoids are usually bound to proteins and/or protected by natural antioxidants. Therefore, when stored properly, retinoids in plasma or serum and other tissues are usually safe for 15 mo to 8 y *(6)*. However, repeated freezing and thawing should be avoided. Once extracted from blood or tissues, the freed retinoids are very susceptible to rapid decomposition by any or all three of the above-mentioned agents. Therefore, extracted samples should be analyzed immediately. Overnight storage of extracted samples is not recommended.

The most frequently analyzed human sample is blood. The handling of blood is of crucial importance. Immediately after withdrawal, blood should be protected from light, and centrifuged to remove the red blood cells. Hemolysis should be avoided. Alternatively, the blood can be allowed to clot in a cool, dark place. After centrifugation, the plasma or serum should be carefully removed from the red blood cell mass, and transferred to a clean dry tube. The size of the tube should be such that the sample of plasma or serum nearly fills it. Thus, air will be displaced, preventing oxidation during storage. Thereafter, the tube should be sealed securely and kept frozen at or below –20°C until time of analysis. Some investigators prefer to flush the tube of plasma or serum with nitrogen or argon before sealing it. However, analysis of the sample immediately after collection is ideal.

2.1.1. Radiolabeled Retinoids

Radiolabeled retinoids are more labile than their unlabeled analogs because of radiolytic autolysis. Usually radiolabeled retinoids are stored dissolved in toluene containing an antioxidant such as vitamin E or 2,5-di-*tert*-butyl hydroxyquinone (DBHQ) under an atmosphere of nitrogen or argon at –70°C. (Although it is chemically inert, benzene is not a satisfactory solvent for low-temperature storage because of its high freezing point, 5°C.) Even when all these precautions are taken, radioactive retinoids degrade, and the degree of degradation varies from retinoid to retinoid. It is very important to check sample purity and confirm the absence of degraded components in the labeled retinoid before it is used. One way to check this is to subject an aliquot of the solution to high-performance liquid chromatography (HPLC), and analyze the collected fractions for radioactivity. Alternatively, the sample can be analyzed by thin layer chromatography (TLC) (general procedure given below). If plastic-backed TLC plates are used, the plate can be cut into small segments that are placed in individual vials and the radioactivity in each segment can be determined by liquid-scintillation counting. The degraded components are usually more polar, and either elute with the solvent front during reverse-phase HPLC, or stay at the origin on TLC. If necessary, the fraction containing the retinoid of interest should be collected, the solvent removed under an inert gas, and the residue reconstituted in an appropriate solvent. The resulting preparation should be used immediately.

2.2. Solubility

All retinoids are fat soluble. The solubilities of individual retinoids in organic solvents depend on the terminal group of the side chain. Retinol (which has an alcoholic group) and retinoic acid (which has a carboxylic group) are soluble in alcoholic solvents such as methanol and ethanol, and also in acetonitrile. On the other hand, retinyl palmitate, which has an esterified long-chain fatty acid, is slightly soluble in methanol or ethanol, but highly soluble in hexane. Therefore, if a mixture of retinoids, e.g., retinoids in a liver extract, is to be dissolved, it is desirable to use a mixture of solvents such as isopropanol/hexane *(7)*. Chlorinated hydrocarbons (dichloromethane, BP 40°C; dichloroethane, BP 83°C; chloroform, BP 61°C) all dissolve retinoids well; however, chloroform should be avoided because it may be acidic or may produce free radicals. Dichloromethane is a very good solvent, but care must be taken when this solvent is used because retinoids tend to isomerize in chlorinated solvents, especially in the presence of light. Acetone is also a good solvent for the retinoids, but should not be used to work with retinaldehyde, which reacts with it (aldol condensation to form a C-23 ketone). Dimethylsulfoxide is often used to introduce retinoids into tissue culture systems; it is appropriate for this use, but its low volatility (BP 189°C) precludes its use when solvent must be evaporated. Ethers such as tetrahydrofuran and diethyl ether dissolve retinoids well, but may contain peroxides that attack retinoids. Ethyl acetate has a convenient boiling point (77°C): it is sufficiently volatile to be removed readily, but not so volatile as to be troublesome. It dissolves retinoids well, and is a good diluent for solutions of retinoids in vegetable oils. If a solvent is to be removed by evaporation, a solvent of low boiling point is preferred, i.e., methanol is preferred over ethanol.

In summary: solutions or tissue extracts containing retinoids should always be kept under nitrogen or argon, protected from light, preferably in amber-colored glass containers, in the cold at 4°C while working, and at –20°C while in storage.

2.3. Exposure to UV Light, Acids, and Alkalis

When it is exposed to UV radiation, retinol exhibits strong yellow-green fluorescence, while undergoing destruction. Measurement of absorption at 325 nm before and after destruction by exposure to UV light was an old method of quantitation of retinol. When retinol is treated with Lewis acids, including sulfuric acid or phosphorus pentoxide, it passes through purple- or blue-colored phases while undergoing rapid destruction (this is the basis of the traditional Carr-Price assay for vitamin A). Anhydrous solvents containing even traces of acid cause structural changes of retinoids. For example, retinol, when treated with methanol containing 0.03 *N* HCl, is dehydrated to anhydroretinol (a hydrocarbon with a *retro*-double bond structure) in a matter of a few min. Thus, use of strong acid should be avoided. Alkali usually is not harmful to retinoids. Indeed, samples containing esters of retinol (e.g., retinyl palmitate in liver or milk) are routinely saponified in the presence of alkali to hydrolyze the esters to free retinol. However, prolonged contact with alkali should be avoided. Retinoid carboxylic acids form salts with alkali. If the free acids are to be analyzed, it is necessary to convert such salts by treatment with dilute acid (preferably acetic acid) to the free carboxylic acid. Retinoyl β-glucuronide (a metabolite of retinoic acid, which is soluble in water) is easily hydrolyzed to retinoic acid if it is treated with hydrochloric acid, but not with acetic acid.

Retinoids in solution degrade faster than in the solid state. It is customary to add antioxidants such as vitamin E, butylated hydroxytoluene (BHT), *tert*-butyl hydroxyanisole (BHA), pyrogallol, or ascorbic acid during workup and purification of retinoids. If the sample to be extracted does not contain an antioxidant naturally, it is recommended that an antioxidant is added during extraction of retinoids from the sample. For example, some investigators add pyrogallol or ascorbic acid to breast milk during extraction and analysis of retinoids. The antioxidant selected should not interfere with the retinoids during analysis. In general, a lipid-soluble antioxidant (BHA, BHT, α-tocopherol) is preferred; however, BHT may interfere with analysis of retinol under some reversed-phase HPLC conditions.

2.4. Ultraviolet-Visible Spectroscopy

Because they have multiple, conjugated carbon-carbon double bonds, retinoids absorb strongly in the ultraviolet-visible region of the spectrum *(1,4,6)*. For example, all-*trans* retinol, which possesses five conjugated double bonds, absorbs maximally ($\lambda_{\max}$) at 325 nm (**Fig. 1**), whereas all-*trans* retinoic acid, with an additional double bond (C=O of the carboxyl group) in conjugation, and 3,4-didehydroretinol (vitamin A_2 alcohol) with the additional carbon–carbon double bond in conjugation in the cyclohexene ring, absorb at 350 nm in ethanol (**Table 1**). 5,6-Epoxy- and 5,8-epoxyretinoids, conversely, have fewer conjugated double bonds and hence have lower-absorption maxima (**Table 1**). Note that retinol, retinyl esters, and retinyl ethers have the same absorption spectra because the polyene chromophore is not disturbed. In an analogous manner, esters of retinoic acid (methyl retinoate, retinoyl β-glucuronide) have absorption spectra similar to that of unionized retinoic acid. It is to be noted that the absorption spectrum of ionized compounds, such as retinoic acid at pH above 7.0, is affected by the carbonyl oxygen that is conjugated with the polyene system. The absorption spectra of all-*trans* retinoic acid at different pH are shown in **Fig. 2**. It is common to observe retinoic acid

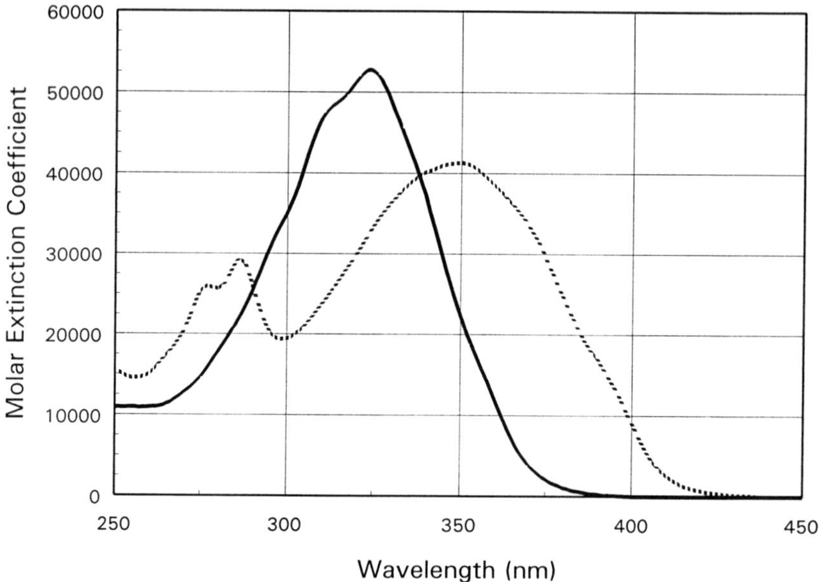

Fig. 1. Absorption spectra of all-*trans* retinol (vitamin A$_1$ alcohol, ——) and all-*trans* 3,4-didehydroretinol (vitamin A$_2$ alcohol, •••) in methanol.

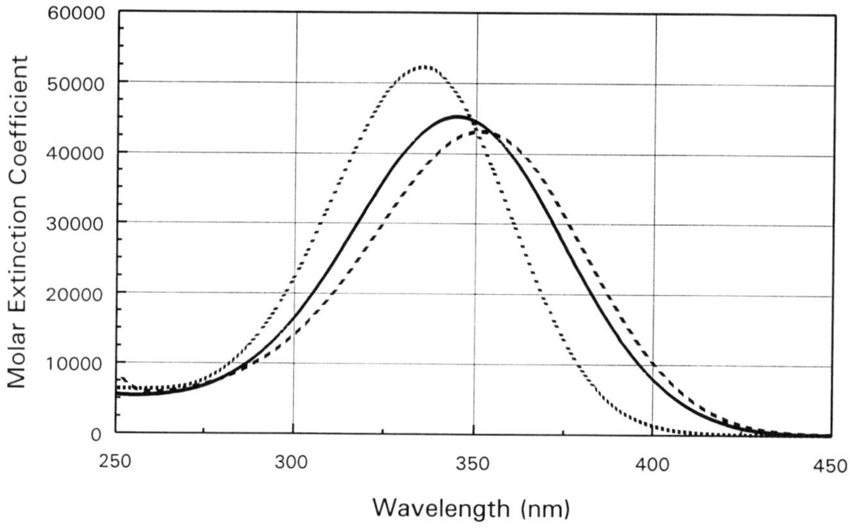

Fig. 2. Absorption spectra of all-*trans* retinoic acid under neutral (——), alkaline (•••), and acidic (– – –) conditions.

showing λ_{max} at 337 nm instead of 350 nm owing to traces of alkali present in the test tube or cuvet. Axerophthene (vitamin A hydrocarbon) has an absorption spectrum nearly identical to that of retinol; anhydroretinol, which has a *retro* double-bond structure, has an absorption spectrum reminiscent of that of β-carotene (**Fig. 3**). The wavelengths of maximum absorption and the molar-extinction coefficients of aporetinoids and apocarotenoids increase almost linearly with an increasing number of double bonds for each chemical class (alcohols, carboxylic acid, and aldehydes/ketones). Light-absorption spectroscopy has also been used in conformational studies of retinals, other polyolefinic retinoids, and aromatic retinoids *(3)*. In general, *cis* isomers have lower molar-extinction coefficients (ε) and lower λ_{max} than the all-*trans* conformers (cf., **Fig. 4** for spectra of 9-*cis* and all-*trans* retinoic acid). The aromatic analogs TMMP-retinol and TMMP-retinoic acid have absorption spectra similar to those of the natural compounds because of a similar polyene structure (**Fig. 5**).

The intensity of absorption, expressed as either $E^{1\%}_{1cm}$ or molar-extinction coefficient (ε), is characteristic for each retinoid. Molar-extinction coefficients and $E^{1\%}_{1cm}$ values of some of the commonly used retinoids are given in **Table 1**. For example, $E^{1\%}_{1cm}$ (that is, the absorption at λ_{max} of a 1% solution measured in a cuvet of 1 cm length) is 1845 for retinol at 325 nm, and 1510 for retinoic acid at 350 nm, both measured in ethanol. The corresponding molar-extinction coefficients (calculated absorbance of a solution of concentration 1 mol/L, in a cuvet of 1-cm pathlength) are 52770 and 45300, respectively. (The use of molar units instead of mass units is encouraged.) This enables one to quantify the amount of retinoid present in a solution of known volume. This is very useful for quantification of amounts of retinoids too small to be weighed. Moreover, retinoids are usually present in very small quantities (picomoles or nanomoles) in biological extracts. Even when milligram or gram quantities can be weighed accurately, and the concentration can be determined by dissolving accurately weighed amounts of retinoids in known volumes of solvent, the quantification may not be accurate, because of the presence of degraded compounds, which usually would not absorb light at the wavelength of maximum absorption of the retinoid. It is customary, therefore, to determine the exact concentration from the extinction coefficient of the retinoid in solution in a particular solvent. Full-absorption spectra (typically 250 nm–400 or 450 nm) should be scanned; determination of absorbance at only a single wavelength will not disclose the presence of decomposition products. For example, if a decomposed sample of retinol is scanned from 400 to 250 nm, it will be observed that although the solution absorbs considerably at 325 nm (typical for retinol), the spectrum is not characteristic of retinol, but shows more absorbance at lower wavelengths. Hexane and methanol or ethanol are useful sol-

Table 1
Light Absorbances of Selected Retinoids

Retinoid	Solvent	λ_{max}	ε	$E^{1\%}_{1cm}$	Ref.
All-*trans* retinol	Ethanol	325	52770	1845	*30*
	Hexane	325	51770	1810	*30*
13-*cis* retinol	Ethanol	328	48305	1689	*31*
11-*cis* retinol	Ethanol	319	34890	1220	*30*
	Hexane	318	34320	1200	*30*
9-*cis* retinol	Ethanol	323	42300	1477	*31*
11,13-di-*cis* retinol	Ethanol	311	29240	1024	*32*
9,13-di-*cis* retinol	Ethanol	324	39500	1379	*31*
All-*trans* retinyl acetate	Ethanol	325	51180	1560	*33*
	Hexane	325	52150	1590	*34*
All-*trans* retinyl palmitate	Ethanol	325	49260	940	*34*
All-*trans* retinal	Ethanol	383	42880	1510	*30*
	Hexane	368	48000	1690	*30*
13-*cis* retinal	Ethanol	375	35500	1250	*30*
	Hexane	363	38770	1365	*30*
11-*cis* retinal	Ethanol	380	24935	878	*30*
	Hexane	365	26360	928	*30*
9-*cis* retinal	Ethanol	373	36100	1270	*35*
11,13-di-*cis* retinal	Ethanol	373	19880	700	*32*
9,13-di-*cis* retinal	Ethanol	368	32380	1140	*32*
All-*trans* retinal oxime (syn)	Hexane	357	55600	1850	*36*
(anti)	Hexane	361	51700	1723	*36*
11-*cis* retinal oxime (syn)	Hexane	347	35900	1197	*36*
(anti)	Hexane	351	30000	1000	*36*
All-*trans* retinoic acid	Ethanol	350	45300	1510	*31*
13-*cis* retinoic acid	Ethanol	354	39750	1325	*31*
9-*cis* retinoic acid	Ethanol	345	36900	1230	*31*
11,13-di-*cis* retinoic acid	Ethanol	346	25890	863	*34*
9,13-di-*cis* retinoic acid	Ethanol	346	34500	1150	*31*
all-*trans* Methyl retinoate	Ethanol	354	44340	1415	*32*
13-*cis* Methyl retinoate	Ethanol	359	38310	1220	*32*
All-*trans* retinoyl β-glucuronide	Methanol	360	50700	1065	*23,37*
13-*cis* retinoyl β-glucuronide	Methanol	369	*a*		*14*
9-*cis* retinoyl β-glucuronide	Methanol	353	*a*		*14*
All-*trans* retinyl β-glucuronide	Methanol	325	44950	973	*24*
α-Retinol	Ethanol	311	47190	1650	*31*
α-Retinal	Ethanol	368	48800	1720	*35*
α-Retinoic acid	Ethanol	340	33000	1100	*31*

Table 1 *(continued)*

Retinoid	Solvent	λ_{max}	ε	$E^{1\%}_{1cm}$	Ref.
All-*trans* 3,4-didehydroretinol	Ethanol	350	41320	1455	*32*
13-*cis* 3,4-didehydroretinol	Ethanol	352	39080	1376	*38*
9-*cis* 3,4-didehydroretinol	Ethanol	348	32460	1143	*38*
9,13-di-*cis* 3,4-didehydroretinol	Ethanol	350	29950	1030	*38*
All-*trans* 3,4-didehydroretinal	Ethanol	401	41450	1470	*32*
All-*trans* 3,4-didehydroretinoic acid	Ethanol	370	41570	1395	*32*
13-*cis* 3,4-didehydroretinoic acid	Ethanol	372	38740	1300	*38*
9-*cis* 3,4-didehydroretinoic acid	Ethanol	369	36950	1240	*38*
9,13-di-*cis* 3,4-didehydroretinoic acid	Ethanol	366	32990	1107	*38*
5,6-Epoxyretinol	Ethanol	310	73140	2422	*34*
5,6-Epoxyretinal	Ethanol	365	45330	1511	*34*
5,6-Epoxyretinoic acid	Ethanol	338	45280	1442	*39*
	Hexane	338	48860	1556	*39*
5,8-Epoxyretinol	Ethanol	278	53390	1768	*34*
5,8-Epoxyretinal	Ethanol	331	43800	1460	*34*
5,8-Epoxyretinoic acid	Ethanol	298	39470	1257	*39*
	Hexane	306	45590	1452	*39*
14-Hydroxy-4, 14-retroretinol	Ethanol(?)	348	53960	1785	*40*
All-*trans* 4-oxoretinoic acid	Ethanol	360	58220	1854	*41*
	Hexane	350	54010	1720	*41*
13-*cis* 4-oxoretinoic acid	Ethanol	361	39000	1242	*42*
All-*trans* 4-oxoretinoyl β-glucuronide	Methanol	364	*a*		*14*
13-*cis* 4-oxoretinoyl β-glucuronide	Methanol	367	*a*		*14*
9-*cis* 4-oxoretinoyl β-glucuronide	Methanol	356	*a*		*14*
Anhydroretinol	Ethanol	371	97820	3650	*43*
Anhydrovitamin A$_2$	Ethanol	370	79270	2980	*44*
Axerophthene	Hexane	326	49950	1850	*34*
Retinyl methyl ether	2-Propanol	328	49800	1660	*45*
All-*trans* N-(4-hydroxyphenyl) retinamide	Methanol	362	47900	1225	*46*
All-*trans* acitretin (TMMP-retinoic acid)	Ethanol	361	41400	1270	*42*
13-*cis* acitretin	Ethanol	361	40450	1241	*42*
TMMP-retinol	Ethanol	325	49800	1596	*42*

Adapted with permission from **ref. 6**.

*a*Molar extinction coefficients may be presumed to be the same as those of the parent retinoic-acid analog.

Where the geometric isomer is not indicated, the preparation is assumed to be predominantly all-*trans*. For many compounds, molar extinction coefficients have been calculated from $E^{1\%}_{1cm}$ values in the original references or vice versa.

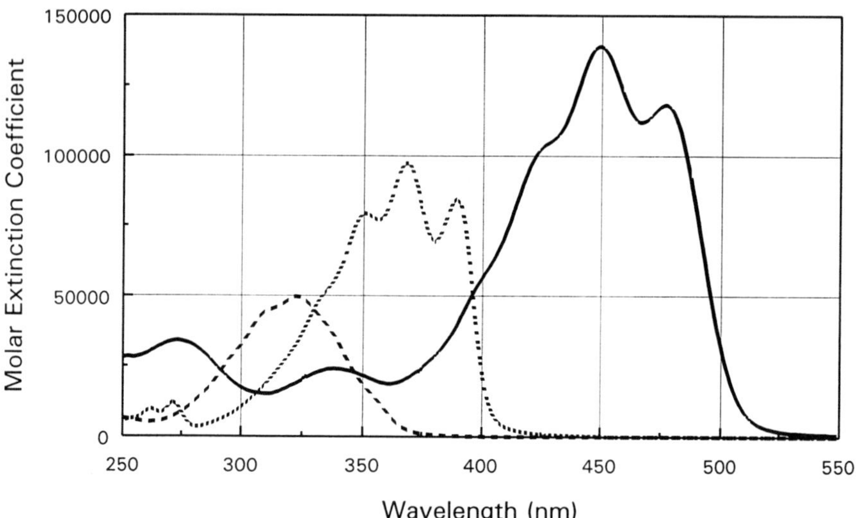

Fig. 3. Absorption spectra of all-*trans* β-carotene (———), all-*trans* anhydroretinol (•••), and all-*trans* axerophthene (– – –) in methanol.

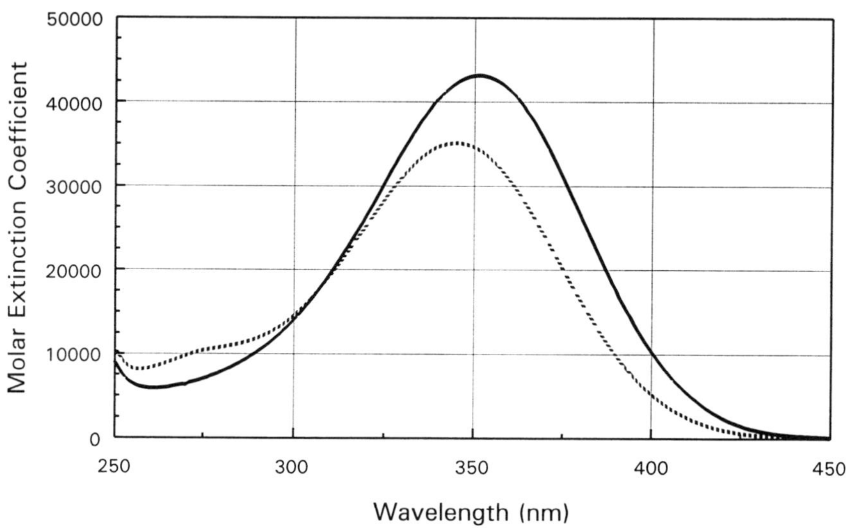

Fig. 4. Absorption spectra of all-*trans* (———) and 9-*cis* (•••) retinoic acid, in acidic methanol.

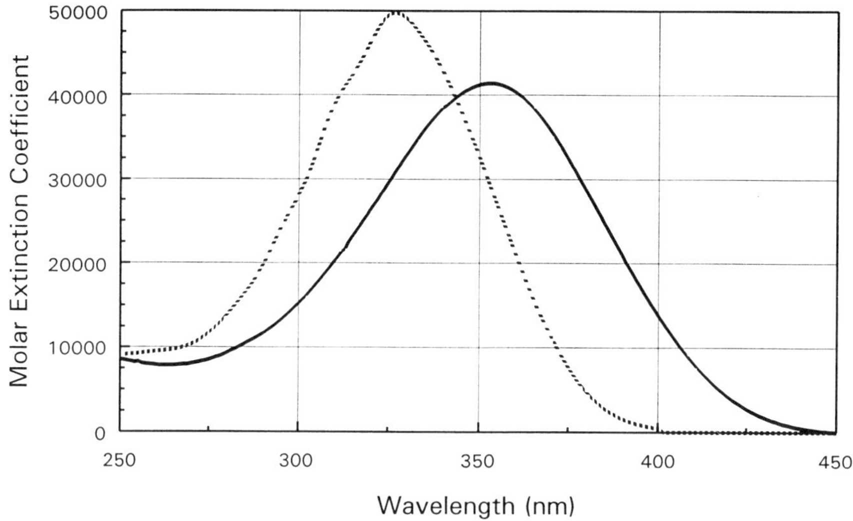

Fig. 5. Absorption spectra of all-*trans* TMMP-retinoic acid (——) and all-*trans* TMMP-retinol (•••) in methanol.

vents for absorption spectroscopy because they are generally good solvents for retinoids and because they are transparent to low wavelengths.

2.4.1. Examples for Quantitation of All-trans Retinoic Acid and Retinol

Retinoic acid: Transfer some crystals (or powder) (approx 2–3 mg) of all-*trans* retinoic acid into a volumetric flask (for example 10-mL capacity). Dissolve the solid in ethanol, and make up the volume (this is the stock solution). Pipet out 1.0 mL of the stock solution and dilute to 10.0 mL by adding 9.0 mL of ethanol. Dilute 1.0 mL of this solution to 10.0 mL. Scan the absorption spectrum from 250 to 400 nm in a UV-visible spectrophotometer. Confirm that the spectrum looks like a typical retinoic acid spectrum (*see* **Fig. 2**). Read the absorbance (optical density, OD) of the solution at the maximum (350 nm). **Note:** If the stock solution is too concentrated (absorbance greater than 1.5 OD for many spectrophotometers), it will be necessary to dilute the solution as described. If the concentration is very low (less than 0.1 OD for many spectrophotometers), it will be necessary to read the absorbance of a less-diluted solution.

Once the volumes of the solutions and the OD reading of the final solution are known, it is easy to calculate the concentration of retinoic acid in the stock solution. Calculate as follows:

Step	Volume	Volume of solution	OD reading at λmax (350 nm for retinoic acid)
1	10 mL	10 mL	Not readable, very high (offscale)
2	10 mL	100 mL (after 1 dilution)	Not readable
3	10 mL	1000 mL (after 2 dilutions)	0.453

The concentration of retinoic acid in **step 3** is:

$$\frac{0.453}{1510} \times 10{,}000 = 3.0\ \mu g/mL\ (1510 = E^{1\%}_{1cm}\ \text{for retinoic acid})$$

(The factor of 10,000 is used to convert from units of percent to units of $\mu g/mL$.) Thus the concentration of retinoic acid in the stock solution $= 3.0\ \mu g/mL \times 1000\ mL = 3000\ \mu g$, or 3.0 mg in the original 10 mL.

Alternatively, the concentrations of retinoids in solutions can be expressed in mol/L by dividing the OD by the molar extinction coefficient ε of the retinoid. For example, in the above exercise:

$$\frac{0.453}{45{,}300} = 0.00001\ mol/L = 10\ \mu mol/L$$

Retinol: Suppose we read the OD for a solution of retinol as 0.922 at 325 nm (the absorption maximum for retinol). Then:

$$\frac{0.922}{52770} \times 10{,}000 = 4.997\ \mu g/mL$$

The concentration of retinol in mol/L can be calculated as follows:

$$\frac{0.922}{1845} = 0.0000175\ mol/L = 17.5\ \mu mol/L$$

2.5. Nuclear Magnetic Resonance (NMR) Spectroscopy

For many retinoids, e.g., retinol and retinyl esters, or retinoic acid and methyl retinoate, the UV-visible absorption spectra are so similar that it is impossible to distinguish them or to confirm identity simply from absorption spectroscopy. The geometric isomers of any retinoid are not only difficult to separate, but also exhibit similar absorption spectra. Under these circumstances, proton-nuclear magnetic resonance (^{1}H-NMR) spectroscopy is very useful in the identification and characterization of closely related retinoids *(1,3)*. Normally, the retinoid is dissolved in an aprotic solvent (such as $CDCl_3$) with a standard such as tetramethylsilane (TMS) for calibration. With older instruments, several milligrams of the compound are necessary, but with advances in technology it is now possible to record a NMR spectrum with a few micrograms of the retinoid. A large number of ¹H-NMR and ¹³C-NMR spectra of geometric isomers of the naturally occurring retinoids and of their synthetic derivatives have been published *(8–13)*. By recording ¹H-NMR and ¹³C-NMR spectra and comparing the spectra with the data available in the data banks, it is possible in most cases to determine the configuration of a novel retinoid structure rapidly and reliably. (For a more detailed description, *see* Frickel in **ref. *1***.)

As some examples, the C-13 (methyl) group proton in 13-*cis* retinoyl glucuronide and 4-oxo-13-*cis* retinoyl glucuronide show signals at 2.11–2.13 ppm, whereas their all-*trans* and 9-*cis* isomers show these signals at 2.33–2.38 ppm *(14)*. Similarly, in the NMR spectra of 9-*cis* retinoyl glucuronide and 4-oxo-9-*cis*-retinoyl glucuronide, the C-7 protons show signals at 6.72 and 7.0 ppm, respectively, and are very different from the C-7 proton signals of the all-*trans* (6.32 ppm) and 13-*cis* isomers (6.26 ppm) *(14)*. Thus by examining the NMR spectra, it is possible to distinguish the isomers.

¹H-Decoupled and ¹³C-NMR spectroscopy have proved particularly useful spectroscopic methods for the rapid yet certain determination of the steric arrangements in retinoids *(10)*. ¹H-NMR and ¹³C-NMR spectroscopy have been found very powerful in studying the visual pigments rhodopsin and bacteriorhodopsin, which contain the protein opsin. In place of the protein in the natural pigments, a simple alkyl amine, retinylidene iminium salt, is usually employed in the model system. The preparation and characterization of some of the retinylidene iminium salts using high field ¹H and ¹³C-NMR spectroscopy has been described in detail *(15)*. By use of ¹³C-NMR spectroscopy, differences in the interaction of ¹³C-retinoic acid with the homologous cellular retinoic acid-binding proteins CRABP I and CRABP II have recently been demonstrated *(16)*.

2.6. Resonance Raman and Infrared (IR) Spectroscopy

Infrared spectroscopy (IR) has not been extensively used in retinoid analysis *(17,18)*. However, the newer techniques of Resonance Raman and infrared-difference spectroscopy have been applied to retinal proteins in rhodopsin and bacteriorhodopsin. By use of these techniques, it is possible to determine the structures of the chromophores in the visual pigments and in the intermediates of their photoreactions. Also, it is possible to study the interactions between the chromophores and the protein, and the structural changes evoked in the protein by the photoreaction *(1,19)*.

2.7. Mass Spectroscopy

The mass spectrum gives the molecular mass of the retinoid under investigation and is therefore particularly useful in structure elucidation *(1,4)*. Gas chromatography-mass spectroscopy (GC-MS) has been most used to date because satisfactory equipment for high-performance liquid chromatography-mass spectroscopy (HPLC-MS) has not been routinely available previously. Only small amounts (ng to μg quantity) of the test compound, which should be volatile enough to vaporize for gas chromatography, are needed to acquire information about the structure of the compound.

Gas chromatography *per se* has not been used extensively for analysis of retinoids because, in general, they are too labile to withstand the elevated temperatures used in this technique. However, it has been possible to chromatograph a number of underivatized retinoids by using cold on-column injection with capillary columns *(20)*.

Retinol and retinyl esters tend to dehydrate under the conditions of electron-impact ionization mass spectroscopy, so that the most prominent ion is that of anhydroretinol (m/z 268) instead of retinol (m/z 286), with small but detectable amounts of the parent ion *(20)*. Methyl retinoate and the trimethylsilyl derivative of retinol usually give prominent molecular ions by electron-impact ionization. Chemical ionization usually gives an identifiable adduct to the molecular ion, thus providing useful information on molecular weight.

Indeed, characterization of retinoic acid as an endogenous compound of human blood has been demonstrated by methylation of retinoic acid present in plasma extract, followed by GC-MS of the methylated product *(21,22)*. Similarly, retinoyl-, and retinyl-β-glucuronides have been characterized in human blood *(23,24)*. Mass spectra of the geometric isomers of retinal, for example, are almost identical, and a distinction cannot be made between the isomers by examination of the mass spectra. On the other hand, valuable information can be gathered by examining the fragmentation pattern in the mass spectra of closely related retinoids, because the fragmentation patterns of different retinoids can be quite different and characteristic. The recent availability of benchtop HPLC-MS offers great promise for the analysis of retinoids, as does further development of tandem mass spectrometry (MS-MS) techniques.

3. Simple Methods of Preparation of Some of the Commonly Used Retinoids

Before trying to prepare some of the commonly used retinoids as standards and reference compounds during analysis, the investigator should be familiar with the following techniques.

3.1. Liquid Chromatographic Techniques for Analysis and Purification

3.1.1. Thin-Layer Chromatography (TLC)

TLC *(4)* is very useful for quickly examining product formation during a chemical reaction, e.g., reduction of retinal to retinol as described in **Subheading 3.2.1.**

Commercial precoated silica gel plates (4 × 8 cm, 5 × 20, or 20 × 20 cm) are convenient, and those with plastic backs can be cut to smaller sizes. If a mini-TLC tank is not available, a glass jar (provided with a cover) that can hold the plate upright inside can be used. By means of a capillary micropipet, apply two spots side by side. (Apply no more than 5 µL at each application to avoid excessive spot broadening. If a greater volume is needed, allow applications to dry before repeating.) The first spot applied should be the starting retinoid (e.g., retinal in the preparation of retinol as in the example below). The second spot should be the product retinoid (e.g., retinol) solution. The spots should be applied at one end of the plate so that when the end of the plate is immersed in the developing solvent, the sample spots are just above the solvent. Develop the chromatogram with a suitable solvent mixture; for many analyses, a mixture of hexane and acetone (4:1) is suitable. When the solvent front has almost reached the top of the plate, remove the plate from the tank and mark the solvent front. If the retinoid-sample solution is concentrated, yellow spots can be seen by the unaided eye. If the retinoid solution is too dilute, no spots will be seen. In any case, examine the plate under a UV lamp in the dark. Retinol and retinyl esters are seen as fluorescent spots under long-wavelength UV light (366 nm). Commercial TLC plates that incorporate a fluorescent binder may be examined under a 254-nm mercury UV lamp; retinoids and impurities will appear as dark spots on a bright fluorescent background. These visualization techniques are sensitive to approx 10 ng retinoid. An even more general technique for visualizing compounds on TLC plates is to place the developed TLC plate in a closed jar containing iodine crystals; iodine vapor dissolves into hydrophobic compounds on the TLC plate, producing brown spots. This technique is sensitive to about 0.1 µg retinoid. If several spots are seen, the preparation is not pure. Relative mobility of compounds (R_f, distance compound moved from origin divided by distance solvent front moved) is not highly reproducible from one brand of TLC plates to another, but is useful for compound identification within a given laboratory *(25)*. Confirm the identity of a spot by analysis of a known standard if possible.

For semipreparative TLC, apply the concentrated solution as a strip to the plate rather than as a small spot. The compounds will now separate as a band. After development of the plate, scrape off the required spot with a spatula, and pour the silica-gel powder containing the desired band immediately into a test tube that has about 0.5 mL of methanol or a mixture of methanol and dichloromethane. Shake the sample well, vortex it, centrifuge, and pipet out the solution. Evaporate the solvent under an inert gas, and reconstitute the sample in an appropriate solvent. Confirm purity by HPLC.

Reversed-phase TLC on small commercial C-18 plates (1 × 3 in) is convenient for solvent scouting when one is preparing reversed-phase HPLC conditions for a new retinoid. Mobile phase compositions that give R_f in the range 0.3 to 0.7 on TLC will usually give capacity factors (k') in the range 1–10 on HPLC.

3.1.2. High-performance Liquid Chromatography (HPLC)

Details of HPLC of retinoids *(4–6)* can be found in Chapter 2 of this book. Here we give a description of purification of a retinoid to be used as a standard. For HPLC purification, a concentrated solution of the retinoid is injected. How much of the solution is to be injected and appropriate concentration of the solution will depend on the impurities present in the sample, their resolution during HPLC, the column size, the solvent used, and other conditions. These can be determined by trials. A reasonably volatile solvent or solvent mixture that can be removed easily under argon or nitrogen or in a rotary evaporator should be selected. Depending on the amount of the retinoid required, several injections may be performed, and the appropriate peak collected each time. Solvent is evaporated from the pooled fractions, the sample reconstituted in an appropriate solvent, and the concentration determined by recording the absorption spectrum. Small quantities of HPLC standards or retinoids for tissue culture studies may be readily purified by this procedure, using standard analytical-scale columns.

3.2. Preparation of Retinol and Retinyl Esters

Retinol or its solutions do not keep well even when kept under an inert gas at low temperature. Commercial retinol often contains substantial amounts of impurities because of its instability. Retinal and retinyl acetate are commercially readily available and are quite stable when kept under an inert gas and at low temperatures. If a pure standard of retinol is not available, it can be prepared easily from either retinal or retinyl acetate as follows.

3.2.1. Preparation of Retinol from Retinal

Transfer a few crystals (or powder or oil) of retinal, 1–10 mg as required, into a test tube. Dissolve the retinal in about 1 mL of methanol or ethanol (the reduction reaction does not occur efficiently in nonhydroxylic solvents such as hexane or acetonitrile). Dilute this solution (for example, touch the retinal solution with a Pasteur pipet, and then immerse the pipet in about 1 mL of methanol) and scan its absorption spectrum, diluting or concentrating this solution as necessary to obtain a representative spectrum (λ_{max} 370–380 nm, **Fig. 6**). Add a small quantity (typically 10–25 mg, need not be exact) of sodium borohydride to the concentrated-retinal solution, and shake the solution gently. The reaction should be rapid. Note the change in color from bright yellow to pale yellow and the rapid evolution of H_2 bubbles. Record the absorption spectrum. If the absorbance maximum is at 325 nm and the spectrum is characteristic of retinol as shown in **Fig. 6**, the reduction of retinal to retinol is complete.

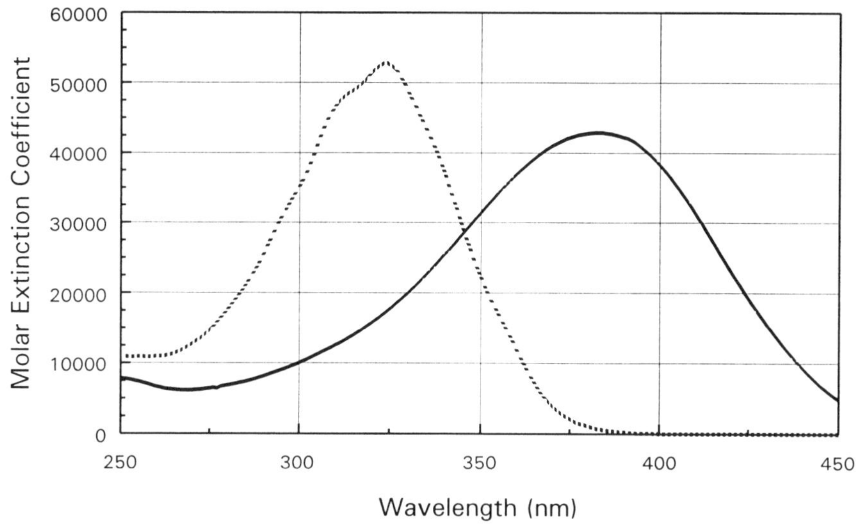

Fig. 6. Absorption spectra of all-*trans* retinal (——) and all-*trans* retinol (•••) in methanol.

If reduction is not complete, shake the solution gently for some more time. If necessary, add more sodium borohydride (excess will not be harmful). Extract the retinol by adding about 1 mL of water and about 2 mL of hexane. Vortex the sample briefly, and centrifuge it. Remove the upper-hexane layer and evaporate it under a gentle stream of nitrogen or argon; dissolve the residue in a small volume of methanol. Confirm purity of the retinol by HPLC or TLC. Purify the retinol by HPLC or TLC as previously described, if necessary.

3.2.2. Preparation of Retinol from Retinyl Acetate

Retinyl acetate is readily available commercially and is relatively inexpensive. Dissolve a few crystals (or a drop, if oily), 1–10 mg as appropriate, of retinyl acetate in methanol (about 1–2 mL). Add about 100 µL of NaOH solution (prepared by adding a drop of water to 1 or 2 pellets of NaOH, and then adding about 1 mL methanol; shake until clear) to the retinyl-acetate solution. Make sure that the solution is alkaline to litmus paper. If necessary, add more NaOH solution (excess will not be harmful). Reflux this solution at 60°C for 15–30 min, or keep warm at 50°C for 1–2 h. Analyze the product by TLC, developing the plate as previously described, and look for the fluorescent retinol spot under a UV lamp (366 nm). When no more retinyl acetate is seen, extract the retinol from the solution by adding water (about 1 mL), and hexane (about 2 mL); vortex and centrifuge the sample. Remove the hexane layer, and wash it by adding 1 mL of water; vortex and centrifuge the sample. Remove this hexane layer, and evaporate the solvent under an inert gas. Dissolve the residue in a small volume of methanol, and confirm its purity by absorption spectroscopy, HPLC and/or TLC. Purify the retinol by HPLC or TLC, if necessary.

3.2.3. Preparation of Retinyl Esters

Acyl esters of retinol may be prepared conveniently by reacting retinol with the appropriate fatty-acid chloride *(26)* or fatty acid anhydride *(27)*. Retinol (1–10 mg, as appropriate) is dissolved in 1 or 2 mL triethylamine. Fatty-acid chloride or fatty acid anhydride (approx 0.5 mL or 0.5 g) is added, and the reaction mixture is allowed to stand in the dark at room temperature. For synthesis of the acetate or propionate esters, the reaction is rapid and exothermic, and should be cooled on ice initially. Progress of the reaction is monitored by TLC (the esters are fluorescent, as is the retinol-starting material; the esters migrate more rapidly on silica-gel TLC). Complete synthesis of short-chain esters such as acetate is rapid (usually within 30 min); synthesis of long-chain esters such as palmitate and stearate is quite slow (often not complete after standing overnight). The product may be purified by HPLC or by TLC, as described earlier.

3.3. Preparation of Methyl Retinoate

3.3.1. Preparation of Methyl Retinoate from Retinoic Acid and Diazomethane

Caution: Diazomethane is a poisonous gas. Proper care should be taken and adequate ventilation should be provided. Perform the reactions in a hood, and wear a protective mask. Prepare an ethereal solution of diazomethane by established procedures (e.g., *see* **refs.** *28* or *29*). Aldrich Chemical Company (Milwaukee, WI) sells several sizes of kits for the preparation and use *in situ* of diazomethane in quantities of 1–100 mmol.

Dissolve retinoic acid in diethyl ether, and cool on ice. To this solution, add a slight excess (>1:1 molar ratio) of cold ethereal solution of diazomethane. Allow the solution to warm to room temperature, and evaporate the solvent under an inert gas or in a rotary evaporator. Dissolve the residue of methyl retinoate in an appropriate solvent. Confirm the formation of methyl retinoate by TLC (methyl retinoate migrates more rapidly than does retinoic acid); add more diazomethane if not all of the retinoic acid has reacted. Check the purity of the methyl ester by HPLC or TLC; confirm purity and determine concentration by absorbance spectrophotometry.

Retinoic acid in serum or tissue extracts can be similiarly methylated to methyl retinoate for further analysis by GC-MS *(21,22)*.

3.3.2. Preparation of Methyl Retinoate from Retinoic Acid and Iodomethane

Methyl retinoate may also be prepared by refluxing retinoic acid in ethylacetate solution with anhydrous K_2CO_3 and CH_3I for 2 h (ratio of methyl retinoate/ K_2CO_3/CH_3I, 1/2/3, w/w/v). After allowing the solution to cool, wash it three times with water, dry it over anhydrous sodium sulfate, and evaporate the solvent. Purify the methyl retinoate product as appropriate (by TLC or HPLC or by crystallization from pentane at $-20°C$).

3.4. Preparation of Methyl 4-Oxoretinoate and Methyl 5,6-Monoepoxyretinoate from Methyl Retinoate

Two common metabolites of retinoic acid are 4-hydroxyretinoic acid and 4-oxoretinoic acid. In addition, 5,6-monoepoxyretinoic acid is formed as a side product in the chemical synthesis described below.

Dissolve methyl retinoate (about 10 mg) in a mixture of dichloromethane/hexane (1:1, 10 mL). Note that the solution is yellow in color. Add 10 mg manganese dioxide (preferably British Drug Houses precipitated variety; British Drug Houses, Poole, UK) and stir the mixture with a magnetic stirrer at room temperature in an enclosed container for about 2 h. Analyze the reaction mixture by TLC (develop in hexane/acetone, 4:1) and examine any change of methyl retinoate. If there is no change in several hours, add 10 mg more of MnO_2. Repeat the addition of MnO_2 one more time if necessary for completion of the reaction. Do not add more than this, because too much manganese dioxide can completely destroy the retinoid, resulting in a colorless solution. Stirring overnight at room temperature may be necessary. When two or three additional spots are seen, filter the solution, and evaporate the solvent. Dissolve the residue in a minimal volume of dichloromethane (or ether), and perform semipreparative TLC as described earlier. Collect the bands, elute with methanol as previously described, and examine the absorbance spectra of the fractions. Methyl 4-oxoretinoate, which is obtained as the major product, absorbs at 360 nm, with a secondary smaller peak at 280 nm (**Fig. 7**). Methyl 5,6-monoepoxyretinoate, which is obtained as a minor product, absorbs maximally at 340 nm. Methyl retinoate moves fastest on TLC, followed by the 5,6-epoxy derivative, which is followed by the 4-oxo derivative.

3.5. Preparation of Methyl 4-Hydroxyretinoate from Methyl 4-Oxoretinoate

Dissolve methyl 4-oxoretinoate in methanol. Reduce with sodium borohydride as previously described for preparation of retinol from retinal. The

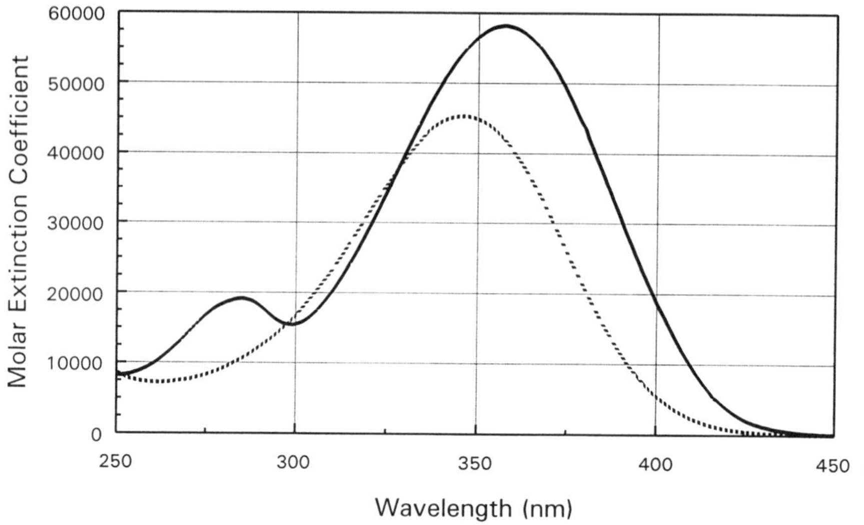

Fig. 7. Absorption spectra of all-*trans* 4-oxoretinoic acid (——) and all-*trans* 4-hydroxyretinoic acid (•••) in methanol.

reduced product is methyl 4-hydroxyretinoate. Confirm its purity and concentration by TLC and/or HPLC, and by absorbance spectroscopy (**Fig. 7**).

3.6. Saponification of Methyl Esters to 4-Oxoretinoic Acid, 4-Hydroxyretinoic Acid, and 5,6-Epoxyretinoic Acid

Dissolve the appropriate methyl ester in methanol, and saponify as previously described for preparation of retinol from retinyl acetate. After saponification, add water, and then acidify with dilute glacial-acetic acid: make sure the solution is acidic to litmus paper. (In aqueous-alkaline solution, retinoid carboxylic acids remain as sodium salts, and are not extracted by organic solvents.) Extract the retinoid-carboxylic acid with diethyl ether two or three times. (Note that hexane is not a good solvent for these polar retinoids.) Then wash the ether extract with water, and dry it over anhydrous sodium sulfate. Alternatively, if the volume is small, vortex and centrifuge the sample, remove any water, and evaporate the solvent. The retinoid-carboxylic acids usually are obtained as yellow solids. Do not add any (not even a trace) HCl to 5,6-epoxy retinoids, because they instantaneously undergo isomerization to 5,8-epoxy retinoids; this change in structure is readily confirmed by the change in absorption spectrum (**Table 1**).

3.7. Reduction of Methyl Retinoates to Retinol Derivatives

The following procedure is generally useful for conversion of methyl retinoate analogs to the corresponding retinol analogs. Dissolve the appropriate methyl retinoate analog in cold dry diethyl ether. With stirring, add an equal-molar quantity of lithium aluminum hydride. (**Caution:** Lithium aluminum hydride ignites spontaneously in the presence of even traces of water. Commercial suspensions of $LiAlH_4$ in oil may be used more safely.) The solution should immediately become pale. Add crushed ice to the solution; remove the ether layer and wash it with water, and then evaporate the solvent. The retinoid alcohol product may be dissolved in an appropriate solvent and purified as appropriate (TLC, HPLC, or conventional-column chromatography).

Acknowledgments

We particularly appreciate the encouragement of James Olson for our research in the chemistry of retinoids. Preparation of this review was partially supported by NIH-DK39733 and USDA-ISU/CDFIN-94-34115-0269 (AB) and the Storrs Agricultural Experiment Station (HF). Various retinoids (9-*cis* retinoic acid, TMMP-retinol, TMMP-retinoic acid) were generously provided by Hoffmann-La Roche. We thank Dr. Pam Duitsman for helpful revisions of the manuscript.

References

1. Frickel, F. (1984) Chemistry and physical properties of retinoids, in *The Retinoids*, 1st ed., vol. I (Sporn, M. B., Roberts, A. B., and Goodman, D. S., eds.), Academic, Orlando, FL, pp. 8–145.
2. Sporn, M. B. and Roberts, A. B. (1994) Introduction, in *The Retinoids*, 2nd ed. (Sporn, M. B., Roberts, A. B., and Goodman, D. S., eds.), Raven, New York, NY, pp. 1–3.
3. Dawson, M. I. and Hobbs, P. D. (1990) Synthetic retinoic acid analogs: handling and characterization. *Methods Enzymol.* **189,** 15–50.
4. Frolik, C. A. and Olson, J. A. (1984) Extraction, separation, and chemical analysis of retinoids, in *The Retinoids*, 1st ed., (Sporn, M. B., Roberts, D. S., and Goodman, D. S., eds.), Academic, Orlando, FL, pp. 181–233.
5. Furr, H. C., Barua, A. B., and Olson, J. A. (1992) Retinoids and carotenoids, in *Modern Chromatographic Analysis of the Vitamins*, 2nd ed. (De Leenheer, A. P., Lambert, W. E., and Nelis, H. J., eds.), Marcel Dekker, New York, NY, pp. 1–71.
6. Furr, H. C., Barua, A. B., and Olson, J. A. (1994) Analytical methods, in *The Retinoids*, 2nd ed. (Sporn, M. B., Roberts, A. B., and Goodman, D. S., eds.), Raven, New York, NY, pp. 179–209.
7. Radin, N. S. (1981) Extraction of tissue lipids with a solvent of low toxicity. *Methods Enzymol.* **72,** 5–7.
8. Schwieter, U., Englert, G., Rigassi, N., and Vetter, W. (1969) Physical organic methods in carotenoid research. *Pure Appl. Chem.* **20,** 365–420.
9. Englert, G. (1975) A ^{13}C-NMR study of *cis-trans* isomeric vitamin A, carotenoids, and related compounds. *Helv. Chim. Acta* **58,** 2367–2390.
10. Englert, G., Weber, S., and Klaus, M. (1978) Isolation by HPLC and identification by NMR spectroscopy of 11 mono-, di-, and tri-*cis* isomers of aromatic

analog of retinoic acid, ethyl all-*trans* 9-(4-methoxy-2,3,6-trimethylphenyl)-3,7-dimethyl-nona-2,4,6,8-tetraenoate. *Helv. Chim. Acta* **61,** 2697–2708.

11. Halley, B. A. and Nelson, E. C. (1979) High-performance liquid chromatography and proton nuclear magnetic resonance spectrometry of eleven isomers of methyl retinoate. *J. Chromatogr.* **175,** 113–123.

12. Halley, B. A. and Nelson, E. C. (1979) Solvent effects on the time-dependent photoisomerization of methyl retinoate. *Int. J. Vitam. Nutr. Res.* **49,** 347–351.

13. Vetter, W., Englert, G., Rigassi, N., and Schwieter, U. (1971) Spectroscopic methods, in *Carotenoids* (Isler, O., ed.), Birkhauser Verlag, Basel, Switzerland, pp. 189–266.

14. Barua, A. B., Huselton, C., and Olson, J. A. (1996) Synthesis of novel glucuronide conjugates of retinoid carboxylic acids. *Synth. Commun.* **26,** 1355–1361.

15. Shaw, G. S. and Childs, R. F. (1990) Characterization of retinylidene iminium salts by high field ^{1}H and ^{13}C nuclear magnetic resonance spectroscopy. *Methods Enzymol.* **189,** 112–122.

16. Norris, A. W., Rong, D., d'Avignon, D. A., Rosenberger, M., Tasaki, K., and Li, E. (1995) Nuclear magnetic resonance studies demonstrate differences in the interaction of retinoic acid with two highly homologous cellular retinoic acid binding proteins. *Biochemistry* **34,** 15,564–15,573.

17. Rockley, N. L., Halley, B. A., Rockley, M. G., and Nelson, E. C. (1983) Infrared spectroscopy of retinoids. *Anal. Biochem.* **133,** 314–321.

18. Rockley, N. L., Rockley, M. G., Halley, B. A., and Nelson, E. C. (1986) Fourier transform infrared spectroscopy of retinoids. *Methods Enzymol.* **123,** 92–101.

19. Sieberts, F. (1990) Resonance Raman and infrared difference spectroscopy of retinal proteins. *Methods Enzymol.* **189,** 123–135.

20. Furr, H. C., Clifford, A. J., and Jones, A. D. (1992) Analysis of apocarotenoids and retinoids by capillary gas chromatography-mass spectrometry. *Methods Enzymol.* **213,** 281–290.

21. De Leenheer, A. P. and Lambert, W. E. (1990) Mass spectrometry of methyl ester of retinoic acid. *Methods Enzymol.* **189,** 104–111.

22. Napoli, J. L. (1986) Quantification of physiological levels of retinoic acid. *Methods Enzymol.* **123,** 112–124.

23. Barua, A. B. and Olson, J. A. (1986) Retinoyl β-glucuronide: an endogenous compound of human blood. *Am. J. Clin. Nutr.* **43,** 481–485.

24. Barua, A. B., Batres, R. O., and Olson, J. A. (1989) Characterization of retinyl β-glucuronide in human blood. *Am. J. Clin. Nutr.* **50,** 370–374.

25. Singh, H., John, J., and Cama, H.R. (1973) Separation of β-apocarotenals and related compounds by reversed-phase paper and thin-layer chromatography. *J. Chromatogr.* **75,** 146–150.

26. Ross, A. C. (1981) Separation of long-chain fatty acid esters of retinol by high-performance liquid chromatography. *Anal. Biochem.* **115,** 324–330.

27. Ross, A. C. (1986) Separation and quantitation of retinyl esters and retinol by high-performance liquid chromatography. *Methods Enzymol.* **123,** 68–74.

28. Furniss, B. S., Hannaford, A. J., Smith, P. W. G., and Tatchell, A. R. (eds.) (1989) *Vogel's Textbook of Practical Organic Chemistry*. Longman Scientific & Technical, Essex, UK, pp. 430–433.

29. Fieser, L. F. and Fieser, M. (1967) *Reagents for Organic Synthesis*, vol. 1., Wiley, New York, NY, pp. 191–195.

30. Hubbard, R., Brown, P. K., and Bownds, D. (1971) Methodology of vitamin A and visual pigments. *Methods Enzymol.* **18C,** 615–653.

31. Robeson, C. D., Cawley, J. D., Weisler, L., Stern, M. H., Edinger, C. C., and Checkak, A. J. (1955) Chemistry of vitamin A. XXIV. The synthesis of geometric isomers of vitamin A via methyl β-methylglutaconate. *J. Am. Chem. Soc.* **77,** 4111–4119.

32. von Planta, C., Schweiter, U., Chopard-dit-Jean, L., Ruegg, R., Kofler, M., and Isler, O. (1962) Physikalische Eigenschaften von Isomeren vitamin-A und vitamin-A_2 Verbindungen. *Helv. Chim. Acta* **45,** 548–561.

33. Olson, J. A. (1990) Vitamin A, in *Handbook of Vitamins* (Machlin, L. J., ed.), Marcel Dekker, New York, NY, pp. 1–57.

34. Schweiter, U. and Isler, O. (1967) Vitamins A and carotene: Chemistry, in *The Vitamins*, 2nd ed., vol. 1 (Sebrell, Jr., W. H. and Harris, R. S., eds.), Academic, New York, NY, pp. 5–101.

35. Robeson, C. D., Blum, W. P., Dieterle, J. M., Cawley, J. D., and Baxter, J. G. (1955) Chemistry of vitamin A. XXV. Geometrical isomers of vitamin A aldehyde and an isomer of its α-ionone analog. *J. Am. Chem. Soc.* **77,** 4120–4125.

36. Groenendijk, G. W. T., Jensen, P. A. A., Bonting, S. L., and Daemen, F. J. M. (1980) Analysis of geometrically isomeric vitamin A compounds. *Methods Enzymol.* **67,** 203–220.

37. Barua, A. B. (1990) Analysis of water-soluble compounds: glucuronides. *Methods Enzymol.* **189,** 136–145.

38. Koefler, M. and Rubin, S. H. (1960) Physiochemical assay of vitamin A and related compounds. *Vitam. Horm.* **18,** 315–339.

39. John, K. V., Lakshmanan, M. R., and Cama, H. R. (1967) Preparation, properties and metabolism of 5,6-monoepoxyretinoic acid. *Biochem. J.* **103,** 539–543.

40. Buck, J., Derguini, F., Levi, E., Nakanishi, K., and Hammerling, U. (1991) Intracellular signaling by 14-hydroxy-4,4-retro-retinol. *Science* **254,** 1654–1656.

41. Rao, M. S. S., John, J., and Cama, H. R. (1972) Studies on vitamin A_2: Preparation, properties, metabolism and biological activity of 4-oxoretinoic acid. *Int. J. Vitam. Nutr. Res.* **42,** 368–370.

42. Vahlquist, A., Torma, H., Rollman, O., and Andersson, E. (1990) High-performance liquid chromatography of natural and synthetic retinoids in human skin samples. *Methods Enzymol.* **190,** 163–174.

43. Shantz, E. M., Cawley, J. D., and Embree, N. D. (1943) Anhydro (cyclized) vitamin A. *J. Am. Chem. Soc.* **65,** 901–906.

44. Shantz, E. M. (1948) Isolation of pure vitamin A_2. *Science* **108,** 417–419.

45. Hanze, A. R., Conger, T. W., Wise, E. C., and Weisblat, D. I. (1948) Crystalline vitamin A methyl ether. *J. Am. Chem. Soc.* **70,** 1253–1256.
46. Moon, R. C., Thompson, H. T., Becci, P. J., Grubbs, C. J., Gander, R. J., Newton, D. L., Smith, J. M., Philips, S. O., Henderson, W. R., Mullen, L. T., Brown, C. C., and Sporn, M. B. (1979) N-(4-hydroxyphenyl)retinamide, a new retinoid for prevention of breast cancer in the rat. *Cancer Res.* **39,** 1139–1146.

2

Quantitative Analyses of Naturally Occurring Retinoids

Joseph L. Napoli and Ronald L. Horst

1. Introduction

It is widely believed that the concentrations of retinoids that activate the two classes of nuclear-retinoid receptors are crucial to the effects of the receptors, and ultimately the actions of the retinoid-humoral system. It is therefore truly important to know the precise concentrations and exact nature of the retinoids present at definite times and specific loci during development, and indeed, during any event mediated by retinoids. To determine conclusively the mechanisms of retinoid action, it is not sufficient to localize only the receptors; eventually we must establish which retinoids are present and the concentrations in which they are present, because several naturally occurring retinoids can stimulate receptor action, albeit with different ED_{50} values. For example, retinoids that can activate retinoic acid (RA) receptors (RARs) include, in addition to all-*trans* RA and 9-*cis* RA, 4-hydroxy RA, 4-oxo RA, and 18-hydroxy RA. 13-*cis* RA also binds to RARs, but with a K_d value higher than the former group. Retinol can also induce responses in RA-sensitive systems; generally at doses ~200-fold higher than RA. In RA-dependent responses, retinol probably functions through conversion into RA (both enzymatic and artifactual conversion may contribute, depending on the circumstances). For example, uses of exogenous retinol, especially in higher concentrations (mM) could generate small amounts of retinoids by oxidation or by the actions of enzymes that normally are denied access to retinol in vivo. The techniques described in this chapter offer sensitive (<2 pmol), specific quantification of a variety of retinoids. A gas chromatography/mass spectrometic (GC-MS) system has been described previously with greater sensitivity (~0.25 pmol), which

From: *Methods in Molecular Biology, Vol. 89: Retinoid Protocols*
Edited by: C. P. F. Redfern © Humana Press Inc., Totowa, NJ

can be coupled with high-performance liquid chromatography (HPLC) to enhance specificity *(1–3)*. Additional HPLC systems for quantifying RA in blood have been summarized previously *(4,5)*.

The methods described here for recovering retinoids from biological samples are fast, reproducible, and have high recoveries (>80%). They do, however, require practice for optimal results. It is important to emphasize that the quality of the sample in any retinoid-sample work-up, e.g., the degree of artifactual isomerization and/or oxidation, will depend on the time it takes to extract the retinoids and prepare the extract for analysis. Some of the extraction methods described below rely on extreme pH values. Extreme pH values, as well as nucleophiles (in the biological milieu?), oxygen and light, can cause isomerization and oxidation of retinoids. Therefore, the amount of time between initiating the extractions by changing the pH and resolubilizing the sample in the HPLC-mobile phase in preparation for injection, will impact substantially on the sample. There is no need for haste; conscientiousness and uninterrupted attention during the extraction should reward the analyst with a sample free of artifactually created isomers and minimize oxidation (*see* **Note 1**).

2. Materials

2.1. General Considerations and Extraction of Biological Samples

1. Solvents: There are "horror stories" frequently about specific solvents, causing some investigators to declare certain solvents "bad" for retinoids. In fact, it is more likely that a badly handled or "old" solvent was used. Solvents of poor quality can damage retinoids. The quality of most solvents can be affected adversely by mishandling or merely aging, especially if partially used, and therefore exposed to oxygen or to white light for surprisingly short periods. Ethyl acetate can accumulate relatively high concentrations of acid, ethers (dioxane, tetrahydrofuran) can accumulate peroxides, chlorinated hydrocarbons (chloroform, methylene chloride, dichloroethane) can accumulate hydrochloric acid, and hydrocarbons can get "wet," especially during high humidity. Therefore, freshly opened bottles of HPLC-grade solvents should be used for extraction and analysis of retinoids. Portions not used should be stored for relatively brief times only, in brown bottles in minimal light after purging with an inert gas (helium, nitrogen, argon). Helium has the advantage that it drives oxygen out of solvents, whereas argon is heavier than air and will prevent the re-entry of oxygen into the container. Nitrogen has neither of these advantages, but costs less than argon. Ethers should be purchased in small containers; any remaining opened portions should be discarded within a few days. Chlorinated hydrocarbons should never be placed in a colorless-container and exposed to strong light, especially sunlight, even for a few hours.

2. Lab conditions: Sunlight should be excluded from the retinoid-analytical lab. It is also advisable to maintain the lab under yellow or gold lighting. Although not strictly required, lower temperatures curb retinoid isomerization and oxidation. Samples kept on ice or at cooler temperatures may fare better than those worked-up at ambient temperature *(4)*.

3. General supplies: Tygon tubing leaches plasticizers, even when a stream of nitrogen blows through it. To avoid problems that may be caused by retinoids or organic solvents contacting plastics, materials used in retinoid assays should be glass or stainless steel. Use Hamilton syringes or Pasteur or glass pipets throughout the assay. Use rubber or Teflon tubing, not Tygon tubing, to reduce sample volumes with streams of inert gas or to purge solvents. Retinoids stick to Teflon. Use stainless-steel lines after the detector, not Teflon, to recover retinoids from HPLC columns.

4. Internal standards: the synthetic retinoids used as internal standards, Ro-13-4306 (*see* **Note 2**) and Ro-23-5525 can be obtained from Hoffmann-La Roche (Basel, Switzerland or Nutley, NJ); retinyl acetate is obtainable from Sigma.

5. Ethanol.

6. 0.025 N KOH in ethanol.

7. Hexane.

8. 4 N aqueous HCl.

9. A nitrogen-gas supply.

10. Isopropanol.

11. Large-reservoir capacity Varian $C_{18}OH$ (500 mg) cartridge.

12. 5% acetic acid in 1:1 isopropanol:chloroform.

13. Methanol.

14. 1% v/v acetic acid.

15. 1% v/v acetic acid in 3:2 methanol:water.

16. 1% ammonium hydroxide in 1:1 methanol:water.

17. 92.5:7.5 Hexane:chloroform.

18. 0.25% v/v acetic acid in hexane.

19. 0.25% v/v acetic acid in 99:1 hexane:isopropanol.

2.2. Additional Solvents and Materials for HPLC Analysis

1. Normal-phase Dupont Zorbax-Sil Reliance cartridge column (0.4 × 4 cm) or Reliance-3 cartridge column (3-μm beads, 0.6 × 4 cm).

2. 4% v/v/tetrahydrofuran and 15% v/v tetrahydrofuran in hexane.

3. Dupont Zorbax-Sil column (0.62 × 25 cm, i.e., semi-preparative; although an analytical column would work as well).

4. 5% acetone in hexane.

5. Reversed-phase Waters ODS column (1.2 × 10 cm).

6. 15% v/v water in methanol.

7. 20% v/v water in 1:1 v/v 2-propanol:methanol.

8. 0.35% v/v acetic acid in 9:1 v/v 1,2-dichloroethane:hexane.

3. Methods

3.1. Recovery of Retinoids From Biological Samples

*3.1.1. Recovery of Retinol, Retinyl Esters and RA from Blood,
Tissue-Culture Medium, or Incubations In Vitro
and Retinal from Incubations In Vitro*

1. Add to each sample as internal RA standard Ro-13-4306 (50–100 pmol) in ethanol (10–100 mL).
2. To 1 vol of tissue homogenate (1–6 mL, up to 25% homogenate) or cell-culture incubation medium (6 mL at pH 7.4) or plasma/serum (0.5–1 mL), add 1 vol of 0.025 *N* KOH in ethanol. To an in vitro incubation done at pH 7.4 (usually 0.5 mL total volume; if incubations are done on a smaller scale, adjust the volume to 0.5 mL with water), quench with two volumes of 0.025 *N* KOH in ethanol (*see* **Notes 3** and **4**).
3. With the exception of in vitro incubations, extract twice by vortexing at least 1 min with 2 volumes of hexane. For incubations in vitro, extract once with 2.5 vol of hexane. Brief spinning in a desktop centrifuge helps effect neat separation of the layers. Remove the hexane (hexane-1), which contains the neutral retinoids: retinyl esters, retinol, and any retinal generated by incubation in vitro. To recover RA, adjust the pH of the remaining alkaline aqueous-ethanol phase to <2 with 4 *N* aqueous HCl (e.g., 75 µL for the incubation in vitro; 0.5 mL for a tissue extract with an initial volume of 6 mL) and re-extract the now-acidified aqueous phase once with a second portion of hexane. Separate this hexane phase (hexane-2) from the aqueous phase. Remove the solvents from the two hexane phases (hexane-1 and hexane-2) by blowing a gentle stream of nitrogen over each sample, while heating at 30°C.

3.1.2. Recovery of Retinol and Retinyl Esters from Plasma

In the procedures described above, the hexane extraction of the alkalinized-aqueous phase serves not only to recover the neutral retinoids, but to remove many neutral lipids that could interfere with the RA assay. If retinol and retinyl esters are the only target analytes, a simpler procedure can be used: Extract 0.25 mL of human plasma by adding 0.5 mL ethanol, 100 ng of retinyl acetate (as internal standard), and then vortexing twice with 2 mL of hexane each time *(6)*. Evaporate the solvents from the combined hexane extracts to dryness under a gentle stream of nitrogen.

3.1.3. Recovery of RA Only from Plasma

In this procedure, the neutral lipids and the RA fraction are separated by eluting the sample through a short column *(7,8)*.

1. To 1 vol of plasma, add as internal standard (Ro-23-5525 or Ro-13-4306, 50–100 pmol) in 50 mL of ethanol. Precipitate plasma proteins with 2 vol of isopropanol, vortex, then centrifuge the mixture at 2000*g* for 15 min.

2. While samples are spinning, prewash a large-reservoir capacity $C_{18}OH$ (500 mg) cartridge (Varian), successively with 5 mL of hexane, 5 mL of 5% acetic acid in isopropanol/chloroform (1/1), 5 mL of methanol, add 5 mL of water. Solvents may be eluted in each step under reduced pressure by attaching the cartridge to VacElut SPS24 (Varian).
3. After centrifugation, transfer the supernatant into a glass tube containing 2 vol of 1% acetic acid. Apply the supernatant to the prewashed $C_{18}OH$ and wash the cartridge successively with 5 mL of 1% acetic acid in methanol/water (3/2), 2 mL of 1% ammonium hydroxide in methanol/water (1/1), 10 mL of hexane/chloroform (92.5/7.5), 8 mL of 0.25% acetic acid in hexane, and then elute the RAs with 8 mL of 0.25% acetic acid in hexane/isopropanol (99/1) (*see* **Note 5**).
4. Remove the solvent from the RA sample either at room temperature under vacuum, or by blowing a gentle stream of nitrogen over the sample, while heating at 35°C.

3.2. HPLC Analysis of Retinoids

3.2.1. Quantification of Retinal and Retinol

Retinol and retinal can be separated and quantified by normal-phase HPLC with a linear tetrahydrofuran to hexane gradient (**Fig. 1**), with a total run time of 10–12 min *(9)*. Resolve retinoids using a normal-phase Dupont Zorbax-Sil Reliance cartridge column (0.4 × 4 cm) eluted at 2 mL/min with a linear gradient from 4% tetrahydrofuran to 15% tetrahydrofuran in hexane for 5 min, followed by 5 min of 4% tetrahydrofuran in hexane. Detect the retinoids with a computer-controlled tunable absorbance-detector set for the first 5 min at 370 nm (retinals) and at 325 nm (retinols) for the remainder of the run. Quantification can be achieved by comparing integrated peak areas from samples to those of standards run immediately before the samples. This system resolves (elution time): 13-*cis*-retinal (1.35 min), 9-*cis*-retinal (1.5 min), all-*trans*-retinal (2 min), 3,4-didehydroretinal (2.3 min), 9-*cis*-retinol (7.4 min), all-*trans*-retinol (7.7 min), and 3,4-didehydroretinol. Retinyl esters are eluted close to the void volume with this system.

As an alternative to the system described above, samples can be loaded onto a Dupont Zorbax-Sil column (0.62 × 25 cm, i.e., semipreparative; although an analytical column would work as well), and retinol isomers resolved by elution with a mobile phase of 5% acetone in hexane. At a flow rate of 3 mL/min retinol isomers are well-resolved: 13-*cis*-retinol (17 min); 9-*cis*-retinol (19.5 min); all-*trans*-retinol (25.3 min).

3.2.2. Analysis of Retinol and Retinyl Esters

The spectrum of neutral retinoids present in samples can be resolved by reversed-phase HPLC (**Fig. 2**; **ref. *10***). Elute samples from a reversed-phase

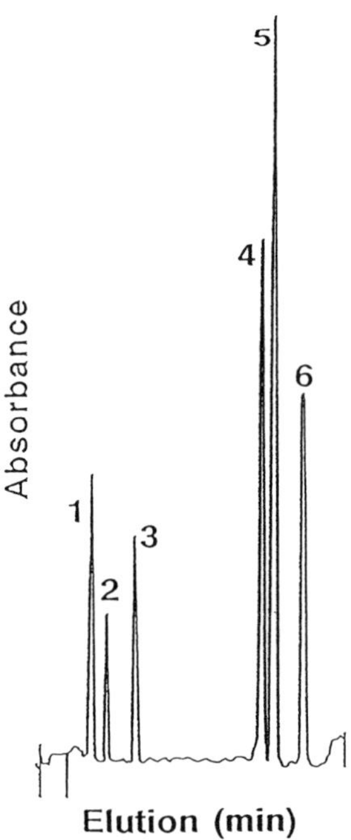

Fig. 1. Quantification of retinal and retinol by HPLC. Retinoids were resolved by a normal-phase Dupont Zorbax-Sil Reliance cartridge column (0.4 × 4 cm) eluted at 2 mL/min with a linear gradient from 4% tetrahydrofuran to 15% tetrahydrofuran in hexane for 5 min, followed by 5 min of 4% tetrahydrofuran in hexane: 13-*cis*-retinal (not shown); 1, 9-*cis*-retinal; 2, all-*trans*-retinal; 3, 3,4,-didehydroretinal; 4, 9-*cis*-retinol; 5, all-*trans*-retinol; 6, 3,4,-didehydroretinol. Retinoids were detected by UV: 370 nm for retinals; 325 nm for retinols.

Waters ODS column (1.2 × 10 cm) at 2 mL/min with a linear gradient of 15% water in methanol to methanol over 20 min, followed by 30 min of methanol. This water/methanol-based system resolves 5,6-epoxyretinol, all-*trans*-retinol, all-*trans*-retinal, anhydroretinol, and the esters retinyl docosahexanoate, retinyl palmitoleate, retinyl linoleate, and retinyl stearate, whereas retinyl palmitate and retinyl oleate elute as a single peak (**Fig. 2**).

An alternative system for measurement of retinol and retinyl esters is to use a reversed-phase column as described above (e.g., for plasma samples), but a different mobile phase. A mobile phase of 20% water in 2-propanol/methanol (1/1) run at 3 mL/min resolves RA (5 min), retinol (13 min), and retinal (15 min) (*see* **Note 6**). Replacing the aqueous-mobile phase with 2-propanol/ methanol (1/1) after 1 min elutes retinyl acetate (21 min), retinyl linoleate (25 min), retinyl palmitate and retinyl oleate together (27 min), and retinyl

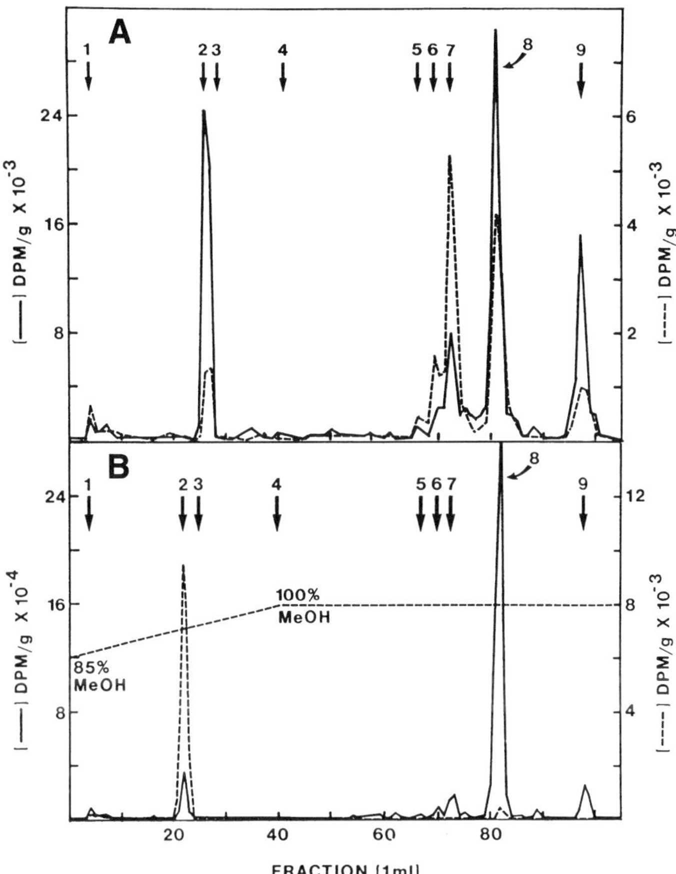

Fig. 2. HPLC analysis of neutral retinoids in tissues and plasma. Samples were eluted from a reverse-phase Waters ODS column (1.2 × 10 cm) at 2 mL/min with a linear gradient of 15% water in methanol to methanol over 20 min, followed by 30 min of methanol: 1, 5,6-epoxyretinol; 2, all-*trans*-retinol; 3, all-*trans*-retinal; 4, anhydroretinol; 5, retinyl docosahexanoate; 6, retinyl palmitoleate; 7, retinyl linoleate; 8, retinyl palmitate and retinyl oleate; 9, retinyl stearate. These specific examples illustrate the analyses of neutral-tissue retinoids equilibrated with orally fed [³H]retinol and extracted from: **(A)** rat small-intestine mucosa (dashed line) or kidney (solid line); **(B)** plasma (dashed line) or liver (solid line). The concentrations of retinol and most retinyl esters in tissues and blood, however, are sufficiently large for detection by UV at 325 nm.

stearate (28.5 min). It is not necessary to replicate these conditions exactly for successful resolution of retinol from its esters. Simply use a combination of methanol/2-propanol (usually 1/1), rather than methanol alone, to elute retinyl esters in a timely manner, and add sufficient water to retain nonester retinoids of interest. When the last nonester retinoid elutes, convert to a mobile phase of just the alcohols to elute the retinyl esters.

3.2.3. Quantification of RA

Quantify RA by normal-phase HPLC (Dupont Zorbax-Sil Reliance-3 cartridges, 3-μm beads, 0.6 × 4 cm) with a mobile phase of 0.35% acetic acid in 1,2-dichloroethane/hexane (9/1) at a flow rate of 2 mL/min. The exact proportions of solvents in these mobile phases are adjusted periodically to respond to differences in solvent batches, relative humidity, and specific columns, or to achieve more or less resolution of isomers, according to the following guidelines: acetic acid suppresses ionization of RA and modifies the elution volume; the amount of 1,2-dichloroethane changes the selectivity. Increasing the acetic acid decreases the elution volume and increasing the 1,2-dichloroethane increases the resolution of all-*trans*-RA from its isomers, such as 13-*cis*-RA, and decreases the resolution between all-*trans*-RA and the internal standard, Ro-13-4306. This HPLC system can detect <2 pmol and has been used regularly to quantify RA generated in vitro *(11–16)* or to measure tissue RA in embryos *(17)*. Quantification of RA with this HPLC system provides the same results as quantification with GC-MS *(18)*. Examples of the use of this system have been summarized previously *(2,5)*.

3.2.4. Quantification of RA and Its Isomers

9-*cis*-RA and RA are not baseline-resolved by the HPLC system described in **Subheading 3.2.3.**, but they can be distinguished easily by the ratios of their elution times to that of the internal standard Ro-13-4306 of 0.79 and 0.84, respectively. Alternative systems for resolving RA isomers are listed below (the columns and reagents are not listed in **Subheading 2.**):

1. A normal-phase HPLC system consisting of an Econosphere 3 mm silica column (0.45 × 15 cm, Altech) eluted at 1 mL/min with 0.35% v/v acetic acid in dichloroethane/hexane (5/95 v/v) resolves 13-*cis*-RA (9.5 min), 9-*cis*-RA (10 min), and all-*trans*-RA (10.6 min).
2. Alternatively the normal-phase HPLC system, with the Econosphere 3-mm column eluted with hexane/methylene chloride/acetic acid (95/5/0.2 v/v), also resolves 13-*cis*-RA and RA from 9-*cis*-RA *(7,8)*. With these three HPLC systems, however, 9-*cis*-RA and 9,13-di-*cis*-RA are co-eluted (**Fig. 3A**).
3. A reversed-phase HPLC system of a Suplex pkb-100 5 mm column (0.46 × 25 cm) eluted with acetonitrile/methanol/water/chloroform/acetic acid resolves all four isomers, and is therefore especially useful to distinguishing 9-*cis*-RA and its much more abundantly occurring isomer in vivo, 9,13-di-*cis*-RA (**Fig. 3B**).
4. RA isomers can be methylated, e.g., by treatment with ethereal diazomethane, and quantified with the Econosphere normal-phase column eluted with 0.25% v/v isopropyl ether in hexane (**Fig. 4A**).

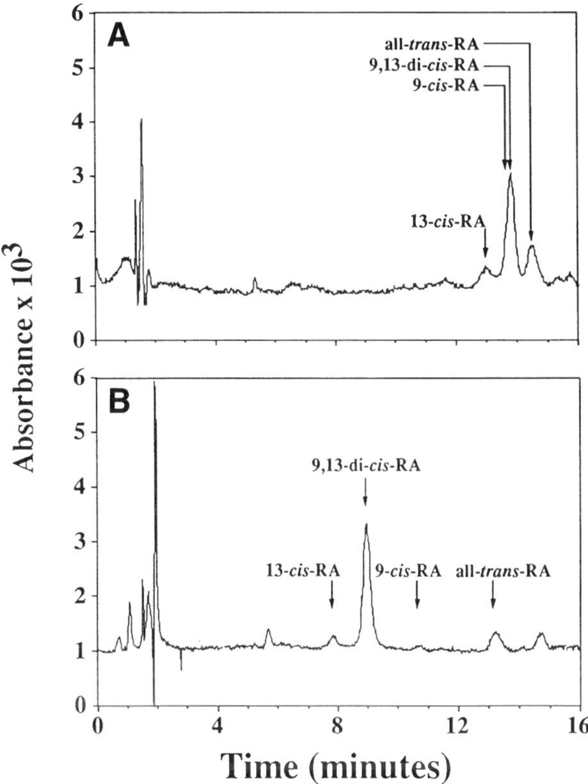

Fig. 3. Resolution of RA isomers by normal- and reverse-phase HPLC. (**A**) A normal-phase HPLC column (Econosphere 3 mm, 0.45 × 15 cm) was eluted with hexane/methylene chloride/acetic acid (95/5/0.2 v/v) at 2 mL/min. (**B**) A reverse-phase HPLC column (Suplex pkb-100, 5 mm, 0.46 × 25 cm) was eluted with acetonitrile/methanol/water/chloroform/acetic acid (17/68/10/5/0.5 v/v) at 2 mL/min. RAs were detected at 340 nm. The example depicts a typical analysis of 2 mL of 2-d-old calf plasma *(8)*.

5. A reversed-phase HPLC system also resolves the four methyl esters (**Fig. 4B**) (*see* **Note 7**).

4. Notes

1. These adaptable methods provide the possibility of quantifying specific retinoids directly. They can be applied to measuring tissue or plasma retinoids during various stages of embryonic development or disease progression. They afford specificity, as well as quantitative evaluation, for determining the kinetic characteristics of proteins involved in retinoid metabolism. An example of the requirement for specificity arose in the study of a purified rat-liver retinal dehydrogenase *(12)*. The enzyme apparently recognized 13-*cis*-retinal as substrate,

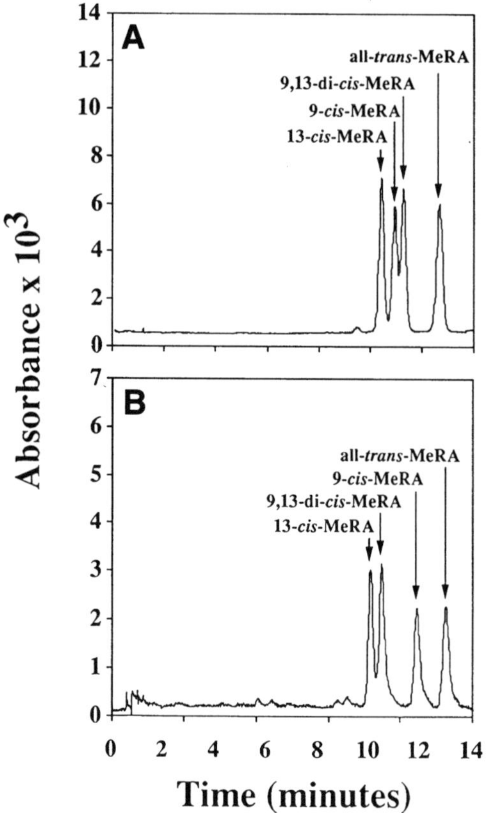

Fig. 4. Resolution of RA isomer methyl esters by normal- and reverse-phase HPLC. (A) A normal-phase HPLC column (Econosphere 3 mm, 0.45 × 15 cm) was eluted with 0.25% isopropyl ether in hexane at 2 mL/min. (B) A reverse-phase HPLC column (Suplex pkb-100, 5 mm, 0.46 × 25 cm) was eluted with acetonitrile/methanol/water/ acetic acid (60/15/25/0.5) at 2 mL/min. RAs were detected at 340 nm *(7)*.

seemingly as efficiently as all-*trans*-retinal. The product of 13-*cis*-retinal metabolism, however, was unmistakably all-*trans*-retinoic acid after HPLC analysis, not the expected 13-*cis*-retinoic acid! In contrast, 9-*cis*-retinal was converted efficiently into 9-*cis*-retinoic acid. Assays done with other than rigorous analytical measurements would not have revealed this discriminating aspect of enzyme specificity and its potential implications for retinoid-signal transduction.

2. Ro-13-4306 was referred to as TIMOTA in a previous publication *(2)*.

3. For a more detailed discussion of working up in vitro incubations, *see* **ref. 5**.

4. To recover retinal efficiently, the pH must be at least 12.0; to prevent isomerization of retinoids, the base used in any of these extractions should not exceed 0.25 *N* KOH.

5. The $C_{18}OH$ columns can be regenerated with the prewash procedure.
6. Note that RA and retinal elutions from biological samples were determined with standards only; in actual serum samples, retinal has not been detected, and RA occurs below limits of detection by UV with small sample sizes.
7. The methyl ester of 9-*cis*-RA elutes before the methyl ester of 9,13-di-*cis*-RA during normal-phase HPLC, whereas the methyl ester of 9,13-di-*cis*-RA elutes before the methyl ester of 9-*cis*-RA during reversed-phase HPLC.

Acknowledgments

Joseph L. Napoli was supported by research grants from the NIH (DK36870, DK47839, AG13566) during the course of this work.

References

1. Napoli, J. L., Pramanik, B. C., Williams, J. B., Dawson, M. I., and Hobbs, P. D. (1985) Quantification of retinoic acid by gas-liquid chromatography/mass spectrometry: total vs. all-*trans*-retinoic acid in human plasma. *J. Lipid. Res.* **26,** 387–392.
2. Napoli, J. L. (1986) Quantification of physiological levels of retinoic acid. *Methods Enzymol.* **123,** 112–124.
3. DeLeenheer, A. P. and Lambert, W. E. (1990) Mass spectrometry of methyl ester of retinoic acid. *Methods Enzymol.* **189,** 104–111.
4. DeLeenheer, A. P. and Nelis, H. J. (1990) High-performance liquid chromatography of retinoids in blood. *Methods Enzymol.* **189,** 50–59.
5. Napoli, J. L. (1990) Quantification and characteristics of retinoid synthesis from retinol and b-carotene in tissue fractions and established cell lines. *Methods in Enzymol.* **189,** 470–482.
6. Patel, P., Hanning, R. M., Atkinson, S. A., Dent, P. B., and Dolovitch, J. (1988) Intoxication from vitamin A in an asthmatic child. *Can. Med. Assoc. J.* **139,** 755–756.
7. Horst, R. L., Reinhardt, T. A., Goff, J. P., Nonnecke, B. J., Gambhir, V. K., Fiorella, P. D., and Napoli, J. L. (1995) Identification of 9-*cis*, 13-*cis*-retinoic acid as a major circulating retinoid in plasma. *Biochemistry* **34,** 1203–1209.
8. Horst, R. L., Reinhardt, T. A., Goff, J. P., Koszewski, N. J., and Napoli, J. L. (1995) Retinoic acid is the major circulating geometric isomer of retinoic acid during the periparturient period. *Arch. Biochem. Biophys.* **322,** 235–239.
9. Boerman, M. H. E. M. and Napoli, J. L. (1995) Characterization of a microsomal retinol dehydrogenase: a short-chain alcohol dehydrogenase with integral and peripheral membrane forms that interacts with holo-CRBP (type I). *Biochemistry* **34,** 7027–7037.
10. Williams, J. B., Pramanik, B. C., and Napoli, J. L. (1984) Vitamin A metabolism: analysis of steady-state neutral metabolites in rat tissues. *J. Lipid Res.* **25,** 638–645.

11. Boerman, M. H. E. M. and Napoli, J. L. (1996) Cellular retinol-binding protein-supported retinol dehydrogenation in cytosol and microsomes: relative roles in retinoic acid synthesis. *J. Biol. Chem.* **271,** 5610–5616.

12. El Akawi, Z. and Napoli, J. L. (1994) Rat liver cytosolic retinal dehydrogenase: comparison of 13-*cis*-, 9-*cis*-, and all-*trans*-retinal as substrates and effects of cellular retinoid-binding proteins and retinoic acid on activity. *Biochemistry* **33,** 1938–1943.

13. Napoli, J. L. (1993) Prostaglandin E and phorbol diester are negative modulators of retinoic acid synthesis. *Arch. Biochem. Biophys.* **300,** 577–581.

14. Posch, K. C., Burns, R. B., and Napoli, J. L. (1992) Biosynthesis of all-*trans*-retinoic acid from retinal: recognition of retinal bound to cellular retinol binding protein (type I) as substrate by a purified cytosolic dehydrogenase. *J. Biol. Chem.* **267,** 19,676–19,682.

15. Posch, K. C. and Napoli, J. L. (1992) Multiple retinoid dehydrogenases in testes cytosol from alcohol dehydrogenase negative or positive deermice. *Biochem. Pharmacol.* **43,** 2296–2298.

16. Posch, K. C., Enright, W. E., and Napoli, J. L. (1989) The synthesis of retinoic acid from retinol by cytosol from alcohol dehydrogenase negative deermice. *Arch. Biochem. Biophys.* **274,** 171–178.

17. McCaffery, P., Posch, K. C., Napoli, J. L., Gudas, L., and Drager, U. C. (1993) Changing patterns of the retinoic acid system in the developing retina. *Dev. Biol.* **158,** 390–399.

18. Napoli, J. L. (1986) Retinol metabolism in LLC-PK$_1$ cells: characterization of retinoic acid synthesis by an established mammalian cell line. *J. Biol. Chem.* **261,** 13,592–13,597.

3

Detection and Measurement of Retinoic Acid Production by Isolated Tissues Using Retinoic Acid-Sensitive Reporter Cell Lines

Michael Wagner

1. Introduction

Retinoids are diffusible signaling molecules important for the normal development, growth, and physiology of vertebrate organisms. A role for retinoids in early embryonic development has been suggested from studies in which exogenous retinoids were administered to developing vertebrate embryos *in situ*. These treatments resulted in morphological alterations in a number of diverse tissues including craniofacial structures, limbs, the hindbrain of the central nervous system, and the vertebral column *(1–5)*. The nature and extent of these alterations were found to be influenced by the developmental stage at which retinoids were administered. Together, these observations led to the proposal that retinoids play a critical role in the patterning, organization, and growth of embryonic tissues. Moreover, the ability of retinoids to carry out these functions appears to depend on their precise spatial and temporal distribution in embryos.

To understand how the differential distribution of cellular retinoids contributes to normal development, it is first necessary to establish the profile of retinoid levels in developing embryos. Direct attempts at measuring retinoids in embryonic tissues have relied on organic extraction of these compounds, followed by their separation and quantitation using high pressure liquid chromatography (HPLC) *(6)*. Whereas HPLC may be the most precise way of identifying and quantitating retinoid isomers, its limited sensitivity precludes easy measurement of retinoids in the minute amounts of tissue obtained from embryos.

From: *Methods in Molecular Biology, Vol. 89: Retinoid Protocols*
Edited by: C. P. F. Redfern © Humana Press Inc., Totowa, NJ

To detect and measure relative retinoid levels in limiting amounts of tissue, we developed a highly sensitive biological assay that uses the retinoid signaling pathway present in vertebrate cells *(7)*. Retinoid receptors are ligand-activated transcription factors present in cell nuclei. Upon binding retinoids, these receptors become associated with response elements in the promoter regions of retinoid responsive genes in such a way as to activate transcription of these genes. We used the retinoid signaling pathway to create a reporter assay for retinoids that consists of a retinoic acid-inducible reporter gene construct transfected into cells expressing retinoic acid receptors. The reporter construct contains a known retinoid-response element, in this case a retinoic acid (RA) response element or RARE *(8)*, that is used to drive the expression of a reporter gene whose product can be easily detected and measured, e.g., β-galactosidase or luciferase. A selectable drug-resistance gene was also engineered into this construct to facilitate selection of transfected cells and to establish a reporter cell line (**Fig. 1**)(*see* **Note 1**). In essence, the reporter cells act as passive detectors of retinoids in their microenvironment. Retinoids, either released from tissue explants or applied as a cell extract, diffuse across the plasma membrane of the reporter cells triggering the retinoid response pathway and expression of the reporter gene. Detection of released retinoid can be qualitative or quantitative depending on the ability either simply to detect or to measure reporter-gene activity. Characterization of reporter cell lines show that they respond specifically to retinoids with sensitivity limits to RA at $5 \times 10^{-9}M$ (*see* **Note 2**).

While this reporter system affords a relatively simple and direct way of measuring retinoids released from small amounts of tissue, it is important to realize the limitations of this assay in order to accurately interpret its results. First, unlike HPLC analysis, the reporter cell lines do not discern the particular isomer of retinoid being measured, for example, whether it is all-*trans* retinoic acid, 3,4-didehydroretinoic acid, 9-*cis* retinoic acid, or any other retinoid isomer(s) that may trigger the reporter construct. Retinoid receptor homodimers and/or heterodimers, e.g., RAR/RAR, RXR/RXR, or RAR/RXR, that specifically bind to a given retinoid isomer, together with different response elements such as RAREs or RXREs, may be able to impart a greater degree of specificity to this assay *(11,12)*. Second, when tissue explants are used in this assay only retinoids released from the tissue are being measured. Retinoids that are not released, owing either to sequestration by cellular retinoid-binding proteins or to intracellular turnover, will not be detected *(13,14)*. Thus, only those retinoids that act as intercellular signaling molecules are detected. Because detection is by the biological response pathway used by cells, levels of released retinoids revealed by this reporter system are likely to be physiologically relevant.

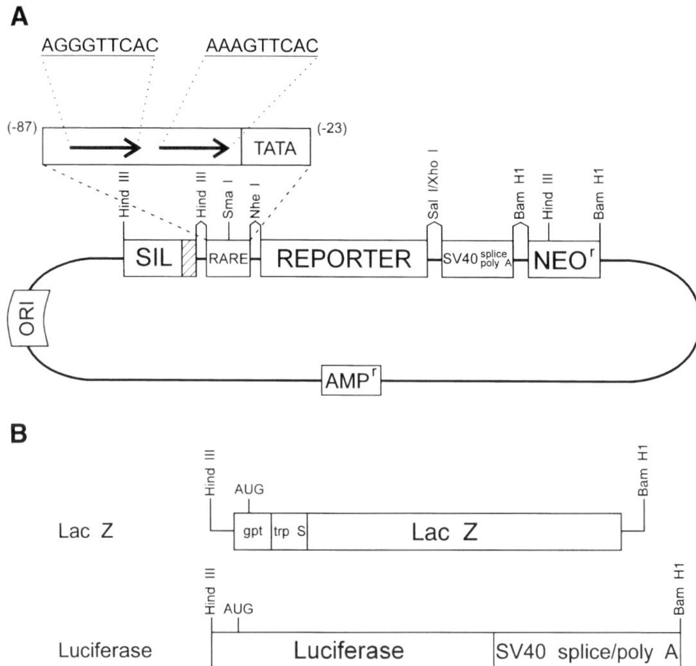

Fig. 1. (**A**) Schematic diagram of the basic reporter plasmid construct used in these studies. (**B**) schematic diagram of the two reporter genes, *E. coli lac*Z and firefly luciferase.

This retinoid reporter cell assay has been successfully used to determine the distribution of retinoids in embryonic tissues *(7,15,16)*. Retinoid reporter cell lines similar to those described here have been constructed by other laboratories *(17,18)*.

2. Materials

2.1. Basic Reporter Construct

The reporter construct used to establish stable reporter cell lines is shown in **Fig. 1**. The construct consists of a reporter gene, in this case *lac*Z or luciferase, driven by a minimal promoter containing a TATA motif and a RARE. The RARE/TATA promoter is derived from the human β-retinoic acid receptor gene *(8)* and functions as an inducible enhancer that responds to all known RAR subtypes *(19)*. Immediumtely 5' to the RARE is a "silencer" cassette consisting of a trimer of SV40 polyadenylation signal sequences that act to dampen spurious transcriptional read-through originating from upstream vector sequences *(20)*. A NEO[r] gene cassette consisting of the SV40 early promoter, the NEO[r] gene, and SV40 early splice-polyadenylation signal sequences is situated 3' to the reporter gene. Inclusion of the NEO[r] gene in the reporter construct ensures the presence of the reporter gene in cells exhibiting geneticin (G418) resistance.

2.2. Normal Passaging of Cells

The two reporter cell lines used in these studies are derived from F9 teratocarcinoma cells and L cells. The F9 reporter cell line was established by transfection with a *lac*Z reporter gene construct and the L cell reporter line was established with luciferase as reporter gene. After selecting for the highest responders to retinoic acid (5×10^{-8} *M*), clonal cell lines were established and maintained under G418 selection.

1. Normal passaging medium for F9 reporter cells: Base medium is L15-CO$_2$ Modification (Specialty Medium, Lavallette, NJ) with "1:1:2" and "FVM" additives added (*see* **Note 3**). Add each premeasured additive to 1 L of L15-CO$_2$ Modification medium and mix well. Add heat-inactivated fetal calf serum (FCS) to 20% (v/v) and G418 (Gibco-BRL, NY) to a final concentration of 0.8 mg/mL (*see* **Note 4**). Adjust pH with a dilute solution of NaOH and filter sterilize.
2. Normal passaging medium for L cell reporter line: Base medium is Dulbecco's modified Eagle's Medium (DMEM) containing 10% (v/v) heat-inactivated FCS and 0.4 mg/mL G418.
3. Gelatin solution: Prepare a 0.1% (w/v) gelatin solution by dissolving gelatin in phosphate buffered saline (PBS). Gentle heating and shaking are needed to thoroughly solubilize the gelatin. Filter sterilize and store at 4°C.
4. Gelatin-coated culture plates or flasks: Coat culture plates or flasks for 2 h with 0.1% gelatin/PBS solution. Aspirate off and wash once with PBS. Cells can be plated immediumtely or, alternatively, the plates or flasks can be stored in PBS until needed.
5. 0.25% v/v Trypsin in 1 m*M* EDTA.

2.3. F9-lacZ Reporter Gene Assay

1. Assay medium for F9 reporter cells: Base medium is L15-CO$_2$ Modification containing 1:1:2 and FVM with 1X concentrated N3 supplement (**ref. *21***, *see* **Table 1**) and glucose (8 mg/mL final concentration) added and filter sterilized (*see* **Note 5**).
2. Cell and tissue fixative: 1% glutaraldehyde, 0.1 *M* sodium phosphate buffer, pH 7.0, 1 m*M* MgCl$_2$. An alternative fixative is 0.2% glutaraldehyde, 0.1 *M* phosphate buffer, pH 7.3, 2% formaldehyde, 5 m*M* EGTA, 2 m*M* MgCl$_2$.
3. X-gal (5-bromo-4-chloro-3-indolyl-β-D-galactopyranoside) solution: 0.2% X-gal, 10 m*M* sodium phosphate buffer, pH 7.0, 150 m*M* NaCl, 1 m*M* MgCl$_2$, 3.3 m*M* K$_4$Fe(CN)$_6$ · 3H$_2$O, 3.3 m*M* K$_3$Fe(CN)$_6$.

2.4. L Cell Luciferase Reporter Gene Assay

1. L cell reporter assay medium: Base medium is DMEM with N3 supplement and glucose added.
2. Terasaki 60-well microculture plates.
3. AG1-X2 ion exchange beads, 100–250 μm in diameter (Bio-Rad Laboratories, Hercules, CA).
4. All-*trans* [11,12-^{3}H]-retinoic acid (Dupont NEN Research Products, Wilmington, DE) or similarly labeled retinoid isomer.
5. Luciferase Assay System (Promega, Madison, WI).

3. Methods

3.1. Normal Passaging of Cells

F9 and L cell reporter cell lines are normally passaged in T-75 flasks. The F9 cells require gelatinized flasks.

1. When cells are confluent, wash once with PBS. Lift cells off flask surface by treating for 5 min with a 0.25% trypsin/1 mM EDTA solution. Triturate the cells well in order to break up cell aggregates.
2. Transfer cell suspension to a 15-mL tube and allow aggregated cell clumps to settle out for approx 5 min.
3. Dilute 2 mL of cell suspension from **step 2** into 20 mL of fresh medium (1:10 split) and place in a T-75 flask. Cells are usually confluent in approx 3 d (*see* **Note 6**).

3.2. F9-lacZ Reporter Assay

The F9-*lacZ* reporter assay provides a qualitative indication of retinoid release from tissue explants maintained in short term coculture with reporter cells. There are two important requirements that must be kept in mind in order to successfully use the assay, particularly when tissue samples are taken from developing embryos. The first requirement is that the tissue being studied must be viable under the culture conditions required for maintaining the reporter cell monolayer (*see* **Note 7**). The second requirement is that the development of the embryos must be timed in accordance with the plating of the reporter cells to ensure that tissue from the desired developmental stage is plated onto near-confluent reporter cell monolayers.

Visualization of retinoid-induced reporter gene expression is dependent on the type of reporter gene. For this assay using *lacZ* as reporter gene, a histochemical procedure using X-gal as substrate allows visualization of *lacZ* gene expression in the responding reporter cells. The procedure used is essentially that of Lim and Chae *(23)*.

Scale the following procedures according to your plate/flask surface area. Starting with a confluent T-75 flask:

1. Aspirate off the normal passaging medium for F9 reporter cells and wash cells once with PBS.
2. Add 5 mL of 0.25% trypsin, 1 mM EDTA to flask; let stand for 5–10 min.
3. Triturate cells off the plastic and pipet up and down repeatedly to give a near or complete single-cell suspension.
4. Add 5 mL of normal passaging medium containing serum and transfer to a 15-mL tube (*see* **Note 8**). Allow aggregated cell clumps to settle for 5–10 min.
5. Using a 2.0-mL pipet, add 40–50 drops of cell suspension to 40 mL of normal passaging medium for F9 reporter cells. Mix well and seed 2 mL of this cell suspension per 35-mm culture plate. This volume (40 mL) is sufficient to seed 20 such plates. Ensure an even plating of cells.
6. Incubate at 37°C in a 5% CO_2 atmosphere for 2 d. Cells should be confluent or near confluent after 2 d.

For coculturing of tissue explants:

7. Approximately 1 mm^3 of tissue from embryos or other tissues under investigation are dissected in any convenient dissection medium, e.g., L15-Liebowitz.
8. Using a fire-polished Pasteur pipet, transfer the tissues to a 35-mm culture dish containing assay medium (L15-CO_2 Modification with N3 and glucose supplements) on ice.
9. Immediumtely before placing tissue explants onto the reporter-cell monolayer, remove the normal-passaging medium containing G418, wash once with PBS, and replace with approx 1.0 mL of assay medium.
10. Transfer tissue pieces onto the cell monolayer using a fire-polished Pasteur pipet (*see* **Note 9**).
11. Use one culture as a positive control to check the response of the cells to retinoids. For RA, a concentration of 5×10^{-8} M all *trans*-retinoic acid (Sigma) should induce a response in the majority of cells.
12. Culture at 37°C in 5% CO_2 for 1–2 h in order to allow the tissue explants to adhere to the reporter-cell monolayer. Once the tissues adhere, gently add more medium to ensure complete coverage of both explant and cell monolayer and incubate overnight.
13. After incubation, fix and develop the explant coculture for *lac*Z expression in reporter cells as outlined in **steps 14–19**.
14. Gently remove medium from the explant cocultures taking care not to dislodge the tissue explants. Gently wash the cocultures twice with PBS.
15. Fix the tissue and cells by incubating in fixative solution for 15 min at room temperature.
16. Remove the fixative and gently wash twice with PBS.
17. Incubate the fixed cells with the 0.2% X-gal solution for a few hours to overnight at 37°C. Periodically check the cultures for development of the typical blue color associated with β-galactosidase activity on the X-gal substrate.
18. After sufficient development of the blue reaction product, remove the X-gal solution and gently wash twice with PBS.

19. Place a thin layer of a 50% glycerol solution in PBS over the cocultures and cover slip, taking care not to dislodge the tissue explants. Storage at 4°C is recommended and often intensifies the blue reaction product.

3.5. L Cell-Luciferase Reporter Assay

In certain instances, it may be critical to know the relative amount of retinoid released from individual tissues or dissected regions of the embryo. To obtain a quantitative measure of retinoids released from tissues, we developed a reporter cell line using luciferase as the reporter molecule. When luciferase oxidizes its substrate luciferin, a photon is emitted that can be detected and quantified by use of a luminometer or scintillation counter. By relating known concentrations of retinoids to the amount of luciferase activity that they induce, the concentration of retinoids released by tissue samples can be determined. For quantifying retinoid release, a standard curve relating known concentrations of released retinoids to induced luciferase activity must be established. Known concentrations of "released retinoids" can be achieved by the use of ion-exchange beads that are loaded with retinoic acid and act as a point source of retinoid released onto reporter cell monolayers.

1. L cell-reporters are passaged in T 75 flasks in normal passaging medium for L-reporter cells.
2. Wash cells once with PBS and trypsinize cells with 5.0 mL of 0.25% trypsin, 1 mM EDTA.
3. Add 10 mL of L cell-passaging medium to resuspend the cells and count a 1:5 dilution of this cell suspension using a hemocytometer. Dilute the cells to a final concentration of 25 cells/μL.
4. For the reporter assay, plate the cells in Terasaki microculture wells by placing 20 μL of a 25 cell/μL dilution into each well.
5. Culture the cells for 3 d in L cell-reporter passaging medium to achieve a near confluent monolayer.
6. Aspirate off the medium and wash once with PBS. Change the medium in the Terasaki wells to L cell-reporter assay medium. Prepare tissue explants by resuspending them in L cell-reporter assay medium and place on the monolayers.
7. Load AG1-X2 ion-exchange beads with known concentrations of ^{3}H-retinoic acid *(24)*. Place single ion-exchange beads (200 μm in diameter) in 250 μL of retinoic-acid solution and shake for 20 min at room temperature. Remove the retinoic-acid solution and wash beads twice in PBS for 1 and 10 min, followed by a final wash in serum-free assay medium for 10 min. Choose a range of RA concentrations between 1 and 25 nM. Prepare two sets of beads over this concentration range.
8. Place one set of beads in Terasaki wells containing only assay medium and the other set into Terasaki wells containing the L cell-reporter monolayer. Culture in a humidified atmosphere for the same period of time (usually 20 h) as the tissue

cocultures. Given the small volume of medium in each Terasaki well, it is important to not let this evaporate.

9. At the end of the culture period, remove the medium from the first set of Terasaki wells containing beads in medium alone. Measure released ^{3}H-retinoic acid by scintillation counting. Convert cpms to dpms and determine the amount of RA released using the known specific activity of the labeled RA.

10. To assay the response of the reporter cells to RA released from the AG1-X2 beads or tissue explants, remove beads and tissue explants from the reporter cell monolayers without disturbing the monolayer and wash once with PBS (*see* **Note 10**).

11. Remove wash PBS and immediumtely add 10 μL of Promega "luciferase lysis reagent."

12. Let stand at room temperature for 10 min.

13. Triturate cells with a pipetman set at 7 μL in order to avoid foaming and transfer the cell lysate to a 1.5 mL microcentrifuge test tube on ice.

14. Wash the culture well again with an additional 5 μL of lysis reagent and add to the first lysate.

15. The luciferase assay is carried out using a commercially available kit (Luciferase Assay System) from Promega. Light production can be detected using a luminometer or a scintillation counter. With the scintillation counter, the coincidence circuit between photomultiplier detectors should be turned off, if possible *(25)*. For all samples, counting should proceed at a set time after cell lysate and luciferase cocktail are mixed together. Count for at least 3 min.

16. Plot a standard curve relating luciferase activity of the L cell-reporter monolayer with RA released into the medium from beads. Determine the amount of luciferase activity induced by each tissue explant and determine the amount of RA released from the tissue using the standard curve.

3.6. Assay of Retinoids in Tissue or Cell Homogenates Using lacZ and Luciferase Reporter Cells

Whereas the reporter assays described thus far measure retinoids released from tissue explants, these assays can also be adapted to measure retinoid levels within tissue samples. McCaffery et al. have modified the F9 reporter cell assay to give a measure of tissue retinoid content *(16)*.

1. F9 reporter cells are cultured and prepared for plating as described in **Subheading 3.2.** Cells are plated into 96-well plates. (Volumes and cell concentrations must be scaled down to accomodate the smaller plating area.)

2. Culture the reporter cells in normal passaging medium to near confluence.

3. Tissue samples for assay are obtained by dissection and, if possible, the wet-weight of the tissue piece determined.

4. Homogenize the tissue sample in 60–90 μL of F9 assay medium at 4°C.

5. Freeze the homogenate briefly, then thaw and fractionate by differential centrifugation:

    a. For complete homogenate minus cellular debris, spin at $1000g$ for 1 h at 4°C.

    b. For cytosolic fraction containing microsomes, spin at $12,000g$ for 1 h at 4°C.

    c. For cytosol free of microsomes, spin at $105,000g$ for 1 h at 4°C.

6. Remove the normal passaging medium from the reporter cells and wash the cells once with PBS.

7. Directly pipet the tissue homogenate onto the reporter cells in the 96-well plates and incubate at 37°C overnight in a 5% CO_2 atmosphere.

8. For quantitation, set up in duplicate a standard curve relating known amounts of retinoic acid to β-galactosidase activity. Add known concentrations of RA in F9 assay medium to the reporter cells over a concentration range of 5×10^{-9} to 10^{-7} M.

9. After overnight incubation, remove the homogenates and RA-containing F9 assay medium from the cells, wash twice with PBS, and develop the reporter cells for detection of *lacZ* expression as described in **Subheading 3.2., steps 14–19**.

10. The intensity of the blue reaction product is measured using a microtiter plate reader at 405 nm. Determine the amount of retinoid in the tissue homogenates by comparing their colormetric readings (at 405 nm) to that of the standard curve relating RA to β-galactosidase activity (light absorption at 405 nm). If increased sensitivity is required, the assay may be performed using the luciferase reporter-cell line (*see* **Note 11**).

4. Notes

1. F9 cells were transfected according to the method of Espeseth et al. *(9)*; L cells were transfected as described *(7)*. After selection in G418-containing medium, single colonies were grown up and tested for their response to RA. The highest responders were established as continuous reporter cell lines. The use of stably transfected reporter cell lines in place of transiently transfected cells provides a degree of stability and uniformity in the response of these cells that allows for a reproducible and reliable reporter assay.

2. The sensitivity of the response to RA of the first F9 reporter cell lines established was characterized using an assay for measurement of β-galactosidase in cell lysates *(10)*. F9 reporter cell clones detected RA at a concentration of 10^{-9} M, with diminution of the response between 10^{-9} and 10^{-10} M RA. All clones respond to retinol concentrations as low as 10^{-6} M; the response at this concentration of retinol is similar to the response to RA at 5×10^{-9} M. No response was detected at a concentration of 10^{-6} M for other steroid hormones, such as dexamethasone, β-estradiol, progesterone, testosterone, L-thyroxine, Vitamin D_3, and d-aldosterone, or to activators of other signal-transduction pathways such as forskolin or phorbol 12-myristate 13-acetate (PMA). For reporter cells in the presence of 10^{-8} M RA, the initial appearance of β-galactosidase expression usually takes about 4 h. In general, F9 reporter cells passaged up to 10 times were used for assays before resorting to fresh cells obtained from storage in liquid N_2.

3. "1:1:2" Supplement contains D-glucose, L-glutamine, penicillin, and streptomycin sulfate. FVM supplement contains glutathione and ascorbic acid. (Specialty Medium, Lavallette, NJ.)

4. The amount of FCS in this medium is relatively high at 20%. According to some reports, high FCS concentrations act to prevent the differentiation of the F9 teratocarcinoma cell line.

5. The N3 supplement is used as a serum substitute to avoid retinoids in serum triggering the response of reporter cells. N3, however, can be arduous and expensive to prepare (*see* **Table 1**). Given the short-term nature of this assay, it may be advisable first to try the assay in serum-containing medium or in medium using serum that has been stripped of retinoids by treatment with charcoal followed by sterile filtration. Alternatively, serum-free medium supplement containing insulin, transferrin, and selenite can be obtained from Boehringer Mannheim (Indianapolis, IN). Whichever medium or serum treatment is chosen, it is important that the viability of both the tissue explant and reporter cell monolayer is ensured over the duration of the assay.

6. When passaging F9 cells, it is important to plate out using single-cell suspensions. Plating aggregates of F9 cells gives a patchy monolayer and leads to poor cell growth. Care should be taken in the passaging of the reporter cell lines, particularly the F9 reporter cells, in order to maintain the high response of these cells to retinoids. Should the response of these reporter cells diminish with passaging, the cell line can be subcloned to regain a 95% or greater response to test concentrations of 5×10^{-8} *M* retinoic acid. This can be achieved by limiting dilution cloning of G418-resistant colonies or, if available, by fluorescence-activated cell sorting using the procedure of Nolan et al. *(22)*.

7. The tissues of certain species require different conditions for maintaining viability in vitro. For instance, tissue explants from *Xenopus* are best maintained at 25°C, whereas reporter cells must be at 37°C. Cocultures of *Xenopus* tissue and reporter cells at 37°C are therefore not feasible. This limitation can be circumvented by using a tissue homogenate (**ref. *16*** and **Subheading 3.6.**) or extracting retinoids directly from amphibian tissues and assaying the homogenate or extract using reporter cells maintained at 37°C *(18)*.

8. Serum contains inhibitors of trypsin that will terminate trypsinization of the cells.

9. It is important for the tissue explants to adhere to the monolayer of reporter cells. This is best achieved by using minimal amounts of medium to avoid "turbulence" that might otherwise dislodge tissue explants before they have had a chance to adhere. Adherence to a single spot on the monolayer is required for the reporter cells in the immediumte area of the tissue explants to detect locally released retinoids.

10. Before removing explants, check each culture well under a dissecting microscope or low-magnification microscope. Assay only those wells in which tissue explants successfully adhered to the reporter cell monolayer. Direct contact between tissue explant and cell monolayer appears crucial to the detection of released retinoids.

**Table 1
N3 Supplement Composition**

Additive	Additive stock solution concentration	Volume of additive required, mL[a]
Hank's balance salt solution (HBSS)	(with calcium and magnesium)	4.9
Bovine serum albumin	10 mg/mL in HBSS	1.0
Human transferrin	100 mg/mL in HBSS	2.0
Putrescine dihydrochloride	80 mg/mL in HBSS	0.4
Sodium selenite	1.04 mg/mL in HBSS	1.0
Triiodothyronine, sodium salt	20 mg/mL in 0.01 M NaOH	0.1
Insulin (bovine pancreas, 24 I.U./mg)	25 mg/mL in 20 mM HCl	0.4
Progesterone	12.5 mg/100 mL absolute ethanol	0.1
Corticosterone grade A	200 mg/100 mL absolute ethanol	0.02

[a]Total volume is 9.92 mL. Filter sterilize, aliquot into 2-mL aliquots, and store at –80°C. All additives may be purchased from Sigma (St. Louis, MO).

11. The luciferase reporter cell line may also be used to measure retinoids in tissue homogenates. The luciferase reporter cells can detect as little as 0.1 fmol of RA, and their response is linear up to 10 fmol of RA *(7)*.

Acknowledgments

I would like to thank Dr. Thomas Jessell (Howard Hughes Medical Institute, Columbia University) for his support during the generation and use of these reporter cell lines in his laboratory. I also thank Barbara Han for expert technical assistance in the production of these reporter cells and Ira Schieren for his expertise in computer-generated graphics.

References

1. Abbott, B. D., Harris, M. W., and Birnbaum, L. S. (1989) Etiology of retinoic acid-induced cleft plate varies with the embryonic stage. *Teratology* **40,** 533–553.
2. Rutledge, J. C., Shourbaji, A. G., Hughes, L. A., Polifka, J. E., Cruz, Y. P., Bishop, J. B., and Generoso, W. M. (1994) Limb and lower-body duplications induced by retinoic acid in mice. *Proc. Natl. Acad. Sci. USA* **91,** 5436–5440.
3. Papalopulu, N., Clarke, J. D. W., Bradley, L., Wilkinson, D., Krumlauf, R., and Holder, N. (1991) Retinoic acid causes abnormal development and segmental patterning of the anterior hindbrain in *Xenopus* embryos. *Development* **113,** 1145–1158.
4. Ruiz i Altaba, A. and Jessell, T. (1991) Retinoic acid modifies mesodermal patterning in early *Xenopus* embryos. *Genes and Dev.* **5,** 175–187.
5. Kessel, M. and Gruss, P. (1991) Homeotic transformations of murine vertebrae and concomitant alteration of *Hox* Codes induced by retinoic acid. *Cell* **67,** 89–104.

6. Thaller, C. and Eichele, G. (1987) Identification and spatial distribution of retinoids in the developing chick limb bud. *Nature* **336,** 775–778.
7. Wagner, M., Han, B., and Jessell, T. M. (1992) Regional differences in retinoid release from embryonic neural tissue detected by an in vitro reporter assay. *Development* **116,** 55–66.
8. de The, H., del Mar Vivanco-Ruiz, M., Tiollais, P., Stunnenberg, H., and Dejean, A. (1990) Identification of a retinoic acid responsive element in the retinoic acid receptor β gene. *Nature* **343,** 177–180.
9. Espeseth, A. S., Murphy, S. P., and Linney, E. (1989) Retinoic acid receptor expression vector inhibits differentiation of F9 embryonal carcinoma cells. *Genes and Dev.* **3,** 1647–1656.
10. Reynolds, A. and Lundblad, V. (1989) Yeast vectors and assays for expression of cloned genes, in *Current Protocols in Molecular Biology* (Ausubel, F. A., Brent, R., Kingston, R. E., Moore, D. D., Seidman, J. G., Smith, J. A., and Struhl, K., eds.), Greene Publishing and Wiley-Interscience, pp. 13.6.1–13.6.4.
11. Leid, M., Kastner, P., Lyons, R. Nakshatri, H., Saunders, M., Zacharewski, T., Chen, J-Y., Staub, A., Garnier, J-M., Mader, S., and Chambon, P. (1992) Purification, cloning, and RXR identity of the HeLa cell factor with which RAR or TR heterodimerizes to bind target sequences efficiently. *Cell* **68,** 377–395.
12. Mangelsdorf, D. J., Umesono, K., Kliewer, S. A., Borgmeyer, U., Ong, E. S., and Evans, R. M. (1991) A direct repeat in the cellular retinol-binding protein Type II gene confers differential regulation by RXR and RAR. *Cell* **66,** 555–561.
13. Boylan, J. F. and Gudas, L. J. (1991) Overexpression of the cellular retinoic acid binding protein-I (CRABP-I) results in a reduction in differentiation-specific gene expression in F9 teratocarcinoma cells. *J. Cell Biol.* **112,** 965–979.
14. Boylan, J. F. and Gudas, L. J. (1992) The level of CRABP-1 expression influences the amounts and types of all-trans-retinoic acid metabolites in F9 teratocarcinoma stem cells. *J. Biol. Chem.* **267,** 21,486–21,491.
15. Kelley, M. W., Xu, X-M., Wagner, M. A., Warchol, M. E., and Corwin, J. T. (1993) The developing organ of Corti contains retinoic acid and forms supernumerary hair cells in response to exogenous retinoic acid in culture. *Development* **119,** 1041–1053.
16. McCaffery, P., Lee, M., Wagner, M. A., Sladek, N. E., and Drager, U. C. (1992) Assymetrical retinoic acid synthesis in the dorso-ventral axis of the retina. *Development* **115,** 371–382.
17. Colbert, M. C., Linney, E., and LaMantia, A. S. (1993) Local sources of retinoic acid coincide with retinoid-mediumted transgene activity during embryonic development. *Proc. Natl. Acad. Sci. USA* **90,** 6572–6576.
18. Chen, Y., Huang, L., and Solursh, M. (1994) A concentration gradient of retinoids in the early *Xenopus laevis* embryo. *Dev. Biol.* **161,** 70–76.
19. Sucov, H. M., Murakami, K. K., and Evans, R. M. (1990) Characterization of an autoregulated response element in the mouse retinoic acid receptor type β gene. *Proc. Natl. Acad. Sci. USA* **87,** 5392–5396.

20. Maxwell, I. H., Harrison, G. S., Wood, W. M., and Maxwell, F. (1989) A DNA cassette containing a trimerized SV40 polyadenylation signal which efficiently blocks spurious plasmid-initiated transcription. *Biotechniques* **7**, 276–280.
21. Romijn, H. J., Habets, A. M., Mud, M. T., and Wolters, P. S. (1981) Nerve outgrowth, synaptogenesis and bioelectric activity in fetal rat cerebral cortex tissue cultured in serum-free, chemically defined medium. *Brain Research* **254**, 583–589.
22. Nolan, G. P., Fiering, S., Nicolas, J. F., and Herzenberg, L. A. (1988) Fluorescence-activated cell analysis and sorting of viable mammalian cells based on β-D-galactosidase activity after transduction of *Escherichia coli lacZ*. *Proc. Natl. Acad. Sci. USA* **85**, 2603–2607.
23. Lim, K. and Chae, C. B. (1989) A simple assay for DNA transfection by incubation of the cells in culture dishes with substrates for beta-galactosidase. *Biotechniques* **7**, 576–579.
24. Eichele, G., Tickle, C., and Alberts, B. M. (1984) Microcontrolled release of biologically active compounds in chick embryos: Beads of 200-μm diameter for the local release of retinoids. *Anal. Biochem.* **142**, 542–555.
25. Nguyen, V. T., Morange, M., and Bensaude, O. (1988) Firefly luciferase luminescence assays using scintillation counters for quantitation in transfected mammalian cells. *Anal. Biochem.* **1781**, 404–408.

II

RETINOID-BINDING PROTEINS

4

Immunohistochemistry for CRBPs and CRABPs

Ulf Eriksson and Anne-Lee Gustafsson

1. Introduction

The existence of two classes of intracellular retinoid-binding proteins, the cellular retinol-binding proteins (CRBP I and II) and cellular retinoic acid-binding proteins (CRABP I and II), have been known for a long time (*1*). Yet, the precise roles of these proteins in retinoid physiology remains poorly defined. One of the major reasons for this has been the difficulties of generating high-quality antibodies to facilitate their detection in tissues and cells by immunohistochemical techniques.

The structures of the cellular retinoid-binding proteins are highly conserved during evolution, rendering them poorly immunogenic, and only a few reports on the use of polyclonal antibodies (PAb) raised against the intact proteins have been published. Given the abundance of the structurally related class of fatty acid-binding proteins, and the problems associated with generation of conventional antibodies to the cellular retinoid-binding proteins, we have developed procedures for the generation of polyclonal antibodies to synthetic peptides derived from CRBP I and CRABP I and the use of such antibodies in the immunohistochemical localization of these two proteins in various tissues and cells. In this chapter, we describe how we generate the polyclonal antipeptide antibodies to CRBP I and CRABP I, and give detailed protocols for the use of these antibodies in the localization of these two proteins in mouse embryos using regular paraffin sections, vibratome sections, and in whole-mount preparations. Similar protocols may be used to generate and use antipeptide antibodies to CRBP II and CRABP II.

From: *Methods in Molecular Biology, Vol. 89: Retinoid Protocols*
Edited by: C. P. F. Redfern © Humana Press Inc., Totowa, NJ

2. Materials

2.1. Preparation of Peptide Conjugate and Generation of Antipeptide Antibodies to CRBPs and CRABPs

1. 100 mM sodium phosphate buffer, pH 7.5, containing 100 mM NaCl.
2. Keyhole limpet hemocyanin (KLH, Sigma or Calbiochem): dissolve 10 mg KLH per mL phosphate buffer.
3. N-succimidyl-3-(2-pyridyldithio)-propionate (SPDP, Pharmacia, Uppsala, Sweden), 40 mM solution in ethanol: dissolve 12.5 mg SPDP per mL of ethanol at 37°C, store at –80°C for up to 2 wk.
4. NAP-25 gel-filtration columns (Pharmacia).
5. Synthetic peptides derived from the amino acid sequences of the retinoid-binding proteins. Select hydrophilic-peptide segments. In many cases the extreme amino- and carboxy-terminal peptides of a given protein will generate antipeptide antibodies that crossreact with the native protein. However, for all retinoid-binding proteins those portions are relatively hydrophobic. We have successfully generated antipeptide antibodies to hydrophilic-peptide segments around amino acids 70–80 in CRBP I and CRABP I *(2)*. The amino acid sequences of those peptides are: GKEFEEDLTGIDDRKC (amino acid residues 67–82 in rat CRBP I) and GEGFEEETVDGRKC (amino acid residues 68–81 in bovine CRABP I). In general, the synthetic peptides should be 10–15 amino acids long and they should contain at least one terminal cysteine residue for coupling to the carrier protein via SPDP. In some cases, it has been shown that coupling of a peptide to the N- or C-terminal regions generates antipeptide antibodies with different characteristics, and to increase the likelihood of generating antipeptide antibodies that will crossreact with the corresponding native protein, it is desirable to have added cysteine residues in both N- and C-terminal ends of the peptides.

2.2. Immunohistochemical Staining

1. Phosphate buffered saline (PBS): 140 mM NaCl, 13 mM Na_2HPO_4, 3 mM KCl, 1.5 mM KH_2PO_4, final pH 7.4.
2. Dipping solution: A. Dissolve 0.1 g $KCrSO_4$ in 150 mL H_2O. B. Let 1 g gelatin swell in 50 mL H_2O for 15 min, warm the solution (50°C) for the gelatin to melt. Mix A and B.
3. Paraffin wax (melting point 52–54°C): e.g., Histovax™ (Histolab, Gothenburg, Sweden).
4. PBS containing Triton X-100 (PBST): 0.3% Triton X-100 in PBS, pH 7.4.
5. 1% H_2O_2 in PBST: Make up fresh immediately before use.
6. Xylene.
7. Ethanol: 100%, 95% v/v, 70% v/v.
8. Blocking solution. Stock solution: Dissolve 10 g bovine serum albumin (BSA) (type IV) in PBS, final volume 100 mL. Store as 1-mL aliquots at –20°C. Working solution (4%): Add 1.5 mL PBS to 1 mL of the stock solution.

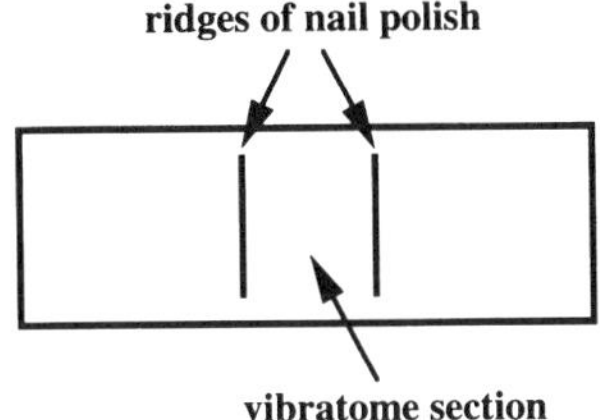

Fig. 1. Preparation of slides for mounting of vibratome sections.

9. AEC-solution: Dissolve 170 mg 3-amino-9-ethylcarbazole (AEC) in 100 mL DMSO; store as 5-mL aliquots at –20°C (*see* **Note 1**).
10. Acetate buffer (20 m*M*), pH 5.0; store as 42-mL aliquots at –20°C.
11. Glycerol-gelatin: Let 5 g gelatin swell in 100 mL H_2O for 15 min, add 90 mL glycerol, warm the solution at 50°C so the gelatin melts and a homogenous solution is achieved, then filter the warm solution through glass wool, and store in 10-mL aliquots at 4°C.
12. Tris-HCl (50 m*M*), pH 7.2.
13. Peroxidase-conjugated goat-antirabbit Ig (Dakopatts AB, Älvsjö, Sweden).
14. Mounting solution: Dissolve 50 mg *p*-phenylenediamine (PPD) in 5 mL PBS, pH 9.0, and mix with 45 mL cooled glycerol. Protect this solution from light and store at –20°C. PPD is easily oxidized: if a change in color of the solution is seen, discard and prepare a new solution.
15. DAB-solution: Solution A. Dissolve 5 mg 3,3-diaminobenzidine tetrahydrochloride (DAB) in 10 mL 50 m*M* Tris-HCl. This solution should be prepared fresh. Solution B. Dilute A 1:3 in 50 m*M* Tris-HCl (*see* **Note 1**).
16. Fluorescein isothiocyanate (FITC)-labeled donkey-antirabbit Ig (Jackson Immunoresearch, West Grove, PA) or any other secondary antibody labeled with a suitable fluorochrome.
17. Preparation of slides for mounting of vibratome sections: Take a slide and create two narrow ridges by means of applying nail polish (**Fig.1**).
18. 30% H_2O_2.
19. Myer's hematoxylin.

3. Methods

3.1. Coupling of the Synthetic Peptides to Carrier Protein and Immunization of Rabbits with the Peptide Conjugates

1. Add 50 µL of the 40 m*M* SPDP solution to 1 mL of KLH dissolved in 100 m*M* phosphate buffer and incubate for 1 h at room temperature. Vortex gently every now and then.
2. Remove uncoupled SPDP by gel filtration using a NAP-25 column equilibrated in the phosphate buffer. Elute the sample with the phosphate buffer in ten 500-µL

fractions and determine the absorbance at 280 nm for each fraction. The first eluted peak is KLH with coupled SPDP and the second peak is uncoupled SPDP. Pool the KLH-SPDP containing fractions.

3. Dilute 10 µL of the KLH-SPDP fraction with 90 µL of the phosphate buffer and measure the absorbance at 343 nm before and after the addition of 2 µL of dithiothreitrol. Calculate the coupling efficiency from the following formula using the molar-extinction coefficient for the release pyridine-2-tion, $8.08 \times 10^3 \ M^{-1}$.

$$\text{Amount of coupled SPDP} = [\text{pyridine-2-tion}] = \frac{\Delta A_{343} \times \text{dilution factor}}{8.08 \times 10^3} \ M$$

Under the conditions specified here, 8–12 molecules of SPDP are coupled per 1×10^5 Dalton of KLH.

4. To the SPDP-activated KLH fraction, add a 5X molar excess of the synthetic peptide and incubate for 1 h at room temperature or overnight at 4°C.

5. To determine the coupling efficiency, remove 10 µL of the peptide conjugate, dilute with 90 mL of the phosphate buffer, and measure the absorbance at 343 nm. Proceed as outlined in **step 3** and calculate the coupling efficiency. Most peptides have a coupling efficiency higher than 80%. If this number is not reached, incubate the peptide conjugate for a longer time. The peptide conjugate may be stored in aliquots at –20°C for up to several years.

6. For each rabbit to be immunized, emulsify peptide conjugate corresponding to 300–500 µg of KLH in Complete Freunds Adjuvant in a maximal volume of 2–3 mL. Subcutaneous injections at multiple sites on the back of the rabbit are recommended. Make sure to obtain some preimmune serum from the rabbits used for immunization.

7. Booster injections with the same amount of peptide conjugate emulsified in Incomplete Freunds Adjuvant is given every second or third week. Blood can be collected following the second booster injection and serum prepared by standard procedures. Normally, high-titer antipeptide antisera are obtained after two booster injections.

3.1.1. Characterization of antisera

A large number of techniques are available for testing the generated antipeptide antisera, including immunoblotting, enzyme-linked immunoassays, precipitation of radiolabeled proteins, and so on. Generally, the immune sera should be characterized using a technique relevant for the further usage. In many cases, we have found that affinity purification of antipeptide antisera using their intact, relevant antigens coupled to Sepharose beads might be necessary, and that they give excellent antibodies suitable for a broad range of analytical and preparative techniques. It is outside the scope of the chapter to describe in more detail such procedures. However, several excellent handbooks on those techniques are available (for example, *see* **ref. 3**).

3.2. Fixation, Paraffin-Embedding, and Sectioning of Embryos

1. Embryos are dissected free from decidual tissue and surrounding membranes, rinsed in PBS, and fixed in 4% formaldehyde in PBS (10 times the volume to be fixed). Gentle shaking is recommended. If embryos older than d 12 p.c. (post coitum) are to be used, we always divide the embryo into a head and a body part. To ensure rapid fixation of older embryos, tear the skin at multiple sites. Embryos are stored in 4% formaldehyde in PBS at 4°C (*see* **Note 2**).
2. Transfer the fixed embryos to 50% ethanol, and rotate the beaker at intervals. After 3–7 h, put the embryos in 70% ethanol and let them incubate overnight at room temperature. Transfer the embryos to 95% ethanol (4 h) and finally to 100% ethanol (4 h). Then the embryos are cleared in xylene until transparent. This takes about 10–15 min for a d-10 p.c. mouse embryo, whereas longer clearing time is needed for older embryos. Put the embryos in melted paraffin and infiltrate them overnight at 52–54°C (*see* **Note 3**).
3. Coating of slides: Dip the slides into the dipping solution (37°C) and air-dry, first at room temperature and then at 37°C overnight.
4. Orientation of embryo and paraffin sectioning are performed according to standard procedures *(4)*, and 5-µm sections are attached to the coated slides by incubation at 37°C overnight. The sections can be stored for extensive periods of time at room temperature.

3.3. Immunohistochemistry Using the Avidin–Biotin Complex Method

1. Select slides to be used and encircle the sections with a diamond pen, which will create surface tension, thereby minimizing the volume of solution needed to cover the sections.
2. Deparaffinize the sections in xylene overnight, and rehydrate them by incubating 5 min each in the following solutions: xylene, twice in 100% ethanol, twice in 95% ethanol, twice in 70% ethanol, and once in PBS.
3. Endogenous peroxidase activity is inactivated by incubating the slides in freshly made 1% H_2O_2 in PBST for 30 min (dark).
4. Rinse the sections once each in PBST and then in PBS.
5. Carefully wipe away excess buffer with a filter paper.
6. Nonspecific binding of antibodies is minimized by incubating the sections with blocking solution at room temperature in a humidity chamber for 1 h.
7. Tap off the blocking solution and wipe away excess solution with a filter paper.
8. Apply the affinity-purified primary rabbit antibody diluted in blocking solution, enough to cover the section (about 20–30 µL/section) and incubate the sections overnight in a humidity chamber at 4°C (*see* **Note 4**).
9. Dilute the biotinylated secondary antibody in blocking solution (we use a biotinylated swine antirabbit Ig [Dakopatt], diluted 1:500), and prepare the avidin–biotin complex by adding 4 µL of avidin and biotinylated peroxidase

to 500 µL of PBS (we use Dakopatts ABComplex/HRP). The avidin–biotin complex solution should be prepared 30 min before use and is stable for 3 d at 4°C.

10. Mix 42 mL acetate buffer with 5 mL AEC-solution and filter. Protect the AEC-solution from direct sunlight.

11. Tap off the primary antibody and rinse the slides three times in PBST and two times in PBS. This is best accomplished by putting the slides in a Hellendahl jar containing the washing buffer and vigorously shaking the jar over a sink.

12. Take up one slide at a time and wipe away excess buffer with a filter paper; then apply the biotinylated-secondary antibody and incubate the slides in a humidity chamber for 30 min at room temperature.

13. Wash as in **step 11**. Then wipe away as much buffer as possible from the slide with a filter paper and apply the avidin–biotin complex. Incubate for 30 min at room temperature in a humidity chamber.

14. Wash as in **step 11**.

15. Activate the filtered AEC-solution by adding 5 µL of 30% H_2O_2 to the filtrate. Pour off the PBS from the Hellendahl jar containing the slides and replace it with the activated AEC-solution. Develop for 15 min at room temperature, and protect from direct light.

16. Rinse the slides in PBS, counterstain in Myer's hematoxylin for 3 min, rinse in tap water for 10 min, and mount with glycerol-gelatin. A representative section showing CRBP I-staining of a mouse embryo limb bud is shown in **Fig. 2** (*see* **Note 5**).

3.4. Fluorescence Immunohistochemistry

1. Follow **steps 1** and **2, Subheading 3.3.**

2. Incubate the sections for 10 min in PBST (in a Hellendahl jar), and rinse the sections in PBS.

3. Carefully wipe away excess buffer with a filter paper.

4. Nonspecific binding of antibodies is minimized by incubating the sections with blocking solution at room temperature in a humidity chamber for 1 h.

5. Tap off the blocking solution and use filter paper to wipe away excess liquid.

6. Apply the primary rabbit antibody (anti-CRBP I or anti-CRABP I) diluted in blocking solution, enough to cover the section (about 20–30 µL per section), and incubate the sections overnight in a humidity chamber at 4°C (*see* **Note 6**).

7. Tap off the primary antibody and wash the slides 3 times in PBST, once for 10 min in 50 m*M* Tris-HCl and once for 10 min in PBS at 4°C on a shaker.

8. Remove excess PBS from slide with filter paper and apply the FITC-labeled donkey-antirabbit Ig (Jackson Immunoresearch), diluted 1:50 in PBS, pH 8.6, and incubate the sections for 2 h in a humidity chamber. Protect from excessive light from here on.

9. Tap off the antibody and rinse the slides twice with PBST, then continue washing the slides at 4°C on a shaker and protected from light, 10 min each in the follow-

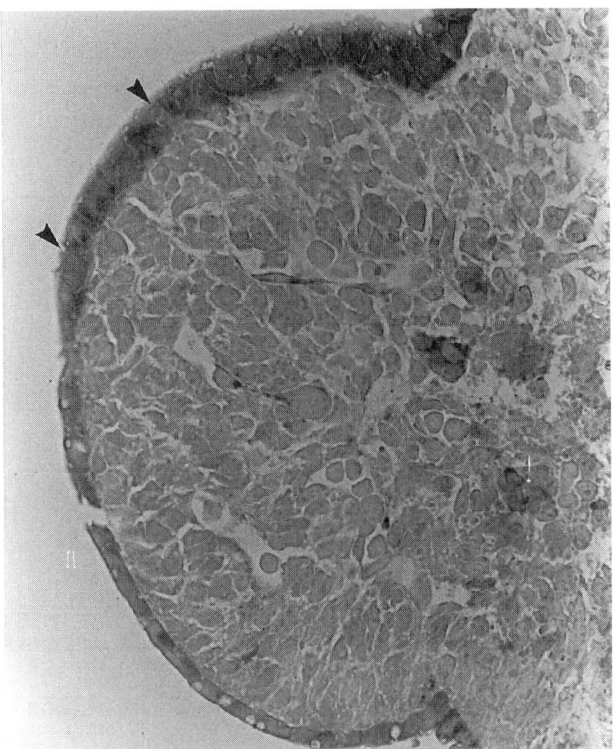

Fig. 2. Immunohistochemical localization of CRBP I in a paraffin section of a mouse-embryo limb bud (d 10 p.c.). Note the intensive staining in the surface ectoderm surrounding the limb bud (arrowheads).

ing buffers: 50 m*M* Tris-HCl, PBS, 50 m*M* Tris-HCl, and PBS. The slides are then washed overnight in 50 m*M* Tris-HCl.

10. Mount the sections using the PPD-mounting media. The cover slips are applied by using instant glue.

3.5. Whole-Mount Immunohistochemistry

Embryos younger than 10.5 p.c. are stained as whole embryos, whereas for older embryos, vibratome sections are used (*see* **Subheading 3.5.1.**). Gentle shaking is recommended during all incubations and washing steps.

1. Rinse fixed embryos 2×1 h in PBST.
2. Block endogenous-peroxidase activity by incubating the embryos 24 h at 4°C in PBST containing 0.09% H_2O_2.
3. Wash with PBST 4×2 h at 4°C.

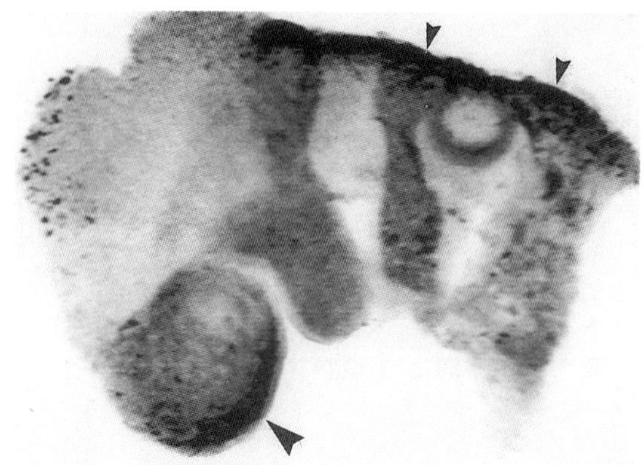

Fig. 3. Whole-mount immunohistochemical localization of CRABP I in a mouse-embryo (d 9 p.c). Specific staining is seen in migrating neural-crest cells, in the frontonasal mesenchyme (large arrowhead) and in several populations of neurons, particularly in the neural tube (small arrowhead).

4. Incubate the embryos with the primary antibody, diluted in fetal bovine serum (FBS) containing 0.3% Triton X-100 and 0.1% NaN$_3$, 4 d at 4°C (*see* **Note 6**).
5. Wash 4 × 2 h in PBST at 4°C.
6. Incubate embryos with peroxidase-conjugated goat-antirabbit Ig (Dakopatt), diluted 1: 400 in 2% BSA in PBST, 2 d at 4°C.
7. Wash 4 × 2 h at 4°C in PBST and 1 × 2 h in PBS.
8. Incubate embryos in 50 mM Tris-HCl, at 4°C for 30 min.
9. Replace the buffer with DAB-solution A, and incubate at 4°C for 3 h.
10. Color is developed by incubating the embryos in DAB-solution B containing 0.008% H$_2$O$_2$. The reaction is allowed to develop in the dark (3–10 min) and stopped by rinsing in PBS.
11. Clear the tissue by incubating the sections in increasing concentrations of glycerol up to 80%.
12. Embryos are stored in 80% glycerol containing 0.1% NaN$_3$. A representative whole-mount mouse embryo stained for CRABP I is shown in **Fig. 3.**

3.5.1. Gelatin-Embedding and Vibratome Sectioning.

1. Rinse fixed embryos 2 × 1 h in PBST, transfer to a 10–30% w/v gelatin in PBS solution (higher concentrations of gelatin are needed for older embryos), incubate at 37°C for 1 h in a tissue-culture dish, and then allow the gelatin to stiffen at 4°C.
2. Cut out a block containing the whole embryo or a part of it. The height of the block intended for sectioning should be kept rather low.
3. 40-μm Sections are cut using a vibratome. All cutting is done in ice-cold PBS. The sections can then either be transferred to 4% formaldehyde in PBS and stored at 4°C, or put in PBS and further processed for immunohistochemistry as described earlier (*see* **Note 7**).

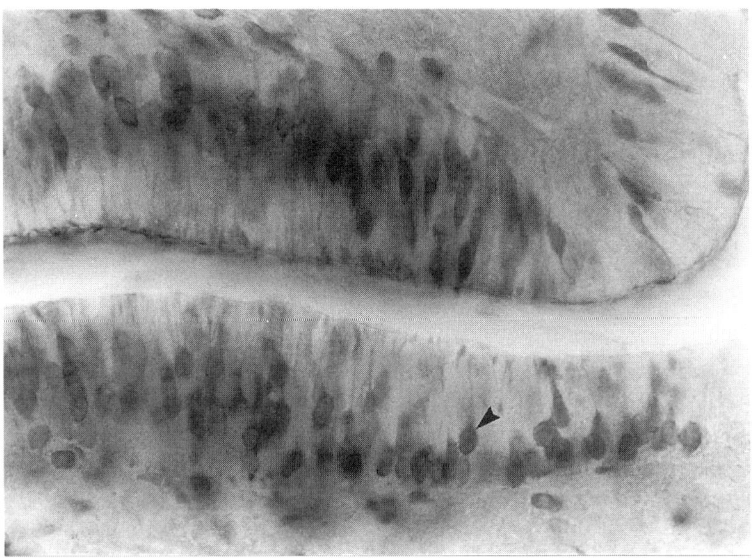

Fig. 4. Immunohistochemical localization of CRABP I in a vibratome section of the olfactory epithelium of a mouse embryo (d 13 p.c). An intense staining of the olfactory-receptor cells is seen (arrowhead).

4. After the staining the sections are mounted in 80% glycerol containing 0.1% NaN$_3$ on prepared slides (*see* **Subheading 2.** and **Fig. 1**). The cover slip is fixed to the ridges by applying more nail-polish. Store the slides horizontally at 4°C. A representative vibratome section of the olfactory epithelium of a mouse embryo stained for CRABP I expression is shown in **Fig. 4.**

4. Notes

1. AEC and DAB are suspected carcinogens and should be handled with great care!
2. We have had embryos stored for prolonged time periods (for up to several months) in fix-solution without seeing any effects on the staining pattern of CRBP I and CRABP I.
3. The dehydration and clearing step are done using a 10-fold excess of solution compared to the volume of the tissue. All incubations as well as the paraffin embedding are done in 5-mL polypropylene beakers (embryos up to d 10 p.c.) or in 20-mL glass beakers (older embryos), so that the clearing process can be checked easily. For transferring embryos, we use the large end of a Pasteur pipet. Because it is critical to keep the temperature below 56°C during the paraffin infiltration, we use Histovax, which melts at 52–54°C (the paraffin wax used for traditional paraffin sectioning melts at 56°C).
4. By using a long incubation time we can use very low concentrations of the antibodies; for CRABP I usually 0.5 µg/mL, and for CRBP I 0.3 µg/mL.

5. Valuable general information regarding immunohistochemical staining methods and good help with troubleshooting is given in Dakopatts booklet: Immunohistochemical staining methods *(5)*.
6. We use the affinity-purified anti-CRBP I- and anti-CRABP I-antibodies at concentrations of 3 and 5 µg/mL, respectively.
7. For storage of vibratome sections, as well as for the immunohistochemical procedures, we have found it very convenient to use 16-microwell plates.

References

1. Ong, D. E., Newcomer, M. E., and Chytil, F. (1994) Cellular retinoid-binding proteins, in *The Retinoids: Biology, Chemistry and Medicine* (Sporn, M. B., Roberts, A. B., and Goodman, D. S., eds.), Raven, New York, NY, pp. 283–317.
2. Busch, C., Saksena, P., Funa, K., Nordlinder, H., and Eriksson, U. (1989) Tissue distribution of cellular retinol-binding protein and cellular retinoic acid-binding protein: Use of monospecific antibodies for immunohistochemistry and cRNA for *in situ* localization of mRNA. *Methods Enzymol.* **189,** 315–324.
3. Harlow, E. and Lane, D. (1988) *Antibodies. A laboratory manual.* Cold Spring Harbor Laboratory, Cold Spring Harbor, New York.
4. Gordon, K. C. and Bradbury, P. (1990) Microtomy and paraffin sections, in *Theory and Practice of Histological Techniques*, 3rd ed. (Bancroft, J. D. and Stevens, A., eds.), Churchill Livingstone, Edinburgh, Scotland, pp. 61–80.
5. Boenisch, T., Farmilo, A. J., and Stead, R. H. (1989) *Handbook: Immunohistochemical staining methods* (Naish, J. L., ed.), DAKO Corporation, Carpinteria, CA.

5

Whole-Mount *In Situ* Hybridization of Mouse Embryos Exposed to Teratogenic Levels of Retinoic Acid

Sally D. Lyn

1. Introduction

In situ hybridization (ISH) allows the localization and relative quantification of mRNA expression for a given gene. Typically, tissue is fixed, permeablized, and then incubated with a labeled riboprobe. Once excess probe has been removed, the hybridized riboprobe that remains is detected by means appropriate to the label. Variations of the technique have been extensively used to analyze the expression patterns of many genes in the tissues of a wide variety of organisms. Radiolabeled probes are commonly used, but necessitate the use of sectioned tissue since the signal must be detected by autoradiography. Whereas this allows precise localization of signal in two dimensions, the reconstruction of three-dimensional expression patterns in complex structures is often difficult. In addition, it is very difficult to compare changes in expression patterns between samples owing to differences in the orientation of each sectioned tissue sample, as well as variations in tissue-section thickness and photoemulsion depth.

Recent improvements in nonradioactive labels have allowed the development of whole-mount ISH. As well as eliminating some of the variation inherent in sectioned tissue, whole-mount ISH allows three-dimensional expression patterns to be viewed directly. This is a particular advantage when expression patterns are examined in complex structures such as embryos. In fact, the whole-mount technique is ideal for the analysis of gene expression during early development. A relatively large number of embryos may be processed in parallel, allowing the expression of a given gene to be closely followed through several stages of embryonic development. In addition, the effects of a com-

From: *Methods in Molecular Biology, Vol. 89: Retinoid Protocols*
Edited by: C. P. F. Redfern © Humana Press Inc., Totowa, NJ

pound (such as a teratogenic dose of a retinoid) on expression levels and patterns can be analyzed in a number of embryos and compared to expression in control embryos at the same developmental stages.

The protocol presented in this chapter is derived from previously published methods *(1,2)*. It has been used to investigate the effects of retinoic acid (RA) on the expression of cellular retinoic-acid binding protein (CRABP) -I and -II in relatively early mouse embryos (6.5–9.5-d postcoitum [p.c.]; E6.5–9.5; *[3]*). Unfortunately, the size of the tissue sample that can be used for whole-mount ISH is limited, since both a riboprobe and an antibody-enzyme conjugate must penetrate the specimen and the unbound excess of both must be removed. Early embryos are small enough to allow penetration and removal of probe and antibody. This procedure should also work well for isolated pieces of larger embryos.

The embryos are dissected from pregnant mice following a teratogenic dose of RA. Digoxigenin-labeled riboprobes are hybridized to the fixed embryos, excess probe is removed, and the embryos are incubated with an anti-digoxigenin antibody-alkaline phosphatase conjugate. Once the unbound antibody-enzyme conjugate is removed, an alkaline phosphatase-dependent chromogenic reaction is performed. The resulting blue precipitate allows the three-dimensional expression patterns to be viewed directly in the cleared embryos.

Specific whole-mount ISH protocols for the embryos, larvae, or small pieces of tissue of a number of other organisms have been published, including *Xenopus (4,5)*, sea urchin *(6,7)*, zebrafish *(8)*, *Drosophila (9)*, chicken *(10,11)*, axolotl *(12)*, and the leech *(13)*. Protocols for the simultaneous hybridization and subsequent detection of two probes with different nonradioactive labels are also available *(14,15)*.

2. Materials

Precautions should be taken to ensure that all solutions, spatulas, plasticware, and glassware are sterile and free of any dust throughout the procedure. All materials must be RNase-free for all stages between the initial fixation and the post-hybridization washes (*see* **Note 1**).

2.1. Preparation of Digoxigenin-Labeled Riboprobes

1. Template DNA: 250–1500 bp *(2)* of the coding region of the gene of interest should be cloned into an appropriate plasmid as described in **Subheading 3.1.** Plasmids may be purified by either the Qiagen® method or centrifugation in CsCl-Ethidium bromide gradients *(16)*.
2. Restriction enzyme and buffer: Preferably a restriction enzyme which does not produce termini with 3′ overhangs (*see* **Subheading 3.1.1.**).

3. Bacteriophage T4 DNA polymerase or the Klenow fragment of *E. coli* DNA polymerase I and appropriate buffer: Only required if it is necessary to remove 3′ overhangs (*see* **Subheading 3.1.1.**).
4. TE buffers: 10 mM Tris-HCl, 1 mM EDTA adjusted to pH 8.0 (TE, pH 8.0) and pH 7.6 (TE, pH 7.6).
5. Phenol equilibrated with TE, pH 8.0 or 100 mM Tris-HCl, pH 8.0: Prepare from unequilibrated liquefied phenol *(16)* or purchase already equilibrated (Sigma, Boehringer Mannheim). Store at 4°C.
6. Phenol/chloroform: Mix equilibrated phenol 1:1 v/v with 24:1 v/v chloroform: isoamyl alcohol *(16)*. Phenol/chloroform is also commercially available (Sigma, Boehringer Mannheim). Store at 4°C.
7. Chloroform.

The following must be RNase-free (*see* **Note 1**):

8. Ethanol.
9. Microcentrifuge tubes: Purchase sterile and RNase-free (Boehringer Mannheim), or DEPC-treat (*see* **Note 1**).
10. DIG RNA Labeling Kit (Boehringer Mannheim).
11. 0.2 M EDTA, pH 8.0.
12. 4 M LiCl.

2.2. Preparation of Mouse Embryos

1. Mice: Both males and females should be at least 6 wk old before they are bred.
2. Gastric lavage (gavage) needles (optional; *see* **Subheading 3.2.1.**).
3. All-*trans* retinoic acid (RA; Sigma): Limit exposure of RA to light as much as possible. On the day of treatment (optional; *see* **Subheading 3.2.1.**), make a stock solution of 25–50 mg/mL RA in dimethyl sulfoxide (DMSO). Dilute 1:10 in corn oil shortly before administering. Reseal the opened ampoule of desiccated RA with Parafilm M® and do not reuse more than 2 wk after the date of opening. Store ampoule at –20°C.
4. Sterile disposable plastic 60-mm diameter Petri dishes (Corning, Falcon).
5. Dumont or Jewelers' forceps: at least two pairs each of straight forceps are required in sizes 3 and 5 (George Tiemann & Co., Hauppauge, NY, or Fine Science Tools, Foster City, CA, and Haverhill, Suffolk, UK) for microdissection.
6. Binocular dissection microscope with incident light source and transillumination base.
7. 1X Phosphate-buffered saline without Ca^{2+} or Mg^{2+} (PBS; *see* **Note 2**) at 4°C. To make 10X stock, dissolve 0.2 g KCl, 8 g NaCl, 0.2 g KH_2PO_4, and 1.15 g Na_2HPO_4 (or 2.89 g $Na_2HPO_4 \cdot 12\ H_2O$) in water to make 1 L. PBS for dissection need only be sterile filtered. For all other uses, it must be RNase-free (*see* **Note 1**). (A 20X stock is required to make the formaldehyde fix.)
8. Glass scintillation vials with foil lined caps, or 10-mL Reacti-Vials (Pierce) with Teflon-lined rubber-cap liners.

9. Formaldehyde fix (FPBS): Heat 190 mL water to 60°C. Add 8 g electron micro-scopy (EM) grade paraformaldehyde and a few drops of 1 *N* NaOH, and stir in a fumehood to dissolve. Allow the solution to cool to room temperature once the paraformaldehyde has dissolved, and add 10 mL 20X PBS. Formaldehyde fix should be made the day it is to be used; always mix and use in a fumehood.

The following must be RNase-free (*see* **Note 1**):

10. PBT: 1X PBS, 0.1% Tween-20.
11. Methanol.
12. Methanol/PBT series: 25, 50, and 75% methanol:PBT.
13. Methanol/peroxide bleach: 5:1 methanol:30% hydrogen peroxide. Prepare on the day of use.

2.3. Whole-Mount ISH

Precautions should be taken to ensure that the following solutions, with the exception of FGPBS, are RNase-free (*see* **Note 1**).

1. Methanol/PBT series and PBT: *see* **Subheading 2.2.**
2. RIPA: 150 m*M* NaCl, 1% NP-40, 0.5% sodium deoxycholate, 0.1% SDS, 1 m*M* EDTA, 50 m*M* Tris-HCl, pH 8.0.
3. FGPBS fix: make FPBS as described in **Subheading 2.2.**, then add EM grade glutaraldehyde (Sigma Grade 1) to 0.2% v/v final, i.e., add 1/40th volume 8% glutaraldehyde (*see* **Note 3**).
4. 20X SSC: 3 *M* NaCl, 0.3 *M* sodium citrate, pH 7.0.
5. 50 mg/mL heparin.
6. 10% Tween-20.
7. Hybridization Buffer: 50% formamide (*see* **Note 3**), 5X SSC, 50 µg/mL heparin, 0.1% Tween-20. Bring to pH 5.0 with 1 *M* citric acid.
8. ssDNA: Stir 10 mg/mL salmon-sperm DNA (Sigma type III sodium salt) in water at room temperature until dissolved. Add 1/50th volume of 5 *M* NaCl and extract with phenol and phenol/chloroform. Force the aqueous phase through a 17-gage needle 12 times to shear the DNA. Extract again with phenol/chloroform: use an RNase-free pipet tip and take care to avoid the interphase above the organic layer when removing the aqueous phase. Transfer the aqueous phase to a sterile, RNase-free tube and add 2 vol of ethanol at 4°C to precipitate the DNA. Pellet the DNA by centrifugation, rinse with 70% ethanol/RNase-free water, and redis-solve in RNase-free water at approx 10 mg/mL. Use a spectrophotometer to determine the exact DNA concentration. Heat the solution in a boiling water bath for 10 min. When cool, aliquot into RNase-free tubes and store at –20°C. Dena-ture the DNA shortly before use by heating for 5 min in a boiling water bath, then transferring immediately to an ice-water bath for 5 min (adapted from Sambrook et al. *[16]*).
9. tRNA: Prepared similarly to ssDNA (*see* **step 8**) but not sheared or denatured. Dissolve yeast tRNA to 10 mg/mL in sterile TE, pH 7.6 containing 0.1 *M* NaCl.

Extract twice each with equilibrated phenol and chloroform, precipitate at room temperature with 2.5 volumes ethanol. Spin at 5000g for 15 min at 4°C to pellet the RNA. Redissolve at 10 mg/mL, aliquot into RNase-free tubes and store at –20°C *(16)*.

The following solutions should NOT be DEPC-treated or made with DEPC-treated water.

10. 10X TBS: dissolve 40 g NaCl and 1.0 g KCl in 125 mL 1 M Tris-HCl, pH 7.5, and water to 500 mL. Autoclave or pass through a 0.2 µm filter to sterilize.
11. TBST: 1X TBS with 0.1% Tween-20. Filter to sterilize.
12. Goat or sheep serum: thaw, aliquot, and store at –20°C until day of post-hybridization washes.
13. Embryo-acetone powder, prepared from embryos in the E11.5 to E13.5 range by standard methods *(17)*: Place freshly dissected embryos in a minimum of PBS (without added Ca^{2+} or Mg^{2+}) on ice. Homogenize the embryos on ice, add four volumes of cold acetone to the homogenate, and vortex or shake vigorously. Keep the mix on ice for 30 min; shake occasionally. Centrifuge for 10 min at 4°C and 10,000g. Discard the supernatant, resuspend the precipitate with cold acetone, shake vigorously, and place on ice for 10 min. Repeat centrifugation as before and remove the supernatant. Use a spatula to spread the pellet onto a piece of filter paper at room temperature and mince continuously until it dries to a powder. Store at 4°C in an airtight container.
14. Anti-digoxigenin-alkaline phosphatase conjugate (Boehringer Mannheim).
15. Substrate buffer (SB): 100 mM NaCl, 50 mM $MgCl_2$, 0.1% Tween-20, 100 mM Tris-HCl, pH 9.5. Make from stock solutions on day of use (*see* **Note 4**).
16. Levamisole (Sigma; optional; *see* **Note 4**).
17. NBT: dissolve 75 mg/mL nitro blue tetrazolium chloride (Pierce, Sigma) in 70% N,N-dimethyl formamide (DMF). Store at –20°C.
18. BCIP: dissolve 50 mg/mL 5-bromo-4-chloro-3-indolyl phosphate (Pierce, Sigma) in 50% DMF. Store at –20°C.
19. NBSB: dilute 4.5 µL/mL NBT and 3.5 µL/mL BCIP in SB just before use.
20. PBTE: PBT containing 1 mM EDTA.

3. Methods

3.1. Preparation of Digoxigenin-Labeled Riboprobes

A fragment of the coding region of the gene of interest should be subcloned into the polylinker of a plasmid containing at least one promoter (T7, SP6, or T3) for RNA polymerase. A plasmid containing a "head to head" pair of these promoters may be used, such as pGEM-4 (T7 and SP6; Promega) or pBluescript (T7 and T3; Stratagene). Digoxigenin-labeled run-off transcripts may then be prepared from each promoter to produce both sense and antisense probes. Alternatively, prepare separate plasmids for the production of sense and antisense probes by subcloning the same fragment in opposite orientations relative to one type of promoter. This will eliminate polymerase-specific size differences between sense and antisense probes (owing to premature termination of transcription). In either case, the yield and size of the labeled transcripts should be double-checked by agarose gel electrophoresis *(16)*.

3.1.1. Preparation of Plasmid Templates

1. Each plasmid should be linearized by restriction enzyme digestion; choose a unique restriction site at the end of the insert farthest from the promoter to be used. Run a small fraction of the digest on an agarose gel to confirm that the plasmid templates are completely linearized *(16)*.
2. If a restriction enzyme that leaves protruding 3' termini is used, the overhang should be removed with bacteriophage T4 DNA polymerase or the Klenow fragment of *E. coli* DNA polymerase I *(16)*.
3. Extract the restriction digest once each with phenol/chloroform and chloroform. After transferring the aqueous phase to an RNase-free microcentrifuge tube, ethanol precipitate, wash and dry by standard methods *(16)*. Resuspend the pellet to between 0.25 and 1 mg/mL.

3.1.2. Transcription of Riboprobes

Digoxigenin-labeled riboprobes are both transcribed (using the DIG RNA Labeling Kit [Boehringer Mannheim]) and quantitated as described by the manufacturer (*see* **Note 5**). Store digoxigenin-labeled riboprobes at –70°C (*see* **Note 6**).

3.2. Preparation of Mouse Embryos

3.2.1. Breeding and Gavage

1. House each male overnight with one or two female mice in estrous.
2. Check the female mice the next morning; a white vaginal "plug" can be observed in females that have mated (*see* **Note 7**).
3. If the effects of a teratogenic dose of RA are to be investigated, 20–50 mg/kg (maternal body weight) RA in 0.2 mL 1:10 DMSO:oil (or 0.2 mL 1:10 DMSO:oil alone for controls) may be administered by gastric lavage at one or more time points (e.g., at 4 h [for immediate effects] to 20 h [for longer term effects]) before dissection (*see* **refs.** *1, 3,* and *18*; *see* **Note 8**).

3.2.2. Dissection

1. Sacrifice each mouse by cervical dislocation at 6.5–9.5 d p.c. (E6.5–9.5; *see* **Note 9**).
2. Remove the uterine horns and transfer to a Petri dish containing cold PBS. Open the uterine horns carefully with the #3 forceps and transfer the decidua to a fresh dish of cold PBS *(19)*.
3. Use the finer #5 forceps to snip open each deciduum and free the embryo. Once free of the decidual tissue, remove the ectoplacental cone of embryos up to

approximately E8; these embryos need no further dissection. More advanced embryos should be dissected from the remaining extraembryonic membranes *(19)*. If the neural tube has formed (about E9 and older), snip a small hole through the thinnest region in the roof of the rhombencephalon with fine forceps (*see* **Note 10**) and puncture the heart chambers.

3.2.3. Fixation and Pretreatment I

1. Rinse the dissected embryos in PBS and transfer to a scintillation vial or Reacti-Vial filled with cold FPBS (*see* **Note 11**). Slowly rotate the vials about the long axis at 4°C overnight or for a minimum of 2 h (E6 embryos) to 5 h (E9.5 embryos). Unless noted otherwise, the vials should be slowly rotated or gently rocked throughout the remaining washes and incubations.
2. Wash the embryos three times in PBT (*see* **Note 12**) and dehydrate by washing with the methanol/PBT series (5–10 min each in 25, 50, 75% methanol and twice in 100% methanol [*see* **Note 13**]).
3. Bleach the embryos for 3–5 h at room temperature in freshly mixed 5:1 methanol/30% hydrogen peroxide.
4. Wash the embryos three times in 100% methanol and store at –20°C until ready to hybridize.

3.3. Whole-Mount ISH

3.3.1. Fixation and Pretreatment II

1. Place the vials at room temperature until warm, then rehydrate by washing through the methanol/PBT series (about 10 min each in 75, 50, and 25% methanol).
2. Wash the embryos three times with PBT, three times for 30 min per wash in RIPA, then fix at room temperature in FGPBS for exactly 20 min.
3. Perform three 5-min washes with RIPA and three 5-min washes with PBT.
4. Transfer groups of embryos into separate tubes (*see* **Note 14**) for hybridization with different probes.

3.3.2. Prehybridization and Hybridization

1. Wash the embryos at room temperature in 1:1 PBT:hybridization buffer until they no longer float (5–10 min).
2. Wash for another 5–10 min in undiluted hybridization buffer, then prehybridize for 1–5 h without rocking at 70°C in hybridization buffer containing 100 µg/mL ssDNA and 100 µg/mL tRNA.
3. Heat an aliquot of riboprobe to 95°C for approx 3 min to denature, then place on ice for at least 3 min.
4. Replace prehybridization buffer with hybridization buffer containing 100 µg/mL ssDNA, 100 µg/mL tRNA, and 0.1–2 µg/mL of denatured digoxigenin-labeled riboprobe. Seal well, rotate very slowly until well mixed, then incubate overnight at 70°C without rocking.

3.3.3. Posthybridization Washes, Blocking, and Antibody Binding

1. Begin the posthybridization embryo washes (*see* **step 2**). In the meantime, place an aliquot of normal goat or sheep serum at room temperature to thaw. Resuspend several milligrams of embryo acetone powder in 1 mL of TBST. Heat-inactivate both the thawed serum and the embryo powder/TBST for 30 min at 70°C (*see* **Note 15**). Once removed from the 70°C water bath, the serum should be vortexed continuously until close to room temperature as it has a tendency to congeal while cooling. Pulse-spin the embryo powder, remove the supernatant, and cool on ice. Resuspend the cooled precipitate with a 1:500 dilution of anti-digoxigenin-alkaline phosphatase conjugate in cold 1% freshly inactivated serum/TBST. Slowly rotate or agitate for 1 h at 4°C, then centrifuge at approximately 10,000*g* for 10 min at 4°C. Dilute the resulting supernatant in cold 1% freshly inactivated serum/TBST to a final antibody concentration of 1:2000 to 1:5000.
2. Wash the embryos two times for 10 min each at 70°C with prewarmed hybridization buffer (without ssDNA, tRNA, or riboprobe).
3. Next, wash twice for 5 min and three times for 30 min in 2X SSC/50% formamide/0.1% Tween-20 at 65°C. After the third 30-min wash, place the embryos at room temperature.
4. Once cool, they should be washed three times with TBST.
5. Incubate the embryos for 1 h at room temperature in 10% freshly heat-inactivated serum in TBST to block nonspecific antibody binding.
6. Replace blocking solution with the dilute preadsorbed antibody-conjugate in serum/TBST (*see* **step 1** of this section) and rotate embryos overnight at 4°C.

3.3.4. Color Development and Clearing

1. Three 5-min washes should be followed by at least three 30-min washes or up to five 1-h washes (depending on the size of the embryos), all in TBST (*see* **Note 16**).
2. Follow with three 10-min washes in SB.
3. Transfer the embryos to a small glass dish, replace SB with NBSB, and protect from strong light. Swirl or rock to mix well, then cover and allow color to develop in the dark at room temperature without rocking.
4. After approx 15 min (and occasionally thereafter if necessary), briefly inspect the embryos under a dissecting microscope for the development of purple staining (*see* **Note 17**).
5. Rinse and then wash the embryos three times each in PBTE to stop the color reaction, and transfer them to a sealable tube or vial (*see* **Note 18**).
6. Wash the embryos three times in TBST at room temperature.
7. Dehydrate through the methanol/TBST series (30, 50, and 70% methanol/TBST followed by two changes of 100% methanol), and then rehydrate back through the series to TBST.
8. Slowly rotate the embryos for 1 h each in 50 and 80% glycerol/PBTE at room temperature (*see* **Note 19**).

9. Use a binocular dissecting microscope to analyze and photograph the finished embryos (*see* **Note 20**).

4. Notes

1. The possibility of RNase contamination can be reduced by ensuring that disposable gloves are worn and changed frequently. Unfortunately, RNases are not irreversibly denatured by autoclaving alone. Heat-resistant materials such as glassware and metal spatulas should be baked at 180°C for at least 8 h. Commercially packaged sterile plasticware is generally RNase-free and should be used when possible. There are several methods that may be used to inhibit or remove RNases from other labware, depending on its composition:
   a. Most heat-resistant plasticware may be soaked for at least 2 h at 37°C in 0.1% diethyl pyrocarbonate (DEPC; *see* below), rinsed with sterile DEPC- or DMPC-treated water (*see* below) and heated to 100°C for 15 min or autoclaved for 15 min at 121°C (15 lb/sq. in.) on a liquid/slow exhaust cycle *(16)*.
   b. Rinse with chloroform *(16)*.
   c. Soak equipment in 0.1 N NaOH, 0.1% EDTA overnight and then rinse thoroughly with RNase-free water *(20)*.
   d. Wash equipment (such as electrophoresis tanks) with detergent, rinse extensively with water followed by 95% ethanol. When dry, soak in 3% hydrogen peroxide for 10 min at room temperature, then rinse thoroughly with sterile DEPC- or DMPC-treated water (*see* **ref. *16***; *see* below).

   Chemicals should be reserved exclusively for RNA work, and handled with gloves and RNase-free spatulas or disposable sterile plastic pipets. Liquid chemicals reserved for RNA work may be considered RNase-free and should not be DEPC-treated.

   The majority of aqueous solutions suitable for autoclaving (*see* below for exceptions) may first be pretreated with 0.1% DEPC for at least 12 h at 37°C *(16)*. Unless noted otherwise, solutions unsuitable for autoclaving should be made up with DEPC-treated water and passed through a 0.2-μm filter to sterilize. DEPC should always be used in a fume hood, as it is a powerful acylating agent *(20)* and a suspected carcinogen *(16)*. Never add DEPC to aqueous solutions containing ammonia, as this will result in the formation of a potent carcinogen. Reagents with amines and sulfhydryl groups, such as Tris and DTT, cannot be directly treated with DEPC. These solutions should be made up in DEPC-treated water and autoclaved or sterile filtered *(20)*.

   Water may be treated with DEPC (as above) or dimethyl pyrocarbonate (DMPC; Sigma), a less-toxic alternative. If DMPC is chosen, dissolve to 1% in a 1:1 ethanol:water mixture, then further dilute 1:10 in distilled deionized water. Incubate this 0.1% DMPC solution at room temperature for 30 min, then autoclave (Boehringer Mannheim). Note that "DEPC- or DMPC-treated water" always refers to water that has been autoclaved following the treatment.

2. PBS with added Ca^{2+} and Mg^{2+} should not be used or the embryos will become adhesive.

3. Formamide and glutaraldehyde should be aliquoted after opening, sealed tightly, and stored at –20°C. Use each aliquot only once.

4. Levamisole may be used if endogenous alkaline phosphatase activity has been observed in control embryos which were not exposed to anti-DIG-antibody-enzyme conjugate. Levamisole does not affect the activity of mammalian intestinal alkaline phosphatase (the form used in most antibody-enzyme conjugates), but including 1 m*M* freshly dissolved levamisole in the substrate buffer will inhibit the form of alkaline phosphatase that is present in a number of other mammalian tissues *(21)*.

5. DNase digestion to remove the template DNA after transcription is not necessary. However, ethanol precipitation of the stopped reaction is required in order to remove unincorporated NTPs, which can increase background staining of hybridized embryos. Avoid extracting digoxigenin-labeled nucleic acids with phenol, as the probe tends to partition with the organic phase.

6. Digoxigenin-labeled riboprobes stored at –70°C are stable for at least 1 yr provided that RNase-free conditions are maintained (Boehringer Mannheim).

7. Since mice generally mate near the middle of the dark cycle, the day the plug is observed should be considered as 0.5 d p.c. (E0.5).

8. Gene expression in untreated embryos should be analyzed first. Gastric lavage should only be performed by those trained in the procedure.

9. Dissect one mouse at a time so that the embryos are placed in fix (FPBS) as soon as possible. It is important to avoid bruising or otherwise damaging the embryos during the dissection.

10. These openings are necessary for the removal of excess probe and antibody from these cavities during later wash steps. The roof of the rhombencephalon includes a thin area which is easily opened immediately after dissection with minimal bruising or disruption of the morphology of the embryo. If gene expression in the roof of the rhombencephalon is of particular interest, an opening may be made by puncturing or snipping the thicker tissue anterior to the rhombencephalon instead.

11. Cut the end from a disposable pipet tip with a razor blade or scalpel, ensuring that the opening is larger than the diameter of the embryos. A pipettor can then be used to transfer embryos from dish to vial. About 10 E9.5 or 20 smaller embryos can be transferred to each vial.

12. Unless otherwise noted, the minimum time for each wash is 5–10 min, depending on the size of the embryos. For multiple washes in the same solution, a short rinse will suffice for the first wash. Use sterile disposable plastic pipets, a pipettor with RNase-free tips, or a baked Pasteur pipet and gentle vacuum to remove solutions from the vials. Great care must be taken to avoid aspirating embryos. Tilt the vial and leave enough liquid to barely cover the embryos until the next solution is added. The embryos should never be allowed to begin to dry.

13. The vials may be stored at –20°C once the embryos are in 100% methanol. Before bleaching, place the vials at room temperature until warm.

14. Small sterile plastic tubes such as cryotubes with external threads (Nalgene) are suitable for hybridizing small groups of embryos. One milliliter of hybridization

solution will leave little void volume in these tubes, so the effects of evaporation are negligible. Alternatively, glass conical-bottomed 10-mL Reacti-Vials (Pierce) may be used for the hybridizations. The larger opening, conical bottom, greater volume, and transparency of Reacti-Vials make them ideal, particularly for groups of larger embryos. If glass vials are to be introduced at this stage, they should be siliconized *(16)*, acid washed, and treated with DEPC before use. These larger vials also require a humidity chamber (containing 50% formamide in water) during hybridization *(2)*.

15. Heat treatment inactivates endogenous alkaline phosphatases *(21)*.
16. The embryos require extensive washing to remove unbound antibody-conjugate.
17. Limit exposure to light as much as possible; control embryos should be exposed to as much light as those hybridized with antisense probe. The amount of time needed for color development can range from 20 min to overnight depending on the abundance of the RNA species being probed. The NBT/BCIP reaction proceeds at a steady rate, so the length of time which will be required can be estimated from early observations. Longer incubations can result in higher background and should be avoided if unnecessary.
18. The embryos can be stored at 4°C in PBTE before proceeding.
19. These steps are performed to enhance the visibility of the signal. Dehydration changes the purple stain to dark blue, whereas clearing the embryo with glycerol makes the tissue more transparent than translucent. After clearing, the embryos may be stored at least overnight in 80% glycerol/PBTE at 4°C.
20. Experiment with different combinations and intensities of light from standard or dark-field transilluminators and incident light sources set at different angles to optimize photographic quality. Sterile 24-, 12-, or 6-well tissue culture plates offer a convenient way to keep groups of embryos separate and organized as they are compared under the microscope. If embryos are to be stored in multiwell plates, be sure to seal the edges with Parafilm M® to minimize evaporation.

Acknowledgments

I would like to thank Vincent Giguère for his support. I am also grateful to Andrew Reaume for critical reading of the manuscript and Ron Conlon for providing whole-mount *in situ* protocols and many practical suggestions. This research was conducted at The Hospital for Sick Children, Toronto, Ontario, and supported by grants from the National Cancer Institute of Canada and the Medical Research Council of Canada.

References

1. Conlon, R. A. and Rossant, J. (1992) Exogenous retinoic acid rapidly induces anterior ectopic expression of murine *Hox-2* genes in vivo. *Development* **116,** 357–368.
2. Rosen, B. and Beddington, R. S. P. (1993) Whole-mount in situ hybridization in the mouse embryo: gene expression in three dimensions. *Trends Genet.* **9,** 162–167.

3. Lyn, S. and Giguère, V. (1994) Localization of CRABP-I and CRABP-II mRNA in the early mouse embryo by whole-mount in situ hybidization: implications for teratogenesis and neural development. *Dev. Dyn.* **199,** 280–291.

4. Hemmati-Brivanlou, A., Frank, D., Bolce, M. E., Brown, B. D., Sive, H. L., and Harland, R. M. (1990) Localization of specific mRNAs in *Xenopus* embryos by whole-mount *in situ* hybridization. *Development* **110,** 325–330.

5. Harland, R. M. (1991) *In situ* hybridization: an improved whole-mount method for *Xenopus* embryos. *Methods Cell Biol.* **36,** 685–695.

6. Harkey, M. A., Whiteley, H. R., and Whiteley, A. H. (1992) Differential expression of the msp130 gene among skeletal lineage cells in the sea urchin embryo: a three dimensional *in situ* hybridization analysis. *Mech. Dev.* **37,** 173–184.

7. Di Bernardo, M., Russo, R., Oliveri, P., Melfi, R., and Spinelli, G. (1995) Homeobox-containing gene transiently expressed in a spatially restricted pattern in the early sea urchin embryo. *Proc. Natl. Acad. Sci. USA* **92,** 8180–8184.

8. White, J. A., Boffa, M. B., Jones, B., and Petkovich, M. (1994) A zebrafish retinoic acid receptor expressed in the regenerating caudal fin. *Development* **120,** 1861–1872.

9. Tautz, K. and Pfeifle, C. (1989) A non-radioactive *in situ* hybridization method for the localization of specific RNAs in *Drosophila* embryos reveals translational control of the segmentation gene *hunchback*. *Chromosoma* **98,** 81–85.

10. Ros, M. A., Lyons, G., and Fallon, J. F. (1993) Spatial and temporal analysis of homeobox genes expressed in chick limb buds by whole-mount *in situ* hybridization. *Prog. Clin. Biol. Res.* **383A,** 79–87.

11. Yutzey, K. E., Rhee, J. T., and Bader, D. (1994) Expression of the atrial-specific myosin heavy chain AMHC1 and the establishment of anteroposterior polarity in the developing chicken heart. *Development* **120,** 871–883.

12. Sun, H. B., Neff, A. W., Mescher, A. L., and Malacinski, G. M. (1995) Expression of the axolotl homologue of mouse chaperonin t-complex protein-1 during early development. *Biochim. Biophys. Acta* **1260,** 157–166.

13. Nitabach, M. N. and Macagno, E. R. (1995) Cell- and tissue-specific expression of putative protein kinase mRNAs in the embryonic leech, *Hirudo medicinalis.* *Cell Tissue Res* **280,** 479–489.

14. Hauptmann, G. and Gerster, T. (1994) Two-color whole-mount in situ hybridization to vertebrate and *Drosophila* embryos. *Trends Genet.* **10,** 226.

15. Jowett, T. and Lettice, L. (1994) Whole-mount *in situ* hybridizations on zebrafish embryos using a mixture of digoxigenin and fluorescein-labelled probes. *Trends Genet.* **10,** 73,74.

16. Sambrook, J., Fritsch, E. F., and Maniatis, T. (1989) *Molecular Cloning: A Laboratory Manual.* Cold Spring Harbor Laboratory Press, Cold Spring Harbor, NY.

17. Harlow, E. and Lane, D. (1988) *Antibodies: A Laboratory Manual.* Cold Spring Harbor Laboratories, Cold Spring Harbor, NY.

18. Rossant, J., Zirngibl, R., Cado, D., Shago, M., and Giguère, V. (1991) Expression of a retinoic acid response element-*hsplacZ* transgene defines specific domains of transcriptional activity during mouse embryogenesis. *Genes Dev.* **5,** 1333–1344.

19. Hogan, B., Costantini, F., and Lacy, E. (1986) *Manipulating the Mouse Embryo.* Cold Spring Harbor Laboratory Press, Cold Spring Harbor, NY.
20. Titus, D. E. (1991) *Promega Protocols and Applications Guide.* Promega Corporation, Madison, WI.
21. Ponder, B. A. and Wilkinson, M. M. (1981) Inhibition of endogenous tissue alkaline phosphatase with the use of alkaline phosphatase conjugates in immunohistochemisty. *J. Histochem. Cytochem.* **29,** 981–984.

6

Reverse Transcriptase-Polymerase Chain Reaction (RT-PCR) for Cellular Retinoid-Binding Proteins

Andrew D. Loughney and Christopher P. F. Redfern

1. Introduction

Cellular retinoid-binding proteins were first identified in the mid-1970s *(1)*. Subsequent studies established the existence of two forms of cellular retinoic acid-binding proteins (CRABP I and II) *(2,3)* and two forms of cellular retinol-binding proteins (CRBP I and II) *(4)*, each encoded by a separate gene. These cellular retinoid-binding proteins are members of a family of related lipid-binding proteins in which the ligand is enclosed within a deep hydrophobic pocket formed by a 10-stranded β-barrel polypeptide structure *(5,6)*. CRBPs and CRABPs are expressed in a tissue-specific manner *(7–10)*, with CRBP II most limited in expression, being confined mainly to the small intestine *(4,11)*. Although the ligand-binding properties of these proteins are well-known, largely as a result of biochemical studies on protein purified from tissues or expressed in *Escherichia coli* (*see* Chapters 8–10), their functions and interactions with other cellular proteins are still uncertain. Recombinant cellular retinoid binding proteins, expressed in *E. coli* or in eukaryotic cells, are important tools for understanding their role in the cellular uptake of retinol, the sequestration and metabolic degradation of retinoic acid or the transfer of retinoids to nuclear receptors. cDNA clones for the complete coding regions of the various binding proteins can be prepared by reverse transcriptase-polymerase chain reaction (RT-PCR), using available sequence information to design the forward and reverse primers. In addition, the regulation of expression of these genes may be studied by quantitative RT-PCR using primers optimized for specificity, as an alternative to conventional Northern blotting. This chapter is based on our experience in cloning human CRABP II and CRBP II via RT-PCR, but similar conditions will be applicable to CRBP I and CRABP I.

From: *Methods in Molecular Biology, Vol. 89: Retinoid Protocols*
Edited by: C. P. F. Redfern © Humana Press Inc., Totowa, NJ

An alternative to cloning these genes via RT-PCR is to search the Expressed Sequence Tag (EST) database with the cellular retinoid binding protein sequence of interest via the Internet. EST clones from the I.M.A.G.E. Consortium (LLNL) with exact matches can be obtained from GenomeSystems (St Louis, MO), and sequenced to verify their identity and length. This may also prove to be a useful "armchair cloning" strategy to identify new members of this gene family.

2. Materials

1. Primers for polymerase chain reaction: the design of primer depends on the purpose of the RT-PCR study. To clone the full coding region, the primers will (but this depends on the vector/expression system employed) include the initiation and termination codons, and cloning is facilitated by adding restriction-enzyme sites at the ends. For qualitative or quantitative studies on gene expression, primers may be designed for optimal specificity and additional restriction sites will be unnecessary. Primers designed for cloning the full-length coding regions of human (h) CRBP II, CRABP I, and CRABP II, and internal primers for hCRBP I, hCRABP I, and hCRABP II are described below. For each cellular retinoid binding protein type, primer 1 (forward) refers to the 5' end and primer 2 (reverse) to the 3' end of the mRNA; initiation and termination codons are in bold, and restriction sites are underlined. Alignments of primer sequences are shown in **Fig. 1** (*see* **Note 1**).
   a. hCRBP I: Torma et al. *(12)* have described internal primers for hCRBP I, giving an expected product size of 391 bp:
      Primer 1: $^{5'}$GTCGACTTCACTGGGTACTGGA$^{3'}$.
      Primer 2: $^{5'}$TTGAATACTTGCTTGCAGACCACA$^{3'}$.
   b. hCRBP II (*see* **Note 2**): Primers incorporating initiation and termination codons, and additional restriction sites, are as follows:
      Primer 1: $^{5'}$AT**TCTAGATG**ACGAGGGACCAGAATGG$^{3'}$ (*Xba*I site).
      Primer 2: $^{5'}$TC**GAATTCA**CTTCTTTTTGAACACTTG$^{3'}$ (*Eco*RI site). Expected product size is 418 bp.
   c. hCRABP I: Degenerate primers for RT-PCR of hCRABP I incorporating the initiation and termination codons were designed by Astrom et al *(13)* on the basis of CRABP sequence data from other species. The forward and reverse primers are 48 and 51 bases long, respectively, and are as follows; degeneracy has been removed by correcting the primers with the hCRABP I cDNA sequence.
      Primer 1: $^{5'}$GAAT**CTAGA**CTGCCACC**ATG**CCCAACTTCGCCGG CACCTGGAAGATG$^{3'}$, (*Xba*I site).
      Primer 2: $^{5'}$CACT**GGATCC**AAGCTGGCCACCT**TTC**ATTCCCGGA CATAAATTCTGGTGCA$^{3'}$ (*Bam*HI site). The expected product size is 436 bp.
   d. Internal hCRABP I primers without additional restriction sites and giving an expected product size of 371 bp *(12)* are:
      Primer 1: $^{5'}$CGGCACCTGGAAGATGCGCA$^{3'}$.
      Primer 2: $^{5'}$CCACGTCATCGGCGCCAAAC$^{3'}$.

5' (amino-terminal) end

```
hCRBP  I:   ATG cCa gtc GAC ttc AcT GGg taC TGG aAG ATG ctG
hCRBP  II:  ATG aCg agg GAC cag AaT GGa acC TGG gAG ATG gaG


hCRABP I:   ATG CCC AAC TTC gCc GGC AcC TGG Aag Atg cgC aGc
hCRABP II:  ATG CCC AAC TTC tCt GGC AaC TGG Aaa Atc atC cGa
```

3' (carboxy-terminal) end

```
hCRBP  I:   TCA CTg Cac cTT ctt gAa Tac ttG ctt gca gac cac a
hCRBP  II:  TCA CTt Ctt tTT gaa cAc Ttg acG gca cac ctg gtc


hCRABP I:   TCA tTC cCG GAC aTA aAt tCT GGT GCA gAC cAC GTC ATC gGC gcc aAa c
hCRABP II:  TCA cTC tCG GAC gTA gAc cCT GGT GCA cAC aAC GTC ATC cGC cgt cAt g
```

Fig. 1. Alignment of primer sequences: underlined bases are primers for full length coding region amplification, bases in bold are internal primers designed by Torma et al. *(12)*. Mismatched bases are in lower case letters, sequences are in conventional 5'–3' arrangement, starting from the initiation codon for 5'-end primers and the termination codon for 3'-end primers. The 3'-end sequences are the reverse complement of the cDNA sequence.

    e. hCRABP II: Primer sequences *(14)* for amplifying the full-coding region are based on the published human CRABP II sequence *(15)*:
Primer 1: 5'ATTCTAGATGCCCAACTTCTCTGGCAACTG3' (*Xba*I site).
Primer 2: 5'CTGGATCCTCACTCTCGGACGTAGACCCTG3' (*Bam*HI site) (*see* **Note 3**). The expected product size is 432 bp.

    f. Alternatively, internal hCRABP II primers lacking restriction sites and giving a 411-bp product *(12)* are:
Primer 1: 5'CCCAACTTCTCTGGCAACTGGA3'.
Primer 2: 5'CTCTCGGACGTAGACCCTGGT3'.

      Note that these two primers are very similar to the full-coding region primers, and lack just the initiation and termination codons and extend two bases further into the coding sequence.

      Primers should be supplied without terminal-trityl groups. For our work, we used these without purification. The primers should be dissolved in 10 mM Tris-HCl, pH 7.5, 1 mM EDTA buffer and the concentration estimated from the OD$_{260}$.

2. Total or poly (A)+ RNA from cells or tissues of interest (*see* **Note 4**).
3. Sterile, RNase-free water.
4. Oligo d(T) primer: 1 μg/μL poly d(T)$_{12-18}$ 5'-phosphate (Pharmacia, Uppsala, Sweden).
5. Mixed deoxynucleotide 5'-triphosphate (dNTP) stock: 10 mM of each 2-deoxynucleoside 5'-triphosphate, deoxyadenosine 5'-triphosphate (dATP),

deoxyguanosine 5′-triphosphate (dGTP), deoxycytidine 5′-triphosphate (dCTP), and deoxythymidine 5′-triphosphate (dTTP) (Ultrapure grade, Pharmacia), at neutral pH.

6. Superscript™ II reverse transcriptase (Life Technologies, Paisley, UK). This enzyme is a recombinant RNase H⁻ reverse transcriptase from Moloney Murine Leukaemia Virus (MMLV) and is supplied with tubes of 0.1 *M* dithiothreitol (DTT) and 5X reverse transcriptase reaction buffer (250 m*M* Tris-HCl, pH 8.3, 375 m*M* KCl and 15 m*M* $MgCl_2$).
7. *Taq* polymerase (Promega, Southampton, UK, or equivalent enzyme from other suppliers).
8. 10X *Taq* buffer (*see* **Note 5**): as supplied with the *Taq* polymerase.
9. 25 m*M* $MgCl_2$ (supplied with the *Taq* polymerase).
10. Light white mineral oil (Sigma, Molecular Biology grade).
11. Thermal cycler.

3. Methods

3.1. cDNA Synthesis: Reverse-Transcription Reaction

1. Add 5 μg of RNA (*see* **Note 4**) to 1 μg oligo d(T) primer to give a total volume of 11 μL in sterile, RNase-free water.
2. Warm the mixture to 70°C for 10 min; then cool briefly on ice.
3. Add the following reagents to the mixture (*see* **Note 6**): 2 μL of 0.1 *M* DTT, 1 μL of mixed dNTP stock, and 4 μL of 5X reaction buffer.
4. Mix the contents of the tube by vortexing gently and spin briefly in a microcentrifuge to return all droplets to the bottom of the tube.
5. Warm the mixture to 37°C for 2 min; then add 400 U (about 2 μL) of Superscript II reverse transcriptase. If necessary, make the volume up to a total of 20 μL with sterile, RNase-free water. Mix the enzyme with the reaction mixture by flicking the tube gently, spin briefly in a microcentrifuge to return all droplets to the bottom of the tube, and incubate for 1 h at 37°C.
6. At the end of the incubation, the reaction mixture can be used directly for PCR or stored at –20°C.

3.2. PCR reaction

1. Add 2 μL of the cDNA reaction (**Subheading 3.1.**, **step 6**) to a thin-walled 0.5-mL microcentrifuge tube containing (*see* **Note 5**): 100 pmol of primer 1, 100 pmol of primer 2, 2 μL of mixed dNTP stock, 10 μL of 10X *Taq* buffer, 5–8 μL of 25 m*M* $MgCl_2$, and sterile water to a total volume of 99 μL.
2. Mix the components together and add 1–5 U of *Taq* DNA polymerase in a volume of 1 μL. Mix the tube contents by flicking with a finger, spin briefly to return all droplets to the bottom of the tube, and overlay with 100 μL light white mineral oil to reduce evaporation during PCR.
3. Using a thermal cycler, amplify the cDNA with 30 cycles of the following conditions (*see* **Note 7**): 1 min denaturation at 95°C, 1 min annealing at 55°C, 1 min

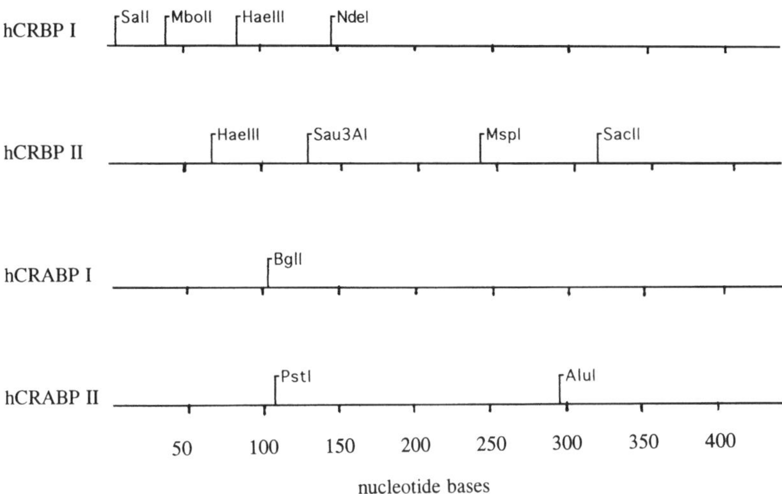

Fig. 2. Restriction enzyme maps for hCRBP I, hCRBP II, hCRABP I, and hCRABP II coding sequences. Length is given in nucleotide bases (*see* hCRABP II map). Although each map is represented as the same length, the coding-region length varies by two to three codons and the scales of the maps therefore differ slightly.

extension at 72°C. After the final cycle, incubate for 10 min at 72°C to ensure completion of product termini.

4. As a negative control run a parallel reaction with all components except the 2-µL aliquot of the cDNA synthesis reaction.

5. Analyze 10 µL of each PCR reaction on an agarose gel to ensure that the amplification has worked (*see* **Note 8**). A band of approx 0.4 kb should be visible on an ethidium bromide-stained gel. The sequence of the PCR product should be verified, either by direct sequencing after elution from the gel, or by sequencing after cloning the product into a suitable vector. Alternatively, digesting the PCR products with restriction enzymes can provide a useful preliminary characterization (*see* **Fig. 2**). For example, using the internal primers, cutting the hCRBP I product with *Nde*I yields 139 and 249 bp fragments, cutting CRABP I with *Bgl*I yields 85 and 285 bp fragments and cutting the CRABP II product with *Pst*I gives 108 and 305 bp fragments *(12)*.

6. Cloning the PCR product can be achieved by various strategies. For example, by digesting with *Xba*I and *Bam*HI (or enzymes corresponding to the restriction sites used) and cloning into Bluescript, or by blunt-ending with T4 DNA polymerase and cloning into PCR-Script vector (Stratagene, La Jolla, CA) cut with *Srf*I restriction enzyme.

4. Notes

1. Over the regions illustrated in **Fig. 1**, there is only approx 30% identity in nucleotide sequence between CRBPs and CRABPs at the 5' and 3' ends of the cDNA

sequence. Therefore, lack of specificity of these primers with respect to these two classes of CRBPs should not be a problem with the RT-PCR conditions described. For the CRABPs, there is approx 70% nucleotide-sequence identity between CRABP I and II at both 5' (amino-terminal) and 3' (carboxy-terminal) ends. To ensure discrimination between CRABP I and CRABP II, it may be worthwhile for some projects to change the design of the primers so that they end at a mismatched base. The CRBPs are more variable, ranging from 55% nucleotide-sequence identity between CRBP I and II at the 5' end to only 24% at the 3' end of the cDNA. Thus, the ability of primers to discriminate between CRBP I and CRBP II should not be a problem. cDNA sequences for hCRBP I, hCRBP II, hCRABP I, and hCRABP II can be found in the Genbank/EMBL database, accession numbers M36809, U13831, S74445, and M68867, respectively. The cDNA sequence for hCRBP I is an incomplete 5' end sequence; the gene sequence for exons 1–4 are in M97814 and M97815.

2. These primer sequences are similar to ones used to clone human CRBP II *(16)* and were based on the rat CRBP II nucleotide and amino acid sequences *(17)* and an N-terminal human CRBP II amino acid sequence *(18)*. In the original primers *(16)*, redundancy was incorporated at base 13 of primer 1 (A or G) and bases 13 and 22 of primer 2 (both C or T). The nonredundant primers described here are based on the sequence of hCRBP II cDNA clones obtained in that study (Genbank accession number U13831).

3. Primer sequences consist of a 20- to 23-base section complementary to the relevant strand of the cDNA, with a restriction site added to the 5' end of each primer followed by a short 2-base extension to facilitate restriction enzyme activity. The 5' ends of each primer may be modified easily to subclone into different restriction sites of alternative vectors. For example, to clone hCRABP II into the pGEX-2T expression vector, new primers to incorporate *Bam*HI and *Eco*RI sites (5' end:
    $^{5'}$ACT<u>GGATCC</u>**ATG**CCCAACTTCTCTGGCAAC$^{3'}$, 3' end,
    $^{5'}$GAT<u>GAATTCC</u>**TCA**CTCTCGGACGTAGAC$^{3'}$,
restriction sites underlined) were used to subclone into the vectors in-frame with the glutathione-*S*-transferase (GST) coding sequence *(14)*.

4. RNA should be extracted from the cells or tissue of interest using an appropriate method, such as a miniprep method *(19)* to prepare cytoplasmic RNA, or extraction of total cellular RNA with guanidine thiocyanate *(20)* or guanidine thiocyanate-phenol *(21)*. For cultured cells we routinely use a miniprep method *(19)*. Although CRABP I, CRABP II, and CRBP I are expressed in a range of tissues, CRBP II is only expressed in the intestine and fetal liver *(4,22)*. To clone human CRBP II we have used the human intestinal-cell line CaCo-2, which expresses CRBP II *(23)* when grown to confluence and maintained as stationary phase cultures for 14 d in DMEM supplemented with 10% fetal calf serum (FCS). We have amplified CRABP II cDNA from total RNA from human retinoic acid-treated dermal fibroblasts *(14)* and neuroblastoma cells. Before proceeding with RT-PCR, the quality of the RNA should be checked by denaturation and electrophoresis through 1.2% agarose (*see* Chapter 15). We have found that total cellular

RNA, total cytoplasmic RNA or poly (A)+ cytoplasmic RNA (prepared using oligo-d[T]-dynabeads, Dynal, UK) work equally well.

5. Buffers for *Taq* polymerase are usually provided with the enzyme. The 10X buffer provided with the Promega enzyme is supplied Mg^{2+}-free, and $MgCl_2$ is supplied as a separate 25-mM solution. The optimal Mg^{2+} concentration for PCR may vary according to the primer-template pair used. We find that a final Mg^{2+} concentration of 1–2 mM gives good results.

6. The precise reaction conditions are flexible. The protocol described here is based on that supplied with the Superscript II reverse transcriptase. In earlier studies *(14)* using 15 U AMV reverse transcriptase (Promega, Southampton, UK) we used 2.5-fold lower concentrations of dNTPs and one-third the amount of oligo d(T) primer, in reactions containing 2.5 U RNasin and 5 mM $MgCl_2$ together with reaction buffer supplied with the enzyme. Incubation conditions for this enzyme were 15 min at 42°C.

7. Thermal cycling conditions will need to be optimized for a particular machine, source of enzyme, and primers. We have also obtained good results with shorter cycles consisting of 30 s denaturation at 95°C, 30 s annealing at 55°C, and 42 s extension at 72°C *(14)*. For amplifying with the internal primers given in the Materials section, Torma et al. *(12)* have used 60 s denaturation at 94°C, 90 s annealing at 60°C, and 120 s extension at 72°C.

8. Failure to obtain a PCR product of the correct size can be due to a number of problems relating to the quality of RNA, absence of target message, failure of the cDNA synthesis reaction, or inappropriate PCR conditions. Ideally, each stage of the procedure should be checked, preferably using as a test cloned cDNA known to be expressed in the RNA sample of interest. Start by verifying that PCR components are working: amplify by PCR a small amount of cloned cDNA for the test gene with the appropriate primers. If the PCR gives the expected product, repeat using an aliquot of cDNA synthesis reaction, taking stringent precautions to prevent contamination by the cloned cDNA used as a test.

References

1. Chytil, F. and Ong, D. E. (1984) Cellular retinoid binding proteins, in *"The Retinoids,"* vol. 2 (Sporn, M. B., Roberts, A. B., and Goodman, D. S., eds.), Academic, Orlando, FL, pp. 89–123.
2. Bailey, J. S. and Siu, C. H. (1988) Purification and partial characterization of a novel binding-protein for retinoic acid from neonatal rat. *J. Biol. Chem.* **263**, 9326–9332.
3. Bailey, J. S. and Siu, C. H. (1990) Unique tissue distribution of 2 distinct cellular retinoic acid binding-proteins in neonatal and adult-rat. *Biochim. Biophys. Acta* **1033**, 267–272.
4. Ong, D. E. (1984) A novel retinol-binding protein from rat-purification and partial characterization, *J. Biol. Chem.* **259**, 1476–1482.
5. Cowan, S. W., Newcomer, M. E., and Jones, T. A. (1993) Crystallographic studies on a family of cellular lipophilic transport proteins-refinement of P2 myelin

protein and the structure determination and refinement of cellular retinol-binding protein in complex with all-*trans*-retinol. *J. Mol. Biol.* **230,** 1225–1246.

6. Winter, N. S., Bratt, J. M., and Banaszak, L. J. (1993) Crystal-structures of holo and apo-cellular retinol-binding protein-II. *J. Mol. Biol.* **230,** 1247–1259.

7. Dollé, P., Ruberte, E., Leroy, P., Morrisskay, G., and Chambon, P. (1990) Retinoic acid receptors and cellular retinoid binding-proteins. 1. a systematic study of their differential pattern of transcription during mouse organogenesis. *Development* **110,** 1133–1151.

8. Ruberte, E., Dollé, P., Chambon, P., and Morriss-Kay, G. (1991) Retinoic acid receptors and cellular retinoid binding-proteins. 2. their differential pattern of transcription during early morphogenesis in mouse embryos. *Development* **111,** 45–60.

9. Ruberte, E., Friederich, V., Morrisskay, G., and Chambon, P. (1992) Differential distribution patterns of CRABP-I and CRABP-II transcripts during mouse embryogenesis. *Development* **115,** 973–987.

10. Ruberte, E., Friederich, V., Chambon, P., and Morriss-Kay, G. (1993) Retinoic acid receptors and cellular retinoid-binding proteins. 3. their differential transcript distribution during mouse nervous- system development. *Development* **118,** 267–282.

11. Schaefer, W. H., Kakkad, B., Crow, J. A., Blair, I. A., and Ong, D. E. (1989) Purification, primary structure characterization, and cellular-distribution of 2 forms of cellular retinol-binding protein, type-II from adult-rat small-intestine. *J. Biol. Chem.* **264,** 4212–4221.

12. Torma, H., Lontz, W., Liu, W., Rollman, O., and Vahlquist, A. (1994) Expression of cytosolic retinoid-binding protein genes in human skin biopsies and cultured keratinocytes and fibroblasts. *Br. J. Derm.* **131,** 243–249.

13. Astrom, A., Tavakkol, A., Pettersson, U., Cromie, M., Elder, J. T., and Voorhees, J. J. (1991) Molecular-cloning of 2 human cellular retinoic acid-binding proteins (CRABP)—retinoic acid-induced expression of CRABP-II but not CRABP-I in adult human skin *in vivo* and in skin fibroblasts *in vitro*. *J. Biol. Chem.* **266,** 17,662–17,666.

14. Redfern, C. P. F. and Wilson, K. E. (1993) Ligand-binding properties of human cellular retinoic acid-binding protein-II expressed in *Escherichia-coli* as a glutathione-s-transferase fusion protein. *FEBS Letts.* **321,** 163–168.

15. Eller, M. S., Oleksiak, M. F., McQuaid, T. J., McAfee, S. G., and Gilchrest, B. A. (1992) The molecular-cloning and expression of 2 CRABP cDNAs from human skin. *Exp. Cell. Res.* **198,** 328–336.

16. Loughney, A. D., Kumarendran, M. K., Thomas, E. J., and Redfern, C. P. F. (1995) Variation in the expression of cellular retinoid-binding proteins in human endometrium throughout the menstrual-cycle. *Hum. Reprod.* **10,** 1297–1304.

17. Li, E., Demmer, L. A., Sweetser, D. A., Ong, D. E., and Gordon, J. I. (1986) Rat cellular retinol-binding protein-II—use of a cloned cDNA to define its primary structure, tissue-specific expression, and developmental regulation. *Proc. Natl. Acad. Sci. USA* **83,** 5779–5783.

18. Inagami, S. and Ong, D. E. (1992) Purification and partial characterization of cellular retinol-binding protein, type-2, from human small-intestine. *J. Nutr.* **122,** 450–456.
19. Wilkinson, M. (1988) RNA isolation - a mini-prep method. *Nucl. Acids Res.* **16,** 10,933.
20. Chirgwin, J. M., Przybyla, A. E., McDonald, R., and Rutter, W. G. (1979) Isolation of biologically-active ribonucleic acid from sources enriched in ribonuclease. *Biochemistry* **18,** 5294–5299.
21. Chomczynski, P. and Sacchi, N. (1987) Single step method of RNA isolation by acid guanidinium thiocyanate phenol chloroform extraction. *Anal. Biochem.* **162,** 156–159.
22. Ong, D. E. and Page, D. L. (1987) Cellular retinol-binding protein (type-2) is abundant in human small-intestine. *J. Lipid Res.* **28,** 739–745.
23. Levin, M. S. (1993) Cellular retinol-binding proteins are determinants of retinol uptake and metabolism in stably transfected caco-2 cells. *J. Biol. Chem.* **268,** 8267–8276.

7

Methods for Producing Recombinant Human Cellular Retinaldehyde-Binding Protein

John W. Crabb, Yang Chen, Steve Goldflam, Karen West, and James Kapron

1. Introduction

The cellular retinaldehyde-binding protein (CRALBP) is expressed at high levels in vertebrate visual tissue, where it may serve to modulate the interaction of 11-*cis*-retinol with visual-cycle enzymes in the retinal pigment epithelium (RPE) *(1)*. This 36-kDa protein is remarkably stereo selective in recognizing the isomer of retinaldehyde involved in vision, and carries 11-*cis*-retinaldehyde or 11-*cis*-retinol as physiological ligands *(2)*. Notably, retinoid bound to CRALBP is less susceptible to photo-isomerization than when bound to rhodopsin *(3)*. In addition to RPE, the protein is expressed in Müller cells of the neural retina *(4)*, ocular ciliary epithelium *(5,6)*, and transiently in iris *(7)*. Recent investigations show the protein is also expressed at lower levels in glia of the optic nerve and brain *(8)*. While in vitro evidence suggests a substrate-routing function for CRALBP in RPE *(1)*, the physiological role of CRALBP in vivo remains unconfirmed.

The primary structure of bovine CRALBP has been determined directly *(9)*, and the cDNAs encoding the bovine and human protein *(10)* and the human CRALBP gene *(11)* have been cloned and sequenced. Topological analysis has suggested buried and exposed regions in the three-dimensional structure of bovine CRALBP *(12)*. Modest sequence identity (25–33% over 209–259 amino acids) has been reported with the putative substrate targeting domain of cytoplasmic protein tyrosine phosphatase *(13)*, SEC14p, a yeast protein involved in secretion from the Golgi complex *(14),* and α-tocopherol transfer protein, a protein that appears to transfer α-tocopherol between membranes in rat liver

From: *Methods in Molecular Biology, Vol. 89: Retinoid Protocols*
Edited by: C. P. F. Redfern © Humana Press Inc., Totowa, NJ

```
GHHHHHHHHH  HSSGHIDDDD  KHMSEGVGTF  RMVPEEEQEL  RAQLEQLTTK   5
DHGPVFGPCS  QLPRHTLQKA  KDELNEREET  REEAVRELQE  MVQAQAASGE  10
ELAVAVAERV  QEKDSGFFLR  FIRARKFNVG  RAYELLRGYV  NFRLQYPELF  15
DSLSPEAVRC  TIEAGYPGVL  SSRDKYGRVV  MLFNIENWQS  QEITFDEILQ  20
AYCFILEKLL  ENEETQINGF  CIIENFKGFT  MQQAASLRTS  DLRKMVDMLQ  25
DSFPARFKAI  HFIHQPWYFT  TTYNVVKPFL  KSKLLERVFV  HGDDLSGFYQ  30
EIDENILPSD  FGGTLPKYDG  KAVAEQLFGP  QAQAENTAF               33
```

Fig. 1. Fusion Human rCRALBP Amino Acid Sequence. The 23 residue N-terminal extension and histidine-tag sequence is underlined; Ser^{24} is the N-terminus of nonfusion human rCRALBP. In fusion rCRALBP, residues R^{141} to N^{336} have been identified as a retinoid-binding fragment *(16)* and K^{244} identified as a component of the hydrophobic retinoid-binding pocket *(17)*.

(15). No significant sequence homology exists with other retinoid-binding proteins; however, the primary structures of bovine and human CRALBP are 92% identical *(10)*. The limited availability of human-retinal tissue for protein purification prompted us to develop a recombinant source of human CRALBP. Structure–function studies of recombinant human CRALBP (rCRALBP) have identified the retinoid-binding domain *(16)* and revealed, through site-directed mutagenesis, residues important for retinoid binding *(17,18)* and autosomal recessive retinitis pigmentosa *(19)*. This chapter describes methods for the expression, purification, and characterization of milligram amounts of human rCRALBP.

2. Materials

1. The pET-expression system *(20)* was used for the recombinant production of human CRALBP in *Escherichia coli* strain BL21(DE3)LysS. The human-CRALBP cDNA *(10)* was engineered by PCR to contain *Nde*I sites at each end of the coding region and cloned into the *Nde*I site of pET vectors. Vector pET3a was used to express a nonfusion rCRALBP and pET19b to express a fusion rCRALBP with a 23-residue N-terminal extension containing a histidine-tag sequence (**Fig. 1**). The pET vectors and host-bacterial strains are available from Novagen (Madison, WI). The nucleotide sequence of bovine and human-CRALBP cDNA can be found in the GenBank/EMBL Data Bank with accession numbers J04213 and J04214, respectively.
2. Standard Luria-Bertani (LB) medium, pH 7.0: 10 g/L bacto-tryptone (Difco), 5 g/L bacto-yeast extract (Difco), 5 g/L sodium chloride, 100 mg/L ampicillin (US Biochemicals) and 25 mg/L chloramphenicol (Calbiochem) is used for growing both small-scale and large-scale cultures.
3. IPTG (isopropyl-β-D-thiogalactopyranoside, US Biochemicals): 1 *M* solution in water for small cultures or dry powder for large cultures.
4. Bacterial cells are harvested by centrifugation in polypropylene bottles using a GS3 rotor and Sorval RC2B centrifuge (10,000*g*/20 min), washed with phosphate buffered saline (PBS) and stored at –70°C.

5. pET 3a Cell-lysis buffer (for purifying nonfusion rCRALBP): 25 m*M* Tris acetate, pH 7.0, 10 m*M* sodium acetate, 0.5 m*M* phenylmethanesulfonyl fluoride (PMSF), 0.5 µ*M* leupeptin, 10 µ*M* benzamidine, and 1 m*M* ethylene diamine tetraacetate-dithiothreitol (EDTA-DTT).

6. pET 19b Cell-lysis buffer (for purifying fusion rCRALBP): 50 m*M* sodium phosphate, pH 8.0, 300 m*M* sodium chloride.

7. DNase (Sigma): fresh solution in deionized water or used as a dry powder (*see* Methods).

8. Chromatography matrices: DEAE-cellulose (DE-52, Whatman); hydroxylapatite (Bio-gel HTP, Bio-Rad); Mono Q anion exchange columns (Pharmacia Fine Chemicals) and Ni-NTA nickel affinity chromatography support (Qiagen).

9. DEAE chromatography buffer: 25 m*M* Tris-acetate, pH 7.0, 10 m*M* sodium acetate, 0.1 m*M* DTT.

10. HTP chromatography buffer: 25 m*M* Tris-acetate, pH 7.0, 300 m*M* sodium acetate, 0.1 m*M* DTT.

11. Mono Q chromatography buffer: 20 m*M* MOPS [3(*N*-morpholino)propane sulfonic acid], pH 7.0, 50 m*M* sodium acetate, 0.1 m*M* DTT.

12. Ni-NTA chromatography buffer and pET19b-lysis buffer: 50 m*M* sodium phosphate pH 8.0, 300 m*M* sodium chloride. For Ni-NTA chromatography, add imidazole to 40 m*M* for eluting contaminants, then add imidazole to 250 m*M* for eluting the target protein.

13. For sodium dodecyl sulfate-polyacrylamide gel electrophoresis (SDS-PAGE) use a Mini-Protein II slab-gel system (Bio-Rad) or equivalent.

14. Hitachi U-2000 spectrophotometer or equivalent for spectral analysis.

15. Supplies, reagents, and instrumentation for amino-acid analysis (for example Applied Biosystems, Perkin-Elmer, model 420H/130/920 Amino acid analyzer and model 470/120/900 protein sequencer).

16. Liquid chromatography mass spectrometry (LCMS) may be performed with a PE Sciex API 300 electrospray triple-quadrupole mass spectrometer, and an Applied Biosystems model 130 HPLC system with a 5-µ Vydac C18 column (1 × 250 mm, using aqueous trifluoroacetic acid/acetonitrile solvents), or equivalent.

17. Retinoids: 11-*cis*-retinaldehyde can be obtained through the National Eye Institute, Fundamental Retinal Processes Research Program, National Institute of Health, Bethesda, MD. 9-*cis*-retinaldehyde is obtainable from Sigma Chemical Company. Both retinoids are stored dry, under argon at −70°C in the dark, and dissolved in ethanol for labeling purposes.

18. A dark work room with dim-red illumination (e.g., a photography dark room) is required to prevent photoisomerization of the CRALBP ligand to the all-*trans*-isomer. An inexpensive dark room may be set up by covering laboratory windows with foil and covering several clamp-on lamps (available at discount department stores such as Wal-Mart) with a red plastic filter such as Roscoe #27, catalog number RC5029 (20 × 24-in. sheet), from Calumet Photographic (Bensenville, Il).

19. A dark cold room or chromatography refrigerator with dim-red illumination is needed for rCRALBP purification.

20. Vibra cell sonifier (Sonics & Materials).
21. Amicon concentrator or centricon centrifugal concentrators (10-kDa molecular-weight cutoff).

3. Methods

3.1. Bacterial Expression of rCRALBP

The same culture conditions are used for bacterial expression of rCRALBP with the pET 3a- or pET-19b plasmids. The following procedure is used for growing a 1.2-L culture in a 2.8-L Fernbach flask and can be scaled up or down as needed. Typically six 1.2-L cultures are grown simultaneously in a controlled environment incubator shaker (model G25, New Brunswick Scientific) at 37°C with vigorous shaking.

1. Grow an initial inoculum of 2 mL (started from a frozen glycerol stock of bacterial cells) overnight in LB medium (*see* **Note 1**), transfer to 50 mL of fresh medium in the morning, grow to mid log-phase, and then transfer to 1.2 L of fresh medium. Monitor cell growth spectrophotometrically at 600 nm. Under these conditions, cells reach mid-log-phase after about 2 h of growth.
2. When the culture reaches mid-log-phase ($OD_{600}0.5$), remove a control sample- (e.g., 0.5 mL) for SDS-PAGE, then add IPTG to 0.5 m*M* final concentration to induce production of the target protein.
3. Stop cell growth in late log phase by cooling the culture flasks in ice water (3–4 h after adding IPTG). Remove an induced sample (e.g., 0.5 mL) for SDS-PAGE. Harvest cells by centrifuging at 4°C, 10,000*g* for 10 min, wash cells 2–3 times with PBS (50–100 mL per wash) to remove growth medium, decant, and store wet-cell pellets at –70°C. Under these conditions, the yield is approx 3 g of wet, packed cells per liter of growth medium.
4. Use SDS-PAGE analysis *(21)* of the uninduced and induced culture samples to verify that expression of the target protein was successful and estimate the level of rCRALBP expression (**Fig. 2**).

3.2. Preparation of Cell Lysates and Labeling with Retinoid

1. Resuspend washed, packed cells from a 1.2-L culture in pET 3a- or pET 19b-lysis buffer (~3 mL buffer per gram wet cells), incubate with DNase (2 U/mL) for 30–40 min at room temperature and sonicate on ice for 5 min (e.g., with 20 s on/off bursts at 60% output using a Vibra Cell Sonifier).
2. Centrifuge the lysed cells at 45,000*g* for 30 min at 4°C to remove debris, decant, measure the volume, and determine the protein concentration *(22)* of the clarified-cell lysate.
3. Estimate the amount of rCRALBP in the crude lysate from SDS-PAGE Coomassie blue staining intensity and the protein concentration of the lysate.
4. In the dark, add to the cell lysate a two- to threefold molar excess of 11-*cis* (or 9-*cis*) retinaldehyde over rCRALBP (typically 1.2–1.8 µmol [0.3–0.5 mg] retin-

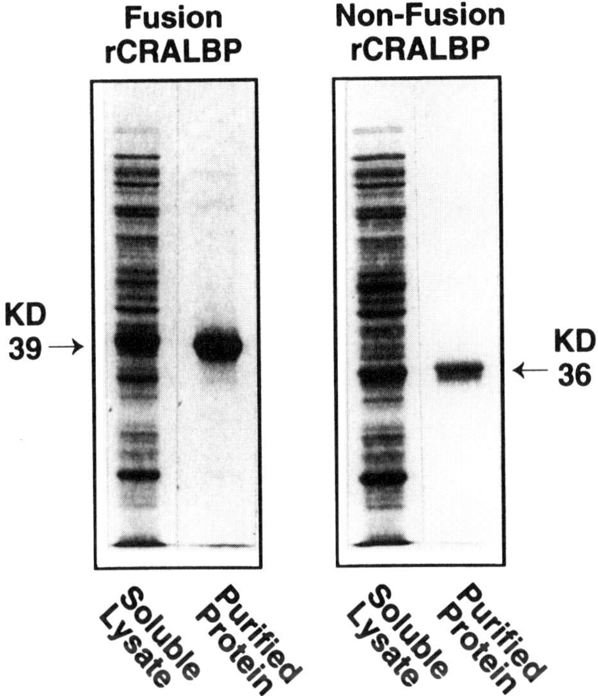

Fig. 2. SDS-PAGE analysis of rCRALBP. Approximately 10 µg of soluble protein from crude lysates containing rCRALBP and 1–2 µg of purified fusion and nonfusion rCRALBP was analyzed according to Laemmli *(21)* on a 12% gel. Detection is by Coomassie blue staining.

oid/1.2-L culture). Incubate in the dark at 4°C for about 1 h (*see* **Note 2**). The retinoid should be added from a concentrated solution (1–2 mg/mL) in ethanol that has been quantified spectrophotometrically (11-*cis*-retinaldehyde, ε^{380nm} = 87.8 [mg/mL]$^{-1}$cm^{-1} in ethanol *(23)*; 9-*cis*-retinaldehyde, ε^{373nm} = 127 [mg/mL]$^{-1}$ cm^{-1} in ethanol *[24]*). The only other retinoid that CRALBP is known to bind is 11-*cis*-retinol *(2)* and this may be used for labeling in place of the retinaldehydes, but also must be added under dark room conditions (*see* **Note 3**).

3.3. Purification of Nonfusion rCRALBP

Purification of the nonfusion rCRALBP produced with the pET-3a vector is accomplished essentially as described by Saari and Bredberg *(25)* for bovine-retinal CRALBP. (**Ref. 25** includes representative chromatography profiles.) All rCRALBP purification steps must be carried out under dim red illumination to prevent photodecomposition of the retinoids.

1. Dialyze the labeled bacterial-cell lysate several hours or overnight at 4°C against pET 3a lysis buffer (1 L); then centrifuge at 45,000g for 20 min at 4°C to remove insoluble material produced during retinoid labeling.

2. Apply the dialyzed and clarified lysate to a DEAE-cellulose column (~2.4 × 30 cm) equilibrated in DEAE-chromatography buffer and wash with the same buffer until the absorbance at 280 nm stabilizes at about 0.1 AU. Elute bound protein with a linear gradient using the DEAE buffer and the same buffer containing 400 mM sodium acetate (total gradient volume = 600 mL; flow rate = 80 mL/h). Collect 5-mL fractions; measure the absorbance of chromatography fractions at 280 nm, and at either 425 nm (for rCRALBP containing 11-*cis*-retinaldehyde) or at 400 nm (for rCRALBP containing 9-*cis*-retinaldehyde); and pool $A_{425/400nm}$ peak fractions (*see* **Note 4**).

3. Apply pooled fractions from the DEAE chromatography (without dialysis) to a hydroxylapatite column (1 × 30 cm) equilibrated in HTP-chromatography buffer and wash with HTP buffer until the absorbance at 280 nm stabilizes. Elute bound protein with a linear gradient using the HTP buffer and the same buffer containing 80 mM sodium phosphate, pH 7.0 (total gradient volume = 600 mL; flow rate = 60 mL/h). Collect chromatography fractions (5 mL), monitor absorbance as above (**step 2**), and pool peak $A_{425/400nm}$ fractions.

4. Dialyze pooled fractions from the HTP chromatography into Mono Q chromatography buffer (1 L × 3 changes of 1 h each), apply the dialyzed pool to a Mono Q anion exchange-HPLC column (0.5 × 5 cm) equilibrated in Mono Q buffer, and wash with the buffer until the A_{280nm} stabilizes. Elute bound protein with a linear gradient of Mono Q buffer and the same buffer containing 0.5 M sodium acetate (total gradient volume = 30 mL; flow rate = 1 mL/min). Collect chromatography fractions (1 mL), monitor absorbance as above (**step 2**), and pool $A_{425/400nm}$ fractions.

5. Pooled fractions from Mono Q chromatography contain the final protein product. The rCRALBP may be concentrated and the buffer exchanged using an Amicon concentrator or by centrifugation using an Amicon Centricon tube (10-kDa mol-wt cutoff) (*see* **Note 5**).

3.4. Purification of Fusion rCRALBP

Fusion rCRALBP produced with the pET 19b vector contains 23 additional amino-terminal residues, including 10 adjacent histidine residues (**Fig. 1**) that facilitate protein purification by nickel-affinity chromatography. Purify with Qiagen Ni-NTA resin according to the supplier's instructions.

1. All fusion-rCRALBP purification steps must be carried out under dim-red illumination to prevent photodecomposition of the retinoids.

2. Centrifuge the labeled bacterial-cell lysate at 45,000g for 20 min at 4°C to remove insoluble material produced during retinoid labeling (*see* **Note 6**).

3. Apply the clarified bacterial-cell lysate to Ni-NTA resin equilibrated in Ni-NTA buffer. We typically apply the lysate to the Ni-NTA resin (~2 mL bed volume for 1.2-L culture) in a tube (e.g., a 50-mL screw-cap, plastic conical tube) and rock gently for 1–2 h or overnight.

4. Wash the Ni-NTA resin-bound protein with the equilibration buffer until the absorbance at 280 nm stabilizes at about 0.02 AU. We use several hundred milliliters of buffer (~400 mL) for this step and caution that it not be cut too short.

5. Next, wash the resin/rCRALBP preparation with Ni-NTA buffer containing 40 mM imidazole until the A_{280nm} again stabilizes near zero. We use about 400 mL of buffer for this step also.

6. Elute bound rCRALBP with Ni-NTA buffer containing 250 mM imidazole. Collect 1-mL fractions. The majority of rCRALBP elutes in fractions 2–5 (with decreasing amounts eluting to about fraction 20 depending on sample load). We generally pool fractions 2–15 from a 1.2-L culture load.

7. The fusion rCRALBP may be concentrated and the buffer exchanged using an Amicon concentrator, or by centrifugation using an Amicon Centricon tube (10-kDa mol-wt cutoff).

3.5. Recombinant Protein Characterization

The identity of any newly expressed recombinant protein should be verified and its structural integrity demonstrated before pursuing structure-function studies (*see* **Note 8**). Several protein-chemistry techniques have proven useful for characterizing rCRALBP.

1. SDS-PAGE on 12% acrylamide gels *(21)* is the method of choice for initially demonstrating over-expression of rCRALBP and for demonstrating purification to apparent homogeneity (**Fig. 2**).

2. Western-blot analysis using CRALBP specific antibodies will rapidly verify that the recombinant product is structurally related to CRALBP. A number of anti-CRALBP antibodies are available *(12)*.

3. Although native CRALBP has an acetylated N-terminus, bacterially expressed rCRALBP has a free N-terminus; therefore, Edman degradation is useful for confirming the identity of the recombinant protein and for defining the amino-terminal structure (*see* **Note 9**). Limited-sequence analysis (10 cycles) can be carried out on a SDS-PAGE/PVDF blot *(12)* of the cell lysate to verify the presence of rCRALBP and on the final purified protein preparation. We usually perform sequence analysis on 1–2 µg rCRALBP samples *(9)*.

4. Amino acid analysis of purified rCRALBP is useful to quantify the protein, and for corroborating the identity of the recombinant protein (**Table 1**). We typically perform phenylthiocarbamyl (PTC) amino acid analysis on about 1 µg rCRALBP samples (20–30 pmol protein per analysis) using Applied Biosystems instrumentation *(26)*.

5. Mass spectral analysis of the final recombinant products is an excellent method for demonstrating that a complete polypeptide chain of known composition has been expressed (**Figs. 3** and **4**). The average isotopic mass of human nonfusion rCRALBP (M_r = 36,343) and fusion rCRALBP (M_r = 39,110) can be measured by LCMS with sufficient accuracy to define the number of residues in the protein (e.g., within about 4 Dalton of the calcu-

Table 1
Amino Acid Composition of Human rCRALBP

| | Residues per molecule | |
| | Fusion rCRALBP | Nonfusion rCRALBP |
Amino acid	$M_r = 39{,}110$	$M_r = 36{,}343$
Asx (D+N)	29	25
Glx (E+Q)	57	57
Ser (S)	17	15
Gly (G)	21	19
His (H)	17	5
Arg (R)	20	20
Thr (T)	15	15
Ala (A)	23	23
Pro (P)	13	13
Tyr (Y)	10	10
Val (V)	20	20
Met (M)	7	6
Ile (I)	14	13
Leu (L)	30	30
Phe (F)	24	24
Lys (K)	16	15
Cys (C)	4	4
Trp (W)	2	2
Total	339	316

lated rCRALBP mass). Usually, the retinoid must be removed from rCRALBP in order to obtain a useful electrospray-mass spectrum. Reverse-phase HPLC (RP-HPLC) in aqueous trifluoroacetic acid/acetonitrile solvents effectively strips the retinoid from rCRALBP (*see* **Note 10**).

6. The retinoid binding functionality of rCRALBP is measured spectrophotometrically with a scanning UV-Vis spectrophotometer (**Fig. 5**). In the absence of bleaching illumination, bovine CRALBP complexed with 11-*cis*-retinaldehyde exhibits an absorbance maximum of 425 nm *(26)* and complexed with 9-*cis*-retinaldehyde, a maximum of 405 nm *(3)*. Human rCRALBP exhibits similar spectral properties with subtle differences (e.g., fusion rCRALBP complexed with 9-*cis*-retinaldehyde exhibits a maximum at 400 nm). These characteristic maxima shift to about 380 nm after exposure to bleaching illumination due to the production of all-*trans*-retinaldehyde, which is not bound by rCRALBP. Bleaching of rCRALBP samples is performed for about 10 min at room temperature with four photography lights (150 W) each placed about 40 cm from the sample on four sides. For bleaching, the sample is placed in a glass cuvet ($2 \times 10 \times 35$ mm), covered with parafilm, and set inside a glass beaker.

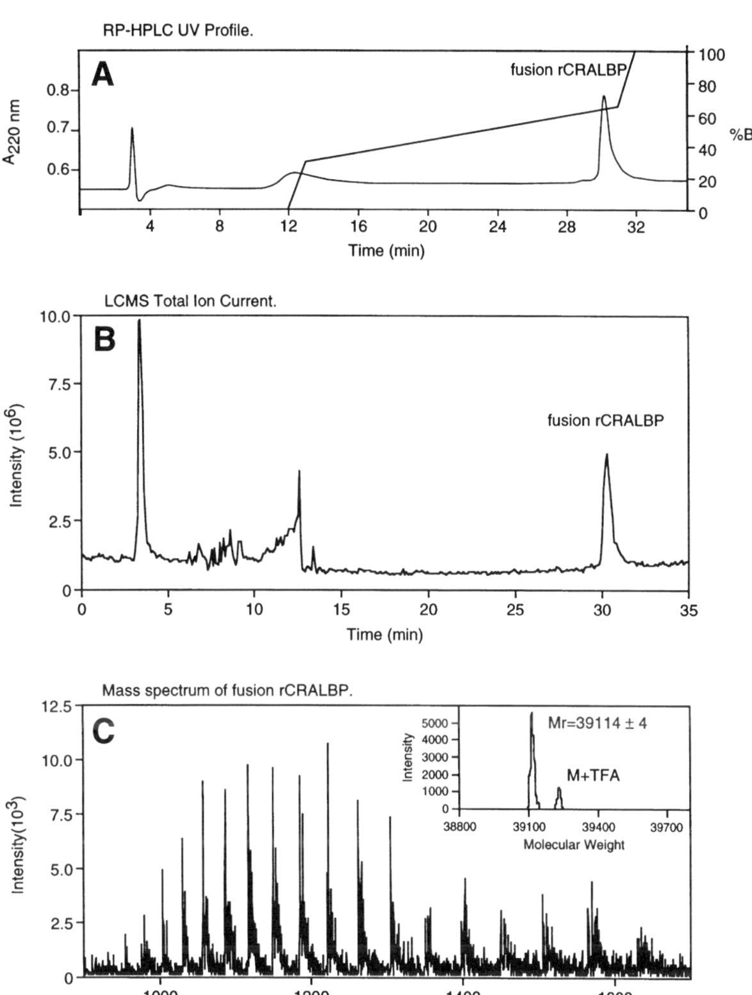

Fig. 3. LCMS Analysis of Fusion rCRALBP. (**A**) RP-HPLC ultraviolet profile (A_{220nm}) from analysis of 1.5 µg fusion human rCRALBP on a 50-µ Vydac C18 column (1 × 250 mm) at 50 µL/min using the indicated gradient. Solvent A was 0.1% trifluoracetic acid (TFA) and solvent B was 84% acetonitrile, containing about 0.07% TFA. (**B**) Reconstructed total ion current from electrospray-mass spectral analysis of the fusion rCRALBP liquid chromatography shown in A. (**C**) Electrospray-mass spectrum of fusion human rCRALBP indicating the presence of a protein of $M_r = 39,114 \pm 4$ (calculated $M_r = 39,110$). The deconvoluted spectrum is shown in the inset.

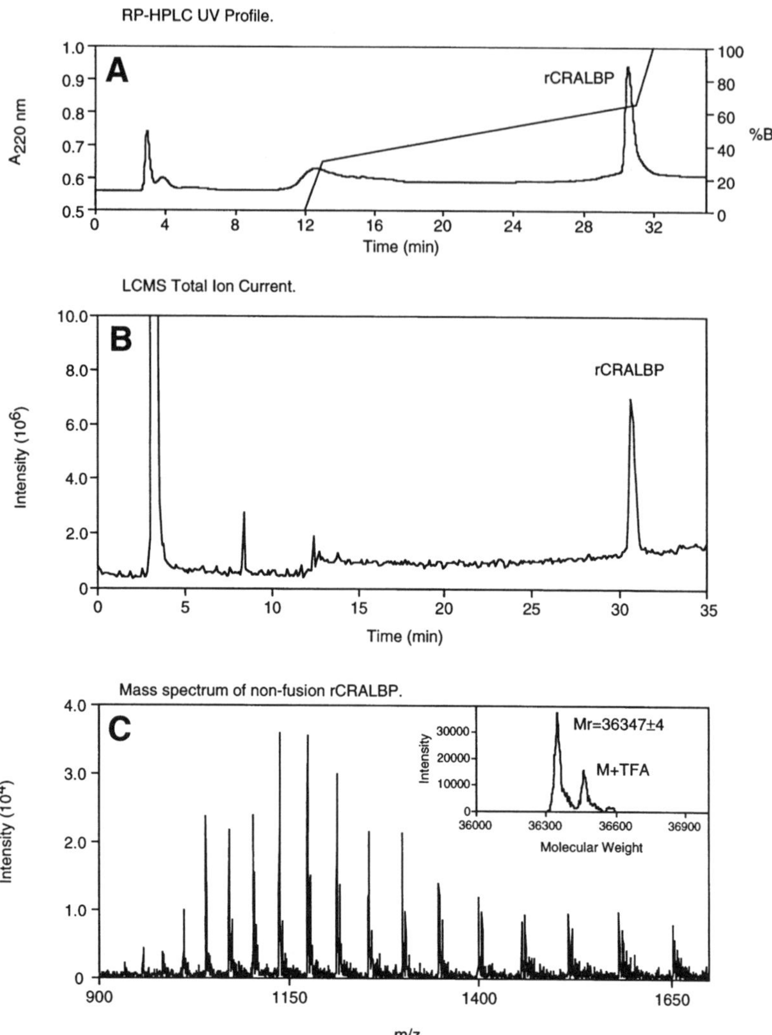

Fig. 4. LCMS Analysis of Non-Fusion rCRALBP. **(A)** RP-HPLC ultraviolet profile, **(B)** reconstructed total ion current, and **(C)** electrospray mass spectrum with the deconvoluted spectrum shown in the inset. LCMS analysis of nonfusion human rCRALBP (1.5 µg) was performed as described in the legend to **Fig. 3**, yielding a measured protein mass of $M_r = 36,347 \pm 4$ (calculated $M_r = 36,343$).

4. Notes

1. Production of wild-type rCRALBP with either pET3a or 19b vectors in *E. coli* strain BL21(DE3)LysS using enriched culture conditions has resulted in increased cell mass (over 10 g/L), but the majority of rCRALBP was insoluble

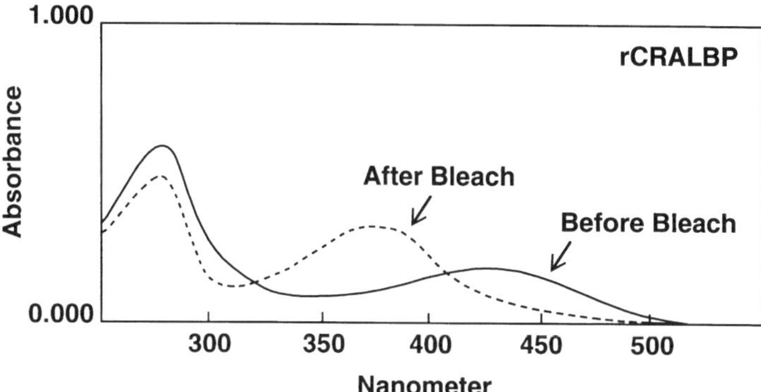

Fig. 5. Spectral Analysis of rCRALBP Before and After Bleaching. Fusion and nonfusion rCRALBP with bound 11-*cis*-retinaldehyde exhibit absorption maximum at 425 nm before bleaching. The proteins undergo a spectral shift to about 380 nm upon exposure to bleaching illumination, owing to the production of free, unbound all-*trans*-retinaldehyde.

and found in inclusion bodies (e.g., in a New Brunswick MPPF-40 L fermentor containing 30 L medium composed of tryptone 6 g/L, yeast extract 3 g/L, glucose 7 g/L, $K_2HPO_4:H_2O$ 2 g/L, KH_2PO_4 2 g/L, $CaCl_2:2H_2O$ 9 mg/L, $MgSO_4$ 6 mg/L, 1 g/L, ampicillin 100 mg/L $[NH_4]_2SO_4$, and chloramphenicol 25 mg/L). Mutant rCRALBPs we have expressed with the pET system (particularly deletion mutants, but also full-length point mutants), have generally been less soluble than the wild-type protein. The solubility of mutant rCRALBPs (produced with pET 19b) has been enhanced slightly (to about 1% of total soluble protein) when the bacteria were grown in a minimal medium containing 10 g/L bacto-tryptone, 1 g/L bacto-yeast extract, and 2 g/L sodium chloride *(18,19,27)*. Minimal media has also allowed productive incorporation of ^{13}C-methionine and 5-fluorotryptophan for solution state NMR analyses of rCRALBP *(27)*.

2. Both 11-*cis*-retinaldehyde and 11-*cis*-retinol have been recovered from purified rCRALBP after labeling of the crude-bacterial lysate with only 11-*cis*-retinaldehyde. This indicates that retinoid modification can occur during the labeling and/or protein-purification steps.

3. Complexed with 11-*cis*-retinol, CRALBP exhibits an absorption maximum at 330 nm *(2)* and fluorescence owing to the bound retinoid (340 nm excitation, 495 nm emission).

4. We have determined molar-extinction coefficients for human rCRALBP with bound 11-*cis*-retinaldehyde ($\varepsilon^{280nm} = 43,400\ M^{-1}cm^{-1}$ and $\varepsilon^{425nm} = 13,400\ M^{-1}cm^{-1}$) and with bound 9-*cis*-retinaldehyde [$\varepsilon^{280nm} = 44,600\ M^{-1}cm^{-1}$ and $\varepsilon^{400nm} = 20,700\ M^{-1}cm^{-1}$]. Similar molar-extinction coefficients have been determined for bovine CRALBP with bound 11-*cis*-retinaldehyde *(28)*.

5. Using the aforementioned culture methods and densitometric quantification of SDS-PAGE Coomassie blue staining, wild-type fusion and nonfusion rCRALBP

typically constitute 15–20% of the total soluble-lysate protein. These methods yield about 7 mg of purified nonfusion rCRALBP per L of pET 3a culture and about 3 mg of purified-fusion rCRALBP per L of pET 19b culture with quantification by amino acid analysis. The relatively lower yield of fusion rCRALBP appears to be associated with very tight binding of the fusion protein with the nickel-affinity chromatography support.

6. Wild-type fusion apo rCRALBP can be readily produced by deleting the retinoid-labeling step (**Subheading 3.2., step 4**) and purifying the protein as usual by Ni-NTA chromatography (**Subheading 3.4.**). Retinoid labeling of purified apo-rCRALBP can be achieved by adding concentrated retinoid (e.g., 5 m*M*) in ethanol to the protein in an aqueous buffer. Add the retinoid in small increments, such as 1 µL retinoid per 1 mL protein solution (with about 0.5-fold molar-excess retinoid over rCRALBP per addition). Add 2–3 increments with mixing between each addition; keep the final volume of retinoid added to 1% of the original sample volume *(29)*. Incubate on ice in the dark for about 10 min, then remove excess retinoid by Ni-NTA or molecular-sieve chromatography. Apo-CRALBP is less stable than the holo protein and tends to fall out of solution.

7. For purifying larger amounts of fusion rCRALBP (e.g., from cultures 7 L), we use an initial DEAE-chromatography step to remove excess retinoid before Ni-NTA chromatography. In this case, follow the protocol for purifying nonfusion rCRALBP from cell lysis through DEAE chromatography, then dialyze the DEAE pool into Ni-NTA buffer and follow the fusion rCRALBP protocol. (*Caution:* Too much retinoid during the labeling process can result in nonspecific association of retinoid with CRALBP, and this can obscure the characteristic spectral properties of the protein. DEAE chromatography is a good way to remove excess retinoid.)

8. Circular-dichroism studies indicate that native bovine CRALBP and human rCRALBP have similar secondary structures, ligand binding, and thermal-stability properties, and support the validity of pursuing structure-function studies of the recombinant protein *(29)*.

9. The N-terminus of bovine CRALBP is sensitive to proteolysis *(12)* and this also appears to be true for nonfusion human rCRALBP. The nonfusion rCRALBP produced with the pET3a vector sometimes has exhibited a frayed N-terminus, with the major component (~90%) starting at position Ser[1], and a minor-truncated species (~10%) starting at Arg[8].

10. Laboratories not set up for structural characterization of proteins can find assistance for amino acid analysis, Edman degradation and mass spectrometry in the Yellow Pages of the Membership Directory of the Association of Biomolecular Resource Facilities (ABRF, 9650 Rockville Pike, Bethesda, MD).

Acknowledgments

This study was supported in part by NIH grant EY06603 and NSF grants DMB 8516111 and BIR 9115824.

References

1. Saari, J. C., Bredburg, D. L., and Noy, N. (1994) Control of substrate flow at a branch in the visual cycle. *Biochemistry* **33,** 3106–3112.
2. Saari, J. C., Bredburg, L., and Garwin, G. G. (1982) Identification of the endogenous retinoids associated with three cellular retinoid-binding proteins from bovine retina and retinal pigment epithelium. *J. Biol. Chem.* **257,** 13,329–13,333.
3. Saari, J. C. and Bredburg, D. L. (1987) Photochemistry and stereoselectivity of cellular retinaldehyde-binding protein from bovine retina. *J. Biol. Chem.* **262,** 7618–7622.
4. Bunt-Milam, A. H. and Saari, J. C. (1983) Immunocytochemical localization of two retinoid-binding proteins in vertebrate retina. *J. Cell Biol.* **97,** 703–712.
5. Eisenfield, A. J., Bunt-Milam, A. H., and Saari, J. C. (1985) Localization of retinoid-binding proteins in developing rat retina. *Exp. Eye Res.* **41,** 299–304.
6. Martin-Alonso, J. M., Ghosh, S., Hernando, N., Crabb, J. W., and Coca-Prados, M. (1993) Differential expression of the cellular retinaldehyde-binding protein in bovine ciliary epithelium. *Exp. Eye Res.* **56,** 659–669.
7. De Leeuw, M., Gaur, V. P., Saari, J. C., and Milam, A. H. (1990) Immunolocalization of cellular retinol-, retinaldehyde- and retinoic acid-binding proteins in rat retina during pre- and postnatal development. *J Neurocytol.* **19,** 253–364.
8. Saari, J. C., Huang, J., Possin, D. E., Fariss, R. N., Leonard, J., Garwin, G. G., Crabb, J. W., and Milam, A. H. (1998) CRALBP is expressed by oligodendrocytes in optic nerve and brain. *Glia,* in press.
9. Crabb, J. W., Johnson, C. M., Carr, S. A., Armes, L. G., and Saari, J. C. (1988) The complete primary structure of the cellular retinaldehyde-binding protein from bovine retina. *J. Biol. Chem.* **263,** 18,678–18,687.
10. Crabb, J. W., Goldflam, S., Harris, S. E., and Saari, J. C. (1988) Cloning of the cDNAs encoding the cellular retinaldehyde-binding protein from bovine and human retina and comparison of the protein structures. *J. Biol. Chem.* **263,** 18,688–18,692.
11. Intres, R., Goldflam, S., Cook, J. R., and Crabb, J. W. (1994) Molecular cloning and structural analysis of the human gene encoding cellular retinaldehyde-binding protein. *J. Biol. Chem.* **269,** 25,411–25,418.
12. Crabb, J. W., Gaur, V. P., Garwin, G. G., Marx, S. V., Chapline, C., Johnson, C. M., and Saari, J. C. (1991) Topological and epitope mapping of the cellular retinaldehyde-binding protein from retina. *J. Biol. Chem.* **266,** 16,674–16,683.
13. Gu, M., Warshawsky, I., and Majerus, P. W. (1992) Cloning and expression of a cytosolic megakaryocyte protein-tyrosine-phosphatase with sequence homology to retinaldehyde-binding protein and yeast SE14p. *Proc. Natl. Acad. Sci. USA* **89,** 2980–2984.
14. Salama, S. R., Cleves, A. E., Malehorn, D.E., Whitters, E. A., and Bankaitis, V. A. (1990) Cloning and characterization of *Kluyveromyces lactis* SEC14, a gene whose product stimulates Golgi Secretory Function in *Saccharomyces cervisiae. J. Bacteriol.* **172,** 4510–4521.

15. Sato, Y., Arai, H., Miyata, A., Tokita S., Yamamoto, K., Tanabe, T., and Inoue, K. (1993) Primary structure of α-tocopherol transfer protein from rat liver. *J. Biol. Chem.* **268,** 17,705–17,710.
16. Chen, Y., Johnson, C., West, K., Goldflam, Bean, M. F., Huddleston, M. J., Carr, S. A., Gabriel, J. L., and Crabb, J. W. (1994) *TECHNIQUES IN PROTEIN CHEMISTRY V* (Crabb, J.W., ed.), Academic, San Diego CA, pp. 371–378.
17. Crabb, J. W., Chen, Y., Kapron, J. T., West, K. A., Bredburg, D. L., and Saari, J. C. (1996) Analysis of the retinoid-binding site of CRALBP. *Invest. Ophthal. Vis. Sci.* **37,** 3691.
18. Crabb, J. W., Roth, K. E., Pradis, S., Luck, L. A., Venters, R. A., and Spicer, L. D. (1997) Mutational mapping of the retinoid binding pocket in the cellular retinaldehyde-binding protein (CRALBP). *Protein Sci.* **6 (suppl. 2),** 67, 76-M.
19. Maw, M. A., Kennedy, B., Knight, A., Bridges, R., Roth, K. E., Mani, E. J., Mukkadan, J. K., Nancarrow, D., Crabb, J. W., and Denton, M. J. (1997) Mutation of the gene encoding cellular retinaldehyde-binding protein in autosomal recessive retinitis pigmentosa. *Nature Genet.* **17,** 198–200.
20. Studier, F. W., Rosenberg, A. H., Dunn, J. J., and Dubendorff, J. W. (1990) Use of T7 RNA polymerase to direct expression of cloned genes. *Methods Enzymology* **185,** 60–89.
21. Laemmli, U. K. (1970) Cleavage of structural proteins during the assembly of the head of bacteriophage T4. *Nature* **227,** 680–685.
22. Lowry, O. H., Rosebrough, N. J., Farr, A. L., and Randall, R. J. (1951) Protein Measurement with the Folin phenol reagent. *J. Biol. Chem.* **193,** 265–275.
23. Brown, P. K. and Wald, G. (1956) The Neo-b isomer of vitamin A and retinene. *J. Biol. Chem.* **222,** 865–877.
24. Robeson, C. D., Blum, W. P., Dieterle, J. M., Cawley, J. D., and Baxter, J.G. (1955) Chemistry of vitamin A. XXV. Geometrical isomers of vitamin A aldhehyde and an isomer of its a-ionone analog. *J. Am. Chem. Soc.* **77,** 4120–4125.
25. Saari, J. C. and Bredburg, D. L. (1988) Purification of cellular retinaldehyde binding protein from bovine retina and retinal pigment epithelium. *Exp. Eye Res.* **46,** 569–578.
26. Crabb, J. W., West, K. A., Dodson, W. S., and Hulmes, J. D. (1997) Amino acid analysis, in *CURRENT PROTOCOLS IN PROTEIN SCIENCE* (Coligan, J. E., Ploegh, H. L., Smith, J. A., and Speicher, D. W., eds.), Unit 11.9 (suppl. 7), Wiley, pp. 11.9.1–11.9.42.
27. Luck, L. A., Barrows, S. A., Venters, R. A., Kapron, J., Roth, K. A., Paradis, S. A., and Crabb, J. W. (1997) NMR methods for analysis of CRALBP retinoid binding, in *TECHNIQUES IN PROTEIN CHEMISTRY VIII* (Marshak, D., ed.), Academic, San Diego, CA, pp. 439–448.
28. Stubbs, G. W., Saari, J. C., and Futerman, S. (1979) 11-*cis*-Retinal-binding protein from retina, isolation and partial characterization. *J. Biol. Chem.* **254,** 8529–8533.
29. Carlson, A., Bok, D., Crabb, J. W., and Horwitz, J. (1993) Circular Dichroism Spectral Analysis of Native and Recombinant Cellular Retinaldehyde-Binding Protein. *Invest. Opthal. Vis. Sci.* **34,** 831.

8

Expression and Purification of CRABPs from *E. coli*

Joseph L. Napoli, Zuzana Sperkova, Marek Sperka,
and Paul D. Fiorella

1. Introduction

Nuclear-retinoid receptors provide a mechanism of retinoid action, but most likely where expressed, the cytosolic cellular retinoic acid-binding proteins (CRABPs) also affect the ability of retinoids to initiate biological signals *(1,2)*. holo-CRABP I sequesters retinoic acid (RA) with a K_d value that may be <1 nM, and serves as a high-affinity (K_m ~2 nM), efficient substrate of RA metabolism, thereby controlling not only the steady-state concentration of RA, but also its availability *(3)*. The impact of CRABP on retinoid metabolism may extend beyond RA. CRABP binds RA metabolites, including 4-OH-RA, 4-oxo-RA, and several metabolites whose structures have not been identified, and could thereby influence their disposition in vivo *(4)*. For example the elimination $t_{1/2}$ of RA in the presence and absence of CRABP I was 35 and 40 min, respectively, in incubations with rat testis microsomes. Unbound 4-OH-RA and 4-oxo-RA had elimination $t_{1/2}$ values of 40 and 6 min, respectively. In contrast, CRABP-bound 4-OH-RA and CRABP-bound 4-oxo-RA were essentially metabolically inert. This shows that CRABP does not affect all retinoids similarly and that it could have a profound influence on the steady-state concentrations of several retinoids in vivo.

The X-ray crystallographic structures of CRABP I and CRABP II have been solved *(5,6)*, high-field NMR studies are available *(7,8)*, and methods to express these proteins and their mutants abundantly have been developed *(3,9–11)*. In the near future, additional site-directed mutagenesis to extend the insight already available from this technique *(12,13)*, and other biochemical approaches should provide further insight into ligand binding by these pro-

From: *Methods in Molecular Biology, Vol. 89: Retinoid Protocols*
Edited by: C. P. F. Redfern © Humana Press Inc., Totowa, NJ

teins, and the purpose of their highly conserved, exterior-projecting amino acids to their functions.

2. Materials

1. Vector: The expression vector pET-3a/CRABP I contains the complete coding region of bovine CRABP I cloned into the plasmid pET-3A, such that the ATG start site of the CRABP I cDNA begins at an *Nde*I site. The CRABP I cDNA was placed downstream and under control of the T7 promotor and upstream from the transcription terminator.
2. Bacteria: pET-3a/CRABP I was used to transform *E. coli* K12 strain BL21(DE3)/pLysS. This strain has the T7 RNA polymerase gene incorporated into its genome under control of the *lac* UV5 promotor and contains the plasmid pLysS, which confers chloramphenicol resistance and produces a modest amount of T7 lysozyme, a T7 polymerase inhibitor. T7 lysozyme prevents constitutive expression of T7 promotor-dependent genes. The inhibition is overcome by isopropyl-β-D-thiogalactopyranoside, which induces synthesis of relatively large amounts of T7 polymerase.
3. Medium preparation: To each of 10 2-L Erlenmeyer flasks, add 400 mL of LB (Luria Broth, Sigma) medium (25 g/L). To a 250-mL Erlenmeyer flask, add 50 mL of LB medium. For agar plates, use LB agar (32 g/L, Sigma). Cover the tops of the flasks with foil and autoclave for 40 min on the slow-exhaust setting.
4. Plate preparation: After autoclaving, gently swirl the contents of the flask containing the agar/medium mixture and allow the mixture to reach 40°C. Add ampicillin (*see* **Note 1**, final concentration 100 mg/mL, made as a 25-mg/mL stock in H_2O by sterilization through a 0.22-μm filter) and chloramphenicol (final concentration 35 μg/mL, made as a 35-mg/mL stock in ethanol). Add 30-mL aliquots of the agar/media/antibiotic mixture to two 100-mm microbiological plates.
5. Isopropyl-β-D-thiogalactopyranoside (IPTG): make a fresh stock solution in water and filter through a 0.22-μm filter. IPTG is added to bacterial-culture medium to a final concentration of 0.6 m*M*.
6. 20 m*M* Tris-HCl, pH 7.4.
7. 20 m*M* Tris-HCl, pH 7.4, 40 m*M* magnesium acetate, 0.5% Triton X-100.
8. DNase I.
9. Bath sonicator (e.g., Branson 1200).
10. Sephadex G-50 column (5 × 115 cm).
11. 20 m*M* Tris-HCl, pH 7.4, 10 m*M* KCl, 2 m*M* 2-mercaptoethanol, and 0.05% w/v sodium azide.
12. Amicon concentrator with type YM5 filter.
13. Buffer A: 20 m*M* Tris-HCl, pH 7.4, 150 m*M* KCl, 2 m*M* 2-mercaptoethanol.
14. 10 m*M* all-*trans* retinoic acid in ethanol.
15. Sephadex G-10 column (0.7 × 3 cm).
16. Buffer B: 20 m*M* *bis*-Tris-HCl, pH 7.0, 2 m*M* 2-mercaptoethanol, 0.05% sodium azide.

17. Mono-Q column type HR5/5.
18. [^{3}H]retinoic acid (1.4 mCi/mmol).
19. Superose-12 column.
20. Centricon 10 filters (Amicon).

3. Methods

3.1. Bacteria Growth

1. *Escherichia coli* growth: after the agar plates have solidified, streak with *E. coli* strain BL21(DE3)/pLysS containing the plasmid pET-3a/CRABP I from a glycerol stock. Incubate overnight at 37°C.
2. Liquid overnight culture: add ampicillin and chloramphenicol to a 250-mL Erlenmeyer flask containing the 50 mL of LB medium. Remove a single bacterial colony from one agar plate and add it to the flask. Shake at 37°C overnight.
3. Set-up for the large-scale induction: add antibiotics (ampicillin 150 µg/mL and chloramphenicol 35 µg/mL) to the 100 2-L flasks, followed by a 100-fold dilution of bacteria from the 50-mL overnight culture, i.e., 4 mL per 400 mL medium. Shake at 250 rpm at 37°C and monitor cell density by spectrophotometric analysis of aliquots at 590 nm. At an optical density of 0.6–1 AU (~4 h), add IPTG to a final concentration of 0.6 mM from a freshly-made stock solution in water. Incubate the cultures for 1.5 h at 250 rpm and 37°C.

3.2. Isolation of CRABP I

1. Processing of bacteria: Centrifuge the contents of the flasks at ~5000g for 12 min in 250-mL bottles and divide the pellets into two aliquots. Place into two 50-mL pre-weighed Falcon tubes. Use ~15 mL of 20 mM Tris-HCl, pH 7.4, buffer to facilitate pellet removal of bacteria from the centrifuge bottles. Centrifuge the 50-mL tubes at 3500g (JA-17 rotor, 5000 rpm) for 10 min. Decant and weigh the tubes to record the pellet weight. Freeze the consolidated pellet at –70°C to facilitate lysis.
2. Preparation of pellet: thaw the pellet in 37°C water. To the lysate, add 4 mL/g wet weight of 20 mM Tris-HCl, 40 mM magnesium acetate, 0.5% Triton X-100, pH 7.4. Add 30 U/g of DNase I and incubate for 30 min at ambient temperature with shaking. Place a plastic pipet tip in the solution, cap, and sonicate 20 min in a bath sonicator (e.g., Branson 1200) (*see* **Note 2**). Centrifuge this reduced viscosity sample at 15,000g for 45 min. The resulting supernatant contains CRABP I.
3. Size-exclusion chromatography of CRABP I: elute the CRABP I-containing supernatant through a Sephadex G-50 column (5 × 115 cm) with 20 mM Tris-HCl, pH 7.4, 10 mM KCl, 2 mM 2-mercaptoethanol, and 0.05% sodium azide, at 4°C at a flow rate of 1 to 1.5 mL/min. Collect 15-mL fractions. CRABP I-containing-fractions are generally preceded by several turbid fractions. A peak (at 280 nm) following the turbid fractions is generally observed. Pool these fractions and take an absorbance spectrum from 400–220 nm. CRABP I shows a

characteristic 280 peak with a small shoulder at 290; a close estimate of the concentration can be made using a molar absorptivity of 28,000 at 280 nm. The yield varies between 50 and 100 mg of CRABP I per 500 mg of supernatant protein.

4. Use of CRABP I without further purification: Concentrate the CRABP I-containing fractions to 2 mg protein/mL with an Amicon concentrator/type YM5 filter while exchanging the buffer to buffer A (*see* **Note 3**). Prepare a 1-to-5 dilution of the concentrate (1 mL) and obtain an absorbance spectrum from 250–450 nm. Add 15 mL of 10 mM all-*trans*-retinoic acid in ethanol, mix inside the cuvet using a Pasteur pipet, and scan again. Remove excess unbound retinoic acid by eluting the solution in the cuvet through a Sephadex G-10 column (0.7 × 3 cm). Collect 0.5-mL fractions. CRABP I will elute in fractions 2 and 3. Obtain an absorbance spectrum on these combined fractions. The ratio A_{350}/A_{280} should be 1.3–1.6.

3.3. Final Purification of CRABP I Expressed in E.coli

1. Final purification of CRABP I by anion-exchange chromatography: omit **step 4** of **Subheading 3.2.** and concentrate the CRABP I-containing fraction to 2 mg protein/mL with an Amicon concentrator/type YM5 filter while exchanging buffer to buffer B. Apply five 1-mL aliquots sequentially to a Mono-Q column type HR5/5 and follow each application with a 5-min wash at 1 mL/min with buffer B. After the last wash, elute CRABP I at 1 mL/min with a linear gradient to 250 mM NaCl in buffer B over 25 min (*see* **Note 4**).
2. Storage: Store CRABP I, which will be used within 1 mo at 4°C in buffer A, otherwise freeze it at –70°C.

3.4. Preparation of Holo-CRABP I

Incubate 150 mg of purified CRABP I in 0.5 mL of buffer B with a 1.5 molar excess [^{3}H]retinoic acid (1.4 mCi/mmol) at room temperature for 1 h. Remove excess unbound [^{3}H]retinoic acid by eluting the solution through a Superose-12 column with buffer B at a flow rate of 0.5 mL/min. Identify the fractions (1 mL each) containing CRABP by liquid scintillation counting of aliquots. Generally, pool four fractions and concentrate in a Centricon-10 filter at 5000g for 30–60 min. The amount of holo-CRABP in the concentrate can be determined by obtaining the absorbance spectrum between 500 nm and 220 nm. Use a molar absorptivity of 50,000 at A_{350} to calculate the amount of holo-CRABP *(14,15)*.

3.5. Expression of CRABP II in E.coli from pET-3a/CRABP II

The expression vector pET-3a/CRABP II is similar in design and function to pET-3a/CRABP I. With pET-3a/CRABP II, CRABP II may be expressed and purified as described above for CRABP I.

3.6. Measurement of K$_d$ and n Values for CRABP I and CRABP II

1. An accurate indication of the number of binding sites for CRABP can be determined by fluorescence titration using an excess of protein (~0.5–1 mM) to all-*trans*-RA (*see* **Note 5**). Such a "high" concentration of protein allows signals to

be read from common fluorimeters, which are relatively insensitive. The data should show a linear relationship between binding and RA concentration that changes abruptly to a slope of zero when the concentration of all-*trans*-RA reaches that of the protein, indicating stoichiometric binding under the experimental conditions, i.e., essentially all of the ligand is bound until saturation is reached, and then no more can be bound.

2. With insensitive instruments, the K_d values of poorly binding ligands, e.g., 9-*cis*-retinoic acid and 13-*cis*-retinoic acid, can also be determined with reasonable accuracy. Be careful not to use so much protein that even poorly binding ligands are substantially bound under the experimental conditions, i.e, show stoichiometric binding.

3. With insensitive instruments, in which concentrations of CRABP are used that exceed the K_d values of tightly binding ligands, K_d values cannot be accurately determined; only upper limits may be evaluated. This is because with stoichiometric binding, the concentration of free ligand is indeterminantly small, i.e., it is not in equilibrium. To overcome this problem, the concentration of CRABP must be lowered so that an equilibrium occurs between it and the ligand. This necessitates use of a sensitive fluorimeter *(16)*.

4. Notes

1. pET-3a confers amplicillin resistance.
2. These steps are necessary to facilitate further purification.
3. A 12.5–15% SDS-PAGE gel can help evaluate quality of CRABP I at this stage.
4. Two major CRABP I molecules with different N-termini are produced by this procedure. Sixty to 100% of the product, depending on the preparation, is authentic CRABP I, with a pI of 4.76. The 40% remaining in some preparations contains a major N-terminal blocked CRABP I variant with a pI of 4.79, and a minor variant with a pI of 4.83. The combination of the two variants binds RA as well as authentic CRABP I.
5. A detailed protocol for fluorescent titration of CRABPs is given in Chapter 10.

References

1. Donovan, M., Olofsson, B., Gustafson, A.-L., Dencker, L., and Eriksson, U. (1995) The cellular retinoic acid binding proteins. *J. Steroid. Biochem. Mol. Biol.* **53,** 459–465.
2. Napoli, J. L. (1993) Biosynthesis and metabolism of retinoic acid: roles of CRBP and CRABP in retinoic acid homeostasis. *J. Nutr.* **123,** 362–366.
3. Fiorella, P. F. and Napoli, J. L. (1991) Expression of cellular retinoic acid binding protein (CRABP) in *Escherichia coli*: characterization and evidence that holo-CRABP is a substrate in retinoic acid metabolism. *J. Biol. Chem.* **266,** 16,572–16,579.
4. Fiorella, P. D. and Napoli, J. L. (1994) Microsomal retinoic acid metabolism: effects of cellular retinoic acid-binding protein (type I) and C18-hydroxylation as an initial step. *J. Biol. Chem.* **269,** 10,538–10,544.

5. Thompson, J. R., Bratt, J. M., and Banaszak, L. J. (1995) Crystal structure of cellular retinoic acid binding protein I shows increased access to the binding cavity due to formation of an intermolecular β-sheet. *J. Mol. Biol.* **252,** 433–446.

6. Kleywegt, G. J., Bergfors, T., Senn, H., Le Motte, P., Gsell, B., Shudo K., and Jones, T. A. (1994) Crystal structures of cellular retinoic acid binding proteins I and II in complex with all-trans-retinoic acid and a synthetic retinoid. *Structure* **2,** 1241–1258.

7. Norris, A. W., Rong, D., d'Avignon, A., Rosenberger, M., Tasaki, K., and Li, E. (1995) Nuclear magnetic resonance studies demonstrate differences in the interaction of retinoic acid with two highly homologous cellular retinoic acid binding proteins. *Biochemistry* **34,** 15,564–15,573.

8. Rizo, J., Liu, Z. P., and Gierasch, L. M. (1994) ^{1}H and ^{15}N resonance assignments and secondary structure of cellular retinoic acid-binding protein with and without bound ligand. *J. Biomol. NMR* **4,** 741–760.

9. Fiorella, P. D, Giguère, V., and Napoli, J. L. (1993) Expression of cellular retinoic acid binding protein (type II) in *Escherichia coli*: characterization and comparison to cellular retinoic acid binding protein (type I). *J. Biol. Chem.* **268,** 21,545–21,552.

10. Redfern, C. P. F. and Wilson, K. E. (1993) Ligand binding properties of human cellular retinoic acid binding protein II expressed in *E. coli* as a glutathione-S-transferase fusion protein. *FEBS Letters* **321,** 163–168.

11. Fogh, K., Voorhees, J. J., and Astrom, A. (1993) Expression, purification, and binding properties of human cellular retinoic acid-binding protein type I and type II. *Arch. Biochem. Biophys.* **300,** 751–755.

12. Chen, L. X., Zhang, Z. P., Scafonas, A., Cavalli, R. C., Gabriel, J. L., Soprano, K. J., and Soprano, D. R. (1995) Arginine 132 of cellular retinoic acid-binding protein (type II) is important for binding of retinoic acid. *J. Biol. Chem.* **270,** 4518–4525.

13. Zhang, J., Liu, Z. P., Jones, T. A., Gierasch, L. M., and Sambrook, J. F. (1992) Mutating the charged residues in the binding pocket of cellular retinoic acid-binding protein simultaneously reduces its binding affinity to retinoic acid and increases its thermostability. *Proteins* **13,** 87–99.

14. Ong, D. E. and Chytil, F. (1978) Cellular retinoic acid-binding protein from rat testis: purification and characterization. *J. Biol. Chem.* **253,** 4551–4554.

15. Ong, D. E. and Chytil, F. (1980) Purification of cellular retinol and retinoic acid-binding proteins from rat tissue. *Methods Enzymol.* **67,** 288–296.

16. Norris, A. W., Cheng, L., Giguère, V., Rosenberger, M., and Li, E. (1994) Measurement of subnanomolar retinoic acid binding affinities for cellular retinoic acid binding proteins by fluorimetric titration. *Biochim. Biophys. Acta.* **1209,** 10–18.

9

Purification and Fluorescent Titration of Cellular Retinol-Binding Protein

Giorgio Malpeli, Claudia Folli, Davide Cavazzini, Giovanni Sartori, and Rodolfo Berni

1. Introduction

The intracellular carriers for all-*trans* retinol are believed to be the homologous cellular retinol-binding protein (CRBP) and cellular retinol-binding protein II (CRBP II) (**ref. *1*** , and references therein). CRBP is present in a wide variety of tissues, including liver, kidney, and testis. The basic feature of CRBP is the retinol-protein recognition, through which a drastically higher stability for CRBP-bound retinol relative to uncomplexed retinol is achieved. In addition, the binding to CRBP permits the solubilization in the aqueous medium of the highly hydrophobic retinol molecule. Evidence has been presented to indicate that the retinol–CRBP complex may serve as substrate of enzymes involved in the metabolism of retinol *(2–4)*. The binding of retinol to CRBP is characterized by a dissociation constant lower than nanomolar *(5,6)*. CRBP exhibits affinity for a variety of retinol analogs *(6,7)*. However, the binding of retinol analogs to CRBP was found to be substantially weaker than that of retinol *(6)*. Based on the three-dimensional structure determination of the retinol-CRBP complex *(8)* and on the results of studies performed in solution *(6)*, it has been suggested that the hydrogen-bonding interaction between the retinol-hydroxyl group and the side chain of Gln 108 contributes remarkably to the strength of binding of retinol to CRBP.

The most widely used method to study retinoid–protein interactions is fluorescence titration. The strong absorbance of retinol and retinol analogs in the region of emission of protein tryptophans is associated with an efficient energy transfer from excited tryptophans to the bound retinoid, and leads to a substantial quenching of intrinsic protein fluorescence when retinoids interact with

From: *Methods in Molecular Biology, Vol. 89: Retinoid Protocols*
Edited by: C. P. F. Redfern © Humana Press Inc., Totowa, NJ

binding proteins. In addition, the fluorescence of retinol and some retinol analogs may provide an intense signal, which allows the interaction of retinoids with binding proteins to be monitored. We report here the preparation of recombinant rat CRBP and the fluorescent retinoid, retinyl methyl ether (RME). RME interacts specifically with the CRBP retinol-binding site, as revealed by the spectral characteristics of the RME-CRBP complex. The fluorescence titration of CRBP with RME is described.

2. Materials

2.1. Synthesis and Purification of All-Trans Retinyl Methyl Ether (RME)

1. Reagents for RME synthesis: All-*trans* retinol (Fluka, Buchs, Switzerland), dimethyl sulfate (Fluka), butyllithium (Aldrich, Steinheim, Germany), sulfuric acid, and ammonium hydroxide.
2. Reaction medium: dry benzene.
3. Mobile phase for thin layer chromatography (TLC): ethanol/hexane (97:3, v/v).
4. Analytical TLC: TLC-Aluminum sheets silica gel 60 F_{254} (Merck, Darmstadt, Germany).
5. Spraying reagent for analytical TLC: 20% w/v $SbCl_3$ in $CHCl_3$.
6. Preparative TLC: TLC-Plates silica gel 60 F_{254} (Merck).
7. High pressure liquid chromatography (HPLC): Waters (Mildford, MA) 6000A solvent delivery system, fitted with a Rheodyne injection valve and equipped with a Waters 484 tunable absorbance detector.
8. HPLC column: reversed-phase Hypersyl 5 ODS column (25 cm × 4.6 mm), eluted isocratically with methanol/water (90:10, v/v) containing 0.01 M ammonium acetate (flow rate: 2.5 mL/min).
9. Diethyl ether.

2.2. Expression and Purification of Rat CRBP

1. Expression system: *E. coli* JM101 cells transformed with a pMON vector containing the full-length CRBP cDNA, the *E. coli* recA promoter (inducible by nalidixic acid) and a Amp^r locus. (*See* **ref. 9** for the preparation of this expression system.)
2. Luria broth: 1% w/v tryptone, 0.5% w/v yeast extract and 1% w/v NaCl. Adjust the pH to 7.9 with 2 M NaOH and autoclave.
3. Nalidixic acid solution: Prepare just before use a 1% w/v nalidixic acid (Fluka) solution in sterile 0.1 M NaOH.
4. Ampicillin solution: 5% w/v ampicillin (Sigma, St. Louis, MO) in sterile water. Store at –20°C.
5. Lysis buffer: 50 mM Tris-HCl, pH 7.9, 10% w/v sucrose, 0.5 mM phenylmethanesulfonyl fluoride. Store at 4°C.
6. Ammonium sulfate solution: 100% sat. ammonium sulfate, 15 mM dithiothreitol, pH 7.1. Store at 4°C.

7. Concentrating cell: Ultrafiltration Amicon Cell equipped with a YM 10 membrane (Amicon, Beverly, MA).
8. Gel filtration column: Ultrogel AcA 54 (Biosepra, Locke Drive, MA) column (5 × 85 cm), equilibrated and developed with 25 mM potassium phosphate, 200 mM NaCl, 3 mM 2-mercaptoethanol, pH 7.2 (flow rate: 0.5 mL/min).
9. Fast protein liquid chromatography (FPLC): Waters 650 solvent-delivery system equipped with a Waters 440 absorbance detector.
10. FPLC column: QMA anion exchange (Waters) column (1.5 × 10.5 cm) equilibrated with 25 mM Tris-HCl, pH 7.3, and eluted with a 100-mL gradient of NaCl (0–200 mM) in 25 mM Tris-HCl, pH 7.3 (flow rate: 2.5 mL/min).

2.3. Binding of RME to CRBP

1. RME: Take to dryness, just before use, an aliquot of the concentrated stock solution of RME purified by means of HPLC and kept in hexane at –80°C (*see* **Subheading 3.1., step 5**). Dissolve RME in ethanol and dilute it to final concentrations of 0.01–1 mM (ε = 49,880 $M^{-1}\mathrm{cm}^{-1}$ at 326 nm in isopropanol *[9]*).
2. Retinol: Prepare just before use ethanolic solutions of 0.5–1 mM all-*trans* retinol (Fluka), using an extinction coefficient of 46,000 M^{-1} cm^{-1} at 325 nm *(10)*.
3. Buffer for binding assays: 20 mM sodium phosphate, 70 mM NaCl, pH 7.3. Store at 4°C.
4. Microsyringe: 5-µL microsyringe from SGE (Austin, Texas).
5. Spectrophotometric measurements: Varian Cary 1E.
6. Spectrofluorometric measurements: Perkin Elmer LS 50B.
7. Analysis of binding data: Computer program Sigmaplot 5.0 (Jandel, Corte Madera, CA).

3. Methods
3.1. Synthesis and Purification of RME

The protocol described here for the synthesis of RME is basically that reported in *(9)*. The lithium derivative of retinol is first prepared in dry benzene from retinol and butyllithium. The lithium derivative of retinol then undergoes a nucleophilic substitution in the reaction with dimethyl sulfate, giving RME in a "one flask" process. All manipulations are carried out under dim light. The structures of retinol and RME are presented in **Fig. 1**.

1. Dissolve 500 mg (1.75 mmol) of crystalline all-*trans* retinol in 30 mL of dry benzene. Add to the solution under nitrogen 1.1 mL of 1.6 M butyllithium (1.76 mmol) in hexane. Upon addition of butyllithium, the color of the solution changes from yellow to deep cherry red. Stir the solution for 5 min at room temperature.
2. Add to the reaction medium 619 mg (4.9 mmol) of freshly distilled dimethyl sulfate dissolved in 20 mL of dry benzene. At this stage the solution changes back to a yellow color. Heat the solution at 70°C for 1 h. After cooling in ice, transfer the reaction mixture to a separating funnel. Wash the solution succes-

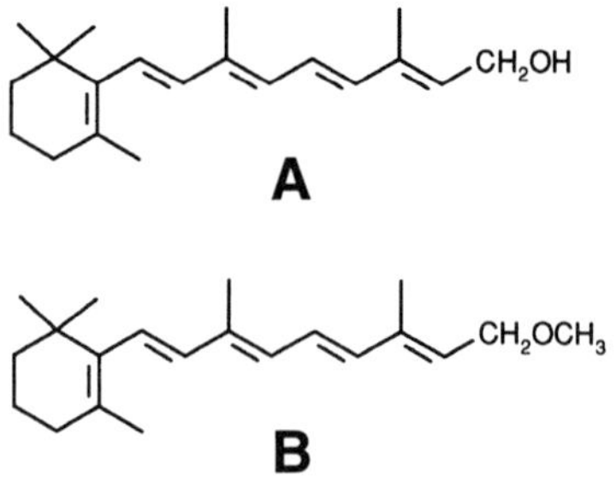

Fig. 1. Structural formulae of all-*trans* retinol (**A**) and all-*trans* retinyl methyl ether (**B**).

sively with cold 0.1 M sulfuric acid, 1 M ammonium hydroxide (twice), and water (three times). RME represents the main reaction product (the yield of the reaction is approx 70%).

3. Analyze reaction products by means of analytical TLC. Recognize RME by the appearance of a blue spot after spraying TLC-Aluminum sheets with $SbCl_3$, which appears to be specific for RME.
4. Purify RME by means of preparative TLC. Recover the fluorescent RME from the silica-gel plates with diethyl ether.
5. Further purify RME by means of reversed-phase HPLC. Take to dryness an aliquot of the RME solution in diethyl ether recovered from the silica-gel plates and containing approx 3 mmol of RME. Dissolve RME in 50 µL of methanol and apply to the HPLC column. Monitor the elution profile at 330 nm, which should reveal only minor components along with the predominant peak of RME. Extract the fractions containing RME with hexane and store the retinoid in hexane at –80°C.

Analytical properties of purified RME are as follows: white solid; mp: 30–33°C; UV λ_{max} (ethanol): 325 nm; MS (m/z): 300 (M^+, 10%), 269 *(8)*.

3.2. Expression and Purification of Rat CRBP

For the expression of recombinant CRBP the following procedure may be used.

1. Grow up overnight, at 37°C, a 20-mL culture of transformed *E. coli* JM101, in freshly prepared Luria broth containing 40 µL of the ampicillin solution.
2. Inoculate 1 L of Luria broth containing 2 mL of the ampicillin solution with the aforementioned culture. Incubate for 60–80 min at 30°C until an absorbance of 0.2 at 600 nm is attained. Add 5 mL of the nalidixic acid solution to induce the expression of CRBP and continue the incubation at 30°C up to an absorbance of 0.85/1.0 at 600 nm.

3. Separate the cells by centrifugation at 9000g for 10 min at 4°C, resuspend the pellet with 2–3 vol of the lysis buffer, sonicate the suspension (nine bursts of 15 s with pauses of 1 min) keeping the temperature below 10°C, and separate the soluble-cell extract by centrifugation at 9000g for 10 min at 4°C.

4. Add slowly to the supernatant an equal volume of ice-cold 100% sat. ammonium sulfate. Stir gently at 4°C for 2 h, centrifuge at 41,000g for 30 min and discard the pellet. Dialyze the supernatant overnight against 5 L of 25 mM potassium phosphate, 200 mM NaCl, 3 mM 2-mercaptoethanol, pH 7.2, and concentrate the protein solution to approx 20 mL with the ultrafiltration Amicon cell.

5. Perform the gel filtration chromatography at 4°C, monitoring the elution profile at 280 nm. Apo-CRBP elutes with a relative retention volume (V_e/V_o) of approx 1.4 and can be unambiguously identified by monitoring the formation of the typical absorption spectrum of the retinol-CRBP complex (*see* **Fig. 2**) obtained for aliquots of the eluted solution in the presence of 5 µM retinol.

6. Concentrate the pool of fractions containing CRBP, dialyze against 25 mM Tris-HCl, pH 7.3, and inject the protein solution into the FPLC-QMA column equilibrated with the same buffer and develop with the linear 0–200 mM NaCl gradient (*see* **Note 1**). Apo-CRBP, which is eluted at about 120 nM NaCl, is nearly completely pure and functionally active at this stage as judged by SDS-PAGE *(11)* and by the formation of the spectrum of an authentic retinol-CRBP complex, upon addition of retinol to aliquots of the purified protein. Quantify apo-CRBP using the extinction coefficient $\varepsilon = 28,080\ M^{-1}\ cm^{-1}$ at 280 nm *(12)*.

3.3. Absorption and Fluorescence Spectra of the RME-CRBP Complex

To establish whether a retinoid interacts specifically with the CRBP retinol-binding site, absorption and fluorescence spectra of the retinoid-CRBP complex may be analyzed. We show here that the RME-CRBP complex exhibits spectral characteristics similar to those of retinol-CRBP, like the vibronic fine structures of the absorption and fluorescence excitation spectra and the quite intense fluorescence emission of CRBP-bound RME (**Figs. 2** and **3**). Absorption and fluorescence spectra of the RME-CRBP complex can be obtained as described below.

1. Add 5 µL of an ethanolic solution of 0.88 mM RME to the spectrophotometer cuvet containing 7.5 µM apo-CRBP in 1 mL phosphate buffer, pH 7.3 (*see* **Note 2**). The excess of CRBP relative to RME (the CRBP/RME molar ratio in the system is approx 1.7) ensures that nearly all the ligand is bound to the protein (a correction for the contribution of free ligand to the spectra of the RME-CRBP complex is, therefore, not needed). Stir gently and let RME bind to CRBP in the dark at 20°C for 15–20 min (*see* **Note 3**). Use this solution to record subsequently both absorption and fluorescence spectra.

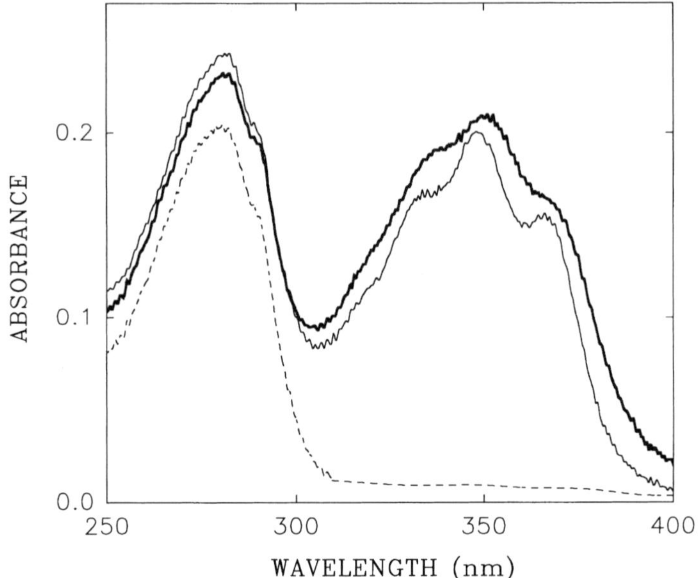

Fig. 2. Absorption spectra of RME-CRBP (bold line) and retinol-CRBP (thin line) complexes, in 20 m*M* sodium phosphate, 70 m*M* NaCl, pH 7.3. The absorption spectrum of apo-CRBP is also shown (dashed line). Upon addition of aliquots of ethanolic solutions of retinol or RME to the solution containing the apoprotein, the reaction medium was incubated for 15 min in the dark before spectra were run. The concentration of both RME and retinol added to the cuvet was 4.4 µ*M*, as compared with a CRBP concentration of 7.5 µ*M*. The temperature was kept at 20 ± 1°C by circulating water through the cell holder. The cuvet path length was 10 mm.

2. Record the absorption spectrum of the RME-CRBP complex in the 250- to 400-nm range and compare it with that of a reconstituted retinol-CRBP complex under the same conditions (*see* **Fig. 2** and **Note 4**).
3. Record the fluorescence excitation and emission spectra (*see* **Note 6**) of the RME–CRBP complex and compare them with those of a reconstituted retinol–CRBP complex under the same conditions (*see* **Fig. 3A, B** and **Note 6**). For excitation and emission spectra, use emission and excitation wavelengths of 460 and 350 nm, respectively, and a reduced excitation slit width (< 3 nm) (*see* **Note 7**).

3.4. Binding of RME to CRBP as Investigated by Fluorescence Titration

To investigate the binding of RME to CRBP, fluorescence titrations can be carried out by following either the quenching of protein fluorescence or the enhancement of RME fluorescence. We report here the method based on the quenching of protein fluorescence (**Fig. 4**), which appears to be more suitable for an accurate analysis of binding properties (*see* **Note 8**).

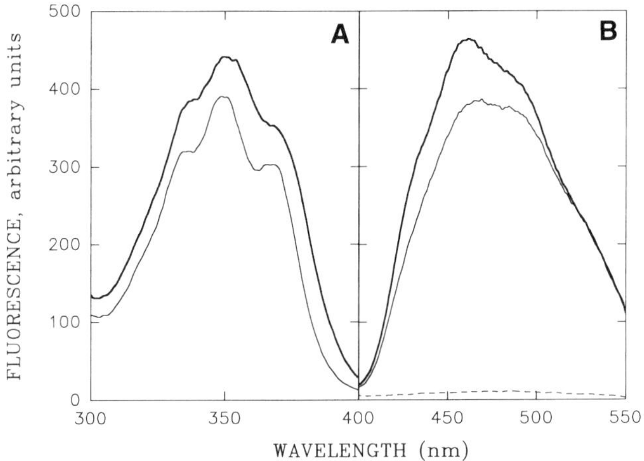

Fig. 3. Fluorescence excitation (**A**) and emission (**B**) spectra of the RME-CRBP (bold line) and retinol-CRBP (thin line) complexes, prepared as described in the legend of **Fig. 2**. The fluorescence emission spectrum of uncomplexed RME in buffer is also shown (dashed line). Conditions were the same as for **Fig. 2**, except for the cuvet path length (3 mm). Excitation and emission spectra were recorded using emission and excitation wavelengths of 460 and 350 nm, respectively. The slit widths for excitation and emission were 2.5 and 3.0 nm, respectively.

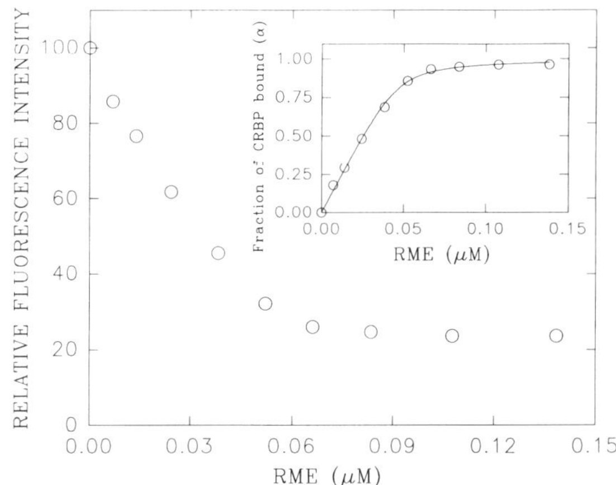

Fig. 4. Fluorescence titration of apoCRBP with RME. The intensity (%) of the intrinsic-protein fluorescence is plotted as a function of RME concentration. Conditions were as follows: 0.05 μ*M* CRBP in 0.4 mL of sodium phosphate buffer, pH 7.3, at 20 ± 1°C; excitation and emission wavelengths: 285 and 345 nm, respectively; excitation and emission slit widths: 2.5 and 10 nm, respectively; cuvet path length: 10 mm. The inset shows the fitting of the experimentally determined values of fractional saturation (α) to **Eq. 7** by means of nonlinear least squares regression for the binding of one mol of RME per mol of CRBP. The estimated apparent dissociation constant (K'_d) is approx 2 n*M* (*see* **Note 14**).

1. Add to the spectrofluorometer cuvet containing 0.05 μM apo-CRBP in 0.4 mL sodium phosphate buffer, pH 7.3, 0.5- to 1-μL aliquots of concentrated ethanolic solutions of RME progressively saturating the protein (*see* **Note 9**). Stir the solution gently after each addition and allow to equilibrate in the dark at 20 ± 1°C, for the time required to obtain a substantially constant fluorescence signal (10–15 min [*see* **Note 10**]).
2. Detect the fluorescence signals using excitation and emission wavelengths of 285 and 345 nm, respectively, and a reduced excitation slit width (< 3 nm).
3. Plot the intensities of intrinsic protein fluorescence as a function of RME concentration (*see* **Notes 11** and **12**).

3.5. Analysis of Fluorescence Titration Data

For the binding of RME to CRBP, possessing one single site for the retinoid, according to the equilibrium expression:

$$CRBP + RME \Leftrightarrow CRBP \bullet RME, \tag{1}$$

the mass law equation is:

$$K_d = [CRBP]\,[RME] / [CRBP \bullet RME], \tag{2}$$

where K_d is the dissociation constant, [CRBP] and [RME] are the concentrations of free protein and free ligand, respectively, and [CRBP $\bullet$ RME] is the concentration of the CRBP-ligand complex. [CRBP] and [RME] present in the system can be expressed as:

$$[CRBP] = [CRBP]_o - [CRBP \bullet RME], \tag{3}$$

$$[RME] = [RME]_o - [CRBP \bullet RME], \tag{4}$$

where $[CRBP]_o$ and $[RME]_o$ are total protein and ligand concentrations, respectively. The fraction of CRBP bound by RME, indicated with α (*see* **Note 13**), is:

$$\alpha = [CRBP \bullet RME] / [CRBP]_o \tag{5}$$

By substituting **Eqs. 3** and **4** in **Eq. 2** and introducing α as defined by **Eq. 5**, the following equation can be obtained from **Eq. 2**:

$$\alpha^2 [CRBP]_o - \alpha\,([RME]_o + [CRBP]_o + K_d) + [RME]_o = 0 \tag{6}$$

By taking the negative square root, the fraction of protein bound can be expressed as:

$$\alpha = \frac{[RME]_o + [CRBP]_o + K_d - \sqrt{([RME]_o + [CRBP]_o + K_d)^2 - 4[CRBP]_o[RME]_o}}{2[CRBP]_o} \tag{7}$$

For each point of fluorescence intensity vs total RME concentrations of the titration curve, the values of α can be calculated using the relationship:

$$\alpha = (F_0 - F) / (F_0 - F_\infty), \tag{8}$$

where F_0 and F_∞ represent the two limiting fluorescence intensities, i.e., in the absence and in the presence of excess saturating RME, respectively, and F represents the fluorescence intensity at a certain concentration of RME. The experimentally determined values of α are fitted to **Eq. 7** by means of non-linear least squares regression (*see* **Fig. 4**, inset; *see* **Notes 14** and **15**).

4. Notes

1. CRBP expressed in *E. coli* consists of two isoforms, differing in their N-terminal amino acid: met$^+$ and met$^-$ CRBPs *(12)*. These forms can be separated by means of the anion exchange chromatography described here (QMA-FPLC), performed at pH 7.9 instead of 7.3. It has been found that *E. coli*-derived met$^+$ and met$^-$ apo-CRBPs interact with retinoids in an identical fashion *(12)*.

2. Retinol and retinoids are particularly unstable in the aqueous medium. However, a certain instability is also present in media (like ethanol and methanol) where they are generally dissolved before binding studies are carried out. Therefore, freshly prepared alcoholic solutions of retinoids, protected from light and kept at 0°C, should be used for such studies.

3. At variance with the relatively fast binding of retinol to apo-CRBP, which takes place in less than 1 min, the kinetics of RME binding to apo-CRBP is rather slow. Therefore, several minutes of incubation after the addition of RME to apo-CRBP are required before absorbance and fluorescence spectra of the CRBP-RME complex are run.

4. The peculiar spectral properties of CRBP-bound retinol and RME are indicative of specific ligand-protein interactions (*see* **Fig. 2**). Instead, other holo-retinoid-binding proteins exhibit less characteristic spectra. For example the absorption spectrum of the complex of retinol with plasma retinol-binding protein (RBP) is characterized by a single, well-shaped peak centered at approx 328 nm. On the other hand, the absorption spectra of some CRBP-bound retinoids, like CRBP-bound all-*trans* retinal *(12,13)*, do not resemble those of CRBP-bound retinol and RME.

5. Fluorescent retinoids exhibit a substantially higher fluorescence emission when they are bound to retinoid-binding protein as compared with uncomplexed retinoids. This feature, typical of retinol bound to retinol-binding proteins, has been found for both CRBP-bound RME (*see* **Fig. 3B**) and RBP-bound RME (data not shown) and it may be attributed to the rigidity of the bound-retinoid molecule and to an apolar environment in the CRBP and RBP retinol-binding sites. Therefore, spectrofluorometric analysis in the case of fluorescent retinoids might prove quite informative in regard to the nature of the interactions with binding proteins.

6. A strong quenching of intrinsic protein fluorescence is usually observed upon binding of retinoids to retinoid-binding proteins, as a result of an efficient energy

transfer from excited tryptophans to bound chromophoric retinoids. The intensity of protein fluorescence of complexes of CRBP with retinol and RME is only 10–20% compared with the intensity exhibited by the apoprotein. A lack of quenching or a limited quenching of intrinsic protein fluorescence of apo retinoid-binding proteins in the presence of chromophoric retinoids may indicate lack of binding or nonspecific binding of such retinoids to the proteins.

7. When fluorescence spectra are run or fluorescence intensities are recorded in the course of fluorescence titrations, the excitation slit widths should be rather small to minimize the possible decomposition of retinoids induced by the excitation light. Likewise, the time of exposure of the spectrofluorometer cuvet to the exciting light should be as short as possible (5–10 s) during measurements.

8. Despite the fact that the enhancement of RME fluorescence associated with RME binding to CRBP might be exploited to perform titrations of CRBP with the retinoid, it is more convenient to monitor the interaction between RME and CRBP by following the quenching of intrinsic protein fluorescence. In particular, the titration method based on the enhancement of RME fluorescence is remarkably less sensitive than the method based on the quenching of protein fluorescence. In addition, a correction for the contribution of the fluorescence of free retinoid present in the medium should be made for each point of a titration based on the enhancement of RME fluorescence.

9. The ethanol/methanol concentrations in the binding-assay buffer should not exceed 1–2% (v/v) in order to reduce to a minimum the possible influence of water-miscible organic solvents on retinoid-protein interactions. Therefore, small aliquots of concentrated alcoholic solutions of retinoids are added to the cuvet containing the protein in the retinoid-binding assays. At the same time the dilution of the protein and, therefore, the change of the signal caused by the dilution are nearly negligible if small aliquots of the solution containing the ligand are added in the course of the titrations.

10. When titrations based on the quenching of protein fluorescence are carried out, it is important to verify that the fluorescence signal remains fairly constant before the ligand is added. A decay of the signal may be observed at low-protein concentrations and might be caused by the protein adsorption to the cuvet walls *(14)*.

11. A negligible inner filter effect by added RME for the titration presented in **Fig. 4** was established by performing a control experiment with L-tryptophan methyl ester (Fluka). RME was added to a solution of the tryptophan derivative exhibiting an absorbance at 285 nm similar to that of CRBP. In the control experiment, wavelengths and slit widths for the excitation and emission were the same as for the titration of CRBP with RME. RME at the same concentrations used for the titrations of CRBP did not cause a significant decrease of tryptophan fluorescence and, therefore, no correction for inner filter effect was made.

12. To evaluate binding affinities by means of fluorescence titrations, the protein concentration in the system should not be substantially greater than the dissociation constant. We have found that under our experimental conditions a protein concentration of 0.05 μM was compatible with a signal of intrinsic protein fluorescence high

enough and stable enough to perform the titration carefully. Because RME binding to CRBP is characterized by an apparent dissociation constant in the nanomolar range (i.e., significantly lower than the protein concentration used for the titration), a substantial portion of the titration shown in **Fig. 4** exhibited a nearly stoichiometric binding, as revealed by the linear change of protein fluorescence as a function of ligand concentration. Binding assays performed at micromolar protein concentrations promoted a fully stoichiometric binding of RME to CRBP throughout the titrations and, actually, this latter condition allowed us to verify that the stoichiometry of binding of RME to CRBP is approx 1:1 (data not shown). The binding stoichiometry can be determined from the titration curve obtained at high protein concentration by extrapolating on the *x*-axis the intersection point of the linear change of protein fluorescence and the plateau value. The extrapolated value is then related to the protein concentration in the system.

13. The described method of calculating the CRBP fractional saturation (α) from titration data is based on the assumption that the quenching of intrinsic-protein fluorescence is proportional to the extent of RME binding. This assumption is verified experimentally by the finding of a linear change of protein fluorescence intensity as a function of ligand concentration under conditions that promote stoichiometric binding of the ligand (*see* **Note 12**).

14. The dissociation constant derived from the titration experiment may represent an apparent constant, indicated with K'_d. In particular, owing to the possible partition of the highly hydrophobic RME molecule between the aqueous phase and small aggregates and micelles, the effective concentration of the retinoid in solution may be lower than its total concentration.

15. **Eqs. 6** and **7** may be rewritten in the form of **Eqs. 9** and **10**:

$$\alpha^2\, n[\text{CRBP}]_0 - \alpha\,([\text{RME}]_0 + n[\text{CRBP}]_0 + K_d) + [\text{RME}]_0 = 0, \tag{9}$$

$$\alpha = \frac{[\text{RME}]_0 + n[\text{CRBP}]_0 + K_d - \sqrt{([\text{RME}]_0 + n[\text{CRBP}]_0 + K_d)^2 - 4n[\text{CRBP}]_0[\text{RME}]_0}}{2n[\text{CRBP}]_0} \tag{10}$$

to include n, the number of equivalent and independent binding sites present on the protein molecule. Thus, the fitting of the experimentally determined values of α to **Eq. 10** allows the determination of both ligand-binding affinity and stoichiometry.

Acknowledgments

We thank Dr. Marc S. Levin (Department of Medicine, Washington University School of Medicine) for the pMON-CRBP plasmid, and our colleague Dr. Simone Ottonello for valuable advice when expression and purification of CRBP were carried out. Giorgio Malpeli and Claudia Folli were recipients of fellowships from the "Associazione Italiana per la Ricerca sul Cancro" (AIRC), Milan, Italy.

References

1. Ong, D. E., Newcomer, M. E., and Chytil, F. (1994) Cellular retinoid-binding proteins, in *The Retinoids: Biology, Chemistry and Medicine* (Sporn, M. B., Roberts, A. B., and Goodman, D. S., eds), Raven, New York, NY, pp. 283–317.
2. Posch, K. C., Boerman, M. H. E. M., Burns, R. D., and Napoli, J. L. (1991) Holocellular retinol-binding protein as substrate for microsomal retinal synthesis. *Biochemistry* **30,** 6224–6230.
3. Herr, F. M. and Ong, D. E. (1992) Differential interaction of lecithin-retinol acyl transferase with cellular retinol-binding proteins. *Biochemistry* **31,** 6748–6755.
4. Ottonello, S., Scita, G., Mantovani, G., Cavazzini, D., and Rossi, G.L. (1993) Retinol bound to cellular retinol-binding protein is substrate for cytosolic retinoic acid synthesis. *J. Biol. Chem.* **268,** 27,133–27,142.
5. Li, E., Quian, S., Winter, N. S., d'Avignon, A., Levin, M. S., and Gordon, J. I. (1991) Fluorine nuclear magnetic resonance analysis of the ligand binding properties of two homologous rat cellular retinol-binding proteins expressed in *E. coli*. *J. Biol. Chem.* **266,** 3622–3629.
6. Malpeli, G., Stoppini, M., Zapponi, M.C., Folli, C., and Berni, R. (1995) Interactions with retinol and retinoids of bovine cellular retinol-binding protein. *Eur. J. Biochem.,* **229,** 486–493.
7. Rong, D., Lovey, A. J., Rosenberger, M., d'Avignon, A., Ponder, J., and Li, E. (1993) Differential binding of retinol analogs to two homologous cellular retinol-binding proteins. *J. Biol. Chem.* **268,** 7929–7934.
8. Cowan, S. W., Newcomer, M. E., and Jones, T. A. (1993) Crystallographic studies on a family of cellular lipophilic transport proteins: the refinement of P2 myelin protein and the structure determination and refinement of cellular retinol-binding protein in complex with all-*trans* retinol. *J. Mol. Biol.* **230,** 1225–1246.
9. Hanze, A. R., Conger, T. W., Wise, E. C., and Weisblat, D. I. (1948) Crystalline vitamin A methyl ether. *J. Am. Chem. Soc.* **70,** 1253–1256.
10. Horwitz, J. and Heller, J. (1973) Interactions of all-*trans*, 9-, 11- and 13-*cys* retinol, all-*trans* retinyl acetate, and retinoic acid with human retinol-binding protein and prealbumin. *J. Biol. Chem.* **248,** 6317–6324.
11. Laemmli, U. K. (1970) Cleavage of structural proteins during the assembly of the head of bacteriophage T4. *Nature (London)* **227,** 680–685.
12. Levin, M. S., Locke, B., Yang, N. C., Li, E., and Gordon, J. I. (1988) Comparison of the ligand binding properties of two homologous rat apocellular retinol-binding proteins expressed in *E. coli*. *J. Biol. Chem.* **263,** 17,715–17,723.
13. MacDonald, P. N. and Ong, D. E. (1987) Binding specificities of cellular retinol-binding protein and cellular retinol-binding protein, type II. *J. Biol. Chem.* **262,** 10,550–10,556.
14. Norris, A. W., Cheng, L., Giguère, V., Rosemberg, M., and Li, E. (1994) Measurement of subnanomolar retinoic acid binding affinities for cellular retinoic acid-binding proteins by fluorometric titration. *Biochim. Biophys. Acta* **1209,** 10–18.

10

Fluorometric Titration of the CRABPs

Andrew W. Norris and Ellen Li

1. Introduction

Fluorescence spectroscopy has long been used to characterize the equilibrium binding of retinoids to proteins *(1)*. Changes in steady-state fluorescence are monitored as the protein is titrated with aliquots of retinoid. The resultant binding curve can be analyzed yielding information about the stoichiometry and affinity of retinoid binding. There are several advantages to this approach. For one, this is a true equilibrium technique. Binding is monitored directly, with no need to physically separate bound from free ligand. A second advantage is the high sensitivity of fluorescence. Titrations can be routinely performed on as little as 2.5 mL of 10^{-6}–10^{-7} M CRABP. Another advantage is that no alteration of the retinoid or protein is necessary. The fluorescence signals are intrinsic to the native protein and retinoid.

Two fluorescence signals are commonly used to monitor retinoid binding. One is intrinsic protein fluorescence, which arises mainly from tryptophans. This signal decreases when retinoid is bound to the protein. This is caused by resonance-energy transfer from the tryptophans to the retinoid, which only occurs when the retinoid is within very close proximity to the tryptophans (i.e., protein bound) *(2)*. One requirement for the resonance-energy transfer is that the absorption spectrum of the retinoid overlaps with the emission spectrum of the protein (~310–360 nm). This is typically the case for retinoids. A second fluorescence signal that can be used to monitor retinoid binding is intrinsic-retinoid fluorescence. The basis for this is that some, but not all, retinoids exhibit increased fluorescence when bound to protein.

Two issues should be kept in mind when determining affinities from retinoid-binding titrations. The first is that binding affinities can only be accurately determined when the protein concentration is near or lower than the dissocia-

From: *Methods in Molecular Biology, Vol. 89: Retinoid Protocols*
Edited by: C. P. F. Redfern © Humana Press Inc., Totowa, NJ

tion constant under study *(3)*. The second is that retinoids are poorly soluble in aqueous buffers. These issues will be discussed in detail later in this chapter.

Long-standing retinoid fluorometric titration methods *(1)* were recently modified for study of the cellular retinoic acid binding proteins (CRABPs) *(4)*. New features of the modified method include independent determination of binding stoichiometry, the study of lower-protein concentrations allowing more accurate determination of strong retinoid-binding affinities, and direct fitting to the binding data. These methods are detailed here. The methods presented here have also been applied to retinoid-binding proteins other than the CRABPs *(5)*.

2. Materials
2.1. Preparation of CRABP

1. Purified CRABP: The large-scale purification of both CRABP I and CRABP II from recombinant expression in *Escherichia coli* is described in Chapter 8. Purified CRABP should be stored at 4°C or on ice.
2. Lipidex-1000 (Packard, Downers Grove, IL) or an equivalent product, hydroxyalkylpropyl dextran (type IV, Sigma).
3. Buffer: We use a simple buffer of 20 mM KHPO$_4$, pH 7.4, 100 mM KCl for most of our studies. Other buffers may be used.
4. Extinction coefficient of CRABP (*see* **Note 1**).

2.2. Preparation of Retinoid Stocks

Retinoids should be handled exclusively under dim or red light. Wrap prepared samples in foil for protection from light. *See* Chapter 1 for more information on retinoid handling.

1. Retinoid: All-*trans*- and 13-*cis*-retinoic acid may be obtained from Kodak. Retinoids should be stored in the dark at –80°C. Some retinoids should be stored under N$_2$ as well.
2. Absolute ethanol (EtOH).
3. Extinction coefficient of retinoid (*see* Chapter 1).

2.3. Data Collection

1. Fluorometer: Important features include sample-stirring assembly, water bath and sample-temperature jacket, and computer interface.
2. Quartz fluorescence cuvets: 1 × 1 cm.
3. Stir bars for the Quartz fluorescence cuvets.

2.4. Titration of High CRABP Concentrations

1. CRABP and retinoid stock solutions (*see* **Subheadings 3.1.** and **3.2.** for preparation of these).
2. Hamilton-style syringe: We prefer the Hamilton 2-, 5-, and 10-μL 7000 Series models.

3. Calibrated pipets for working with aqueous samples from 2 μL to 1 mL.
4. Buffer blank: Save some of the exact buffer created for the dialysis of CRABP (**Subheading 3.1., step 2**).
5. 0.5% (w/v) gelatin: Prepare in buffer solution. We use gelatin Type A, Porcine Skin, 300 Bloom (Sigma, purchase number G-2500) (*see* **Note 2**).

2.5. Titration of Low CRABP Concentrations

The materials required are as for **Subheading 2.4.**

2.6. Measurement of Inner Filter Effect

In addition to the materials listed in **Subheading 2.4.**, a stock solution of ~0.2 m*M* *N*-acetyl-tryptophanamide (Sigma) will be required: measure its absorbance at the excitation wavelength to be used in the fluorescence studies (typically 280 nm, *see* **Subheading 3.3.**).

2.7. Analysis of Data

1. Personal computer.
2. Software package that comes with the fluorometer.
3. Mathematical, spreadsheet, or graphing software package with nonlinear-regression capability. We use Sigma Plot (Jandel, Corte Madera, CA) and Scientist (MicroMath, Salt Lake City, UT).

3. Methods
3.1. Preparation of CRABP

CRABP often copurifies with endogenous-bacterial lipids (unpublished data). These lipids should be removed by delipidation prior to binding studies. Delipidation reduces the shelf life of CRABP, so delipidate only the amount of CRABP expected to be used in the coming few months. Following delipidation, both CRABP I and CRABP II will slowly precipitate over the span of months.

Working stock concentrations of 5–20 μ*M* CRABP are convenient for the fluorescence studies. In general, 10 mL of a 10-μ*M* stock of CRABP will provide enough protein for 20–40 stoichiometry determinations or 100–1000 binding-constant determinations.

1. Delipidate CRABP: This is best accomplished by passage of CRABP through a column of Lipidex-1000 at 37°C (*6,7*). Pool fractions containing protein as demonstrated by A_{280} (*see* **Note 3**).
2. Dialyze the delipidated CRABP against the buffer to be used in the fluorescence studies. Save a portion of the buffer unused as this will serve as the reference buffer. This will be called buffer blank.
3. Centrifuge the CRABP sample. Do this immediately before the fluorescence study in order to remove dust and precipitated protein. A 10 min spin at 5000*g* should be adequate. Transfer the supernatant to a clean tube.

4. Carefully quantify the centrifuged CRABP by measurement of its A_{280}. Use the buffer blank as the reference (*see* **Note 4**).

3.2. Preparation of Retinoid Stocks

Retinoid stocks of the appropriate concentration should be prepared in ethanol. Under conditions in which [CRABP] >> K_d, retinoid-stock concentrations should be ~300- to 600-fold higher than [CRABP]. This is so that there will be less than 1% EtOH in the sample once the titration is completed, and will also minimize volume changes during the titration. If [CRABP] ≤ K_d, then higher-retinoid stock concentrations will probably be necessary. Create retinoid stocks at several concentrations, for example 1000, 200, and 20 μ*M*, allowing titrations under a variety of conditions.

1. Weigh out retinoid for 1–10 mL of high-concentration stock under dim or red light. As a typical example, weigh out ~3 mg of retinoic acid to create 10 mL of a 1000-μ*M* stock.
2. Dissolve the retinoid in EtOH. Make additional stocks by diluting aliquots of the concentrated stock in EtOH.
3. Quantify the retinoid stocks by UV spectroscopy. Use positive displacement pipets to accurately pipet ethanol (*see* **Note 4**). Discard the sample used in the spectrometer, as it may be photo-isomerized.
4. Store retinoids at –80°C in the dark. Storage under N_2 may also be necessary.

3.3. Data Collection

This section describes set up of the fluorometer. Some experimentation will be necessary to determine the settings that maximize signal and minimize noise.

1. Set the excitation and emission wavelengths to 280 and 330 nm, respectively. These correspond closely to the excitation and emission maxima of the CRABPs *(4)*, and will thus maximize signal. Monitoring retinoid fluorescence yields poorer signal to noise than protein fluorescence (*see* **Note 5**).
2. Set the excitation slit widths: Narrow the slits as far as possible without sacrificing signal to noise, in order to minimize light exposure of sample. For our system we use 2 nm (*see* **Note 6**).
3. Set the emission slit widths: Typically we use 5 nm. Opening the emission slits further may improve the signal-to-noise ratio.
4. It is essential that the fluorometer operates on the linear portion of its response curve. If signal levels are too high, the photomultiplier tube may no longer yield linear responses. If this occurs, the slits may be narrowed to decrease the signal. Alternatively a neutral-density filter may be used to decrease the total signal reaching the photomultiplier tube.
5. The fluorometer may have a reference-normalization option. Activating this option may enhance long-term stability as well as increasing the signal relative to noise.

6. We set the instrument to collect one point per second. Our fluorometer integrates the signal during this time period, increasing the signal relative to noise. Optimum settings for this option may vary between fluorometers. In general, collecting data points too fast will waste memory and make the output appear noisier, and collecting too slowly will delay the titration as one waits for data to collect.
7. Sample stirring facilitates mixing of retinoid and protein and may reduce noise. The stirring speed should be adjusted by trial and error so as to produce the smoothest signal.
8. The use of dust-free samples and buffers will improve the signal-to-noise ratio.

3.4. Titration of High CRABP Concentrations

Titrations are first performed at high CRABP concentrations (0.5–2 μM) in order to obtain a rough idea of the strength of binding. For strong binding interactions these titrations will also be used to determine the stoichiometry of binding. The pre-equilibration step with gelatin helps stabilize the fluorescence signal of CRABP, possibly by decreasing adsorption of CRABP to the cuvet walls. Refer to **Fig. 1A** for an example.

1. Set up the fluorometer as described in **Subheading 3.3.** and turn on the water bath. Protein fluorescence signal is generally more stable at 25°C compared with 37°C. Other temperatures may be used as well.
2. Vent the retinoid stock as it warms from –80°C to aid degassing. Repeated shaking and venting may be necessary to complete degassing.
3. Add the appropriate amount of buffer to cuvet. Add 50 µL of 0.5% gelatin. Place sample in fluorometer and monitor the fluorescence signal, which will probably initially drop or rise. After ~5 min, remove the cuvet and gently tap to release any bubbles that may have formed. Return sample to the fluorometer.
4. Monitor the fluorescence signal until it is stable. This will require at least 10–45 min, depending on the degree of stability required (*see* **Note 7**).
5. Start data acquisition and collect ~20 s of baseline signal. Add the appropriate volume of CRABP. Monitor the signal as it stabilizes. Typically this requires 5–20 min.
6. Titrate the sample with retinoid using a Hamilton-style syringe. Perform the titration rapidly to minimize photoisomerization and to keep changes in signal due to instability to a minimum. Aim to titrate one molar equivalent of retinoid (compared to the amount of protein) in ~5–10 additions (*see* **Note 8**). Close the lamp shutter while adding retinoid to minimize photoisomerization. For later reference, record or mark the points of addition.
7. Collect > 12 s of data per retinoid addition once the signal has stablilized. This usually occurs very rapidly, within 1–2 s.
8. Titrate well past apparent saturation of the protein. These titration points are very important to analysis of the binding curve. Note however that there is little reason to titrate beyond 10 μM total retinoid, as inner-filter effects will become quite large (*see* **Subheading 3.6.**).

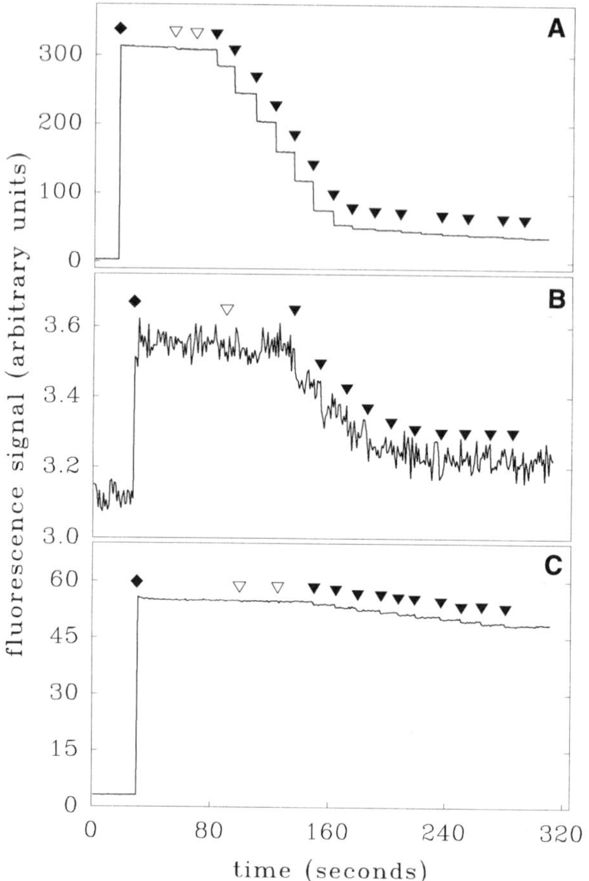

Fig. 1. Collection of raw fluorescence data. (**A**) Titration of high CRABP concentration. (**B**) Titration of low CRABP concentration. (**C**) Titration of NAT to measure inner filter effect. (♦) Addition of CRABP or NAT to buffer/gelatin. (▽) Pause in data collection. (▼) Addition of retinoid.

3.5. Titration of Low CRABP Concentrations

The purpose of these titrations is to measure K_d. The aim is to lower the concentration of CRABP such that $[CRABP] \leq K_d$. Under optimum conditions we have been able to study protein concentrations as low as 5 nM. An example is shown in **Fig. 1B**.

1. These titrations are performed as described for high CRABP concentrations, except that lower protein concentrations are used.
2. *See* **Subheading 3.3.** for methods to optimize signal to noise.

3. A much higher degree of signal stability will be required as the total signal owing to protein will be very small. Equilibration periods of 45 min after adding gelatin are often required at very low CRABP concentrations.
4. For calculation of the K_d it is even more important to titrate well past the apparent saturation of the protein. It is important to obtain as complete a binding curve as possible.

3.6. Measurement of Inner Filter Effect

The inner filter effect arises from absorption of light by the sample. For these studies, each addition of retinoid to the sample artifactually decreases the fluorescence signal. We have chosen an empirical means by which to measure the inner filter effect (*see* **ref. 3** for more detail). Each retinoid will produce an inner filter effect of different magnitude. Measure the inner filter effect for each retinoid stock used. Note that changes in the instrument, such as adjusting the slits, may affect the magnitude of the inner filter effect. An example is shown in **Fig. 1C**.

1. These titrations should be performed exactly as above, including use of gelatin, except that *N*-acetyl-tryptophanamide (NAT) is substituted for CRABP. Use an amount of NAT such that the final A_{280} in the sample is equal to that used in the CRABP titrations.
2. Titrate with amounts of retinoid approximately equal to that used in the CRABP titrations.
3. When large amounts of retinoid (≥ 0.5–$2\ \mu M$) have been titrated into solution, the signal may move upwards after the initial drop, which occurs with each retinoid addition. This is most likely owing to precipitation of retinoid from solution. When this occurs, there is little reason to continue the titration, as these data points will not be useful.

3.7. Analysis of Data

3.7.1. Data Processing

The overall scheme is to average 10 data points for each titration addition. This improves signal to noise and allows calculation of the uncertainty associated with the fluorescence signal. This process can be largely automated.

1. Determine the mean and standard error for 10 data points following each retinoid addition. Do this also for the fluorescence signal immediately prior to the first addition of retinoid.
2. The means and errors are then placed into Sigma Plot, Scientist, or similar program. Two columns are created:
   a. Fluorescence signal.
   b. Retinoid concentration (in nM).

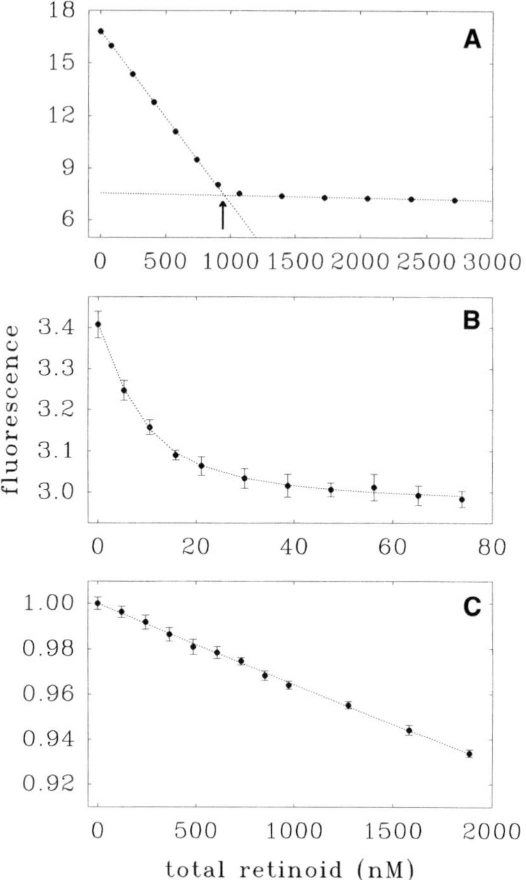

Fig. 2. Processed and fitted data. **(A)** Determination of binding stoichiometry. **(B)** Fitting to determine K_d. **(C)** Fitting to determine the inner filter effect.

3.7.2. Determination of the Inner Filter Effect

A single parameter, here called c, can be used to describe *IFE*, the fractional loss in signal owing to the inner filter effect.

$$IFE = 10^{-c \cdot [retinoid]}$$

See **Appendix A** for additional explanation of this equation. The value of c should be determined for each retinoid stock used. This parameter will be used in the stoichiometry and affinity calculations. An example is shown in **Fig. 2C**.

1. The signal data should be normalized so that the first point (that is, the signal in the absence of retinoid) has a value of one.

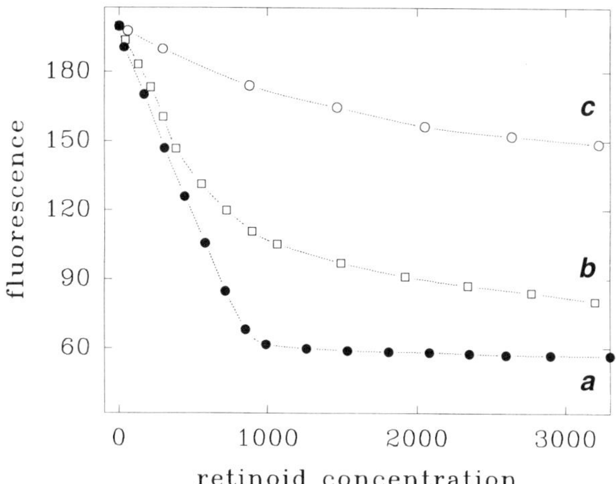

Fig. 3. Titration of 1 μM CRABP with three retinoids of different affinity: *a*, high affinity; *b*, weak affinity; *c*, very weak affinity.

2. Using the nonlinear regression option, fit the data to the following equation.

$$signal = 10^{-c \cdot [retinoid]}$$

See **Appendix C** for an example Sigma Plot fitting file.
3. The best fit value of *c* should be returned. Note that its units will be nM^{-1}.
4. Inner filter-effect data will often deviate from ideality at higher retinoid concentrations. This is owing to the insolubility of retinoids at such concentrations. Data points where insolubility seems to be at play, for example where upgoing signals are noted (*see* **Subheading 3.6.**), should not be included in the fit.

3.7.3. Determination of Binding Stoichiometry

Figure 3 shows the titration of a high concentration of CRABP with three retinoids. Curve *a* shows stoichiometric binding: all of the added retinoid binds to the CRABP until the protein is saturated. Stoichiometric binding occurs when [CRABP] >> K_d. Such binding curves may be used to determine the binding stoichiometry, but may not be used to determine K_d accurately. In order to calculate the K_d for this retinoid, titrations should be performed at lower CRABP concentrations.

Curve *b*, **Fig. 3**, shows titration with a weaker binding ligand. When this is the case it is difficult to calculate the stoichiometry rigorously. The reason for this is that there are likely to be large amounts of retinoid aggregation at the concentrations of unbound ligand required to saturate binding. This makes it difficult to reach saturation of the protein. For the same reason it is difficult to calculate an accurate K_d for such retinoid-CRABP interactions, even at low-protein concentrations.

Curve c, **Fig. 3**, shows little signal change. In these cases it is likely that $K_d >$ than the retinoid-solubility limit, and thus it is impossible to calculate the binding stoichiometry or affinity accurately. We usually report these as "binding not detected" (*see* **Note 9**).

1. The inner filter effect should be corrected. The fluorescence signal should be multiplied by $1/IFE$, using the value of c determined for the relevant retinoid stock, where $IFE = 10^{-c \cdot [retinoid]}$.
2. The binding stoichiometry can be determined graphically as shown in **Fig. 2A** *(8)*. Lines are extrapolated from the two phases of the binding curve, and their intersection represents the total molar amount of retinoid-binding sites. The binding stoichiometry is then simply this value divided by the concentration of protein. This can be done graphically, or within the spreadsheet program (*see* **Note 10**).
3. An alternative method is to fit the binding data as described (*see* **Subheading 3.7.4.**). The value obtained for the protein concentration parameter will represent the total molar amount of retinoid binding sites. This method is only accurate when the stoichiometric binding is observed (Curve a, **Fig. 3**).

3.7.4. Determination of Binding Affinity

As mentioned in the introduction, two factors need to be considered when determining retinoid binding affinities. The first is that retinoids are poorly soluble in aqueous buffers. Retinoic acid has a solubility limit of ~200 nM in aqueous buffers at pH 7.4 *(9)*. This fact profoundly influences the resultant binding curves. For example, in the case of weaker binding, unbound concentrations of retinoid greater than 200 nM are produced before the protein is saturated. These portions of the binding curve will thus be perturbed, and accurate K_d's cannot be determined (*see* **Appendix B** for a more mathematical treatment of this problem). Note though that for strong binding retinoids, the unbound concentrations do not reach 200 nM until after essentially complete saturation of the protein. Because retinoid solubility/aggregation is poorly characterized, its effects on K_d determinations are unknown. For this reason only apparent binding constants can be determined, noted as K_d'.

A second factor that needs to be considered is that the protein concentration used in the titration needs to be near or lower than the K_d under study in order to determine the K_d accurately *(3)*. The hallmark of this condition is that the binding curve will not be stoichiometric. Thus it will be necessary to lower the protein concentration until nonstoichiometric binding curves are obtained. An example is shown in **Fig. 2B**.

There are a few additional considerations. One is that the following analysis requires the amount of change in protein fluorescence to be proportional to the amount of ligand binding. This is true if the decreases in signal observed before saturation of the protein under stoichiometric conditions are linear *(10)* (this is the case for **Fig. 2A**). Also note that the following analysis applies only to single-site binding. Methods that may be used to study multiple binding-site interactions have been developed *(8)*.

1. Correct for the inner-filter effect as (**Subheading 3.7.3., step 1**).
2. Using the nonlinear regression option, fit the data to the following equation (*see* **Appendix B** for derivation).

$$F = F_0 \frac{1 + (P_T + R_T)\,K_a - \sqrt{(P_T - R_T)^2\,K_a^2 + 2(P_T + R_T)K_a + 1}}{2P_T K_a}\,(F_0 - F_\infty)$$

   The variables of the fit are F, the fluorescence signal; and R_T, the corresponding retinoid concentration. The constant of the fit is F_0, the fluorescence signal at the first point of the titration, where [retinoid] = 0. The parameters of the fit are P_T, the total protein concentration; K_a, the association constant of binding; and $F\infty$, the fluorescence signal at complete saturation of the protein. *See* **Appendix C** for an example Sigma Plot Fitting file.
3. Best fit values for the three parameters will be returned. The K_d of the fit is simply $1/K_a$. The fitted curve should agree well with the data when examined graphically. *See* **Fig. 2B**. In the case of a poor fit, the obtained parameters are probably unreliable. We commonly obtain poor fits for weaker K_d's (curves *b* and *c*, **Fig. 3**).
4. At times the fitting routine will converge to impossible values, for example negative protein concentrations. When this happens different starting parameter values may solve the problem. Alternatively, constraints such as $P_T > 0$, may solve the problem.
5. Many nonlinear regression packages will report some measure of error associated with each fitted parameters. When calculating these errors, many nonlinear regression packages assume that the parameters are independent of one another. Because this is not true in this case, such reported errors will be unreliable. Some software packages, including Scientist, do not make this assumption and report reliable confidence intervals (*see* **refs. *11*** and ***12*** for more information).
6. If possible, it is best to constrain the protein concentration to the known value (as determined by A_{280} and the dilution) during the fitting (*see* **Note 11**). This will yield greater confidence in the fitted K_d. We usually perform nonlinear regressions with and without the protein concentration constrained. Comparing the obtained K_d from the two approaches helps define the confidence associated with the results.
7. Several titrations should be performed, analyzed, and compared also to help determine the degree of confidence. If possible, several protein concentrations

should be titrated and compared. The one best-fit K_d should predict the binding curves obtained at the different protein concentrations.

4. Notes

1. Mouse CRABP I has been determined to have an extinction coefficient of 21,270 cm^{-1} M^{-1}; mouse CRABP II, 19,990 cm^{-1} M^{-1} in our buffer *(4)* (*see* **refs.** *13* and *14* for an accurate, convenient method to determine protein-extinction coefficients).
2. It is possible that gelatin may be contaminated with endogenous retinoids, or contaminants which bind RA. Determine the inner-filter effect with and without gelatin. The obtained values should be identical, showing that the gelatin (or contaminants) are not interacting with retinoid. It may be necessary to examine the gelatin for the presence of endogenous retinoid.
3. CRABP is not retained by Lipidex-1000, and will elute in the void volume. 20 mL of packed Lipidex gel should adequately delipidate ~10 mg CRABP (40 mL of 20 µM).
4. The following technique for quantification of samples by UV spectroscopy is convenient, uses little sample, and decreases errors associated with unmatched or inadequately cleaned cuvets. One milliliter of reference solution (buffer blank for proteins, EtOH for retinoids) is added to a standard 1-mL quartz cuvet. This is placed in the spectrometer and the instrument is zeroed. An appropriate amount of sample is then thoroughly mixed into this cuvet and the absorbance recorded. Examples: addition of 100 µL of a 10-µM solution of CRABP will yield a final A_{280} of ~0.02; addition of 10 µL of a 1-mM stock of typical retinoid will yield a final A_{280} of ~0.4. The measurement should be repeated several times, preferably using several different dilutions.
5. If large quantities of retinoid (> 5 µM) are being used, the inner filter effect (*see* **Subheading 3.6.**) can become quite large. These effects may be reduced by choosing wavelengths away from the absorption maximum of the retinoid.
6. The irradiation of the sample with the excitation light can cause photo-isomerization of the retinoid. We and others have extracted retinoic acid from the sample after titration and examined for photo-isomerization by HPLC *(4,15)*. Under our conditions ~5% photoisomerization is observed. Systems with higher intensity lamps or different optics may incur larger amounts of photo-isomerization.
7. The degree of signal stability required depends upon the purpose of the titration, as well as the concentration of protein under study. Quantitative work and titrations at low-protein concentrations will require higher degrees of stability. For stringent work, the amount of signal change owing to instability, which takes place during the time it takes to perform a titration (~10 min), should be drastically smaller than the ligand-dependent signal changes that will take place during the titration. For titration of high concentrations of CRABP, 15 min of equilibration after adding gelatin is often sufficient. For low-CRABP concentrations, 45 min of equilibration may be necessary.

8. Too many additions will increase the total time it takes to complete the titration and thus photo-isomerization. Too many additions will also increase the errors in total delivered volume. Too few additions will result in loss of information from the binding curve.

9. Some retinoids may not cause changes in protein fluorescence when protein bound. This is even possible, but unlikely, for retinoids with an absorption spectrum that overlaps with the protein-emission spectrum.

10. We have estimated total errors in determining the binding stoichiometry at $\pm$ ~20%. The errors are owing to uncertainty in pipetting, in measuring absorbances, and in extinction coefficients.

11. It is necessary to determine whether CRABP added to the cuvet remains in solution. This can be determined by measuring the fluorescence signal of added CRABP over a range of added amounts. We have found a linear relationship under our conditions suggesting that CRABP remains in solution.

Acknowledgments

This work was supported by grants from the National Institutes of Health (DK-40172 and DK-49684). Ellen Li is a Burroughs Wellcome Scholar in Toxicology.

References

1. Cogan, U., Kopelman, M., Mokady, S., and Shinitzky, M. (1976) Binding affinities of retinol and related compounds to retinol binding binding proteins. *Eur. J. Biochem.* **65,** 71–78.

2. Stryer, L. (1978) Fluorescence energy transfer as a spectroscopic ruler. *Ann. Rev. Biochem.* **47,** 819–846.

3. Birdsall, B., King, R. W., Wheeler, M. R., Lewis, Jr., C. A., Goode, S. R., Dunlap, R. B., and Roberts, G. C. K. (1983) Correction for light absorption in fluorescence studies of protein-ligand interactions. *Anal. Biochem.* **132,** 353–361.

4. Norris, A. W., Cheng, L., Giguère, V., Rosenberger, M., and Li, E. (1994) Measurement of subnanomolar retinoic acid binding affinities for cellular retinoic acid binding proteins by fluorometric titration. *Biochim. Biophys. Acta* **1209,** 10–18.

5. Cheng, L., Norris, A. W., Tate, B. F., Rosenberger, M., Grippo, J. F., and Li, E. (1994) Characterization of the ligand binding domain of human retinoid X receptor α expressed in *Escherichia coli. J. Biol. Chem.* **269,** 18,662–18,667.

6. Glatz, J. F. C. and Veerkamp, J. H. (1983) Removal of fatty acids from serum albumin by Lipidex 1000 chromatography. *J. Biochem. Biophys. Meth.* **8,** 57–61.

7. Lowe, J. B., Sacchettini, J. C., Laposata, M., McQuillan, J. J., and Gordon, J. I. (1987) Expression of rat intestinal fatty acid-binding protein in *Escherichia coli. J. Biol. Chem.* **262,** 5931–5937.

8. Lohman, T. M. and Mascotti, D. P. (1992) Nonspecific ligand-DNA equilibrium binding parameters determined by fluorescence methods. *Methods Enzymol.* **212,** 424–458.

9. Szuts, E. Z. and Harosi, F. I. (1991) Solubility of retinoids in water. *Arch. Biochem. Biophys.* **287,** 297–304.
10. Halfman, C. J. and Nishida, T. (1972) Method for measuring the binding of small molecules to proteins from binding-induced alterations of physical-chemical properties. *Biochemistry* **11,** 3493–3498.
11. Straume, M. and Johnson, M. L. (1992) Analysis of residuals: criteria for determining goodness-of-fit. *Methods Enzymol.* **210,** 87–105.
12. Johnson, M. L. and Frasier, S. G. (1985) Nonlinear least-squares analysis. *Methods Enzymol.* **117,** 301–342.
13. Lohman, T. M., Chao, K., Green, J. M., Sage, S., and Runyon, G. T. (1989) Large-scale purification and characterization of the *Escherichia coli* rep gene product. *J. Biol. Chem.* **264,** 10,139–10,147.
14. Gill, S. C. and von Hippel, P. H. (1989) Calculation of protein extinction coefficients from amino acid sequence data. *Anal. Biochem.* **182,** 319–326.
15. Fiorella, P. D., Giguère, V., and Napoli, J. L. (1993) Expression of cellular retinoic acid-binding protein (type II) in *Escherichia coli. J. Biol. Chem.* **268,** 21,545–21,552.

Appendix

A: The Inner Filter Effect

An equation describing the inner filter effect can be derived by integrating total light absorbances over the path length of the cuvet, over both the excitation and emission wavelengths (*see* **ref. 3**. At low absorbances this simplifies to:

$$ife = 10^{-A/W}$$

where *ife* represents the fractional loss of signal owing to the inner filter effect, A is the total absorbance over both the excitation and emission wavelengths, and W is a parameter which depends on instrument optics. This approximation holds well when $A < 0.2$, which roughly corresponds to [retinoid] $\lesssim 5\ \mu M$. Note that:

$$A = P \cdot \varepsilon_p + R \cdot \varepsilon_r$$

where ε_p and ε_r are the extinction coefficients over both the excitation and emission bands, and P and R are the total concentrations of protein and retinoid respectively. Thus we can rewrite our expression as:

$$ife = 10^{-P \cdot \varepsilon_p/W} \times 10^{-R \cdot \varepsilon_r/W}$$

Using the substitution $c = \varepsilon_r/W$ and knowing that P does not change during the titration we can write:

$$ife = k \cdot 10^{-c \cdot R}$$

where k is $10^{-P} \cdot \varepsilon_r / W$. Thus at the first point in the titration, where $R = 0$, $ife = k$. We can normalize the inner filter data against this value, introducing the normalized fractional loss of signal due to the inner filter effect: $IFE = ife/k$. Thus:

$$IFE = 10^{-c \cdot R}$$

Many of the approximations used for these derivations break down at higher absorbances and thus higher concentrations of retinoid. In addition, high unbound concentrations of retinoid are insoluble and precipitate from solution making measurement of the inner filter effect at such concentrations difficult. For these reasons it is best to titrate quantities of protein less than 5 μM, in order to keep inner filter effects to a minimum.

B: Binding Equations

The binding equation used in our studies is simply derived by combining the mass law equation for a single binding site with an expression for conservation of ligand as follows. Start by considering the binding of retinoid, R, to a single site on a protein, P:

$$P + R \rightleftharpoons PR$$

The mass law equation for this equilibrium binding, where K_a is the association constant, can be expressed in the following forms:

$$K_a = \frac{[PR]}{[P][R]} = \frac{\bar{R} \cdot P_T}{(P_T - [PR]) \cdot [R]} = \frac{\bar{R}}{(1 - [PR]/P_T) \cdot [R]} = \frac{\bar{R}}{(1 - R) \cdot [R]}$$

$\bar{R}$ is the degree of binding, defined as $[PR]/P_T$, where P_T represents the total protein concentration. Assuming that retinoid is either free in solution or bound to the macromolecule, conservation of ligand states that:

$$R_T = [R] + RP_T$$

where R_T is the total retinoid concentration. The free-ligand term is removed by substituting $R_T - RP_T$ for $[R]$ in the mass law equation:

$$K_a = \frac{\bar{R}}{(1 - \bar{R}) \cdot (R_T - \bar{R} \cdot P_T)}$$

Rearranging:

$$K_a \cdot R_T - K_a \cdot R_T \cdot \bar{R} - K_a \cdot P_T \cdot \bar{R} + K_a \cdot P_T \cdot \bar{R}^2 = \bar{R}$$

Grouping terms:

$$\bar{R}^2 (K_a \cdot P_T) + \bar{R}(-K_a \cdot R_T - K_a \cdot P_T - 1) + K_a \cdot R_T = 0$$

Solving for $\bar{R}$ using the quadratic equation:

$$\bar{R} = \frac{1 + (P_T + R_T)K_a \pm \sqrt{K_a^2(P_T + R_T + 2P_T \cdot R_T) + 1 + K_a(2P_T + 2R_T) - 4K_a^2 P_T \cdot R_T}}{2P_T \cdot K_a}$$

Only the negative root yields values of R with physical reality. Rearranging:

$$\bar{R} = \frac{1 + (P_T + R_T)K_a - \sqrt{(P_T - R_T)^2 K_a^2 + 2(P_T + R_T)K_a + 1}}{2P_T K_a}$$

If the fluorescence signal changes are proportional to the amount binding and if the fluorescence of free retinoid is zero, the observed fluorescence will follow this relation:

$$F = F_0 - \bar{R}(F_0 - F_\infty)$$

where F is the observed fluorescence, F_0 is the protein fluorescence in the absence of ligand, and F_∞ is the fluorescence signal of the protein once completely saturated with retinoid. The fluorescence signal of each point of a titration of protein with retinoid will be defined as follows:

$$F = F_0 - \frac{1 + (P_T + R_T)K_a - \sqrt{(P_T - R_T)^2 K_a^2 + 2(P_T + R_T)K_a + 1}}{2P_T K_a} (F_0 - F_\infty)$$

This equation has been derived for nonretinoid binding proteins *(3)*.

If the unbound concentration of retinoid exceeds its solubility limit then the conservation of retinoid equation will change to:

$$R_T = [R] + \bar{R}P_T + [R_{sol}]$$

where $[R_{sol}]$ is he concentration of aggregated retinoid. Thus under such conditions the above binding equation no longer holds true.

C: Sigma Plot Fitting Files

Inner Filter Effect Fitting File

Note: The normalized fluorescence signals (s) are placed in column 1, and $r = $ the related total retinoid concentrations (r) are placed in column 2.

```
[IFE.FIT]
[Parameters]
c = 0
[Variables]
s = col(1)
r = col(2)
```

[Equations]
$f = 10\verb|^|(-c*r)$
fit *f* to *s*
[Constraints]
[Options]

Binding Equation Fitting File

Note: The fluorescence signals (*y*) are placed in column 1, and the related total retinoid concentrations (*r*) are placed in column 2.

[Parameters]
$p = 10$
$K_{\mathrm{d}} = 3$
$F_{\mathrm{m}} = 10$
[Variables]
$r = \mathrm{col}(2)$
$y = \mathrm{col}(1)$
[Equations]
$K = 1/K_{\mathrm{d}}$
$n = 1+(r+P)*K-([P\text{-}r]\verb|^|2*K\verb|^|2+2*[P+r]*K+1)\verb|^|0.5$
$d = 2*P*K$
$f = \mathrm{cell}(1,1)-(n/d)*(\mathrm{cell}(1,1)\text{-}F_{\mathrm{m}})$
fit *f* to *y*
[Constraints]
[Options]

11

Expression and Mutagenesis
of Retinol-Binding Protein

Manickavasagam Sundaram, Asipu Sivaprasadarao,
and John B. C. Findlay

1. Introduction

Vitamin A is transported in the plasma as retinol bound to a carrier protein, called retinol-binding protein (RBP), which itself forms a complex with the thyroxine-binding protein, known as transthyretin (TTR). This complex exists in equilibrium with free holo-RBP, which can then interact with a specific cell-surface receptor, thereby inducing the release of its retinol to the target cell. Thus, RBP possesses at least three molecular-recognition properties: it binds retinol and it interacts with both TTR and the cell-surface receptor (for reviews, *see* **refs.** *1* and *2*).

One commonly used approach to define the molecular interfaces involved in these interactions is to mutate those residues that are likely to participate in the interactions, and analyze the consequences of these mutations using appropriate binding assays. A prerequisite for such studies is the production of the protein using a suitable expression system, and its subsequent purification in sufficient quantities for the proposed investigation. Once this is achieved, mutants of the protein can be generated, examined for activity, and interesting forms subjected to structural analysis. This chapter describes the production of RBP in *Escherichia coli*, purification of expressed protein, and the methods used to introduce amino-acid substitutions and deletions.

Because purification can be a time-consuming and problematic process, it is often convenient to reconstruct the protein to facilitate its isolation. The best approach is to employ a system that is distinct from any structure or activity the native protein may have. In this way, disturbance of those native characteristics will not perturb the purification strategy. We describe in the following

From: *Methods in Molecular Biology, Vol. 89: Retinoid Protocols*
Edited by: C. P. F. Redfern © Humana Press Inc., Totowa, NJ

sections methods for the isolation of both the native protein and one engineered to contain an affinity tag—the "strep tag"—which interacts with streptavidin. The presence of this additional feature also allows alternative methods for its detection and quantitation to be employed.

2. Materials

1. Competent cells: XL-Blue *E.coli* cells can be purchased from Stratagene (Cambridge, UK). BL21 (DE3) cells are supplied by Novagen (Madison, WI). Alternatively, both can be prepared by the method of Hanahan *(3)*.
2. LB medium: 1% w/v bactotryptone, 0.5% w/v yeast extract, 0.5% w/v sodium chloride.
3. Ampicillin: 50 mg/mL in water, sterilize by filtering through a 0.22-μm filter.
4. LB-ampicillin agar plates: Add 1.4 g of bacto-agar to 100 mL of LB medium, autoclave at 120°C and 121 psi, cool to about 50°C, then add 100 μL of ampicillin and pour into four 90-mm Petri dishes. Allow the agar to solidify. Store the plates at 4°C.
5. LB-ampicillin/X-gal/IPTG agar plates: Prepared as in **step 4** except that 160 μL of X-gal solution and 200 μL of IPTG are added along with ampicillin.
6. 2% X-Gal (Sigma, Poole, UK): Prepare in dimethyl formamide.
7. 0.1 M Isopropyl-β-thio-galactoside (IPTG, Calbiochem, Beeston, UK) in water: sterilize by filtering through a 0.22-μ filter.
8. Lysis buffer: 20% sucrose w/v in 35 mM Tris-HCl, pH 8.0, 1 mM PMSF.
9. CNBr-activated Sepharose CL-4B (Pharmacia, St. Albans, UK).
10. Coupling buffer: 0.1 M NaHCO$_3$, pH 8.5, 0.5 M NaCl.
11. Human transthyretin (Sigma).
12. TTR-column buffer: 50 mM Tris-HCl, pH 7.4, 0.5 M NaCl.
13. Sintered-glass filter and a Buchner flask.
14. 50 mM all-*trans*-retinol in dimethyl sulfoxide (store in an amber-colored glass container at –20°C under nitrogen).
15. Spiramix (Denley Instruments, Billingshurst, UK).
16. Peristaltic pump.
17. 1 M Ethanolamine, pH 8.0.
18. Streptavidin-agarose (Sigma).
19. 2-iminobiotin (Sigma).
20. Strep-equilibration buffer: 100 mM Tris-HCl, pH 8.0, 1 mM EDTA.
21. Strep-elution buffer I: Dilute strep-elution buffer II 10-fold with deionized water.
22. Strep-elution buffer II: 1 mM 2-iminobiotin, 100 mM Tris-HCl, pH 8.0, 1 mM EDTA.
23. Lipidex-1000 (Sigma).
24. UV spectrophotometer.
25. Thermocycler (e.g., Perkin Elmer).
26. Template DNA : pO-RBP (**Fig. 1**).
27. Universal forward primer, 5'-CCCAGTCACGACGTTGTAAAACG-3' 10 pmol/μL.

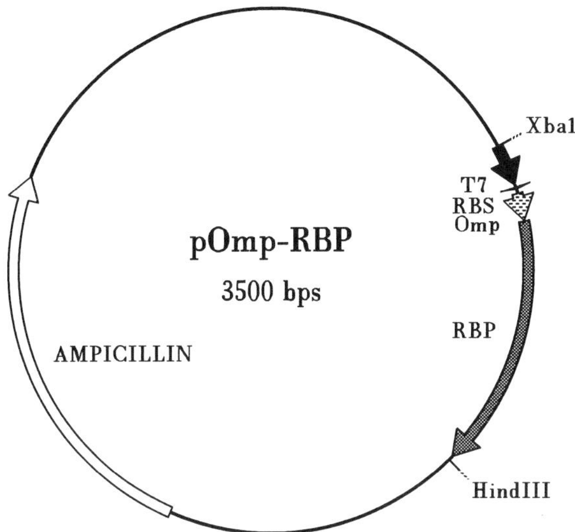

Fig. 1. Schematic diagram of the human RBP expression vector, pOmp-RBP. RBP, cDNA for human retinol-binding protein; Omp, bacterial outer membrane signal sequence; RBS, ribosome-binding site; AMPICILLIN, the β-lactamase gene. The arrow indicates the direction of the coding sequence.

28. Universal reverse primer, 5'-AGCGGATAACAATTTCACACAGG-3' 10 pmol/μL.
29. Mutagenic and deletion primers: Design a primer incorporating the required base change(s). The primer (approx 10 μmol/μL) should have an annealing temperature between 55 and 60°C. The primers can be custom-made by several companies, (e.g., Gibco-BRL, Pharmacia) (for design of primers *see* **ref. 7**).
30. Deoxyribonucleotide triphosphates (dNTPs): Mixture of dATP, dTTP, dGTP, and dCTP, 2 m*M* each. Store as aliquots at –20°C (Stock solutions supplied by Pharmacia).
31. *Pfu* DNA polymerase (Stratagene).
32. 10X *Pfu* PCR buffer (supplied along with *Pfu* DNA polymerase).
33. Mineral oil (Sigma).
34. Agarose (electrophoresis grade).
35. 1% w/v ethidium bromide in water.
36. 50X TAE buffer: 242 g Tris-base, 57.1 mL glacial acetic acid and 100 mL of 0.5 *M* EDTA, pH 8.0, made up to 1L.
37. DNA miniprep kits (commercially available from a number of sources).
38. 10 m*M* ATP (prepare in water, pH to 7.0).
39. 6X loading buffer: 0.25% bromophenol blue, 30% glycerol.
40. 10X ligase buffer (supplied along with the enzyme).
41. T4 DNA ligase.
42. Geneclean kit (Anachem, Luton, UK).

43. pBluescript KS (Stratagene).
44. *Xba*I.
45. *Hin*dIII.
46. T4 Polynucleotide kinase.
47. 10X PNK buffer, obtained from enzyme suppliers.
48. Sequenase version 2.0 DNA sequencing kit (Amersham, Little Chalfont, UK).
49. Lysozyme (Sigma).
50. 0.5 *M* EDTA, pH 8.0.

3. Methods
3.1. Expression of Recombinant RBP

RBP contains three disulfide bonds that are essential for the functional integrity of the protein. It is therefore important to choose an expression system that facilitates not only the expression of the protein but proper folding and disulfide bridge formation. The vector construct, pO-RBP, used in the system described in this section allows RBP to be expressed as a fusion protein in *E. coli* with the OmpA-signal peptide added to its N-terminus. This sequence directs the fusion protein into the periplasm of the cell via the inner-plasma membrane. At some stage the OmpA-signal peptide is cleaved off such that only the mature form of RBP is released into the periplasm. Unlike the cytosol, the periplasm offers the oxidizing environment necessary for the protein to form disulfide bridges. Thus, the expressed RBP would have the best chance of adopting the native structure and thus possessing the required functional properties *(4)*.

The expression of RBP is carried out in three stages. First, BL21-DE3 cells are transformed with the pO-RBP vector construct. Second, a colony containing the construct is grown in a large volume of medium to midlog phase and the expression of RBP induced with IPTG for 4 h. Finally, the periplasmic proteins are recovered and RBP purified by affinity chromatography. The yield of the purified protein is often in the range of 200–300 µg/L of culture. The expression is normally carried out in a scale of three litres in six 2-L flasks, each containing 500 mL of medium. Alternatively, it is possible to use a fermenter where yield is higher.

1. Thaw out 100 µL of competent BL21 DE3 *E.coli* cells in a 1.5-mL microtube on ice.
2. Add 1 µL of the pO-RBP plasmid DNA (approx 1 ng) to the cells and mix gently with a pipet tip.
3. Place in a water bath for exactly 45 s and immediately chill on ice.
4. Add 450 µL of LB medium, and incubate at 37°C with shaking at 200 rpm for 1 h.
5. Spin cells down at 6000*g* in a microcentrifuge.
6. Discard 300 µL of the supernatant.

7. Resuspend the cells in the remaining medium and spread the suspension on an LB-ampicillin-agar plate and incubate the plate (inverted) for 16–20 h at 37°C to allow colony growth.
8. Inoculate 5 mL of LB-broth supplemented with 100 µg/mL ampicillin with a single colony from the plate and grow at 30°C in a shaking incubator at 200 rpm for 12 h (*see* **Note 1**).
9. Add the overnight culture to 500 mL of LB medium containing ampicillin (100 µg/mL) in a 2-L flask and incubate this culture at 30°C in a shaking incubator at 200 rpm.
10. Continue the incubation until the OD_{550} of the culture reaches 0.6 (approx 2–3 h).
11. At this point, add 1 mL of 0.1 M IPTG (final concentration 0.2 mM) and continue the incubation for a further 4 h. Chill on ice.
12. Centrifuge the culture in a 500-mL centrifuge bottle at 4500g for 10 min at 4°C.
13. Decant the supernatant and drain any excess as completely as possible.
14. Resuspend the cell pellet in 50 mL (one-tenth the volume of the original culture) of ice-cold lysis buffer, add 0.5 M EDTA dropwise to a final concentration of 10 mM and chill on ice for 10 min.
15. Release the expressed protein from the periplasm by one of the following methods (*see* **Note 2**):
    a. Osmotic shock: Centrifuge the cell suspension at 10,000g for 10 min at 4°C. Discard the supernatant, and resuspend the resulting pellet in 50 mL of ice-cold deionized water. After incubating on ice for a further 10 min, centrifuge at 10,000g for 30 min at 4°C. Freeze-dry the supernatant.
    b. Lysozyme treatment: To the cell suspension from **step 14**, add 5 mL of a 1 mg/mL stock solution of lysozyme in water and incubate at room temperature for 30 min. Centrifuge at 10,000g for 30 min. Freeze-dry the supernatant.

3.2. Purification of Recombinant RBP

Recombinant RBP is purified from the freeze-dried periplasmic extract by affinity chromatography using a TTR-affinity resin.

3.2.1. Preparation of TTR-Affinity Resin

1. Dissolve 8–10 mg of TTR in 8 mL of ice-cold coupling buffer and keep on ice.
2. Weigh out 2.0 g of CNBr-activated Sepharose CL-4B and resuspend in 1 mM ice-cold HCl.
3. Wash the resin with about 400 mL of ice-cold 1 mM HCl on a sintered-glass filter under gentle suction (do not let the resin dry at any stage).
4. Transfer using a spatula as quickly as possible to the TTR solution. Add coupling buffer so that the final concentration of TTR is approx 0.5 mg/mL.
5. Mix on a Spiramix overnight at 4°C.
6. Add an equal volume of 1 M ethanolamine, pH 8.0, to block the unbound sites. Continue mixing on the Spiramix for 2 h at room temperature.

7. Filter the suspension through a sintered-glass filter under gentle suction (collect the filtrate and store separately to determine coupling efficiency later). Wash the resin on the filter with 500 mL of deionized water followed by 500 mL of column-equilibration buffer.

8. Pack resin into a 10-mL column and equilibrate with ice-cold TTR-column buffer. If the column is not immediately used, wash with column buffer containing 0.02% azide and store at 4°C (the TTR-resin is stable for about 8 mo at 4°C and can be used up to about 50 chromatographic runs).

9. Determination of coupling efficiency: Dialyse the filtrate from **step 7** against a large excess (2–3 L) of water, with 2–3 changes (to remove ethanolamine). Freeze-dry the dialyzed filtrate, dissolve in 1 mL of water and measure the absorbance at 280 nm. Calculate the amount of protein from the absorbance using a molar-extinction coefficient of 76,000. Use this value and the amount of TTR applied to the resin to calculate the amount of TTR coupled to the support.

3.2.2. TTR-Affinity Chromatography

1. All chromatographic steps are performed at 4°C using a flow rate of 10 mL/h.

2. Dissolve the freeze-dried periplasmic extracts, derived from 3 L of culture in 50 mL of TTR-column buffer.

3. To this, add 500 µL of the 50 m*M* retinol solution dropwise (this should be done in dim light or under red light) with swirling. Wrap the flask in aluminium foil and incubate at 37°C for 2 h. All steps henceforth are carried out in diffused light and at 4°C.

4. Centrifuge the retinol-treated extract at 10,000*g* for 30 min. Collect the supernatant into a fresh container and store at 4°C.

5. Wash the TTR column with 10 bed volumes of the TTR-column buffer.

6. Circulate the clear supernatant from **step 3** through the TTR resin overnight using a peristaltic pump.

7. On the following day, stop the circulation, let the excess extract flow out of the column, and then wash the resin with 5 bed volumes of the TTR-column buffer.

8. Replace the TTR-column buffer with deionized water, pH 8.5 (adjusted with a drop of aqueous ammonia). Collect 1-mL fractions and measure their absorbance spectra using a UV-Visible spectrophotometer (for a typical spectrum, *see* **Fig. 2**).

9. Pool fractions showing absorbance peaks at 280 and 330 nm, freeze-dry and dissolve the freeze-dried protein in 500 µL of water or PBS. Check its purity on SDS-PAGE.

3.3. Purification of RBP-Streptavidin Fusion Protein

RBP can be expressed as a fusion protein with a variety of tags added on to its termini. We have expressed RBP as a streptag-fusion protein *(5)*. This has two advantages. First, commercially available streptavidin-affinity resins can be used to purify the protein (instead of TTR-resin). Second, the apo-or inactive

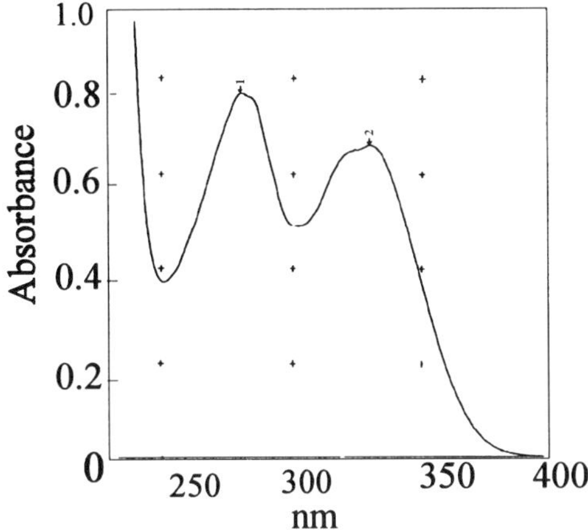

Fig. 2. UV spectrum of purified recombinant RBP.

forms of RBP can be purified. The expression of the protein can be carried out essentially as described for RBP earlier except that BL21 DE3 cells are transformed with the pO-RBP-Streptag construct. This construct contains the nucleotide sequence corresponding to the streptag peptide (Ser-Ala-Trp-Arg-His-Pro-Glu-Phe-Gly-Gly), fused to the 3' end of the RBP-coding sequence before the stop codon. Expression and preparation of periplasmic extracts are carried out as above (**Subheading 3.1.**).

3.3.1. Chromatography with Streptavidin-Affinity Resins

All chromatographic steps are performed using a flow rate of 10 mL/h and at 4°C.

1. Pack 5 mL of streptavidin-agarose into a small column (a 10-mL syringe may be used) and equilibrate it with 10 bed volumes of Strep-equilibration buffer.
2. Apply the concentrated periplasmic protein fraction containing the streptag-RBP fusion protein to the affinity resin.
3. Stop the flow for 30 min after the entire sample has entered the resin.
4. Wash the resin with strep-equilibration buffer until the absorbance at 280 nm drops to base line.
5. Elute the resin-bound protein, first with 3 vol of Strep-elution buffer-I and then with 3 vol of Strep-elution buffer-II. Collect 5-mL fractions (*see* **Note 3**).
6. Dialyze each fraction against 2 L of distilled water overnight and freeze-dry.
7. Dissolve each freeze-dried fraction in 200 µL of 10 m*M* Tris-HCl, pH 7.5. Analyze 20 µL on a 15% SDS-polyacrylamide gel. Pool the fractions containing the purified protein.

3.3.2. Preparation of Holo-RBP-Streptag Fusion Protein

When required, the apo-form of streptag-RBP fusion protein can be converted to the holo form using the following procedure:

1. To 100 µg of RBP in 900 µL of 10 m*M* Tris-HCl, pH 7.5, add 50 µL of 2 m*M* all-*trans*-retinol in ethanol (20-fold molar excess) and incubate in the dark at 37°C for an hour.
2. Apply the mixture to a Lipidex-1000 resin (column bed volume ~2 mL) and elute with three column volumes of 10 m*M* Tris-HCl, pH 7.5. Collect the eluant and freeze-dry. (Excess retinol will be retained by the resin.) Dissolve the protein in a suitable volume of PBS and measure the absorbance spectrum to confirm the binding of retinol to the protein.

3.4. Mutagenesis

Point mutations and deletions can be readily introduced into the cDNA for RBP using the polymerase chain reaction (PCR). The methods described here are carried out using the pO-RBP vector as the template DNA *(6)*. (For details of recombinant DNA procedures, *see* **ref. 3**.)

3.4.1. Point Mutations

In order to introduce point mutations, a two stage-PCR reaction is carried out using a mismatched oligonucleotide primer containing appropriate base change(s) and the universal M13 forward and reverse primers (**Fig. 3**).

3.4.1.1. FIRST PCR (SYNTHESIS OF MEGAPRIMER)

The first PCR is carried out to generate a megaprimer using the M13-reverse primer and the mutagenic primer. The reaction is set up as follows:

1. Into a 0.5-mL microtube, pipet out: 10 µL 10X PCR buffer, 10 µL 2 m*M* dNTPs, 10 µL reverse primer, 10 µL mutagenic primer, 1 µL DNA template (10 ng/µL), 59 µL water.
2. Overlay with 50 µL of mineral oil.
3. Heat to 95°C for 5 min in a PCR thermocycler.
4. Add, while at 95°C in the block, 1 µL of *Pfu* DNA polymerase.
5. Subject to 30 PCR cycles, each cycle consisting of: 40 s denaturation at 95°C, 1 min annealing at 55°C, and 1 min extension at 72°C.
6. Carry out the final extension at 72°C for 5 min.
7. Remove tubes from the block and add 20 µL of 6X loading buffer.
8. To remove the oil, add 200 µL of chloroform, vortex, and centrifuge in a microcentrifuge for 1 min at top speed.
9. Remove the bottom organic phase with a pipet and discard.

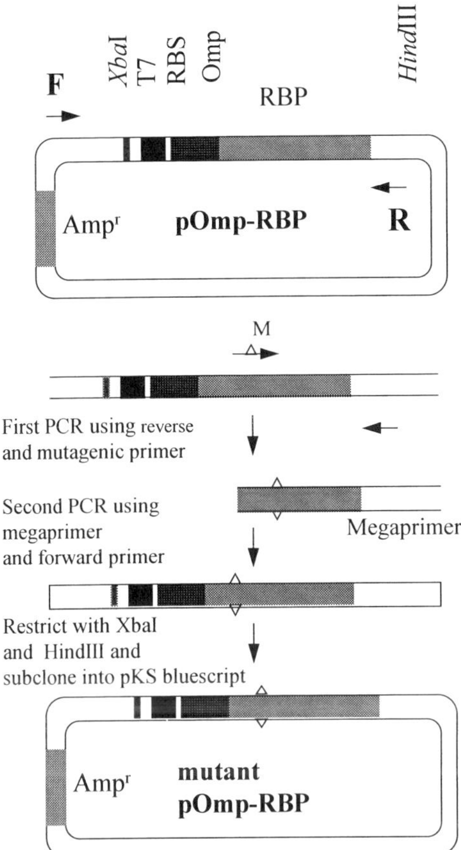

Fig. 3. Site-directed mutagenesis of RBP cDNA. R, reverse primer; F, forward primer; M, mutagenic primer; RBS, ribosome-binding site; Omp, OmpA signal sequence; Amp^r, ampicillin-resistance gene.

3.4.1.2. PURIFICATION OF PCR PRODUCT:

The PCR product is purified by first running the PCR reaction mixture on an agarose gel and then purifying the DNA from the relevant band excised out of the gel.

Agarose gel electrophoresis:

1. Dissolve 1 g of agarose in 1X TAE buffer by boiling, cool to about 50°C, add 5 µL of 1% ethidium bromide, and pour into an appropriate gel-casting tray, making wells that take ≥ 20 µL of sample.

2. Load the first PCR reaction product into 5–6 wells of the gel alongside marker DNA (e.g., λDNA *Hin*dIII marker).
3. Run in an electrophoresis tank at 6 V/cm length of the gel until the blue dye migrates two-thirds of the gel length.
4. View under UV light.
5. Cut out the bands with a razor blade.
6. Isolate DNA from the agarose bands using GeneClean according to the protocol supplied by the manufacturer. Elute the DNA (megaprimer) bound to glass beads into 20 μL of water.

3.4.1.3. Second PCR

1. Into a 0.5-mL microtube, pipet out: 20 μL megaprimer (from **step 6** in **Subheading 3.4.1.2.**), 10 μL 10X PCR buffer, 10 μL 2 m*M* dNTPs, 10 μL forward primer, 1 μL DNA template (10 ng/μL), 49 μL water.
2. Subject to PCR as above (*see* **Subheading 3.4.1.1.**).
3. Remove the oil by extracting with chloroform (**step 8** of **Subheading 3.4.1.1.**).
4. Run the extracted sample on an agarose gel. If a product larger than the megaprimer (650 bp) is seen, it indicates that the PCR has been successful.
5. Excise the band, GeneClean and elute the product into 20 μL of water.

3.4.1.4. Subcloning of Purified PCR Product into pBluescript

3.4.1.4.1. Restriction Digestion of Second PCR Product and pKS Bluescript Vector DNA

1. Into a 0.5-mL microtube, pipet out: 20 μL second PCR product (from **step 5** in **Subheading 3.4.1.3.**), 3 μL restriction buffer B, 5 μL water, 1 μL *Xba*I (10–12 U), 1μL *Hin*dIII (10–12 U).
2. Incubate at 37°C overnight.
3. Set up a parallel restriction digestion of the vector DNA using 20 μL of pKS-Bluescript (2–5 μg) in place of second PCR product.
4. GeneClean the restricted PCR product and the vector DNA; elute each into 10 μL of water.

3.4.1.4.2. Ligation

1. Into a 0.5-mL microtube, pipet out: 1 μL vector DNA, 5 μL insert DNA, 1 μL 10X Ligase buffer, 2 μL water, 1 μL T4 DNA ligase (0.1–0.5 U).
2. Incubate at 16°C overnight or at room temperature for 2 h.

3.4.1.4.3. Transformation

1. Thaw out 100 μL of competent *E. coli* XL-Blue cells in a 1.5-mL microtube on ice.
2. Add 5 μL of the ligated sample to the cells and mix gently with the pipet tip.
3. Place in a water bath for exactly 45 s and immediately chill on ice.

4. Add 450 µL of LB medium, and incubate at 37°C with shaking at 200 rpm for 1 h.
5. Spin cells down at 6000*g* in a microcentrifuge.
6. Discard 300 µL of the supernatant.
7. Resuspend the cells in the remaining medium and spread the suspension on to the LB-agar plate containing X-gal and IPTG and incubate the plate for 16–20 h at 37°C.
8. Blue and and white colonies should appear on the plate after 16–20 h. White colonies should contain the insert cDNA.
9. Prepare DNA from four to six white colonies using DNA miniprep kits (these kits are expensive; methods described in most manuals, e.g., Sambrook et al. *(3)*, give satisfactory results).
10. Perform digestion with *Xba*I and *Hin*dIII as above on 1/10th of the miniprep DNA.
11. Analyze the restriction digests on an agarose gel as above.
12. Presence of a 650-bp band would suggest the presence of RBP cDNA.
13. Sequence 1/4 of the miniprep DNA using the Sequenase DNA sequencing kit and universal primers and check for the mutation.
14. Nearly all clones should be mutants.

3.4.2. Deletion PCR

In order to generate mutants containing deletions of an amino acid residue or a region, two oligonucleotide primers, flanking the DNA segment to be deleted are required. One primer should be made to the sense strand and the other to the antisense strand (*see* **Fig. 4**). PCR is then performed on the template DNA, at the end of which, the amplified DNA should contain the entire DNA sequence (including the vector sequence) except for the fragment to be deleted. The amplified DNA is self-ligated and transformed into *E. coli*. The protocol described in this section was used to delete residues 92–98 of RBP using the primers identified *(6)*. The strategy can be adopted to remove any part of the DNA by designing appropriate primers flanking the sequence to be deleted.

Oligonucleotide primers:

1. 5'-AAAGGAAATGATGACCAC-3' (sense).
2. 5'-CCAGTACTTCATCTTGAA-3' (antisense).

3.4.2.1. Phosphorylation of Oligonucleotides

Phosphate groups are added to the 5' ends of oligonucleotides to allow subsequent ligation of the PCR product.

1. Into a 0.5-mL microtube, pipet out: 20 µL sense oligonucleotide, 20 µL antisense oligonucleotide, 5 µL 10X PNK buffer, 5 µL 10 m*M* ATP, 1–2 µL T4 polynucleotide kinase (10–20 U).
2. Incubate at 37°C for 1 h.

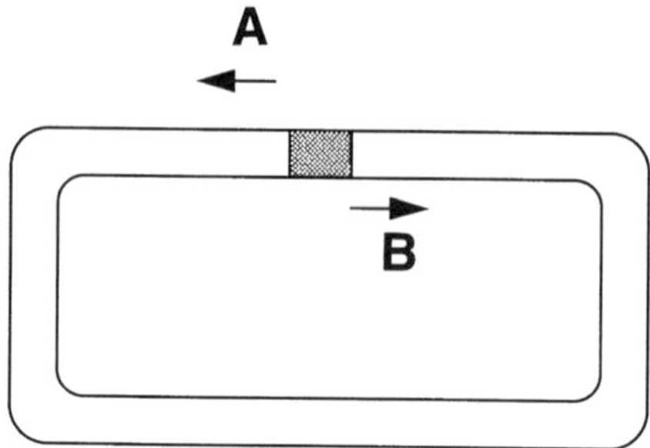

Fig. 4. Schematic diagram for deletion PCR. Primer A, sense primer; primer B, antisense primer. The primers flank the region to be deleted.

3.4.2.2. PCR

1. Into a 0.5-mL microtube, pipet out: 10 μL phosphorylated oligonucleotides (from **step 2**), 10 μL 10X *Pfu* PCR buffer, 10 μL 2 m*M* dNTPs, 1 μL pO-RBP DNA (10 ng) (*see* **Note 4**), 69 μL water.
2. Overlay with 50 μL of mineral oil.
3. Heat at 95°C for 5 min in a PCR block.
4. Add 1 μL of *Pfu* DNA polymerase.
5. Subject to the following 30 PCR cycles: 40 s denaturation at 95°C, 1 min annealing at 46°C, 7 min extension at 72°C, and 10 min final extension at 72°C.
6. Run the PCR product on a 1% agarose gel and GeneClean the band (expected size is about 3.6 kb) as noted earlier. Elute the DNA into 10 μL of water.

3.4.2.3. Ligation

1. Into a 0.5-mL microtube, pipet out: 10 μL eluted DNA (from **step 6** in **Subheading 3.4.2.2.**), 2 μL 10X Ligase buffer, 7 μL water, 1 μL T4 DNA ligase (1 U).
2. Incubate at 16°C overnight and transform as in **Subheading 3.4.1.4.3.**
3. Sequence the miniprep DNA made from the transformed colonies to confirm the deletion.

4. Notes

1. It is important to use colonies from a freshly transformed plate.
2. Both methods are easy to use. With the osmotic-shock procedure, however, some intracellular proteins are also released in addition to the periplasmic proteins. This might result in the appearance of contaminating proteins during the purification of the RBP-streptag fusion protein. The lysozyme method, on the other hand, causes the release of only periplasmic proteins and thus is recommended for the recovery of the expressed protein.
3. 2-Iminobiotin is used to ensure column reusability. Nevertheless, if maximal recovery is desired the same elution buffer containing 1 m*M* D-biotin, instead of 2-iminobiotin, should be used.

4. To reduce the percentage of wild-type DNA, it may be necessary to use the super-coiled version of the plasmid, purified either by CsCl-density gradient centrifugation or from agarose gels by the GeneClean protocol.

References

1. Soprano, D. R. and Blaner, W. S. (1994) Plasma retinol-binding protein, in *The Retinoids: Biology, Chemistry and Medicine*, (Sporn, M. B., Roberts, A. B., and Goodman, D. S., eds.), Raven, New York, NY, pp. 257–283.
2. Sivaprasadarao, A. and Findlay, J. B. C. (1994) The retinol-binding protein super-family, in *Vitamin A in Health and Disease*, (Blomhoff, R. ed.), Marcel Dekker, New York, NY, pp. 87–117.
3. Sambrook, J., Fritsch, E. F., and Maniatis, T. (1989) *Molecular Cloning: A Laboratory Manual*, 2nd ed., Cold Spring Harbor Laboratory, Cold Spring Harbor, NY.
4. Sivaprasadarao, A. and Findlay, J. B. C. (1993) Expression of functional human retinol-binding proteins in *Escherichia coli* using a secretion vector. *Biochem. J.* **296,** 209–215.
5. Müller, H. N. and Skerra, A. (1993) Functional expression of the uncomplexed serum retinol-binding protein in Escherichia coli. *J. Mol. Biol.* **230,** 725–732.
6. Sivaprasadarao, A. and Findlay, J. B. C. (1994) Structure-function studies on human retinol-binding protein using site-directed mutagenesis. *Biochem. J.* **300,** 437–442.
7. Rychlik, W. (1993) Selection of primers for polymerase chain reaction, in *PCR Protocols: Current Methods and Applications* (White, B. A., ed.), Humana, Totowa, NJ, pp. 31–40.

12

Interactions of Retinol-Binding Protein with Transthyretin and Its Receptor

Asipu Sivaprasadarao, Manickavasagam Sundaram, and John B. C. Findlay

1. Introduction

To study the interaction of retinol-binding protein (RBP) with its plasma carrier, transthyretin (TTR), spectrofluorimetry, and circular dichroism have previously been used. Both these techniques require milligram quantities of the proteins and this sets limitations on the use of these techniques for the study of RBP-TTR interactions using recombinant proteins. The *Escherichia coli* expression system described in Chapter 11 does not readily produce milligram quantities of RBP for routine use. For this reason, we have developed a highly sensitive method which employs radioiodinated ^{125}I-RBP (unpublished). The method requires only microgram quantities of protein. This chapter describes a method to radioiodinate RBP without loss of activity and protocols for its use in the study of its interaction with TTR.

The interaction of RBP with its membrane-bound receptor can be investigated using a variety of methods, all employing radioiodinated RBP. However, of all the methods used, the oil-centrifugation protocol described here (*1*) has been found to be the most reliable and sensitive. Methods to assay both the membrane-bound and the detergent-solubilized receptor are provided. In addition, a method for purifying the receptor from membranes is also included.

2. Materials

1. Iodobeads (Pierce, Chester, UK).
2. Carrier-free Na^{125}I (5000 Ci/mmol, Du Pont NEN, Stevenage, UK).
3. Iodination buffer: 100 m*M* sodium phosphate buffer, pH 6.5.
4. Sephadex G-50 (fine) (Pharmacia, St. Albans, UK).

From: *Methods in Molecular Biology, Vol. 89: Retinoid Protocols*
Edited by: C. P. F. Redfern © Humana Press Inc., Totowa, NJ

5. PBS: 20 mM Sodium phosphate, pH 7.4, 150 mM NaCl.
6. Assay buffer: 0.2% w/v ovalbumin in PBS.
7. Polyethylene glycol 8000 (Sigma, Poole, UK).
8. 20% w/v Trichloroacetic acid.
9. Goat γ-globulins in 0.1 M sodium phosphate buffer, pH 7.4.
10. Transthyretin (Sigma).
11. Recombinant RBP (50 μM).
12. Dibutyl phthalate (Sigma).
13. Dinonyl phthalate (Fluka, Gillingham, UK).
14. 0.15 M NaCl.
15. 1% w/v Sodium dodecyl sulfate (SDS).
16. Placental buffer:10 mM Mannitol, 2 mM Tris-HCl, pH 7.1.
17. BCA protein-assay kit (Bio-Rad, Hemel Hempstead, UK).
18. Reacti-Gel 6X (Pierce).
19. Coupling buffer: 0.1 M borate, pH 8.5.
20. 2 M Ethanolamine, pH 8.0.
21. 50 mM Tris-HCl, pH 7.4, 1 M NaCl.
22. Wash buffer: PBS, 2 mM MgCl$_2$, 0.2 mM phenylmethylsufonyl fluoride (PMSF, *see* **Note 1**).
23. Solubilization buffer: 5% w/v octyl glucoside in wash buffer.
24. Buffer A: PBS, 0.5% w/v octylglucoside, 0.2 mM PMSF.
25. Buffer B: 20 mM sodium phosphate, pH 7.4, 1 M NaCl, 0.5% w/v octylglucoside, 0.2 mM PMSF.
26. Buffer C: 0.5% w/v octylglucoside in deionized water, 0.2 mM PMSF.
27. Buffer D: 50 mM Sodium acetate, pH 5.0, 150 mM NaCl, 0.5% w/v octylglucoside, 0.2 mM PMSF.
28. Buffer E: 0.3 M Sodium phosphate, pH 7.4, 0.5% w/v octylglucoside, 0.2 mM PMSF.

3. Methods

Binding of RBP to TTR as well as to the receptor can be studied using RBP radiolabeled with [125]I. Either the native or the recombinant form of RBP can be used for this purpose. However, the recombinant form of RBP is less hetero-geneous and binds the receptor more effectively than the native form *(2)*.

3.1. Radioiodination of RBP

Although RBP can be radiolabeled with ^{125}I using a variety of methods, we have found that the use of harsh iodination conditions, such as those employed in the chloramine-T method, results in a marked loss of biological activity. The use of mild-iodinating reagents, such as Enzymobeads (Bio-Rad) or Iodobeads (Pierce), does not cause any apparent loss of activity. However, Enzymobeads are no longer on the market. We therefore describe the Iodobead method, which can be used to generate ^{125}I-RBP of high specific activity (250–500 Ci/mmol).

1. Just prior to use, wash two iodobeads with 500 µL of iodination buffer. Dry the beads on a filter paper (this wash step removes any loose reagent particles from the beads).
2. To 95 µL of iodination buffer in a small round-bottomed polypropylene tube, add 1 µL (1 mCi) of Na^{125}I solution and then the washed beads. Incubate the mixture at room temperature for 5 min (*see* **Note 2**).
3. Dilute the 50 µ*M* RBP stock solution 10-fold with iodination buffer and add 100 µL (0.5 nmoles) of this to the vial.
4. Incubate at room temperature for 5 min. Then, remove the solution into a fresh 1.5-mL microtube (leaving the beads in the vial).
5. Add 800 µL of ice-cold assay buffer and store on ice. Remove three 10-µL aliquots into separate tubes, dilute them to 1.0 mL with assay buffer and store them on ice to determine specific activity later (**Subheading 3.1.2.**). Separate the unreacted iodine from the ^{125}I-RBP as in **Subheading 3.1.1.**

3.1.1. Separation of ^{125}I-RBP from Free ^{125}I

Gel filtration on Sephadex G-50 is used for this purpose.

1. Pack Sephadex G-50 slurry into a column (60 × 1 cm). Wash with several bed volumes of PBS. Just before use, wash the column with at least two column volumes of assay buffer at a flow rate of 15 mL/h.
2. Remove the excess buffer from the top of the resin and load remainder of the labeled protein (about 970 µL) on to the top of the resin without disturbing the surface. Let the sample run into the bed. Add 2 mL of assay buffer and let it run into the resin.
3. Add 1 mL of assay buffer again and elute the labeled protein with the assay buffer using a flow rate of 12 mL/h. Collect 50 fractions, 1 mL each, into polypropylene tubes. Count 10 µL of each fraction in a γ-counter. Plot cpm against fraction number. Pool the fractions containing the first peak of radioactivity.
4. Determine the protein-bound radioactivity by TCA precipitation (**Subheading 3.1.2.**). If the TCA-precipitable counts are less than 98% of total counts, dialyze the pooled fractions against PBS overnight at 4°C. Store the labeled protein as 100-µL aliquots at –20°C.

3.1.2. Determination of Specific Activity of ^{125}I-Labeled RBP

The specific activity of the labeled RBP is estimated by determining the TCA-precipitable counts (TCA precipitates only protein-bound radioactivity) using ovalbumin as a carrier protein.

1. Into three 1.5-mL microtubes, pipet 10 µL of the diluted labeling reaction mixture (*see* **step 5** in **Subheading 3.1.**). Add 490 µL of ice-cold assay buffer and mix by vortexing. To this, add 500 µL of ice-cold 20% TCA and mix by vortexing.
2. Place the tubes on ice for 20 min, and centrifuge at 10,000*g* in a microcentrifuge for 15 min. Remove 500 µL of the supernatant into separate microtubes (the former set of tubes contains TCA precipitate plus 500 µL supernatant, whereas the latter contains only the supernatant). Count the tubes in a γ-counter.
3. To 20 µL of the final pool (**step 4**, **Subheading 3.1.1.**), add 480 µL of assay buffer and 500 µL of ice-cold 20% TCA and determine TCA-precipitable counts as before.

3.1.2.1. CALCULATIONS

1. TCA preciptable counts = (counts in pellet + supernatant) – (counts in supernatant alone).
2. Specific radioacivity =

$$\text{TCA precipitable counts} \times \frac{1}{\text{counter efficiency}} \times \frac{1}{^{125}\text{I decay factor}} \times \frac{10^4}{0.5} = \text{d.p.m./nmol}$$

3. Specific activity in Ci/mmol = $\dfrac{(\text{d.p.m./nmol} \times 10^6)}{2.2 \times 10^{12}}$

4. Final concentration of ^{125}I-RBP in the final pool is given by:

$$\text{TCA precipitable counts} \times \frac{1}{\text{counter efficiency}} \times \frac{1}{^{125}\text{I decay factor}} \times 50$$

$$= \text{d.p.m. per mL of sample}$$

$$\text{nmole } ^{125}\text{I-RBP/mL} = \frac{\text{d.p.m. per mL of sample}}{\text{specific activity in d.p.m./nmol}}$$

3.2. Binding Assay for RBP–TTR Interaction

The assay described in this section involves incubation of ^{125}I-RBP with TTR until equilibrium binding is achieved. The TTR-bound ^{125}I-RBP is then separated from the unbound ^{125}I-RBP by selective precipitation of the complex with 10% PEG 8000. At this PEG concentration, negligible amounts of free ^{125}I-RBP are precipitated.

1. Dilute ^{125}I-RBP stock with assay buffer to 20 n*M* concentration.
2. Set up reactions (100 µL) in triplicate in 0.5-mL microtubes as shown in **Table 1**.
3. Mix by vortexing gently and incubate at 37°C for 15 min.
4. Chill on ice for 2 min. Add 50 µL of ice-cold solution of goat γ-globulins and mix well.
5. Then add 100 µL of ice-cold 25% PEG 8000 solution. Vortex immediately. Incubate on ice for 25 min (*see* **Note 3**).

Table 1
Binding Assay for RBP–TTR Interaction

Addition	Total binding, µL	Nonspecific binding, µL	Background, µL
^{125}I-RBP	25	25	25
TTR	50	50	
RBP	—	10	—
Assay buffer	25	15	75

6. Centrifuge at 10,000g for 10 min in a microcentrifuge at 4°C to pellet the precipitate.
7. Overlay the mixture with ice-cold dibutyl phthalate and centrifuge again for 5 min. (The aqueous layer containing the unprecipitated ^{125}I-RBP should rise to the top through the denser oil layer.)
8. Freeze the tube in dry ice/ethanol and cut off the bottom of the tube containing the pellet with a sharp razor blade. Invert on a paper towel to drain any excess oil.
9. Place the bottoms of the tubes in suitable containers and count for radioactivity in a gamma counter.
10. Correct all values by subtracting the background. Specific binding is calculated by subtracting the nonspecific from the total binding.

3.3. Preparation of Placental Membranes

Microvillar membranes are prepared from freshly delivered human placentae according to the method of Booth et al. *(3)*. All procedures are carried out at 4°C.

1. Remove the umbilical cord and amniotic membranes from the placenta and cut it into pieces weighing 20–30 g each.
2. Wash with ice-cold 0.15 M NaCl solution and mince the pieces in a mincer.
3. Stir the mince with 1.5 vol of 0.15 M NaCl for 1 h using a magnetic stirrer.
4. Pass the mince through a nylon sieve to remove large pieces of tissue. Centrifuge the filtrate at 800g for 10 min.
5. Centrifuge the resultant supernatant at 10,000g for 10 min.
6. Centrifuge the resultant supernatant at 90,000g for 30 min (Beckmann AL30 rotor at 30,000 rpm).
7. Discard the supernatant.
8. Resuspend the pellet in 50 mL of placental buffer.
9. Add 0.1 g of solid $MgCl_2$ and place on ice for 10 min (this aggregates the nonmicrovillar membranes) with occasional mixing.
10. Centrifuge at 2000g for 15 min (AL30 rotor, 5000 rpm).
11. Centrifuge the supernatant for 30 min at 15,000g (AL30 rotor, 15,000 rpm).
12. Resuspend the pellet in 30 mL of 0.15 M NaCl (about 50–60 mg of membrane protein is normally obtained from one placenta).
13. Assay the protein content by BCA-protein assay (BCA-assay kit) after dissolving a portion of the sample in 1% SDS according to the instructions of the manufacturer.
14. Store as 1-mL aliquots at –70°C.

Table 2
Assay of the Membrane-Bound RBP Receptor

Addition	Total binding, µL	Nonspecific binding, µL
Membranes	50	50
^{125}I-RBP	25	25
RBP	—	10
Assay buffer	25	15

3.4. Assay of RBP Receptor

3.4.1. Assay of the Membrane-Bound Receptor

The binding of RBP to intact cells or membranes is assayed by a minor modification of the oil centrifugation method described earlier. The method used here is to assay the binding activity of isolated membranes but can be readily adapted for intact cells.

1. Thaw out membranes by incubating in a 37°C water bath for 5 min.
2. Chill on ice and then centrifuge in a microcentrifuge at 4°C at 10,000g for 10 min.
3. Discard the supernatant, resuspend the pellet in ice-cold assay buffer, and centrifuge as in **step 2**.
4. Set up reactions (100 µL) in triplicate in 0.5-mL microtubes as shown in **Table 2**.
5. Mix well and incubate at 37°C for 15 min.
6. Centrifuge at 10,000g for 5 min in a microcentrifuge to pellet the membranes.
7. Overlay with 100 µL of a 3:2 (v/v) mixture of dibutyl phthalate and dinonyl phthalate (with intact cells, use a 2:1 mixture).
8. Centrifuge again for 2 min. The oil (phthalate mixture) should go to the bottom and the aqueous phase rise to the top.
9. Freeze the tubes in dry ice/ethanol and cut off the bottoms of the tubes containing the pellets and count for radioactivity.
10. Subtract the nonspecific from the total binding to obtain specific binding.

3.4.2. Assay of the Detergent-Solubilized Receptor

The assay is a modification of the method described by Sivaprasadarao and Findlay *(2)*.

3.4.2.1. Detergent Solubilization of Placental Membranes

1. Centrifuge 1 mL of placental membranes (5 mg protein/mL) in a microcentrifuge at 10,000g at 4°C for 30 min.
2. Resuspend the pellet in 1 mL of wash buffer and centrifuge again as in **step 1**.
3. Resuspend the pellet in 0.6 mL of wash buffer.

Table 3
Assay of the Detergent-Solubilized RBP Receptor

Addition	Total binding, µL	Nonspecific binding, µL	Background, µL
Soluble extract	50	50	
^{125}I-RBP	25	25	25
RBP	—	10	—
Assay buffer	25	15	75

4. Add 0.4 mL of solubilization buffer. Vortex gently and incubate at room temperature for 15 min with occasional mixing.
5. Spin at 4°C in a Ti 50 rotor in a Beckman ultracentrifuge at 120,000g for 1 h.
6. Collect the supernatant into a fresh tube and use it to assay the receptor.

3.4.2.2. BINDING ASSAY

1. Set up the reactions (100 µL) in triplicate in 0.5-mL microtubes as shown in **Table 3**.
2. Mix well and incubate at 37°C for 15 min.
3. Follow **steps 4–10** in **Subheading 3.2.**

3.5. Purification of RBP Receptor

3.5.1. Coupling of RBP to Reacti-Gel

1. Dissolve RBP in ice-cold coupling buffer at a concentration of 1.0 mg/mL in a Universal tube and store on ice.
2. Filter Reacti-Gel through a sintered-glass filter funnel under gentle suction until the moist cake no longer drips. Do not dry the gel.
3. Weigh 3 g of the wet cake and wash the gel (under gentle suction) with several volumes of ice-cold water to remove acetone.
4. Quickly transfer the gel to the RBP solution and resuspend with a clean glass rod.
5. Mix on a Spiramix overnight at 4°C.
6. Add 10 mL of 2 M ethanolamine solution to stop the coupling reaction.
7. Incubate at room temperature for 2 h.
8. Filter the suspension through a sintered-glass filter funnel under gentle suction (store the filtrate to determine the efficiency of coupling).
9. Wash the gel on the filter with several volumes of ice-cold buffers A and B.
10. Resuspend the gel in buffer B and store at 4°C.

3.5.2. Solubilization of the Receptor

1. Centrifuge 60 mL of placental membranes (5 mg protein/mL) in an Al 30 rotor in a Beckman ultracentrifuge at 100,000g at 4°C for 30 min.
2. Resuspend the pellet in 30 mL of wash buffer and centrifuge again as in **step 1**. Discard the supernatant.

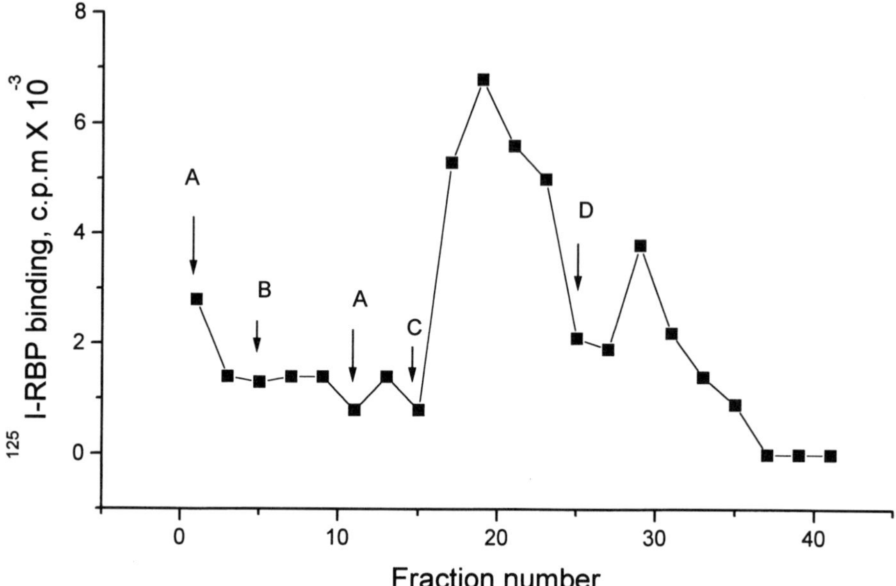

Fig. 1. Affinity chromatography of the RBP receptor on RBP-affinity resin. The RBP-receptor was purified from an octyl-glucoside extract of human placental membranes (120 mg of protein) using an RBP-Gel 6X affinity matrix. Arrows indicate the stages at which various buffers (A–D) were introduced to the resins. The compositions of the buffers are described in the **Subheading 2.**

3. Resuspend the pellet in 14 mL of wash buffer.
4. Add 14 mL of solubilization buffer. Pass the suspension through a 21-gage needle repeatedly (about 20 times) using a 50-mL syringe.
5. Incubate at room temperature for 15 min with occasional mixing.
6. Centrifuge at 4°C in Ti 50 rotor in a Beckman ultracentrifuge at 120,000g for 1 h.
7. Collect the supernatant into a fresh tube and use it to purify the receptor.

3.5.3. RBP-Affinity Chromatography

1. Pack 1 mL of RBP-affinity resin into a small column (~6 mL) and wash with buffer B at 4°C.
2. To 3 mL of RBP-affinity resin taken in a 15-mL Falcon tube, add the supernatant and mix for 15 min at room temperature on a Spiramix.
3. After 15 min, chill the suspension on ice, and tranfer the suspension on to the top of the resin in the column.
4. Wash the column with 10 bed volumes of buffer B at a flow rate of 10 mL/h. Start collecting 1-mL fractions into 1-mL microtubes using a suitable fraction collector.
5. Wash with 3 bed volumes of buffer A.
6. While washing with buffer A is in progress, warm buffers C and D at 37°C.

7. Elute the receptor using 3-bed volumes of the preheated buffer C.
8. Elute any remainder of the bound receptor with 5 vol of preheated buffer D. Collect these fractions into tubes containing 0.5 mL of buffer E.
9. Add 0.1 mL of 10X PBS to fractions eluted with buffer C.
10. Dialyze the fractions eluted with buffer D against buffer A at 4°C. The fractions can be stored at 4°C.
11. Assay all fractions for RBP-binding activity using 50 µL of each fraction in the binding assay (**Subheading 3.4.2.2.**) (*see* **Fig. 1** for a typical profile).
12. Pool fractions containing receptor activity. Dialyze against water, freeze-dry, and analyze about 25% of the freeze-dried protein by SDS-PAGE. A 63-kDa band should be seen (a second band of 55 kDa is sometimes seen).

4. Notes

1. Add PMSF to the solutions just before use from a 0.2 *M* (1000X) stock solution in isopropanol. (All buffers should be stored at 4°C.)
2. The highest specific activity is obtained when 1 mCi of Na^{125}I is used. However, as far as receptor-binding ability is concerned, 0.5 mCi appears to be the optimum amount. Specific activity can also be controlled by changing the number of beads.
3. It is crucial to vortex the mixture as soon as PEG solution is added. Do not leave the precipitation step longer than 25 min to prevent high background and large standard errors.

References

1. Sivaprasadarao, A. and Findlay, J. B. C. (1988) The interaction of retinol-binding protein with its plasma-membrane receptor. *Biochem. J.* **255,** 561–569.
2. Sivaprasadarao, A., Boudjelal, M., and Findlay, J. B. C. (1994) Solubilisation and purification of the retinol-binding protein receptor from human placental membranes. *Biochem. J.* **302,** 245–251.
3. Booth, A. G., Olaniyani, R. O., and Vanderpaye, O. A. (1980) An improved method for the preparation of human placental syncytiotrophoblast microvilli. *Placenta* **1,** 327–336.

13

Detection of Conformational Changes in Cellular Retinoid-Binding Proteins by Limited Proteolysis

Robert S. Jamison, Marcia E. Newcomer, and David E. Ong

1. Introduction

Partial proteolysis of an undenatured protein is a widely used, powerful technique to probe protein conformation in the native state. The basis for this technique is that the more exposed an amino-acid residue is to the solvent, the easier it is for a protease to cleave a peptide bond at that site *(1,2)*. Therefore, regions of a protein with an extended conformation, such as those found in large multidomain proteins, are better substrates for proteolysis than are more tightly folded motifs. Using this technique, it is possible to define protein domains, because the flexible regions between them are more susceptible to proteolysis. It is possible to further define these domains by obtaining N-terminal amino acid sequence of the resulting fragments. Likewise, by monitoring altered susceptibility to proteolysis, changes in protein conformation may be detected. If partial sequence of the resulting proteolytic fragments is obtained, the regions of the protein involved in these conformational changes can be mapped.

Generally, the smaller the protein, the more resistant it is to proteolysis. Changes in protein structure may expose (or hide) residues from attack. A number of factors may induce these structural changes, such as the binding of substrates or cofactors, heat, denaturants, and stabilizing compounds such as DMSO and glycerol. Protease specificity can also affect the rate of proteolysis. Enzymes that recognize sites that include hydrophobic residues are less likely to cleave native structures than proteases that cleave bonds adjacent to charged side chains. The size of both the proteinase and its active site also influence the

From: *Methods in Molecular Biology, Vol. 89: Retinoid Protocols*
Edited by: C. P. F. Redfern © Humana Press Inc., Totowa, NJ

rate of proteolysis. Generally, more compact proteases have a greater degree of access to cleavage sites on a native protein.

The underlying assumption that differential proteolysis can reveal domain structure depends on the protein being in its native state. Great care must be exercised to maintain the native-folded state and avoid introducing artifacts due to denaturation. This is particularly critical when the protease is being inactivated. Most inhibitors do not completely inhibit proteolysis. Strong denaturants such as SDS, urea, or dithiothreitol usually increase the rate of proteolysis, rather than inhibit it. For this reason, protease reactions are commonly stopped by a combination of chemical inhibitors and rapid heat denaturation. Even after boiling however, it is important to ensure that the protease is unable to refold into an active form. A commonly used technique to inactivate proteases is the addition of sodium dodecyl sulfate-polyacrylamide gel electrophoresis (SDS-PAGE) sample loading buffer followed immediately by boiling for at least 2 min.

The retinoid-binding proteins are members of a family of small intracellular proteins which bind retinoids and fatty acids *(3,4)*. The function of these proteins is presumably to protect the ligand from modification by anything other than physiologically relevant enzymes *(5)*. The nine members of this family for which a crystal structure has been reported share the same basic motif, which consists of a 10-stranded up-and-down β-barrel closed at one end by the N-terminal strand and at the other end by a helix-turn-helix motif (*see* **refs. 6 and 7**). The ligand is bound in the interior of the barrel. The binding proteins encapsulate the ligand, such that there are no openings large enough to allow the ligand access to the protein interior. Further, the crystal structures of the unliganded (apo) and the liganded (holo) forms of one of the members of this family, cellular retinoid-binding protein type II (CRBP II), showed no significant conformational difference between the two forms of a magnitude which may allow ligand entry *(8)*. Clearly, some region of the protein must be able to move in order to create a site of ligand entry/exit. Additionally, evidence has suggested that some retinoid-metabolizing enzymes have the ability to discriminate between the apo and holo forms of the binding proteins *(9,10)*.

Comparison of the crystal structures of the apo and holo forms of adipocyte lipid-binding protein (ALBP), another member of the retinoid-binding protein family, has led to the identification of a probable ligand-entry site *(6)*. A portion of the helix-turn-helix motif of the apoprotein appears able to twist away from the barrel. Furthermore, residues at the area of joining between this helix and the barrel appeared relatively mobile, suggesting they may act as a hinge. From this demonstration, one may infer that the helix-turn-helix motif of ALBP is relatively flexible in the apoprotein, and may be able to move away from the barrel to allow ligand access to the binding cavity, thus acting as a movable cap.

We have used partial proteolysis of both the apo (unliganded) and holo (liganded) forms of intracellular retinoid-binding proteins to help identify any regions that undergo conformational change upon ligand binding *(11)*. Identification of these regions combined with the knowledge of the tertiary structure, may indicate likely sites for ligand entry and exit from the binding cavity. Additionally, these regions may be involved in the recognition of the carrier proteins by retinoid-metabolizing enzymes. Because of its structural independence, the α-helical cap of the apoprotein is relatively accessible to proteases. In the holo form this cap makes a number of contacts with the ligand that help to secure it to the barrel, making the α-helical cap less accessible to proteases. Therefore the holo form of the binding protein is less susceptible to proteolysis than the apo form.

We describe methods for generating quantities of binding protein sufficient for proteolysis, screening of proteases to identify those of most use, determination of the relative rates of different binding proteins to proteolysis, identification of the initial site(s) of cleavage, and preparation of proteolytic fragments for partial sequencing. These techniques have also been applied to heart fatty acid-binding protein, and the results suggest that this ligand-induced conformational change is common to all members of this protein family.

2. Materials

2.1. Production and Purification of Retinoid-Binding Proteins

1. LB medium with 0.5% casamino acids: 10 g tryptone, 5 g yeast extract, 10 g NaCl, 5 g casamino acids, water to 1 L. Autoclave 30 min to sterilize. Add 1/1000 vol of antibiotic stocks just before use. For culture plates, add 15 g agar per L of medium.
2. 1000X Antibiotic stocks (Gold Biotechnology):
   a. Ampicillin stock: Dissolve ampicillin in water at a final concentration of 100 mg/mL. Filter sterilize and store in 1-mL aliquots at –20°C.
   b. Chloramphenicol stock: Dissolve 34 mg/mL chloramphenicol in 95% ethanol, store at –20°C.
3. IPTG stock: We prepare a 1-*M* stock of IPTG (Gold Biotechnology) in deionized H_2O. This stock may be stored at 4°C for several months.
4. TEK buffer: 10 m*M* Tris-HCl, pH 8.3, 1 m*M* EDTA, 100 m*M* KCl. Store at 4°C.
5. BL21(DE3)pLysS cells (Stratagene) containing the retinoid-binding protein coding sequences *(12–15)* cloned into the pT7-1 *(16)* or pET (Stratagene) vectors (*see* **Note 1**).
6. A large (approx 1000 mL) G-75 column.
7. A large DEAE column. We use a 90 mL TSK-GEL DEAE-SPW FPLC column (Toso Haas), but any similar column will suffice; however, the elution times may vary somewhat with different column matrices.

8. 330 m*M* Tris-acetate, pH 8.3: Mix 2 vol of 0.5 *M* Tris with 1 vol of 0.5 *M* acetic acid.
9. 0.5 *M* imidazole acetate, pH 6.4 (or 6.6): per 500 mL, mix 250 mL of 1 *M* imidazole (recrystallized from ethyl acetate) with approx 235 mL of 1 *M* acetic acid. Add the acetic acid carefully until the pH reaches 6.4 (or 6.6). Add water to a total volume of 500 mL.

2.2. Comparative Proteolysis of Apo and Holo Forms of the Binding Proteins Using a Battery of Enzymes

1. Proteases. A wide range of enzymes can be tested, because we have found that each binding protein has different levels of cleavage with each enzyme. We have obtained our enzymes from Calbiochem, Promega, and Sigma. Use only preparations of high purity to avoid inconsistent results.
2. Retinoids. Retinoids (Sigma) are suspended in DMSO at a concentration of 2.5 mg/mL. Extreme care must be exercised to avoid isomerization owing to light.
3. Apo-retinoid-binding proteins.
4. Reaction buffers. Each protease has a different pH optimum. Use the buffers recommended by the manufacturer.
5. Reaction termination. The reactions are terminated by the addition of SDS-PAGE sample buffer (10 m*M* Tris-HCl, pH 6.8, 1 m*M* EDTA, 1% DTT, 2.5–5% SDS, 0.01% bromphenol blue, final) followed immediately by boiling. Prepare a 4X stock of SDS-PAGE sample buffer and store in aliquots at –20°C. Quickly boiling the samples is essential to minimize artifactual proteolysis owing to binding-protein denaturation.
6. Rapid Coomassie stain: 0.03% w/v Coomassie brilliant blue R250, 40% v/v methanol, 7% v/v acetic acid. Destain: 40% methanol, 10% acetic acid.

2.3. Time-Course of Proteolysis

1. Materials used are the same as for **Subheading 2.2.**

2.4. Determination of the Initial Cleavage Site

1. Materials used are the same as those in **Subheading 2.2.**

2.5. Sequencing of Proteolytic Fragments

1. Proteolysis. The materials used for proteolysis are identical to those used in **Subheading 2.2.**
2. Polyvinylidine Fluoride (PVDF) membranes. We have had good success with Pro Blott (Applied Biosystems) (*see* **Note 2**).
3. Electrophoretic transfer buffer: 10 m*M* 3-[cyclohexylamino]-1-propanesulfonic acid (CAPS) (Sigma), pH 11.0 (pH with NaOH), 10% methanol.
4. PVDF-membrane stain. Stains used vary with the sequencing system used, check with the chosen sequencing facility (*see* **Note 4**). We have used three different stains: A: 1% Coomassie blue R250 (Sigma) in 100% methanol; B. 0.1% Ponceau S (Sigma), 1% acetic acid; C. 0.1% Coomassie blue R250 (Sigma), 40% methanol, 1% acetic acid.

5. Destain. Either 100% methanol for stain A, 1% acetic acid for stain B, 50% methanol, 1% acetic acid for stain C.

3. Methods

3.1. Production and Purification of Retinoid-Binding Proteins

A very efficient system for production of binding proteins is to express their coding sequences *(12–15)* in the *E. coli* BL21(DE3)p*Lys*S strain. Binding-protein coding sequences must be cloned into the bacterial expression vector pT7-7 *(16)* or equivalent (*see* **Note 1**). Binding-protein expression is induced in culture, and the protein is purified from the cell lysate by a size-exclusion column followed by an ion-exchange column.

1. Prior to induction of binding protein in the cell clones, make a working-stock culture by streaking an LB/ampicillin-chloramphenicol plate from the frozen stock. Incubate the plate overnight at 37°C (if desired, wrap the plate in plastic wrap and store at 4°C for up to a month).
2. Inoculate a 5-mL tube of LB/amp-chlor with an isolated colony from the plate. Grow 4 h to overnight at 37°C with shaking.
3. Put 0.5 mL of the overnight culture into each of two sterile Fernbach flasks (or large baffled flasks) containing 500 mL LB/amp-chlor supplemented with 0.5% casamino acids. Grow cultures at 37°C with rapid shaking until the $A_{600} = 0.6$ (about 3–4 h). Induce protein expression with the addition of 0.5 mM IPTG (final) to each culture. Do the induction overnight at room temperature with rapid shaking (*see* **Note 3**).
4. Following the induction, spin the cells at 5000g for 15 min and discard the supernatant. If desired, the pellet may be stored for extended periods at –70°C. Resuspend the cell pellet in 40 mL of chilled TEK buffer. From now on, all procedures are to be done on ice. Lyse the cells by sonication (5 × 30 s) or by four passes through a French press. Centrifuge the cells 32,000g for 30 min at 4°C.
5. Load the supernatant on a 1000-mL G-75 column (up to 5% column volume). For CRBP I, run the column in 20 mM Tris acetate, pH 8.3. For CRBP II run the column in 8 mM imidazole acetate, pH 6.6, 1 mM 2-mercaptoethanol. For CRABP I and CRABP II, run the column in 20 mM imidazole acetate, pH 6.4, 1 mM 2-mercaptoethanol. Use a flow rate of 1.5 mL/min or less. The protein will elute from the column in approx 70% of the column volume. The protein peak may be determined by A_{280}, fluorescence, or SDS-PAGE. A_{280} is most convenient because the binding protein will be the major protein at 70% of column volume.
6. Load the peak fractions on a DEAE (or equivalent) column, purified binding proteins will be the only major A_{280} peak(s) (*see below*). Use caution during the last steps of purification to avoid contaminating the protein preps with proteins from the hands, such as keratin. Concentration of the fractions before loading is not necessary, but is often done (we use an Amicon YM-3 membrane). We normally use a Toso Haas DEAE-5PW column with a flow rate of 3 mL/min. For

CRBP I, run the DEAE column in a 20–330 mM Tris-acetate, pH 8.3, gradient over 80 min. For CRBP II, run an 80 min gradient from 8 mM imidazole acetate, pH 6.6, 1 mM 2-mercaptoethanol to 50 mM imidazole acetate, pH 6.6, 1 mM 2-mercaptoethanol. For CRABP I and CRABP II, run a stepwise gradient from 20 mM imidazole acetate, pH 6.4, 1 mM 2-mercaptoethanol to 160 mM imidazole acetate, pH 6.4, 1 mM 2-mercaptoethanol. Step 1: 10 min 20–110 mM for 10 min; Step 2: 110–115 mM for 60 min; Step 3: 115–160 mM for 10 min. For CRBP I, the pure protein will elute as three A_{280} peaks owing to differential processing of the N-terminal formyl-methionine (*see* **Note 3**). We have never noticed any difference in the three forms of CRBP I as far as susceptibility to proteolysis, but to avoid any possible inconsistencies in proteolytic assays, we collect only the first (and largest) of the three DEAE peaks, which has the N-terminal methionine removed. This peak will elute at approx 60 min of the gradient. For CRBP II, CRABP I, and CRABP II, only a single major A_{280} peak will be observable at approximately 40–60 min. The pure proteins may be quantitated by dividing the A_{280} of the combined DEAE peak fractions by the molar extinction coefficient of the binding protein multiplied by molecular weight. This will give protein concentration in mg/mL. Molar extinction coefficients of the binding proteins are: 26,700 for CRBP I and CRBP II; 20,940 for CRABP I and CRABP II. Molecular weights of the binding proteins are: 15,700 for CRBP I; 15,500 for CRBP II; 15,600 for CRABP I; and 15,000 for CRABP II. If necessary, concentrate the proteins using centricon-3 concentrators (Amicon). Apo forms of the binding proteins are fairly stable at 4°C, but for long-term storage, aliquot the proteins, quick freeze in dry ice or liquid nitrogen, and store at –70°C.

3.2. Comparative Proteolysis of Apo and Holo Forms of Binding Proteins Using a Battery of Enzymes

It is difficult to predict which proteases will efficiently cleave the retinoid-binding protein. This is partly owing to sequence and structural variations between the members of this family. Another variable is the degree of accessibility of sites to a particular protease. Proteases differ in their sizes and in optimal-reaction conditions, both of which may effect the rate of proteolysis. Larger-sized proteases or those with extended active sites will have greater difficulty attacking some target sites than will smaller enzymes. Certain reaction conditions may cause a change in the accessibility of the target site owing to altered substrate protein conformation. Additionally, the most useful proteases in differential proteolysis are those that show the strongest preference for one form of the substrate protein, in this case, either the apo or the holo form of the binding proteins. Finally, because sequencing of proteolytic fragments can be used to determine the sites of cleavage, it is important to identify proteases that do not rapidly degrade the cleavage products of the binding proteins.

Because of their compactness and resulting resistance to cleavage, it is necessary to use a relatively low binding protein to protease ratio, approx 20:1 w/w. For less compact proteins, the useful substrate to protease ratio is often 2-500:1 w/w. This ratio may vary slightly with both the protease species used, as well as the supplier. The structures of the binding proteins do not seem to be greatly effected by pH, reducing agent, or salt concentration within the optimal conditions for most proteases, so the digestions are done at the supplier's suggested conditions.

1. Generation of holo retinoid-binding protein must be done under safe light. Use a plastic yellow filter over a fluorescent lamp for a light source. Prepare holo forms of the retinoid-binding proteins by adding a 1.2-fold molar excess of ligand to the protein stock. Remove excess ligand by chromatography over a 9-mL sephadex G-25 (medium) column. The progress of the holo-protein over the column may be monitored by its fluorescence during brief illumination with a low-intensity ultraviolet light source. Great care must be exercised to minimize ultraviolet damage to the ligand. Holo-binding protein may be quantitated by the following equation: A280/molar-extinction coefficient $[1 - 0.08R] \times$ mol wt = mg/mL. R equals (A_{350}/A_{280}). If necessary, concentrate the eluted protein using centricon 3 concentrators (Amicon). Holo-binding protein stocks are stable at 4°C for several weeks as long as they are protected from light. For long-term storage, quick freeze and store at –70°C.
2. Preheat a water bath or heating block to 100°C.
3. To two microtubes per reaction add 3 µL of 4X SDS-PAGE sample buffer. Keep the tubes on ice until use.
4. A protease stock is prepared by hydrating the protease in reaction buffer at a 10X concentration, unless otherwise indicated in the supplier's instructions. The hydrated-protease stock is kept on ice until use to avoid self-digestion.
5. Binding protein is added to the reaction solution (optimized for the particular protease being used) at a final concentration of 0.2–0.5 mg/mL. Typical reaction volumes are 20–50 µL. Tubes are kept on ice until after the time-zero sample has been removed.
6. Add 1/10 vol of protease stock and mix by gently flicking the tube. Quickly remove a 9-µL amount for the time-zero sample, add to SDS-PAGE sample buffer, immediately boil for 3 min, and keep the boiled samples on ice.
7. Incubate the remainder of the reaction at 37°C for 60 min. After the incubation, briefly centrifuge the tube, mix, recentrifuge. This minimizes concentration of the protein owing to evaporation. Remove 9 µL and immediately boil in SDS-PAGE sample buffer for 5 min. During this, reboil the time-zero samples for 2 min.
8. Run the samples on a 20% SDS polyacrylamide gel.
9. Stain the gel with Coomassie stain solution for 30 min to overnight (*see* **Note 5**). Destain the gel until the blue background is removed. Coomassie-stained protein

bands are visible when the gel is held over a light box or white paper. The degree of proteolysis may be determined by comparison of the intensity of the 15-kDa binding-protein band in the time-zero and 60-min lanes. Choose the protease that demonstrates the greatest level of differential proteolysis of the apo vs the holo form of each binding protein.

3.3. Time-Course of Proteolysis

Once the most useful, in terms of ability to preferentially cleave either the apo or the holo form of the binding protein, protease(s) has (have) been identified, it is necessary to perform a time course of proteolysis. This aids in both the determination of the order of evolution of cleavage products and in the optimal reaction time for production of the desired proteolytic fragment. The time-course also provides additional information on the relative rates of proteolysis of the apo and holo forms of each of the binding proteins.

1. Prepare holo forms of the binding proteins as described in **Subheading 3.2., step 1**.
2. Heat water to boiling and prepare a series of microtubes each with 3 µL SDS-PAGE sample buffer. Store tubes on ice until needed.
3. Hydrate the protease in reaction buffer and place on ice.
4. Prepare reaction solutions with all reagents except protease.
5. Add protease, mix briefly, remove 9 µL, add this to the time-zero tube, and immediately boil for 3 min.
6. Incubate reaction tubes at 37°C, remove 9-µL aliquots at predetermined time-points, and process as for time zero. Reasonable initial time-points are 15 min, 30 min, 1, 2, 4, 8 h, and overnight (approx 15 h). Before the removal of each time-point sample, briefly centrifuge the reaction tube, mix, and recentrifuge. This minimizes concentration of the protein owing to evaporation. After boiling each sample for 3 min, cool rapidly on ice and store at 20°C until all time points have been taken.
7. Analyze the level of proteolysis in the samples by electrophoresis on a 20% SDS polyacrylamide gel, followed by Coomassie staining as described earlier. Reboil the frozen samples for 2 min before loading samples on the gel.

3.4. Determination of the Initial Cleavage Site

The region of the substrate protein most accessible to the protease active site will be the location of the first site of cleavage. This is helpful in probing binding-protein structure because it identifies extended or accessible conformations. Often after the first cleavage, other sites are made accessible to the protease. Because these are not necessarily representative of extended conformations in the native structure, it is important to differentiate between the cleavage of the native protein and cleavage of proteolytic fragments, as access to these fragments by the proteases is no longer in the context of the native structure. To do this, it is important to determine the site of the initial cleavage of the binding protein by the protease. Determination of the initial site of proteolysis involves very brief reaction times to avoid secondary cleavage of the proteolytic fragments. Additionally, it is advantageous to use relatively high concentrations of binding protein to be able to detect any unstable fragments, which will be present only in low amounts.

The proteolysis is done as in the time-course of proteolysis, except that 0.5–1 mg/mL of binding protein is used, and reaction times are shortened considerably. Depending on the protease used, we have used reaction times as short as 1 min. The more rapid the rate of cleavage, the shorter the reaction time. Great care must be taken when using such high-protein concentrations to avoid artifacts owing to the addition of denaturants to stop the reaction. Use the shortest reaction time that gives detectable levels of proteolysis above the time 0 control.

1. Set up a time-course of proteolysis reaction as in **Subheading 3.3.**
2. Take out 9-μL aliquots of the reaction mixture, add to SDS-PAGE sample buffer and boil as in **Subheading 3.3.** Recommended initial times are 0, 5, 10, 15, 30, 45, and 60 min. These may be varied depending on the protease and substrate protein used.
3. Separate the proteolytic fragments by electrophoresis on a 20% SDS polyacrylamide gel and Coomassie stain. Note the rate of production of each fragment. Those appearing at the shortest reaction times are most likely to represent the initial site(s) of cleavage.

3.5. N-terminal Amino Acid Sequences of the Proteolytic Fragments

It is possible to approximate the sites of cleavage based on the sizes of the proteolytic fragments. However the exact locations of cleavage can best be determined by N-terminal sequencing of the proteolytic fragments. The best sequence is obtained using a protease that cleaves bonds adjacent to only one type of amino acid residue (e.g., arg-C). Proteolytic fragments of a defined molecular weight generated by proteases with a narrow range of substrate specificity will be likely to all have identical N-termini. Less-specific proteases (e.g., papain) will often generate a population of proteolytic fragments of similar molecular weight, each with a different N-terminal residue, from a single protease-sensitive region.

Here is described a simple technique of electrophoretic separation of proteolytic fragments, their transfer to PVDF membrane, staining of the membrane, and excision of protein bands for sequencing.

1. One day before proteolyzing the binding protein, pour a 20% SDS polyacrylamide gel. Drape a damp paper towel on top of the comb and wrap the gel with plastic wrap. Store the gel at 4°C. It is necessary to pour the gel in advance to reduce the amount of oxidizing agents that could interfere with the sequencing process. The paper towel helps keep the gel from drying out.

2. Using the results from the time-course of proteolysis experiment, **Subheading 3.3.**, determine the optimal reaction time in order to obtain the maximal amount of the desired fragment.

3. Some fragments are produced in only small amounts. In these cases, it may be useful to try increasing the concentration of substrate protein in the reaction in order to obtain sufficient quantities for sequencing.

4. Perform and terminate the reaction as described in **Subheading 3.3.**

5. Separate the proteolytic fragments by SDS-PAGE.

6. For electrophoretic transfer of proteolytic fragments from the gel to PVDF membrane use a tank electrophoretic-transfer unit. These are available from various manufacturers including Hoefer and Bio-Rad. Great care must be taken to avoid contamination of the transferred protein with other proteins. Wear gloves at all times, wash all equipment thoroughly, and use disposable materials whenever possible. Soak the gel for a few minutes in CAPS transfer buffer. Prewet the PVDF membrane, cut to a size slightly larger than the gel, first in 100% methanol, followed by CAPS transfer buffer. Cut two pieces of Whatman 3MM paper to a size slightly larger than the gel and prewet in transfer buffer. On the foam sponge of a gel cassette, stack one piece of Whatman paper, the polyacrylamide gel, the PVDF membrane, and the other piece of Whatman paper. Remove all bubbles between each layer by carefully rolling a disposable pipet over the top surface. Close the gel cassette and place it in the tank of the transfer unit with the PVDF membrane between the gel and the positive electrode. Fill the tank with CAPS buffer and transfer the proteins at 1 amp/L of CAPS buffer for approx 60 min. The actual rate of transfer will alter somewhat depending on the molecular weight of the fragment. The smaller the protein, the more rapid the rate of transfer.

7. After the transfer is complete, remove the PVDF membrane and stain with either Coomassie or Ponceau S for 1 h to overnight. The stain used depends on the sequencing method used; check with the chosen sequencing facility (*see* **Note 4**). Destain carefully until all fragments are visible. Use care not to overdestain, resulting in the loss of visible bands. If this happens, restain the blot.

8. Using a clean disposable scalpel, cut out the desired protein band, taking care to minimize the area of the PVDF slice. Wash the slice six times in a clean Eppendorf tube with deionized water. Vortex the tube for the final wash. Leave a little water in the tube after the final wash to keep the membrane wet. Parafilm-seal the tube and store on dry ice.

9. Sequencing facilities. There are a number of commercial facilities that offer protein sequencing services (*see* **Note 4**). Be sure to obtain their recommended protocol for sample preparation before submitting a fragment for analysis.

4. Notes

1. Both plasmids are very similar, but pET is commercially available. The cloning of the coding sequences into these vectors requires the introduction of an *Nde*I site 5' to the translation start codon. This may be done using standard PCR reactions with the cDNA *(12–15)* as template. The 5' primers all have the same basic

motif: 5'-CCC<u>CATATG</u>-[18–20 bases of coding sequence]-3' (*see* **Note 4**). The *Nde*I sequence is underlined, and the translational start codon is in bold.

2. Other PVDF membranes with a relatively low-binding strength (such as Bio-Rad's) may also be used. Immobilon (Millipore) should be avoided because it binds transferred proteins too tightly. This tight binding, although excellent for Western blotting, interferes with protein elution from the membrane during the sequencing process.

3. Induction of protein expression in *E. coli* is normally done for 3 h at 37°C. We have found that in the case of the binding proteins, this rate of induction is too rapid and much of the induced protein (up to 50%) is sequestered into inclusion bodies, making purification difficult. In addition, most of the CRBP I produced during the 37°C induction does not have the N-terminal formyl-methionine removed, which changes its rate of migration on native gels and elution on ion-exchange columns. Inducing protein expression overnight at room temperature greatly reduces these problems. Additionally, the total amount of protein produced per cell is increased overnight vs the 37°C incubation. Finally, we have found that maximal aeration of the cells during induction is essential to obtain high levels of protein expression. Use minimal volumes per flask and shake the cultures as rapidly as possible.

4. Oligonucleotide synthesis and DNA sequencing facilities are commonly advertised in many scientific journals such as *BioTechniques, Nature,* and *Science.* Prices and quality vary widely; check carefully before choosing a particular company. Protein sequencing facilities are found at most universities and research centers. Protein sequencing of excellent quality is performed by Harvard Microchem (Cambridge, MA).

5. The binding proteins generally do not stain well, and the proteolytic fragments stain even less well. We have found that it may be necessary to stain overnight in order to visualize all of the fragments. It may even be useful to silverstain the gels to visualize very faint fragments, but we have not explored this.

Acknowledgments

We would like to thank Bharati Kakkad for technical assistance in the purification of the retinoid-binding proteins.

References

1. Price, N. C. and Johnson, C. M. (1989) Proteinases as probes of conformation of soluble proteins, in *Proteolytic Enzymes: A Practical Approach* (Beynon, R. J. and Bond J. S., eds.), Oxford, pp. 163–179.

2. Mihalyi, E. (1978) Proteolytic enzymes, enzymatic proteolysis—general considerations, in *Application of Proteolytic Enzymes to Protein Structure Studies,* 2nd ed., **1,** 43–149.

3. Sundelin, J., Das, S. R., Eriksson, U., Rask, L., and Peterson, P. A. (1985) The primary structure of bovine cellular retinoic acid-binding protein. *J. Biol. Chem.* **260,** 6494–6499.

4. Matarese, V., Stone, R. L., Waggoner, D. W., and Bernlohr, D. A. (1990) Intracellular fatty acid trafficking and the role of cytosoloc lipid-binding proteins. *Prog. Lipid Res.* **28,** 245–272.

5. Ong, D. E., Newcomer, M. E., and Chytil, F. (1994) Cellular retinoid-binding proteins, in *The Retinoids: Biology, Chemistry and Medicine,* 2nd ed. (Spore, M. B., Roberts, A. B., and Goodman, D.S., eds.), Raven, New York, pp. 283–316.

6. Xu, Z., Bernlohr, D. A., and Banaszak, L. J. (1993) The adipocyte lipid-binding protein at 1.6-Å resolution. *J. Biol. Chem.* **268,** 7874–7884.

7. Cowan, S. W., Newcomer, M. E., and Jones, T. A. (1993) Crystallographic studies on a family of cellular lipophilic transport proteins. *J. Mol. Bio.* **230,** 1225–1246.

8. Winter, N., Bratt, J., and Banaszak, L. J. (1993) Crystal structures of holo and apo-cellular retinol-binding protein (II). *J. Mol Biol.* **230,** 1247–1259.

9. Herr, F. M. and Ong, D. E. (1992) Differential interaction of lecithin-retinol acyl transferase with cellular retinol-binding proteins. *Biochemistry* **31,** 6748–6755.

10. El Akawi, Z. and Napoli, J. L. (1994) Rat liver cytosolic retinal dehydrogenase: comparison of 13-cis-, 9-cis-, and all-trans-retinal as substrates and effects of cellular retinoid-binding proteins and retinoic acid on activity. *Biochemistry* **33,** 1938–1943.

11. Jamison, R. S., Newcomer, M. E., and Ong, D. E. (1994) Cellular retinoid-binding proteins: limited proteolysis reveals a conformational change upon ligand binding. *Biochemistry* **33,** 2873–2879.

12. Li, E., Demmer, L. A., Sweetser, D. A., Ong, D. E., and Gordon, J. I. (1986) Rat cellular retinol-binding protein (II): use of a cloned cDNA to define its primary structure, tissue-specific expression, and developmental regulation. *Proc. Natl. Acad. Sci. USA* **83,** 5779–5783.

13. Sherman, D. R., Lloyd, S. R., and Chytil, F. (1987) Rat cellular retinol-binding protein: cDNA sequence and rapid retinol-dependent accumulation of mRNA. *Proc. Nat. Acad. Sci. USA* **84,** 3209–3213.

14. Shubeita, H. E., Sambrook J. F., and McCormick, A. M. (1987) Molecular cloning and analysis of functional cDNA and genomic clones encoding bovine cellular retinoic acid-binding protein. *Proc. Nat. Acad. Sci. USA* **84,** 5645–5649.

15. Giguere, V., Lyn, S., Yip, P., Siu, C. H., and Amin, S. (1990) Molecular cloning of a cDNA encoding a second cellular retinoic acid-binding protein. *Proc. Natl. Acad. Sci. USA* **87,** 6233–6237.

16. Tabor, S. and Richardson, C. C. (1985) A bacteriophage T7 RNA polymerase/promoter system for controlled exclusive expression of specific genes. *Proc. Natl. Acad. Sci. USA* **82,** 1074–1078.

14

Measurement of Rates of Dissociation of Retinoids from the Interphotoreceptor Retinoid-Binding Protein

Noa Noy

1. Introduction

Information regarding rates of dissociation of ligands from binding proteins is helpful for obtaining insights into the forces that stabilize protein–ligand interactions, as well as for understanding how the ligands distribute between the different cellular compartments, which are their sites of action. Such information is especially useful when the proteins being studied are involved in mediating transport or delivery of ligands to particular cellular locations. The complete scope of the functions of retinoid-binding proteins—either cellular proteins such as the cellular retinol-binding proteins (CRBPs), cellular retinoic acid-binding proteins (CRABPs), and cellular retinal-binding protein (CRALBP); or extracellular proteins like retinol-binding protein (RBP) and interphotoreceptor retinoid-binding protein (IRBP)—is not completely understood at present. However, all of these proteins have been implicated in participating in transport of their hydrophobic ligands across aqueous spaces. Knowledge of the rates by which they release retinoids and of the factors that influence these rates is thus important for understanding how the activities of the proteins as retinoid-carriers are regulated.

The protocols for measurements of the rates of dissociation of retinoid–IRBP complexes detailed below are applicable to studies of other retinoid-binding proteins. Similar procedures have been used to measure the rate constants of the dissociation of complexes of retinoids with RBP *(1)*, CRBP *(2)*, IRBP *(3,4)*, and the retinoid X receptor *(5)*. The method is based on the usage of unilamellar vesicles of phospholipids to induce dissociation of

From: *Methods in Molecular Biology, Vol. 89: Retinoid Protocols*
Edited by: C. P. F. Redfern © Humana Press Inc., Totowa, NJ

retinoids from a binding protein. Introduction of vesicles is necessary because retinoids are poorly soluble in water and will not leave protein-binding sites at measurable amounts unless another phase with high affinity for them is present. Phospholipid vesicles have a twofold advantage as such a phase: (1) Retinoids readily dissolve in the hydrophobic milieu of lipid bilayers, and mixing retinoid-protein complexes with vesicles results in transfer of the ligand from the protein to the vesicles; (2) appropriate fluorescent lipid probes can be readily incorporated into vesicles and serve as a sensitive readout for the arrival of the ligands at the bilayers. By appropriate choices of probes, movement of a variety of retinoids can be followed.

IRBP is a large glycoprotein that is the major protein component of the interphotoreceptor matrix, an aqueous space separating the pigment epithelium from photoreceptor cells in the eye *(6)*. IRBP binds several chemical derivatives and isomeric forms of retinoids and is also known to associate with long-chain fatty acids *(7,8,16)*. It has been suggested that this protein plays an important role in the visual cycle by serving as a vehicle for shuttling retinoids between photoreceptor and pigment epithelium cells *(6,9)*. IRBP possesses two binding sites for retinoids *(7,10,11)*. Binding of retinol in one of these sites results in a significant enhancement of the fluorescence of the ligand, indicating that the site consists of a hydrophobic/restrictive environment. The other retinoid-binding site of IRBP does not seem to provide a hydrophobic environment for binding the isoprenoid moiety of retinoids.

In the experimental protocols detailed in **Subheading 3.4.**, vesicles containing fluorescent probes that are sensitive to the presence of either retinols or retinals are used. The vesicles are mixed with IRBP and the rate of transfer of the retinoid from the protein to the vesicles is followed by monitoring time-dependent changes in fluorescence. It has been shown that, provided that the concentration of the vesicles is sufficiently high, the rate constant of transfer of retinoids from protein to vesicles is independent of the concentrations of either the protein or the vesicles, indicating that transfer does not occur by direct collisions between the two phases. Instead, retinoids move into the vesicles following solvation from the protein and diffusion through the aqueous phase. Because the dissociation of the ligand from the protein is the rate-limiting step of the overall process, the observed rate of transfer reflects the rate of dissociation of retinoids from IRBP *(1–3)*, and the first order rate constant characterizing the dissociation of a retinoid-IRBP complex, k_{diss}, can be directly extracted from the data.

2. Materials

2.1. Retinoids

Solutions of retinoids are prepared in ethanol. Dissolve several grains of the appropriate retinoid in 0.5–2 mL of ethanol in an amber vessel. Determine the concentration of each solution using the extinction coefficients of the ligands: all-*trans*-retinol, $\varepsilon_{325\ nm}$ = 52,770 $M^{-1}cm^{-1}$; 11-*cis*-retinol, $\varepsilon_{319\ nm}$ = 34,890 $M^{-1}cm^{-1}$; all-*trans*-retinal, ε_{383} nm = 42,880 $M^{-1}cm^{-1}$; 11-*cis*-retinal, $\varepsilon_{380\ nm}$ = 24,935 $M^{-1}cm^{-1}$. *(12)*. Keep the retinoid solutions in the dark and on ice while in use. They can be stored at –20°C for up to 1 wk.

2.2. IRBP

2.2.1. Protein Purification and Storage

1. Buffer A: 20 m*M* HEPES, pH 7.0, 150 m*M* NaCl, 0.1 m*M* DTT.
2. Isolate bovine IRBP from frozen bovine retina by a procedure that utilizes 200 retinas to yield 50–80 mg of pure protein (**ref. *13***, *see* **Note 1**). Following elution from the last column, dialyze IRBP against buffer A. Determine the protein concentration from its molar-extinction coefficient $\varepsilon_{280\ nm}$ = 135,432 $M^{-1}cm^{-1}$ *(10)*.
3. Concentrate pure IRBP by ultrafiltration (Amicon, YM-10 membranes) to 10–15 µ*M* (*see* **Note 2**). Add an equal volume of glycerol and store the protein at –20°C (*see* **Note 3**). Prior to use, dialyze the required amount of protein against buffer A.

2.2.2. Protein Delipidation

Delipidate IRBP by hydrophobic chromatography (*see* **Note 4**). Pack 70 mL Lipidex-5000 (Packard Instruments Co.) into a water-jacketed column and equilibrate with buffer A at 37°C. Circulate up to 25 mL IRBP solution through the column using a peristaltic pump for 30 min. The resulting delipidated protein is either stored at 4°C and used within a day, or placed back into a solution containing 50% glycerol and stored at –20°C for up to 3 wk.

2.3. Phospholipid Vesicles Containing Fluorescent Lipid Probes

1. Phospholipids: purchase dioleoyl phosphatidylcholine (DOPC) or egg-yolk phosphatidylcholine (egg PC) (*see* **Note 5**).
2. Fluorescent probes: for experiments with retinols: *N*-(7-nitrobenz-2-oxa-1,3-diazol-4-yl)-1,2-dihexadecanoyl-*sn*-glycero-3-phosphoethanolamine (NBD-DPPE). For experiments with retinals: 1-hexadecanoyl-2-(1-pyrendecanoyl)-*sn*-glycero-3-phosphocholine (PY-PC) 6. Fluorescent probes can be purchased from Molecular Probes (Eugene, OR).
3. Phosphorus standard: 400 µ*M* potassium phosphate.
4. Ammonium molybdate solution: Dissolve 4.4. g ammonium molybdate in 250 mL distilled water. Add 14 mL of concentrated H_2SO_4 followed by water to a final volume of 1 L.
5. Fiske-Subbarow reducing agent: Grind 30 g of sodium bisulfite, 6 g of sodium sulfite, and 0.5 g 1,2,4-aminonaphthol sulfonic acid together using a mortar and pestle until thoroughly mixed. Dissolve the mixture in distilled water in a final volume of 250 mL. Incubate the solution in the dark for 3 h and filter into an amber bottle. Dilute the reducing agent 12-fold prior to use.

6. Chloroform.
7. Argon or nitrogen gas supply.
8. To obtain vesicle suspensions containing a high-lipid concentration, a high-output sonicator is needed. For example, a Heat-System-Ultrasonics sonicator (Farmingdale, New York). The sonicator can be used either in the probe or in the cup-horn mode.
9. Perchloric acid.
10. Washed glass beads, glass test tubes, and glass marble "stoppers."

2.4. Instrumentation

As the rates of dissociation of retinoid-IRBP complexes are of the order of seconds, a rapid-mixing (stopped-flow) apparatus operating in conjunction with a fluorometer is required (*see* **Note 6**). A rapid-mixing device consists of two syringes that are filled with solutions containing the reactants (in this case, holo-IRBP and the probe-containing vesicle suspension). The plungers of the syringes are pressed simultaneously and rapidly, usually by applying pressure from an air tank. The solutions are pushed into connecting tubing where they mix, and the mixed solution is deposited into a cuvet. The apparatus is connected with the fluorometer via a trigger that initiates data collection upon mixing. The "dead-time" (i.e., the mixing time) of the instrument to be used should be short enough to allow measurements of reactions with $t_{1/2}$ of the order of 0.2 s. The procedure detailed in **Subheading 3.4.** assumes that equal volumes of the two solutions are used. Other mixing ratios can be used provided that the final concentrations of reactants are kept similar to those given below.

3. Methods
3.1. Preparation of Vesicles Containing Fluorescent Probes

Unilamellar vesicles are prepared by cosonication of the phospholipids and the fluorescent probes.

1. Dissolve the appropriate fluorescent probe (either NBD-DPPE or PY-PC) in chloroform. Probes are purchased in 1–10 mg-sizes and the entire content of the bottle is dissolved in 0.5–1 mL of chloroform.
2. Mix phospholipids with the fluorescent probe at a mole ratio of 50-100:1 (lipids:probe) in a vessel resistant to chloroform and suitable for sonication. To make 1 mL of a vesicle suspension, use 5–15 μmol of lipids.
3. Evaporate organic solvents under a stream of argon or nitrogen, and place the lipids under vacuum for 1–2 h to eliminate traces of solvents.
4. Add 5–20 mL of buffer A, vortex the mixture, and sonicate the lipid suspension using a high-output sonicator. During sonication, immerse the sample in a water–ice bath and purge continuously with a gentle stream of argon. To avoid over-

heating, carry out the sonication using 30-s bursts at 20-s intervals. Sonicate until the sample is clear and has a transluscent rather than opaque appearance (10–45 min depending on sample concentration and sonication energy).

5. Centrifuge the vesicle suspension at 100,000*g* for 10 min and discard the pellet.

3.2. Determination of Vesicle Concentrations

Determine the lipid concentration of the vesicle suspension by assaying the phosphorus content *(14)*.

1. Prepare phosphate standards in the range of 0–160 nmol phosphate using 0–400 µL of the phosphate standard solution which should be pipetted into glass test tubes.
2. Prepare samples of the vesicle suspension (1–10 µL) in duplicates.
3. Add 400 µL of perchloric acid and a washed glass bead to each of the tubes.
4. Stopper the tubes with glass marbles, place on a heating block, and heat at 190°C until completely clear (1–4 h). Only the bottom of the tubes should be in contact with the heating block to allow for reflux of the solution and to avoid splattering.
5. When cool, add 2.4 mL each of the ammonium molybdate solution and the diluted reducing agent, and vortex the solutions immediately.
6. Heat the test tubes, stoppered with glass marbles, in a boiling-water bath for 10 min.
7. When cool, measure the absorbances of the samples at 830 nm. If absorbances are too high, carry out the measurements at 735 nm.
8. Calculate the phosphorus content of the vesicle suspension by comparing the absorbances of the samples to those of the known standards using the standard curve.

3.3. Verification of Protein Viability

IRBP is susceptible to loss of retinoid-binding activity during storage. Most notably, the hydrophobic-binding site of the protein becomes significantly weakened, whereas binding of retinoids at the second site tends to be completely lost. Consequently, it is important to verify that the protein to be used in an experiment is viable, i.e., possesses high retinoid-binding affinity at both sites. This is readily done by fluorescence titrations of IRBP. Usage of fluorescence titrations for measurements of binding of retinoids to proteins is detailed in two separate chapters in this volume. It should be noted that, in order to ascertain the viability of IRBP, measurements of the characteristics of binding of retinol in the protein hydrophobic site (**Subheading 3.3.1.**), as well as in both binding sites of the protein (**Subheading 3.3.2.**), need to be carried out (*see* **Note 7**).

3.3.1. Binding of Retinol to the Hydrophobic Retinoid-Binding Site of IRBP

The fluorescence of retinol is significantly enhanced upon binding to one of the retinoid-binding sites of IRBP, and this enhancement can be used to monitor the association of this ligand with this site.

1. Place IRBP (0.5–2 μM in buffer A) in a fluorescence cuvet in the sample compartment of a fluorometer. Measure the fluorescence of the sample with excitation and emission monochromators set at 330 and 480 nm, respectively (*see* **Note 8**).
2. Titrate IRBP with all-*trans*-retinol, which should be added from a concentrated solution in ethanol. Usually, 1-μL steps corresponding to a final concentration of retinol of 0.05–0.2 μM are used. Measure the fluorescence of the sample after each addition and continue the titration until saturation is reached (*see* **Note 9**).
3. Correct titration curves for the contribution of unbound retinol *(15)* and analyze either by a linearization method *(15)* or by curve-fitting to an equation derived from binding theory (**Eq. 1**; *see* **Note 10**).

$$F = F_0 - \frac{1 + \dfrac{(P+R)}{K_d} - \sqrt{\dfrac{(P-R)^2}{K_d^2} + 2\dfrac{(P+R)}{K_d} + 1}}{\dfrac{2P}{K_d}} (F_0 - F_{sat}) \tag{1}$$

In **Eq. 1**, P and R are the total concentrations of the protein and the retinoid, respectively; F_0 is the fluorescence in the absence of ligand. F is the fluorescence value following each titration step, and F_{sat} is the value of the fluorescence at saturation.

Data analyses yield the number of retinol-binding sites per mole of protein (in our hands, 0.55–0.95 in different protein preparations), and the equilibrium dissociation constant (K_d) of the retinol-IRBP complex (typically, 40–100 nM).

3.3.2. Binding of Retinol at Both Retinoid-Binding Sites of IRBP

Unlike the hydrophobic site, binding of retinol to the second retinoid-binding site of IRBP is not accompanied by enhancement of the fluorescence of the ligand. However, binding to both sites results in quenching of the intrinsic fluorescence of the protein. Changes in the intrinsic fluorescence of IRBP upon titration with retinol are followed in order to monitor binding of retinol to both sites. The procedure is the same as described in **Subheading 3.3.1.**, except that excitation and emission wavelengths are set at 280 and 340 nm, respectively. Fluorescence signals are lower than those observed when following the fluorescence of retinol because tryptophanes are less efficient fluorophores as compared with retinol. Hence, the widths of the fluorometer's slits should be adjusted to obtain a high enough signal for these measurements. Upon titration of the protein with all-*trans*-retinol, the fluorescence values decrease until saturation is reached (*see* **Note 11**). Since the retinol binding affinities of the two IRBP sites differ by less than an order of magnitude *(4)*, titration curves can be analyzed either by the linearization method *(15)* or by standard curve-fitting to yield the total number of binding sites and the average K_d for the two sites. For a viable protein, the number of binding sites is twice that found by monitoring the fluorescence of retinol (**Subheading 3.3.1.**). Average K_d for the two sites is in the 100–250 n*M* range.

3.4. Rates of Dissociation of Retinoids from IRBP

Rates of dissociation of retinoids from IRBP are measured by monitoring the process of transfer of the ligands from the protein to unilamellar vesicles of phospholipids. The vesicles contain fluorescent lipid probes that serve as readouts for the arrival of retinoids at the bilayers. The data yield information on rates of dissociation of retinoids from both retinoid-binding sites of IRBP.

3.4.1. Rates of Dissociation of Retinol-IRBP Complexes

To follow the dissociation of a retinol (e.g., all-*trans*- or 11-*cis*-retinol) from IRBP, vesicles containing the fluorescent probe NBD-DPPE are used. The absorption spectrum of NBD extensively overlaps with the fluorescence emission spectrum of retinol. Consequently, when both retinol and the probe are present within the same bilayer, energy transfer between the ligand and the NBD moiety of the probe results in quenching of retinol fluorescence. Transfer of retinol from the protein to the vesicles can thus be monitored by following the time-dependent decrease in the fluorescence of the ligand.

1. Set the excitation and emission wavelengths of the fluorometer at 330 and 480 nm, respectively.
2. Place a solution containing 1–3 µ*M* of retinol-IRBP complex in buffer A in one of the stopped-flow syringes of the rapid-mixing apparatus (*see* **Subheading 2.4.**). Pre-mix retinol with IRBP from a concentrated solution in ethanol. The retinol/IRBP mole ratio can be varied in the range of 0.5–2. At low ratios, a larger fraction of the observed signal reflects dissociation from the stronger retinoid-binding site of IRBP. At higher ratios, dissociation from both binding sites are observed (*see* **Note 12**).
3. Place the vesicle suspension containing 1–2 m*M* lipids in buffer A in the second stopped-flow syringe.
3. Mix the two solutions and follow the fluorescence until equilibrium is reached (30–60 s). Typical integration times are 0.01–0.05 s/point and special care should be taken to collect sufficient data points at the initial 0- to 5-s interval. A typical trace is shown in **Fig. 1**.

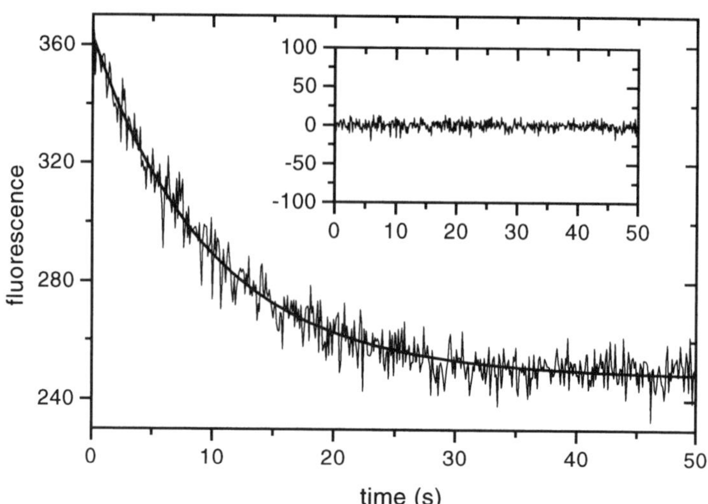

Fig. 1. Transfer of all-*trans*-retinol from IRBP to unilamellar vesicles of DOPC. Vesicles containing 2 mol% NBD-DPPE (2 m*M*) were mixed with all-*trans*-retinol-IRBP complex (1 µ*M*, ligand/protein mol ratio = 2). Transfer of all-*trans*-retinol from IRBP to vesicles was followed by the time-dependent quenching of all-*trans*-retinol fluorescence upon its arrival at the vesicles (excitation, 350 nm; emission, 480 nm). The insets shows the residuals of the fit of the trace to a single first-order reaction.

3.4.2. Rates of Dissociation of Retinal-IRBP Complexes

To monitor dissociation of retinals from IRBP, use vesicles containing the fluorescent lipid probe PY-PC. The fluorescence emission spectrum of PY-PC overlaps with the absorption spectra of retinals and the presence of retinal in close proximity to the probe results in quenching of the probe fluorescence. Consequently, movement of retinal from IRBP to vesicles can be followed by monitoring the time-dependent decrease in PY-PC fluorescence following mixing a retinal-IRBP complex with vesicles.

1. Set excitation and emission wavelengths at 330 and 400 nm, respectively.
2. Fill the stopped-flow syringes with the appropriate solutions: A. 1–3 µ*M* retinal-IRBP complex at the desired retinal/IRBP mole ratio. B. Suspension of PY-PC-containing vesicles (1–2 m*M* lipids).
3. Mix the solutions and follow the fluorescence to equilibrium. A typical trace depicting the time-dependent decrease in PY-PC fluorescence following mixing of IRBP complexed with retinal with probe-containing vesicles is shown in **Fig. 2**.

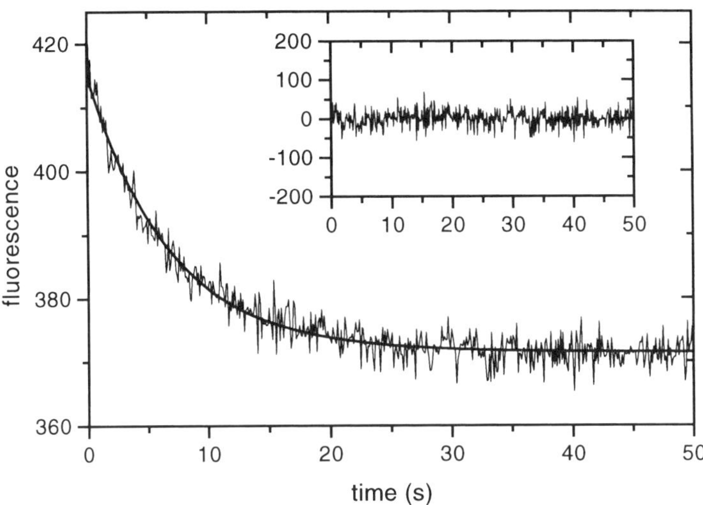

Fig. 2. Transfer of 11-*cis*-retinal from IRBP to unilamellar vesicles of DOPC. Vesicles containing 2 mol% PY-PC (2 m*M*) were mixed with 11-*cis*-retinal-IRBP complex (1 µ*M*, ligand/protein mole ratio = 2). Transfer of the ligand from IRBP to vesicles was followed by the time-dependent decrease in the fluorescence of the probe (excitation, 330 nm; emission, 400 nm). The inset shows the residuals of the fit of the trace to a single first-order reaction.

3.4.3. Calculation of Rate Constants of the Dissociation of Retinoid-IRBP Complexes

Rates of dissociation of retinoids from either of the retinoid-binding sites of IRBP are characterized by a first-order rate constant, k_{diss}, which can be directly extracted from the time-dependent changes in fluorescence obtained as described in **Subheadings 3.4.1.** and **3.4.2.** It should be kept in mind, however, that the data derived from the aforementioned experiments report on rates of dissociation of the ligands from both retinoid-binding sites of the protein. When the rates of dissociation of retinoids from the two sites of IRBP are similar, the data can be fitted to a single first order equation. As can be seen by the good fits of the traces depicted in **Figs. 1** and **2**, the rates of dissociation of retinoids from the two sites of delipidated IRBP are similar and display a $t_{1/2}$ on the order of 5 s.

However, under some conditions, the rates of dissociation of retinoids from the two IRBP sites may be quite different. The equilibrium-dissociation constant that governs the affinity of a protein for a ligand can be expressed as the ratio of the rate constants for dissociation and for association of the complex,

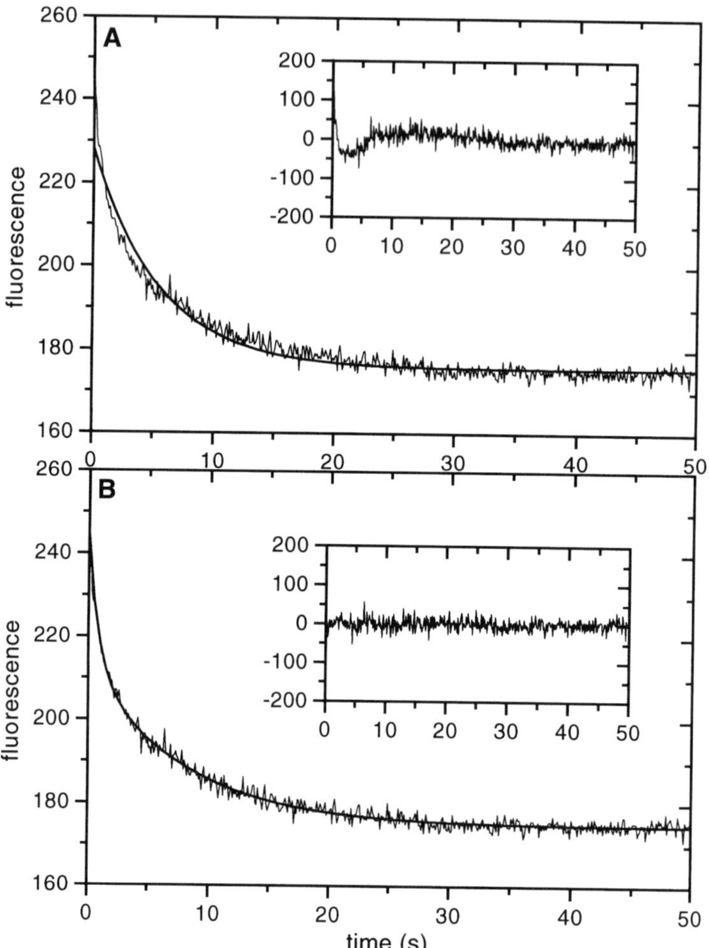

Fig. 3. Transfer of 11-*cis*-retinal from IRBP to unilamellar vesicles of DOPC in the presence of docosahexaenoic acid. Experiments were carried out as described in the Legend to **Fig. 2** in the presence of 10 μ*M* docosahexaenoic acid. Insets show the residuals corresponding to the fit of the traces to a single first-order reaction (**A**), or to two first order reactions (**B**).

$K_d = k_{diss}/k_{assoc}$. Hence, factors that alter the affinity of either of the binding sites of IRBP for a retinoid often also modify k_{diss} for that site. When the rates of dissociation of a retinoid from the two retinoid-binding sites of IRBP differ significantly, the data cannot be fit as a single first-order process. Instead, fits are carried out according to a model including two independent first-order reactions. An example is shown in **Fig. 3**, which depicts data on the dissociation of 11-*cis*-retinal from IRBP in the presence of the polyunsaturated fatty acid docosahexaenoic acid. This fatty acid specifically inhibits binding of 11-*cis*-retinal to one of the retinoid-binding sites of IRBP, an effect that stems, at least partially, from facilitation of the dissociation of this ligand from the site *(16)*. Thus, in the presence of docosahexaenoic acid, the rate constants characterizing the dissociation of 11-*cis*-retinal from the two IRBP binding sites are considerably different, and the overall process cannot be fitted as a single first-order reaction (**Fig. 3A**). A good fit can nevertheless be obtained by assuming that the overall reaction is comprised of two independent first-order reactions that occur in parallel (**Fig. 3B**). The analysis, in this case, indicated that the rate of the dissociation of 11-*cis*-retinal from the two sites differed by about an order of magnitude.

4. Notes

1. Dark-adapted, frozen retina can be purchased in the United States from J. A. and W. L. Lawson (Lincoln, NE).
2. At concentrations higher than 10–15 μM, IRBP tends to precipitate irreversibly.
3. Freezing IRBP, as well as storage of the protein at 4°C, results in rapid loss of the retinoid-binding activity of the protein. Viability of IRBP preparations can be maintained for a period of about 8 wk at –20°C, provided that freezing of protein solutions is prevented by including 50% glycerol in the storage buffer.
4. IRBP purified from retina contains endogenous ligands, mainly retinoids and long-chain fatty acids. Because some long-chain fatty acids interfere with the interactions of retinoids with the protein, it is important to delipidate IRBP prior to use. Delipidation also extracts traces of endogenous retinoids.
5. Dioleoylphosphatidylcholine or egg-yolk phosphatidylcholine are used because their phase transition temperatures are low and they maintain a liquid-crystalline state at physiological temperatures. The former lipid is better defined chemically and the affinity of vesicles made from it for retinoids is very consistent between different preparations. The latter lipid is, however, less costly. Lipids can be purchased from Avanti Polar Lipids (Alabaster, AL).
6. Rapid mixing accessories are available either as a separate device that can be linked to a fluorometer (e.g., Hi-Tech Scientific, Salisbury, UK), or as integrated stopped-flow-fluorometers.
7. It has been our experience that the number of retinoid-binding sites vary between different protein preparations, most likely as a result of deactivation of a fraction of the protein upon delipidation and storage. The presence of a small fraction of inactivated protein does not interfere with the measurements. However, to verify the viability of the protein, it is important to ascertain that the total number of binding sites (measured by monitoring the intrinsic fluorescence of IRBP) is twice the number of hydrophobic-binding sites (measured by monitoring the fluorescence of retinol) in any particular protein preparation.
8. To avoid photo-damage to retinoids and probes, fluorescence measurements should be brief. We routinely measure the signal for 2–5 s, followed by closing of

the shutter to isolate the sample from the light source. To ensure that binding reached an equilibrium after each titration step, i.e., that the fluorescence signal is constant, 3–4 such brief measurements are taken at 1- to 2-min intervals.

9. Upon saturation, depending on measurement wavelengths and fluorometer setting, fluorescence signals either reach a constant value or the slope becomes shallow and linear. In the latter case, the titration curve should be corrected for the nonspecific slope (*see* **ref. *15***). Usually saturation is reached at a mole ratio of ligand/protein of 1–1.5. Titrations should be continued to a mole ratio of about 2.5 ligand/protein to validate the attainment of saturation.

10. Curve-fitting involves fitting the data to several parameters. To increase confidence of attaining a meaningful solution, two of these parameters, the fluorescence in the absence of ligand and the fluorescence at saturation (F_0 and F_{sat} in **Eq. 1**, respectively) can be fixed using the measured values.

11. Saturation is attained at about twice the ligand concentration required for saturation in titrations monitored by following the fluorescence of retinol.

12. If the ligand/IRBP ratio is higher than two, the excess ligand will rapidly partition into the lipid bilayers resulting in quenching of fluorescence signals and loss of sensitivity.

Acknowledgments

Work in the author's laboratory has been supported by grants from the National Institute of Health (EY09296, CA68150, and DK42601).

References

1. Noy, N. and Xu, Z.-J. (1990) The interactions of retinol with binding proteins: Implications for the mechanism of uptake by cells. *Biochemistry* **29,** 3878–3883.

2. Noy, N. and Blaner, W. S. (1991) Interactions of retinol with binding proteins: studies with rat Cellular Retinol-Binding Protein and with Rat Retinol-Binding Protein. *Biochemistry* **30,** 6380–6386.

3. Ho, M.-T. P., Massey, J. B., Pownall, H. J., Anderson, R. E., and Hollyfield, J. G. (1989) Mechanism of vitamin A movement between rod outer segments, interphotoreceptor retinoid-binding protein, and liposomes. *J. Biol. Chem.* **264,** 928–935.

4. Chen, Y. and Noy, N. (1994) Retinoid specificity of interphotoreceptor retinoid-binding protein. *Biochemistry* **33,** 10,658–10,665.

5. Kersten, S., Pan, L., and Noy, N. (1995) On the role of ligand in retinoid signaling. Positive cooperativity in the interactions of 9-*cis* retinoic acid with tetramers of the retinoid X receptor. *Biochemistry* **34,** 14,263–14,269.

6. Pepperberg, D. R., Okajima, T-I. L., Wiggert, B., Ripps, H., Crouch, R. K., and Chader, G. J. (1993) Interphotoreceptor retinoid-binding protein: molecular biology and physiological role in the visual cycle of rhodopsin. *Mol. Endocrin.* **7,** 61–85.

7. Fong, S.-L., Liou, G. I., Landers, R. A., Alvarez, R. A., and Bridges, C. D. B. (1984) Purification and characterization of a retinol binding glycoprotein synthesized and secreted by bovine neural retina. *J. Biol. Chem.* **259,** 6534–6542.
8. Bazan, N. G., Reddy, T. S., Redmond, T. M., Wiggert, B., and Chader, G. J. (1985) Endogenous fatty acids are covalently and noncovalently bound to interphotoreceptor retinoid binding protein in the monkey retina. *J. Biol. Chem.* **260,** 13,677–13,680.
9. Saari, J. C. (1994) Retinoids in photosensitive systems, in *The Retinoids, Biology, Chemistry, and Medicine* (Sporn, M. B., Roberts, A. B., and Goodman, D. S., eds.), Raven, New York, pp. 351–386.
10. Saari, J. C., Teller, D. C., Crabb, J. W., and Bredberg, L. (1985) Properties of an interphotoreceptor retinoid-binding protein from bovine retina. *J. Biol. Chem.* **260,** 195–201.
11. Chen, Y., Saari, J. C., and Noy, N. (1993) Studies on the interactions of all-*trans*-retinol and long-chain fatty acids with interphotoreceptor retinoid-binding protein. *Biochemistry* **32,** 11,311–11,318.
12. Furr, H. C., Barua, A. B., and Olson, J. A. (1994) Analytical methods, in *The Retinoids, Biology, Chemistry, and Medicine* (Sporn, M. B., Roberts, A. B., and Goodman, D. S., eds.), Raven, New York, p.188.
13. Saari, J. C. and Bredberg, L. D. (1988) Purification of cellular retinaldehyde-binding protein from bovine retina and retinal pigment epithelium. *Exp. Eye Res.* **46,** 569–578.
14. Dittmer, J. C. and Wells, M. A. (1969) Quantitative and qualitative analysis of lipids and lipid components. *Methods Enzymol.* **14,** 482–530.
15. Cogan, U., Kopelman, M., Mokady, S., and Shinitzky, M. (1976) Binding affinities of retinol and related compounds to retinol-binding proteins. *Eur. J. Biochem.* **65,** 71–78.
16. Chen, Y. and Noy, N. (1996) Docosahexaenoic acid modulates the interactions of the interphotoreceptor matrix retinoid-binding protein with 11-*cis*-retinal. *J. Biol. Chem.* **271,** 20,507–20,515.

15

Use of Antisense Oligonucleotides to Study the Role of CRABPs in Retinoic Acid-Induced Gene Expression

Paul Nugent and Robert M. Greene

1. Introduction

Antisense methodologies have been used extensively to inhibit the expression of specific genes with a view to elucidating their role in particular cellular processes *(1)*. The technique is based on the ability of mRNA to bind, in a sequence-specific fashion, to a complimentary oligonucleotide sequence (the antisense sequence), via Watson-Crick hydrogen bonding. Binding of the oligonucleotide then prevents efficient translation of the mRNA, either by preventing the ribosome from reading the RNA message, or by activation of RNAse H, an enzyme that specifically cleaves the RNA strand of a DNA-RNA duplex (**Fig. 1**). In either case, the result is the arrest of specific protein synthesis *(2)*. Though primarily used in cell and tissue-culture systems *(1–3)*, efforts are underway to develop therapeutic applications *(4)*.

A large number of variables have been identified as affecting the success of an antisense experiment. The pharmacokinetic properties of the oligonucleotides, including cellular uptake and subcellular distribution and metabolism, the sequence of the mRNA that is targeted, and the type of cell/tissue being studied, all contribute to the efficacy with which protein expression is inhibited *(5)*. Antisense technology is very much an empirical science, even "art," at the moment, and requires much trial-and-error to develop conditions suitable for the particular system under investigation *(6)*. However, certain principles of design have emerged in the past few years that may be used as a starting point for preliminary explorations.

From: *Methods in Molecular Biology, Vol. 89: Retinoid Protocols*
Edited by: C. P. F. Redfern © Humana Press Inc., Totowa, NJ

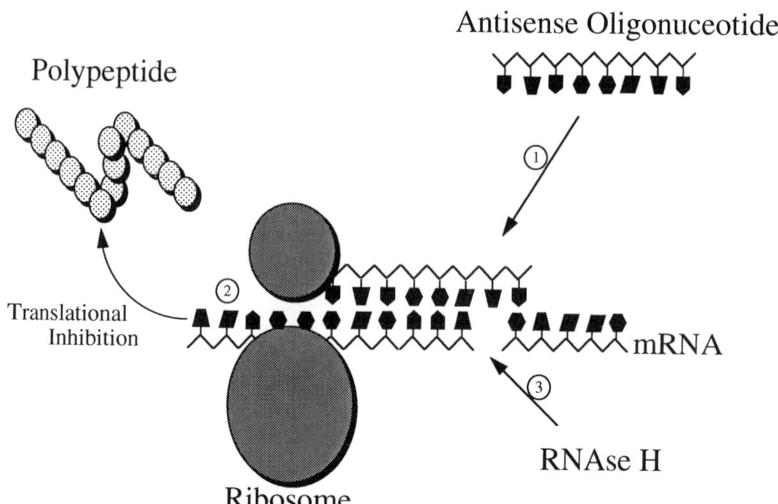

Fig. 1. Mechanism of action of antisense oligonucleotides. After binding to the complimentary mRNA sequence *(1)*, the oligonucleotides inhibit protein translation by interference with movement of the mRNA through the translation machinery *(2)*, or by inducing RNAse H-dependent degradation of the mRNA in the duplex *(3)*.

We have developed an antisense strategy to elucidate the role of the cellular retinoic acid-binding proteins (CRABP) I and II in retinoic acid (RA)-induced changes in gene expression in primary cultures of murine embryonic palate mesenchymal (MEPM) cells *(7)*. These cells basally express high levels of CRABP I mRNA, and much lower levels of CRABP II *(8)*. Treatment of MEPM cells with RA results in downregulation of CRABP I and upregulation of CRABP II mRNA *(8)*. This expression pattern has been taken into account in the design of our antisense strategy: for inhibition of CRABP I, oligonucleotides are added for several days to inhibit CRABP I expression prior to addition of RA; for CRABP II, oligonucleotides are added simultaneously with RA, so that retinoid-induced CRABP II mRNA is inhibited by the oligonucleotides as it appears. The expression of several genes, including retinoic acid receptor-β (RAR-β), transforming growth factor-β (TGF-β3), and tenascin, are regulated by RA in MEPM cells *(7–9)*. Using the expression of these genes to monitor effects of CRABP antisense oligonucleotides on changes in RA-responsiveness in MEPM cells, we have demonstrated that CRABP I and CRABP II play a role in the induction of specific-gene expression by RA in embryonic palate cells (**Fig. 2**) *(7)*.

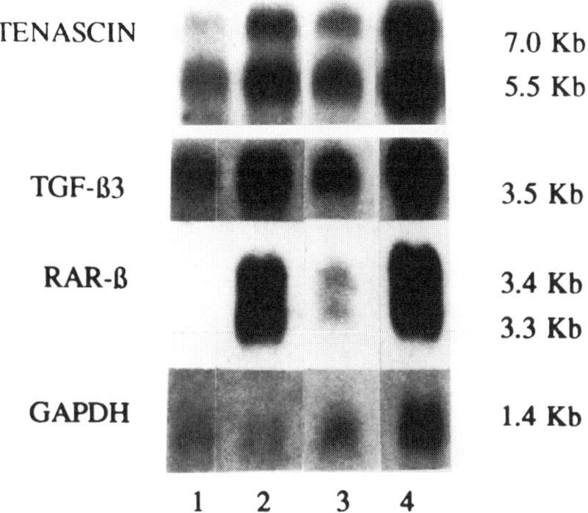

Fig. 2. Northern-blot autoradiogram showing the effect of antisense oligonucle-otides to CRABP I on the expression of TGF-β3, RAR-β, and tenascin mRNA in MEPM cells. Subconfluent cultures of cells were treated with 10 μ*M* oligonucleotide each day for 3 successive days, 3.3 μ*M* RA being added for the last 5 h of treatment. MEPM cells were cultured without oligonucleotide and RA (lane 1); with RA but without oligonucleotide (lane 2); with RA and antisense oligonucleotide (lane 3); with RA and mis-sense oligonucleotide (lane 4). RNA loading was checked by hybridiza-tion of the blot with a GAPDH probe. (*Note:* Lane 4 was slightly overloaded, as indi-cated by the GAPDH signal.)

2. Materials

2.1. Treatment of Cells with Oligonucleotides

2.1.1. Oligonucleotides

The most critical factors in determining the success of an antisense experi-ment are the activity and quality of the oligonucleotides. By activity we mean the ability of the oligonucleotides to seek out and bind to their complimentary mRNA. Not all parts of the mRNA transcript can bind antisense oligonucle-otides, presumably owing to secondary structure, protein binding, and so on. Having identified a suitable target sequence, one must use oligonucleotides of a high quality, i.e., intact and free from contaminating incomplete oligonucle-otide sequences. Many commercial sources of oligonucleotides are available (*see*, for example, the last few pages of any recent issue of the journal *Science*). Oligonucleotides we use are synthesized on a 394 DNA Synthesizer (Applied

Biosystems) and purified on a reversed-phase column (Nensorb Prep column, DuPont) by the DNA Core Facility at Thomas Jefferson University (*see* **Note 1**). The oligonucleotides are supplied as a lyophilized powder, with an estimate of yield. The powder is resuspended in 0.5 mL autoclaved, distilled water, and the exact concentration of DNA determined by spectrophotometry at 260 nm. The volume of this solution (stock solution) that must be added to culture medium (3.0 mL for a 60-mm culture dish) to give a final oligonucleotide concentration of 10 μM (*see* **Note 2**) is then calculated by dividing the amount of oligonucleotide in 3 mL of a 10-μM solution (178 μg of an 18-mer) by the concentration of oligonucleotide in the stock solution. For example: concentration of stock solution = 4.88 μg/mL; required final oligonucleotide concentration in culture medium = 10 μM; oligonucleotide required in 3 mL culture medium to give 10 μM solution =178 μg; volume of stock solution to be added to culture medium = 178/4.88 = 36.5 μL.

In general, the best target sequence for the antisense must be determined empirically, because it is difficult to predict the optimal site from primary sequence of the mRNA, and slight shifts in the oligonucleotide sequence have been reported to produce quite different efficacies. We have used 18-mer oligonucleotides to target the ATG-translation initiation site of the CRABP I and CRABP II mRNAs. This length of oligonucleotide should be sufficient to achieve specificity without resulting in nonspecific binding to mRNA, DNA, or protein, whereas an approximately equal ratio of A/T to G/C is considered desirable for optimal hybridization to the mRNA.

Primary nucleolytic degradation of oligonucleotides takes place at the phosphate center of the phosphodiester backbone. Because digestion of oligonucleotides by nucleases in serum is a major concern in antisense experiments, we use oligonucleotides containing the nuclease-resistant phosphorothioate modification, wherein sulfur is substituted for phosphorous at the diester linkages between adjacent nucleotides. We also use serum-free conditions for treatment of cells (*see* **Note 3**). Such phosphorothioate oligonucleotides are easily taken up by cells and, although they hybridize somewhat less readily than natural phosphodiesters, their nuclease-resistance has made them effective and popular molecular tools. We have used the following sequences in our studies with MEPM cells (translation initiation region underlined):

> 5'..gtg acg gtg <u>gta cgg</u> gtt..3' CRABP I antisense oligonucleotide
> 5'..acg acg gtg a<u>ta cgg</u> att..3' CRABP II antisense oligonucleotide

These are undoubtedly not the only sequences in the CRABPs that will work, and indeed we have had success with another CRABP I sequence *(7)*. Furthermore, these sequences may not work optimally in all cell types.

2.1.2. MEPM Cells

Our cell-culture system comprises mesenchymal cells derived from the developing secondary palate, a tissue sensitive to the teratogenic effects of RA. MEPM cells were derived from palatal tissue dissected from gestational d 13 ICR mouse embryos (date of vaginal plug detection was considered as d 0 of gestation). Palates are minced and dissociated with 0.25% trypsin 1:250/0.1% EDTA in phosphate-buffered saline (PBS) for 10 min at 37°C with constant shaking. Trypsin is inhibited by the addition of Opti-MEM medium (Gibco) containing 5% v/v fetal bovine serum (FBS, Sigma). Cells are plated in 60-mm tissue-culture dishes at an initial density of 2–3×10^5 cells/cm^2 in Opti-MEM containing Earle's salts, 25 mM HEPES buffer and supplemented with 2 mM glutamine, 5% FBS, 100 mg/mL streptomycin, and 100 U/mL penicillin, and grown at 37°C in a 95% air/5% CO_2 atmosphere, with media replaced every other day. The cells are used after 6 d, when approx 80% confluent. At this stage MEPM cells can survive in serum-free conditions, thus obviating the need to use fetal calf serum (FCS) rich in oligonucleotide-degrading nucleases.

2.1.3. Retinoic Acid

All-*trans* RA (Sigma) is dissolved in ethanol to a concentration of 1 mg/mL. The RA is dissolved under low-light intensity, and handled therafter under these conditions (*see* **Note 4**). The stock solution is diluted 1:1000 in serum-free medium; typically 9 µL is added to 9-mL culture medium and distributed to three culture plates at 3 mL/plate. The final concentration of ethanol never exceeds 0.1%, and is not toxic to the cells.

2.2. RNA Extraction

All stock solutions for RNA work should be made with diethylpyrocarbonate (DEPC)-treated water and, where possible, autoclaved.

1. DEPC-treated water.
2. Solution D: 4 M guanidium thiocyanate, 25 mM sodium citrate, pH 7.0, 0.5% sarkosyl, 0.1 M 2-mercaptoethanol.
3. 2 M Sodium acetate, pH 4.0.
4. Water-saturated phenol (molecular-biology grade).
5. Chloroform/isoamylalcohol 24:1 v/v.
6. Isopropanol.

2.3. Northern Blotting

2.3.1. Northern Blotting Materials

1. Agarose (low endo-osmosis, molecular-biology grade).
2. 10X MOPS: 0.2 M MOPS, 5 mM sodium acetate, pH 7.0, 10 mM EDTA.

3. Formaldehyde.
4. Sample buffer: 0.5% SDS, 0.25% bromophenol blue, 25 µ*M* EDTA, 25% glycerol.
5. 20XSSC: 3 *M* sodium chloride, 0.3 *M* sodium citrate, pH 7.0 10X SSC, 2X SSC, and 0.2X SSC are made by diluting the 20X SSC with the required volume of DEPC-treated water.
6. Salmon-sperm DNA (SS DNA): 10 mg/mL.
7. 100X Denhardts solution: 10 g of Ficoll 400, 10 g of polyvinylpyrrolidone and 10 g of BSA in 500 mL DEPC-treated water. Filter and store at –20°C.
8. Prehybridization solution: mix together 10 mL of formamide, 5 mL of 20X SSC, 1 mL of 100X Denhardt's solution, 2 mL of 0.5 *M* sodium phosphate, pH 6.5, 0.5 mL of 10 mg/mL SS DNA, and 1.5 mL of water. Before adding the SS DNA, boil for 10 min with the water and then place on ice for 10 min.
9. Cocktail mix: mix together 25 mL of 20X SSC, 4 mL of 0.5 *M* sodium phosphate, pH 6.5, 10 g of dextran sulfate, and 1 mL of 100X Denhardt's Solution.
10. 2X SSC/0.1% SDS.
11. 0.2X SSC/0.1% SDS.

2.3.2. Probes

Northern blots are probed with fragments of the cDNAs (murine) of genes whose expression in MEPM cells has been shown to be regulated by RA. It is our experience that use of fragments of cDNAs gives superior results to those obtained with linearized labeled plasmid. We have used a 0.74-kb *Eco*RI fragment of pMT-CRABP I encoding the murine CRABP I gene *(10)*, a 0.86-kb *Eco*RI fragment of pKSmCRABP II *(11)*, a 0.61-kb *Eco*RI/*Sma*I fragment encoding the murine TGF-β3 gene *(12)*, a 1.9-kb *Eco*RI fragment of the murine RAR-β cDNA *(13)*, and a 2.5-kb *Bam*HI fragment of the murine tenascin cDNA *(14)*. Other genes may be more suitable for other cell systems; the aforementioned genes are appropriate for MEPM cells. Selected genes should fulfil the following criteria:

1. Detectable expression by Northern blotting.
2. Induction of expression by doses of RA which are not deleterious to the cells.

Variations in amount of RNA loaded per lane is determined by probing the blot with a human cDNA probe for glyceraldehyde phosphate dehydrogenase (GAPDH) (obtained from ATCC).

3. Methods

3.1. Treatment of Cells with Oligonucleotide

We have used the following treatment regimes in our studies on MEPM cells. Precise details may need to be modified for other cell types, because our protocol is based on published accounts of antisense treatment of fibroblastic cells.

1. On the first day of treatment, wash cell monolayers twice with 3 mL serum-free medium (SFM); add CRABP I antisense oligonucleotides in SFM at a concentration of 10 µ*M* (*see* **Subheading 2.1.**).
2. We have treated MEPM cells with 10 µ*M* CRABP I antisense oligonucleotide each day for up to 3 d without observing toxic effects. On the second and third day, add oligonucleotide directly without change of culture medium. For the second and third day, we assume zero concentration of oligonucleotide in the culture medium, owing to cellular uptake and/or endogenous-tissue nucleotidase activity, though no doubt there is some oligonucleotide left over from the previous day.
3. On the third day of oligonucleotide treatment, add 3 µL of RA stock solution (1 mg/mL) (*see* **Note 2**) to each culture plate, giving a final concentration of 3.3 µ*M*.
4. Extract total RNA after 5 h (*see* **Subheading 3.2.**).
5. In the case of experiments utilizing CRABP II, 5 µ*M* antisense oligonucleotides are sufficient to elicit a response. Cells are treated for 2 d, the RA being added for the last 24 h.

The slight differences in treatment protocol for CRABP I and CRABP II antisense oligonucleotides derive from the expression patterns of these genes in MEPM cells. Details of these protocols may need to be modified for other cell-culture models, depending on their patterns of CRABP I and CRABP II mRNA expression, and the effect of RA on that expression.

This antisense strategy has been used to investigate the role of CRABP I and CRABP II in RA-induced gene expression. Alterations in the expression of RAR-β, TGF-β3, and tenascin are determined by Northern-blot hybridization using total-cellular RNA.

3.2. Extraction of RNA

The reader should consult **refs.** *15* and *16* for additional practical details concerning RNA extraction and Northern blotting. The following represents our preferred method.

Manipulations of RNA samples should be done with gloved hands. Extract total RNA from cell monolayers by the method of Chomczynski and Sacchi *(17)* or an equivalent method.

1. Remove the culture medium, wash the cell layer with 5 mL of SFM, and add 0.33 mL of solution D to each plate.
2. Remove the cell contents with a Teflon cell scraper, combine the contents of triplicate plates, and divide equally between two 1.5-mL microcentrifuge tubes.
3. Add 60 µL of 2 *M* sodium acetate (pH 4.0), 600 µL of water-saturated phenol, and 120 µL chloroform/isoamylalcohol (24:1) to each tube and vortex the tubes vigorously for 10 s.
4. After a 15-min incubation on ice, centrifuge the tubes at 4°C for 15 min and transfer the aqueous RNA-containing supernatant to a clean tube.

5. Add an equal volume of isopropanol, vortex the tubes briefly, and precipitate the RNA at –20°C overnight, or –70°C for 2 h.
6. Centrifuge the tubes at 4°C for 15 min and decant the supernatant.
7. Resuspend the pellets in 300 µL of solution D and combine duplicate tubes.
8. Add an equal volume of isopropanol, mix with the vortexer, and precipitate the RNA by incubation overnight at –20°C.
9. Centrifuge the tubes at 4°C for 15 min and decant the supernatant.
10. Wash the pellet three times with 70% ethanol (in DEPC-treated water), lyophilize and resuspend in 23 µL DEPC-treated water.
11. The integrity of the RNA, as indicated by the presence of 28S and 18S RNA bands, is determined by electrophoresis of 1 µL on a 1% w/v agarose gel (**Fig. 3**). Quantify total RNA by spectrophotometry at 260 nm using 1 µL diluted with 99 µL water, and a small quartz cuvet. We generally obtain a total yield of 40–80 µg per sample. At 20 µg/lane on the agarose gel, this is enough to generate several blots (*see* **Note 5**).

3.3. Northern-Blot Assay

Steady-state levels of mRNA are determined by Northern-blot hybridization.

1. The samples are run on a 1.2% w/v agarose gel prepared as follows:
   For a 350-mL gel (sufficient for a 24 × 20 cm gel bed):
   a. Dissolve 4.2 g of agarose in 250 mL water using a microwave oven;
   b. Cool, add 35 mL of 10X MOPS and 65 mL of formaldehyde, and pour into gel frame.
2. Prepare samples by mixing: 20 µg of total RNA (up to 10 µL), 25 µL of formamide, 5 µL of 10X MOPS, 10 µL of formaldehyde, 10 µL of sample buffer.
3. Heat samples to 60°C for 10 min, cool on ice for 10 min and apply to gel wells.
4. We run our gels overnight at 60 volts, by which time the dye front has migrated approx 16 cm.
5. Carefully remove the gel from the frame and trim to remove first and last lanes (which do not contain sample) and gel remaining beyond dye front.
6. Soak the gel in 10X SSC for 1–2 h with one change, and set up a capillary-transfer apparatus as described in **ref. 15** for transfer of RNA to a nitrocellulose filter.
7. Allow transfer to proceed for at least 18 h.
8. Remove the nitrocellulose filter, wash with 2X SSC, and crosslink RNA to the filter with a UV crosslinker (Stratagene).
9. Prehybridize the filter in prehybridization solution at 42°C in a shaking water bath for at least 3 h.
10. cDNA fragments (50–100 ng) may be labeled with $(\alpha\text{-}^{32}P)$-deoxycytidine 5'-triphosphate (dCTP) by primer-extension labeling using a commercially available kit. The labeled probe may be separated from unincorporated isotope by a variety of techniques, most of which are marketed in the form of a kit (*see* **Note 6**).

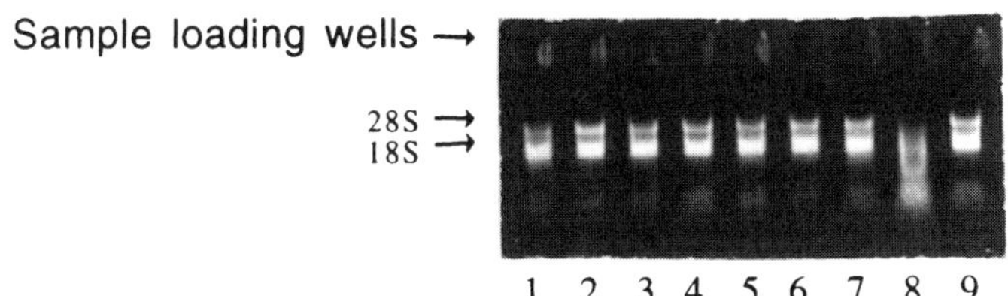

Fig. 3. Assessment of integrity of total RNA extracted from MEPM cells. One microliter of RNA extract was subjected to electrophoresis for approx 30 min at 90 V in a 1% agarose gel containing ethidium bromide. In all lanes, except lane 8, 28S and 18S ribosomal RNA fluoresces as distinct bands, indicating that the RNA is intact. The ribosomal RNA in lane 8 is smeared, indicative of RNA degradation. Because rRNA comprises over 95% of total RNA, degradation of rRNA is indicative of degradation of mRNA.

11. Mix a volume of labeled probe containing $2–4 \times 10^8$ dpm in a 50-mL screwcap-plastic tube with 250 μL salmon-sperm DNA (10 mg/mL), add water to a total volume of 1.7 mL, and boil the solution for 10 min. Place the tube on ice for 10 min.
12. Prepare the hybridization solution by adding 5 mL formamide and 3.3 mL cocktail mix to the probe and mix the contents by inversion.
13. Remove prehybridization solution, add hybridization solution to celluloid bag containing the filter, and incubate at 42°C for at least 18 h in a shaking water bath.
14. Remove filter from bag and wash twice with 2X SSC/0.1% SDS each time for 10 min at room temperature with gentle shaking (*see* **Note 7**).
15. Wash in 0.2X SSC/0.1% SDS for 40 min at 52°C in a shaking water bath (*see* **Note 8**).
16. Remove filters to 3MM blotting paper, wrap moist filters in cling film, and expose to X-ray film in a cassette at –20°C. The exposure time varies depending on the strength of the signal, which can be estimated with a Geiger counter before the filter is put into the cassette. The filters may be prehybridized again and hybridized with a different probe.
17. Autoradiographic signals may be quantified by densitometry and presented as fold (or percentage) increase or decrease relative to the appropriate control, taking into account differences in total RNA loading between lanes (as assessed by hybridization with a probe for GAPDH) (*see* **Note 9**).

3.4. Inhibition of CRABP Protein Expression

Ideally one would like to confirm that the antisense has inhibited CRABP expression. This could be done by: assaying for reduction in protein expression by Western blot or the PAGE method of Siegenthaler *(18)*; or assaying RNAse H-mediated destruction of CRABP mRNA by Northern blotting.

Unfortunately antibodies to the CRABP proteins are not generally available. It is uncertain if the PAGE method would be sufficiently sensitive to detect an antisense-induced inhibition, especially if the effect is less than complete.

4. Notes

1. Purification of the oligonucleotides with the Nensorb-prep column will remove contaminating salts, failure sequences, and synthetic byproducts, which might interfere with the specificity of action of the antisense oligonucleotides *(19)*.
2. Although it is often recommended that oligonucleotide concentrations of less than 5 μM be utilized to minimize nonsequence-specific effects, we have found that in this system a concentration of 10 μM results in optimal efficiency.
3. MEPM cells treated with oligonucleotides when at least 80% confluent survive quite well in serum-free conditions for several days. One should check the effects of oligonucleotide treatment on survival of the cell type of interest growing in serum-free conditions at less than full confluency.
4. Exposure of RA in solution to light produces metabolites that may vary in their biological and biochemical activity. We drop the weighed RA powder into a screw-cap tube wrapped in a paper towel and containing the required volume of ethanol. After capping the tube, it is completely covered and set on a gently rocking surface for 30–60 min. The tube is opened in the tissue-culture hood with the hood light off and only a dim light in the room.
5. The yield of total RNA will vary with cell type, number, and size of plates per treatment. The amount of total RNA loaded in the wells need not be 20 μg; just ensure that the same amount is loaded in every well.
6. We endeavor to use probes labeled to a very high-specific activity since they tend to give a better signal to noise ratio (i.e., good signal, low background), and detect mRNAs expressed at low levels. To generate these probes we try to use the $\alpha\text{-}^{32}P$dCTP before its activity date and certainly before a half-life has elapsed. We also separate labeled probe from unincorporated isotope using Sephadex G50 Nick Columns (Pharmacia Biotech) prior to hybridization. These columns are quick, easy to use, and do not involve the use of a centrifuge (one less piece of equipment to be monitored for radioactive contamination!).
7. This wash solution is usually too radioactive to discard down the sink and must be stored for decay. However, the solution produced by the 52°C wash can be safely discarded down the sink.
8. The stringency of washing depends, among other things, on the sequence similarity between the gene of interest and other members of its family. We have found that this washing regime generally gives a good signal for RAR-β, TGF-β, and tenascin, with little background. It is also our experience that complete removal of the smallest amount of background remaining at this stage requires such stringency that the specific mRNA signal is lost or substantially reduced.
9. Equality of RNA loading between lanes may also be assessed by comparing intensity of ethidium bromide staining of 18S and 28S RNA bands. We have

found this technique to be less sensitive than determination of GAPDH expression, because the ethidium bromide signal may be too intense to evaluate small differences between lanes. It should be noted that all methods of assessment have their drawbacks and should be interpreted carefully. Use of densitometry to quantify differences between lanes is useful when GAPDH expression is variable owing to errors in loading. However, it is our experience that densitometry merely puts a number on patterns observed by the naked eye: if you do not see a significant difference, changes seen by a densitometer should be interpreted with great care.

Acknowledgments

This work was supported in part by NIH grants DE05550, DE08199, and DE09540 to RMG. Paul Nugent is supported by a PHS NRSA F32-DE05633. The authors thank Drs. L. Gudas (Cornell Medical School, New York, NY), V. Giguere (University of Toronto, Ontario, Canada), H. Moses (Vanderbilt University, Nashville, TN), P. Chambon (CNRS, Strasbourg, France), and Y. Saga (Ibaraki, Japan) for the gifts of the cDNA clones.

References

1. Colman, A. (1987) Antisense strategies in cell and developmental biology. *J. Cell Science* **97,** 399–409.
2. Helene, C. (1991) Rational design of sequence-specific oncogene inhibitors based on antisense and antigene oligonucleotides. *Eur. J. Cancer* **27,** 1466–1471.
3. Krieg, A. M. (1993) Uptake and efficacy of phosphodiester and modified antisense oligonucleotides in primary cell cultures. *Clin. Chem.* **39,** 710–712.
4. Heidenreich, O., Kang, S.-H., Xu, X., and Nerenberg, M. (1995) Application of antisense technology to therapeutics. *Mol. Med. Today* **1,** 128–133.
5. Crooke, R. M., Graham, M. L., Cooke, M. E., and Crooke, S. T. (1995) *In vitro* pharmacokinetics of phosphorothioate antisense oligonucleotides. *J. Pharmacol. Exp. Ther.* **275,** 462–473.
6. Wagner, R. W. (1995) The state of the art in antisense research. *Nature Med.* **1,** 1116–1118.
7. Nugent, P. and Greene, R. M. (1995) Antisense oligonucleotides to CRABP I and II alter the expression of TGF-β3, RAR-β, and tenascin in primary cultures of embryonic palate cells. *In Vitro Cell Develop. Biol.* **31,** 553–558.
8. Nugent, P. and Greene, R. M. (1994) Interactions between the TGF-β and retinoic acid signal transduction pathways in embryonic palatal cells. *Differentiation* **58,** 149–155.
9. Nugent, P., Potchinsky, M., Lafferty, C., and Greene, R.M. (1995) TGF-β modulates the expression of retinoic acid-induced RAR-β expression in primary cultures of embryonic palate cells. *Exp. Cell Res.* **220,** 495–500.
10. Stoner, C. M. and Gudas, L. J. (1989) Mouse cellular retinoic acid binding protein: cloning, complimentary DNA sequence, and messenger RNA expression during retinoic acid-induced differentiation of F9 wild type and RA-3-10 mutant teratocarcinoma cells. *Cancer Res.* **49,** 1497–1504.

11. Giguere, V., Lyn, S., Yip, P., Siu, C-H., and Amin, S. (1990) Molecular cloning of cDNA encoding a second cellular retinoic acid-binding protein. *Proc. Natl. Acad. Sci. USA* **87.** 6233–6237.

12. Miller, D. A., Lee, A., Matsui, Y., Chen, E. Y., Moses, H. L., and Derynck, R. (1989) Complementary DNA cloning of the murine transforming growth factor-β3 (TGFβ3) precursor and the comparative expression of TGFβ3 and TGFβ1 messenger RNA in murine embryos and adult tissues. *Mol. Endocrinol.* **3,** 1926–1934.

13. Zelent, A., Krust, A., Petkovich, M., Kastner, P., and Chambon, P. (1989) Cloning of murine α and β retinoic acid receptors and a novel receptor γ predominantly expressed in skin. *Nature* **339,** 714–717.

14. Saga, Y., Tsukamoto, T., Jing, N., Kusakabe, M., and Sakakura, T. (1991) Murine tenascin: cDNA cloning, structure and temporal expression of isoforms. *Gene* **104,** 177–185.

15. Sambrook, J., Fritsch, E. F., and Maniatis, T. (1989) Extraction, purification, and analysis of messenger RNA from eukaryotic cells, in *Molecular Cloning. A Laboratory Manual,* 2nd ed. (Ford, N., ed.), Cold Spring Harbor Laboratory, Cold Spring Harbor, NY, pp. 7.1–7.87.

16. KrumLauf, R. (1991) Northern blot analysis of gene expression, in *Methods in Molecular Biology, Vol. 7: Gene Transfer and Expression Protocols* (Murray, E. J., ed.), Humana, Totowa, NJ, pp. 307–323.

17. Chomczynski, P. and Saachi, N. (1987) Single-step method of RNA isolation by acid guanidium thiocyanate-phenol-chloroform extraction. *Anal. Biochem.* **162,** 156–159.

18. Siegenthaler, G. (1990) Gel electrophoresis of cellular retinoic acid-binding protein, cellular retinol-binding protein, and serum retinol-binding protein. *Meth. Enzymol.* **189,** 299–307.

19. Johnson, B., McClain, S. G., Doran, E. R., Tice, G., and Kirsch, M. A. (1990) Rapid purification of synthetic oligonucleotides, a convenient alternative to HPLC and polyacrylamide gel electrophoresis. *Biotechniques* **8,** 424–429.

III

Nuclear Retinoid Receptors

16

Preparation of Polyclonal Antibodies to Retinoid Receptors

Jacqueline A. Dyck and Vickie J. LaMorte

1. Introduction

Techniques for the preparation of antibodies and their applications are widely established in the literature. Antibodies have proven to be useful reagents in characterizing the molecular activities of the retinoid receptors *(1)*. The use of a retinoid "X" receptor α (RXRα) antibody on Western blots contributed to the identification of the RXR ligand by detection of the receptor in fractions of whole-cell lysates that were able to bind the 9-*cis* retinoic acid ligand *(2)*. Antibodies to the retinoid receptors used in supershifting gel-retardation assays also supported the finding that RXR is the heterodimeric partner of other members of the retinoid/thyroid hormone-receptor superfamily *(3,4)*. Additionally, immunohistochemistry with antibodies to retinoic acid receptor α (RARα) enabled visualization of the promyelocyte-retinoic acid receptor (PML-RAR) fusion oncoprotein in cells from acute promyelocytic-leukemia patients *(5,6)*. This chapter describes basic protocols for polyclonal rabbit-antibody production and affinity purification. Suggestions for potential applications in the study of retinoid-molecular biology are also discussed.

The appropriate selection of the antigen determines the usefulness of the antibodies in the molecular techniques to which they will be applied. For example, to examine a receptor in vivo, a region of the molecule that is likely to be exposed in the three-dimensional or nondenatured state would serve best as a candidate for the antigen. For techniques that detect the denatured form of the protein, such as Western analysis, the epitope location may not be an issue. Thus, not all antibodies are applicable for all techniques. Because of the conservation of sequence between the retinoid receptors, it may be necessary to select a unique sequence of the molecule to ensure specificity of the reagent. In

From: *Methods in Molecular Biology, Vol. 89: Retinoid Protocols*
Edited by: C. P. F. Redfern © Humana Press Inc., Totowa, NJ

general, two types of antigens are employed: a bacterial/ baculoviral-expressed protein or a synthetic peptide. Easily purified, soluble glutathione-*S*-transferase (GST) fusion proteins *(7)* can be synthesized containing either the full-length receptor or discrete regions within it. One may utilize PCR technology to select precise domains such that regions of hydrophobicity can be avoided. Bacterially expressed protein purified from polyacrylamide gels can be emulsified and also be used as an antigen *(8)*. Synthetic peptides can also be generated for coupling to carriers and are typically 10–15 amino acids in length *(9)*. The strategy for selection of an appropriate peptide *(10)* is based on choosing a unique stretch of amino acids, not present in related molecules, that is likely to: (1) represent an exposed hydrophilic surface of the protein/polypeptide *(11)*, (2) contain a predicted β-turn *(12)*, and (3) contain appropriate residues for coupling to the carrier protein. The peptides are conjugated to carrier proteins such as human α-globulin, bovine serum albumin (BSA), or keyhole limpet hemacyanin (KLH) using bisdiazotized benzene or glutaraldehyde *(8,13,14)*. Antiserum exhibiting a high titer is desired because nonspecific interactions pose less of a problem at higher dilutions. Affinity purification of antisera will aid in eliminating background signal and is the method of choice in preparing a high-quality reagent.

2. Materials

2.1. Bacterially Expressed Protein as an Immunogen

1. Bacterial or baculoviral-expressed protein as a full length protein or a GST-fusion protein.
2. Polyacrylamide gel electrophoresis apparatus, gel reagents, Coomassie blue stain.
3. Emulsion supplies and reagents: petri dish, razor blades, 5-mL syringes (18-, 20-, 22-gage needles), sterile physiological saline (0.9% NaCl).

2.2. Synthetic Peptide as an Immunogen

1. Human α-globulin, BSA, or KLH-conjugated synthetic peptide: 0.5 mg for each of the first two injections and 0.25 mg for subsequent injections for each rabbit.
2. Freund's complete adjuvant (CalBiochem).
3. Freund's incomplete adjuvant (CalBiochem).
4. Sterile physiological saline (0.9% NaCl).
5. 10-mL Syringes.

2.3. Immunization, Boosting, and Bleeding Schedule

Animals: 3 New Zealand Rabbits, 6- to 8-wk-old males (*see* **Note 1**). Immunization syringes (glass) and other necessary immunization and blood drawing supplies in accordance with your animal-resource department.

2.4. Titer Analysis

1. 96-Well microtiter plates.
2. Antigen (*see* **Note 2**): 5 µg/mL stock in 0.1 *M* sodium bicarbonate, pH 9.0.
3. Blocking buffer: 10 mg/mL BSA in 0.1 *M* sodium bicarbonate, pH 9.0.
4. PBS: 140 m*M* NaCl, 2.7 m*M* KCl, 6.5 m*M* $Na_2HPO_4 \cdot H_2O$, and 1.5 m*M* KH_2PO_4 in H_2O.
5. Wash buffer: 0.05% v/v Tween-20 in PBS.
6. Serum and preimmune serum: Prepare serial dilutions such as 1:100, 1:1000, 1:10,000, 1:100,000 in 0.05% Tween-20/PBS.
7. Secondary antibody: horseradish peroxidase-coupled secondary antibody (Boehringer Mannheim). The appropriate dilution should be determined. Usually a range of 1 in 1000 to 1 in 5000 works best. The antibody should be diluted in 0.05% Tween-20/PBS.
8. Substrate: ABTS, 2,2'-azino-di-[3-ethylbenzthiozolene sulfonate (6)] diammonium salt (Boehringer Mannheim) in ABTS buffer. 1 mg/mL suggested working concentration.

2.5. Affinity Purification and Antibody Storage

2.5.1. Subtractive Purification (Batch Process)

1. Ammonium sulfate.
2. TBST buffer: 10 m*M* Tris-HCl, pH 8.0, 150 m*M* NaCl, 0.05% v/v Tween-20.
3. Subtractive matrix: Bacterially expressed GST lysate conjugated to CNBr Sepharose (Pharmacia): performed according to manufacturer's instructions.
4. 15-mL Polypropylene conical tubes.
5. 0.2-µm Filter (Millipore).

2.5.2. Affinity Purification (Column Process)

1. Affinity Matrix: CNBr Sepharose (Pharmacia), Affigel 15 (Bio-Rad) or EAH Sepharose 4B (Pharmacia) conjugated to the appropriate protein or peptide.
2. 5-mL Polypropylene disposable columns (Bio-Rad).
3. TBST buffer: 10 m*M* Tris-HCl, pH 8.0, 150 m*M* NaCl, 0.05% v/v Tween-20.
4. 100 m*M* Glycine, pH 2.8.
5. 100 m*M* Tris base, pH 11.0.
6. TBST containing 0.01% w/v sodium azide.
7. Polyacrylamide gel electrophoresis apparatus, gel reagents, Coomassie blue stain.

2.5.3. Storage

1. Centricon 30 concentrator unit (Amicon).
2. BSA.
3. Sodium azide (*see* **Note 3**).

3. Methods

3.1. Bacterially Expressed Protein as an Immunogen

To generate antibodies to RARα, a full-length protein was expressed in bacteria using the pET8c-expression vector *(15)* and partially purified by SDS-PAGE. The protein was cut from the acrylamide gel and emulsified in sterile physiological saline *(8)*. This technique is also valuable for insoluble GST-fusion proteins that cannot be purified by binding on glutathione sepharose beads.

3.1.1. Preparation Of Emulsified Antigen In Acrylamide Gel Slice

1. Load approx 200 mg of bacterial lysate or insoluble bacterial pellet in Laemmli sample buffer *(16)* to a 3-mm preparative SDS-polyacrylamide gel and run according to standard protocol *(17)* including prestained molecular-weight markers on an adjacent track.
2. To visualize the location of the recombinant protein, cut a strip from the resulting gel and stain and destain as usual with Coomassie blue. To compensate for changes in the size of the gel strip, rehydrate with H_2O, as needed. Estimate the location of the band on the remaining unstained gel. Other methods for visualizing the expressed protein are available such as KCl staining, sodium acetate, and copper chloride staining *(8)*.
3. Using a fresh razor cut out the band (1- to 2-cm band) containing the immunogen.
4. Rock the gel strip in three changes of distilled H_2O for 5 min each to reduce the SDS concentration.
5. Finely dice the gel strip in a sterile Petri dish with a fresh razor and transfer the material to a 5-mL syringe (loading from the top with a sterile weighing spatula). Add 1–2 mL of sterile physiological saline.
6. Run the saline-gel mixture through increasingly smaller gage needles starting with 18 gage going up to 22 gage in order to create the emulsion. It may be necessary to add more saline during the procedure. Ideally the final volume will be 3.5 to 5 mL. It is not necessary to use any adjuvants with this technique because the polyacrylamide gel appears to provide the same function.

Assuming that approx 5% of the bacterial lysate or pellet is composed of the recombinant protein, and 200 mg are loaded onto the gel, it can be estimated that approx 10 mg of recombinant protein is present in the excised gel slice. It is suggested that adjustments be made in the amount of protein loaded based on your estimate of the percent of expressed protein in your lysate, such that the band excised contains approx 10 mg protein. The resulting emulsion can be utilized for the immunization of three rabbits, each rabbit receiving approx 3 mg of protein. For the first two immunizations, 3–4 mg of protein/rabbit are appropriate. The boosting immunizations require only half of this amount.

3.2. Synthetic Peptide as an Immunogen

Although the amino and carboxy-termini are frequently chosen sequences for the creation of synthetic peptides, we chose to raise RXRα antibodies against a synthetic 16-mer peptide representing the internal hinge region of the receptor (a.a. 214–229), which is located C-terminally of the DNA-binding domain. The amino acid residues in this sequence (DRNENEVESTSSANED) are primarily polar, suggesting that this region is exposed in the native protein. The peptide was conjugated to human α-globulin *(13)*. For a strong polyclonal response to peptide antigens, it is necessary to accompany them with an adjuvant, e.g., Freund's or some other titer-boosting adjuvant. Typically, the first two injections are 0.5 mg of protein in 1 mL of complete adjuvant. Subsequent boosting injections contain 0.25 mg of protein in 1 mL of incomplete Freund's adjuvant.

For the first two injections:

1. In a 10-mL syringe, with the open tip blocked with parafilm, add 5 mg of synthetic peptide conjugate and bring to a total volume of 5 mL with physiological saline. After this is mixed, add 5 mL of Complete Freund's adjuvant to the syringe.
2. Using a Polytron homogenizer with a "microtip," emulsify the contents of the syringe. It is important to generate a thick emulsion.
3. Each rabbit receives 1 mL (0.5 mg of peptide) and the remaining emulsion can be refrigerated for the subsequent injections.

For subsequent injections:

1. In a 10-mL syringe, add 2.5 mg of synthetic-peptide conjugate and bring to a total volume of 5 mL with physiological saline. After this is mixed, add 5 mL of Incomplete Freund's adjuvant to the syringe.
2. Using a Polytron homogenizer, emulsify the contents of the syringe.
3. Each rabbit receives 1 mL (0.25 mg of peptide). Refrigerate the remaining emulsion for future injection.

3.3. Immunization, Boosting, and Bleeding Schedule

Typically, two to three rabbits are simultaneously injected to ensure successful antibody production. After receiving the rabbits, permit them to acclimate for 1 wk before drawing the preimmune sample. The antigen emulsion may be injected intradermally into multiple (20–30) sites. It is most economical and efficient to exsanguinate the rabbit when a high titer of antibody is attained. Adherence to institutional guidelines and consultation with your animal-resource department will assist you in this phase of production. It may be most practical for your laboratory to contract out the antisera production to a company specializing in antibody preparation (e.g., BABCO, Berkeley, CA).

3.3.1. A Typical Schedule for Antibody Production

1. Wk 1: Rabbits are received and permitted to rest.
2. Wk 2: Draw preimmune serum.
3. Wk 3: Initial injection.
4. Wk 6: First boost injection.
5. Subsequent boosts every 3–6 wk followed by blood collection 7–10 d postinjection.
6. Exsanguination by cardiac puncture 7–10 d after final boost.

3.3.2. Serum Preparation

Serum can be collected 7–10 d after boosting injections. Generally, 6–9 mL of blood/kg of body weight is collected (generally about 50 mL) and permitted to clot for 1 h at 37°C so that the clot is released from the sides of the collection tube (if necessary, ream the sides of the tube with a sterile wooden stick) then transfer to 4°C overnight. The serum is collected after centrifugation at 3000g for 10 min, heat inactivated at 56°C for 30 min, and then stored in aliquots at –20°C.

3.4. Titer Analysis

It is worth noting that because not all antibodies work for all applications, initial assays may be attempted in the immunodetection system for which the antibodies have been designed. Typically, a sample of the serum can be titered by several methods including Western blotting (*see* **Notes 4** and **5**) and ELISA *(8)*. ELISA is the most routine method for determining the titer of the antibody. The following guidelines may be applied in setting up an ELISA:

1. Antigen. Add 50 µL of antigen to each well of a microtiter plate. Cover plate with parafilm and leave overnight at 4°C or 1h at room temperature for antigen to adsorb. Remove solution (*see* **Note 6**).
2. Blocking. To block nonspecific binding, incubate with 150 µL/well of blocking buffer for 1 h at room temperature. Remove blocking solution and wash with wash buffer.
3. Antibody. To each well, add 75 µL of serially diluted test antibody or control preimmune antibody. Incubate 1h at 37°C followed by three washes with wash buffer.
4. Secondary antibody. To each well add 50 µL of horseradish peroxidase-coupled secondary antibody. Incubate 1h room temperature followed by washing.
5. Substrate. To each well add 50 µL of ABTS substrate solution. Leave at room temperature for 5–15 min. Plates may be visually analysed or read using a standard microtiter plate reader at 405 nm.

3.5. Affinity Purification and Antibody Storage

As an initial step in antibody purification, we use a subtractive strategy *(18–20)* to remove antibodies reactive to the GST and nonspecific antigens (*see* **Note 7**). This is achieved by creating a subtractive matrix consisting of bacterially (BL21 strain) overexpressed GST *(7)* bacterial lysate conjugated to CNBr-Sepharose beads. Because the GST protein was not purified from the bacterial lysate, the resulting matrix contains a myriad of bacterial proteins in addition to the GST protein.

3.5.1. Subtractive Purification (Batch Process)

1. Prepare overexpressed GST protein. We recommend using a French Press for bacterial lysis to maximize soluble, bacterial protein lysate yield.
2. To create the subtractive matrix, couple GST-containing lysate to CNBr beads according to manufacturer's specifications (generally 20 mg protein per mL of settled beads).
3. Wash substractive matrix with TBST (*see* **Note 8**).
4. Approximately 10 mL of serum is ammonium sulfate cut (50%) according to Harlow and Lane *(8)*. Spin at 3000*g* for 20 min, remove supernatant and resuspend pellet in 10 mL of TBST buffer.
5. Add 5 mL of sulfate-cut serum to 2.5 mL of washed subtractive matrix in a 15-mL conical tube.
6. Agitate gently on an orbital rotator for 90 min at room temperature.
7. Centrifuge for 5 min to pellet the matrix, and remove the supernatant to a fresh tube.
8. Repeat the subtractive process twice with supernatant utilizing new subtractive matrix each time, for a total of three subtractive treatments.
9. Centrifuge supernatant and filter the resulting supernatant through a 0.2-µm Millipore syringe filter unit to remove all traces of matrix and proceed to affintiy purification.

3.5.2. Affinity Purification (Column Process)

To affinity purify the subtracted serum, an appropriate affinity support must be selected. The antigen is immobilized on a solid phase support. For example, synthetic peptides or proteins can be attached to CNBr Sepharose (Pharmacia) or Affigel 15 (Bio-Rad), which couple through primary amines. If use of an amine is not optimal, one may couple via carbodiimide bonding utilizing EAH Sepharose 4B (Pharmacia) crosslinked with EDC (Pierce). In general, 20 mg of peptide per mL of gel should be sufficient.

1. Affinity matrix can be poured into a 1-mL column (washed with 1 mg/mL BSA) and equilibrated with TBST buffer. Column can be stored in TBST buffer with 0.01% w/v sodium azide at 4°C.

2. Load 10 mL of subtracted serum onto column. Pass flow-through over column four times. Save the final flow-through for SDS-PAGE analysis.
3. Wash the affinity column four times with 10 mL of TBST buffer.
4. Elute 0.5-mL fractions with 10 mL of 150 m*M* glycine, pH 2.8 (*see* **Note 9**), into microcentrifuge tubes containing 100 m*M* Tris, pH 11.0, to neutralize (the number of microliters needed to neutralize should be predetermined).
5. Wash column with TBST containing 0.01% w/v sodium azide for storage.
6. 10 µL of each fraction can be analyzed by SDS-PAGE followed by Coomassie blue staining. The more concentrated eluate fractions are pooled to minimize volume. Antibodies may be dialyzed into appropriate buffers at this point if required.
7. Final concentration can be estimated by comparison to BSA standards by SDS-PAGE analysis or Bradford protein determination assays.

3.5.3. Storage of Affinity-Purified Antibodies

If the concentration of the affinity-purified antibody is low, it is recommended that they be concentrated to approx 0.5–2 mg/mL with a 30-kDa Centricon-filter apparatus prior to storage. Otherwise, include BSA to 1 mg/ml to reduce loss of the antibody by absorption to the walls of the storage tube. Purified antibodies should be aliquoted into 5- to 10-µL aliquots in 500-µL microcentrifuge tubes to minimize repeat freeze-thawing and stored at –80°C. Sodium azide can be added to the thawed antibodies (final concentration 0.01%) to maintain storage at 4°C.

3.6. Applications

Antibodies have a multitude of applications in the retinoid field. In this section, we discuss special considerations for a few of these applications, namely: Western blotting, immunohistochemistry, gel mobility super-shift assays, and microinjection.

3.6.1. Western Blotting

Immunoblotting procedures for the detection of retinoid receptors can be challenging when searching for endogenous protein, which may be present in relatively low concentrations in whole-cell and nuclear extracts. Preparations made from cell lines known to express higher levels of receptors, such as the human 293 cell line and the RXRα-nuclear receptor, are most likely to provide a satisfactory result. Affinity purified antibodies, which are the reagent of choice for immunoblotting, may still be capable of cross reactions, it is thus recommended that, as a control, full-length overexpressed receptor be run to verify the location of the expected band. To resolve issues of nonspecific crossreactivity, absorption of the antisera with purified receptor, fusion protein, or peptide should also confirm the specificity of the antibody. Immunoblotting yields the least ambiguous results when the receptor molecule predominates in the sample; for example, when detecting receptors in preparations of overexpressed protein (e.g., bacterial/baculoviral or COS cell extracts) and for following the stepwise biochemical purification of a receptor.

3.6.2. Immunohistochemistry

Immunohistochemistry utilizing the antibodies to the retinoid receptors demonstrates the expected nuclear localization. Although the results in transfected cells can be impressive, it is also possible to visualize endogenous protein, especially when utilizing cell lines known to express the receptor of interest at an elevated level. It is also possible to detect the PML-RAR fusion protein in acute promyelocytic-leukemia cells *(5,6)*, which localizes to a speckled nuclear pattern. For the visualization of endogenous protein in patient samples, fixation of the cell preparation with analytical grade reagents is critically important. Fixation of cytospun human bone marrow or cell lines carried out utilizing cold acetone-methanol (1:1) and air-drying followed by standard procedures for antibody labeling *(8)* was found to be successful for immunodetection with antibodies to the retinoid receptors. Again, superior results are obtained with affinity-purified antibodies. To reduce background fluorescent signal, the fluorescently conjugated secondary antibody (*see* **Note 10**) can be preabsorbed by incubating with formaldehyde-fixed cells prior to use, then pelleting out the cell debris before applying the supernatant containing the antibody to the specimen. The fixed cells utilized should correspond to, or at least have the same species identity, as the cells to be examined by immunohistochemistry.

3.6.3. Gel-Mobility Supershift Assays

The utilization of antibodies for supershift assays involves a simple additional step in the gel-mobility shift assay *(21)*. Briefly, for each reaction, 5–10 µg of cell extract or 1–2 µL of in vitro-translated protein bound to 1 ng ^{32}P-end labeled double-stranded DNA probes are incubated on ice for 15 min with antibody. It is suggested that titrations of 1–5 µL antisera or diluted affinity-purified antibodies, are utilized when testing a new antibody. The resulting complexes are resolved by electrophoresis on a nondenaturing 5% acrylamide gel in 0.5X TBE at 150 V according to standard laboratory technique *(17)*. For proper interpretation of the autoradiograph, appropriate control-DNA probes are included. When utilizing nonaffinity purified sera, one should also test preimmune sera that is not expected to cause a supershift of retarded probes. It is worth noting that while most of the time a supershift can be visualized as a slowly migrating band, with some antibodies, the band is entirely absent. This may happen if the location of the antigenic site is present in a region critical for DNA binding, causing the release of the probe fragment and thus eliminating complex formation.

3.6.4. Microinjection

Microinjection into mammalian cells provides a means to introduce inhibitory antibodies in molecular excess of the protein of interest in order to discern its function *(22)*. The number of molecules per cell ranges from several hundred for transcription factors to millions for structural proteins. Because only a very minute volume is injected (0.1 pL), affinity-purified antibodies at higher concentrations are necessary (mg/mL range). Antibodies must be in a physiologically compatible buffer that does not contain detergents or sodium azide. Not all antibodies are suitable for microinjection because immuno-neutralization is epitope dependent. To inhibit the function of nuclear-receptor proteins, the antibody must be directly injected into the nucleus of the cell, because antibodies are too large to diffuse through the nuclear envelope. Injected cells can be identified immunohistochemically by labeling with a fluorescently conjugated secondary antibody. Thus, microinjection provides a technical approach through which the phenotypic consequences of inhibiting a receptor can be analyzed in vivo.

4. Notes

1. It is worth noting that different rabbits provide a different immune response. It may therefore be necessary to immunize two or more animals to obtain an antibody with the desired characteristics.
2. In the case of peptide-generated sera, a second conjugated peptide should be used (e.g., if KLH-conjugated peptide was used for immunization, BSA-conjugated peptide would be used) for the ELISA to prevent the detection of the carrier moiety. As a control, the conjugating moiety is also immobilized (e.g., BSA alone).
3. Sodium azide should not be used for in vivo applications and has been shown to inhibit horseradish peroxidase.
4. Western blotting can also be useful to determine the titer and specificity of the antibody. In the case of the utilization of a full-length or truncated receptor as an antigen, or if the synthetic peptide selected contains some identity with a related receptor, it is recommended that the resulting antibodies be tested in a Western blot against other members of the receptor family to determine specificity. This can be accomplished by running either transfected COS-7 cell extracts or overexpressed bacterial/baculoviral lysates by standard PAGE/Western-blotting protocols.
5. To identify the region to which antibodies against a full-length receptor are reactive, one may also take advantage of domain-swapped and truncated-receptor

recombinant-expression plasmids that were constructed to study functional aspects of the receptors. These chimeric and truncated proteins can be expressed by COS-cell transfection and immunoblotted to determine the domains targeted by the antibody.

6. To remove liquid from the microtiter plate, invert plate and give a sharp downward jerk.

7. We found that during the subtractive treatment of the antisera, not only GST antibodies but nonspecific antibodies are also removed. This procedure alone can substantially improve the quality of the antibody reagent. Thus, we now routinely use this process independent of the nature of the immunogen.

8. If affinity-purified antibody will be used in microinjection experiments, it is necessary to perform the affinity purification in the absence of detergents and sodium azide. TBST buffer can be substituted with TEK buffer (20 mM Tris pH 8.0, 1 mM EDTA, 50 mM KCl).

9. It is sometimes necessary to decrease the pH of the 150 mM glycine to 2.5 in order to elute higher-affinity antibodies from the column.

10. We recommend Jackson Laboratories as a source of secondary antibodies for immunodetection because they offer an excellent assortment of affinity-purified reagents with minimal crossreactivity.

Acknowledgments

We are indebted to Drs. Jean Rivier, Wylie Vale, Joan Vaughan, and Carl Hoeger (Salk Institute) for selection and synthesis of hRXRα peptide. Dr. Christopher Wright (Vanderbilt University) was most generous in sharing detailed antibody affinity-purification protocols and he offers to provide these protocols to interested parties upon request. Vickie J. LaMorte is supported by the American Cancer Society and JAD by the Damon Runyon-Walter Winchell Cancer Research Fund. This work was carried out in the laboratory of Dr. Ronald M. Evans (HHMI Investigator, Gene Expression Laboratory, Salk Institute) and was supported by program project (CA54418).

References

1. Mangelsdorf, D. J. and Evans, R. M. (1995) The RXR Heterodimers and Orphan Receptors. *Cell* **83,** 841–850.

2. Heyman, R. A., Mangelsdorf, D. J., Dyck, J. A., Stein, R. B., Eichele G., Evans, R. M., and Thaller, C. (1992) 9-*cis* Retinoic acid is a high affinity ligand for the retinoid X receptor. *Cell* **68,** 397–406.

3. Kliewer, S. A., Umesono, K., Heyman, R. A., Mangelsdorf, D. J., Dyck, J. A., and Evans, R. M. (1992a) Retinoid X receptor-COUP-TF interactions modulate retinoic acid signaling. *PNAS,* **89,** 1448–1452.

4. Kliewer, S. A., Umesono, K., Noonan, D. J., Heyman, R. A., and Evans R. M. (1992b) Convergence of 9-cis retinoic acid and peroxisome proliferator signalling pathways through heterodimer formation of their receptors. *Nature* **358,** 771–774.

5. Dyck, J. A., Maul, G. G., Miller, W. H., Kakizuka, A., and Evans, R. M. A novel macromolecular structure is a target of the PML-RAR oncoprotein. (1994) *Cell* **76,** 333–343.

6. Dyck, J. A., Warrell, R., Evans, R. M., and Miller, W. H. (1995) Rapid diagnosis of acute Promyelocytic leukemia by immunohistochemical localization of PML/RARα protein. *Blood* **86,** 862–867.

7. Smith, D. B. and Johnson, K. S. (1988) Single step purification of polypeptides expressed in *Escherichia coli* as fusions with glutathione transferase. *Gene* **67,** 31–40.

8. Harlow, E. and Lane, D. (1988) *Antibodies: A Laboratory Manual,* Cold Spring Harbor Laboratory, Cold Spring Harbor, NY.

9. Rivier, J., Rivier, C., and Vale, W. (1984) Synthetic Competitive Antagonists of Corticotropin-Releasing Factor: Effect on ACTH Secretion in the Rat. *Science* **224,** 889–891.

10. Ausubel, F. M., Brent, R., Kingston, R. E., Moore, D. M., Seidman, J. G., Smith, J. A., and Struhl, K. (eds.) Immunology, in *Short Protocols in Molecular Biology* (1995) A Compendium of Methods from Current Protocols in Molecular Biology. Wiley, New York, pp. 11.28–11.30.

11. Kyte, J. and Doolittle, R. F. (1982) A Simple Method for Displaying the Hydropathic Character of a Protein. *J. Mol. Biol.* **157,** 105–132.

12. Corrigan, A. J. and Huang, P. C. (1982) A BASIC Microcomputer Program for Plotting the Secondary Structure of Proteins. *Comput. Programs Biomedicine* **15,** 163–168.

13. Vaughan, J. M., Rivier, J., Corrigan, A. Z., McClintock, R., Campen, C. A., Jolley, D., Voglmayr, J. K., Bardin, C. W., Rivier, C., and Vale, W. (1989) Detection and purification of inhibin using antisera generated against synthetic peptide fragments, *Methods in Enzymology,* vol. 168, Conn, P. M., ed., Academic, Orlando, FL, pp. 588–617.

14. Reichlin, M. (1980) Use of Glutaraldehyde as a Coupling Agent for Proteins and Peptides. *Methods Enzymol.* **70,** 159–165.

15. Yang, N., Schule, R., Mangelsdorf, D. J., and Evans, R. M. (1991) Characterization of DNA binding and Retinoic Acid binding Properties of Retinoic Acid Receptor. *Proc. Natl. Acad. Sci. USA* **88,** 3559–3569.

16. Laemmli, E. K. (1970) Cleavage of structural proteins during the assembly of the head of bacteriophage T4. *J. Immunol. Methods* **47,** 303–307.

17. Sambrook J., Fritsch E. F., and Maniatis T. (1989) *Molecular Cloning: A Laboratory Manual,* 2nd ed., Cold Spring Harbor Laboratory, Cold Spring Harbor, NY.

18. Wright, C. V., Schnegelsberg, P., and De Robertis, E. M. (1989) XIH box 8: A novel Xenopus homeo protein restricted to a narrow band of endoderm. *Development* **105,** 787–794.

19. Oliver, G., Wright, C. V., Hardwicke, J., and De Robertis, E. M. (1988) Differential antero-posterior expression of two proteins encoded by a homeobox gene in *Xenopus* and mouse embryos. *Embo J.* **7,** 3199–3209.

20. Wall, N. A., Jones, C. M., Hogan, B. L., and Wright, C. V. (1992) Expression and modification of Hox 2.1 protein in mouse embryos. *Mech. Dev.* **37,** 111–120.
21. Mangelsdorf, D. J., Umesono, K., Kliewer, S. A., Borgmeyer, U., Ong, E. S., and Evans, R. M. (1991) A direct repeat in the cellular retinol-binding protein type II gene confers differential regulation by RXR and RAR. *Cell* **66,** 555–561.
22. LaMorte,V. J., Goldsmith, P. K., Spiegel, A. M., Meinkoth, J. L., and Feramisco, J. R. (1992) Inhibition of DNA synthesis in living cells by microinjection of Gi 2 antibodies. *J. Biol. Chem.* **267,** 691–694.

17

Detection of RARs and RXRs in Cells and Tissues Using Specific Ligand-Binding Assays and Ligand-Binding Immunoprecipitation Techniques

Elizabeth A. Allegretto

1. Introduction

Ligand-binding assays have been used extensively in the past to detect and characterize intracellular receptors. In fact, the initial discoveries of the first known members of this family of proteins were achieved by the demonstration of binding of radioactively labeled hormonal ligands to the target tissues containing receptors for those hormones *(1–5)*. Since then, ligand-binding assays have been used as a tool to characterize receptor–ligand interactions including determination of the affinities and specificities of the receptors for various ligands such as naturally occurring endogenous hormones as well as synthetic compounds including drugs. Ligand-binding assay methodologies include saturation-binding techniques and competition assays. Saturation binding is a method which requires a radiolabeled ligand and its receptor. This method used in combination with Scatchard analysis enables the direct measurement of dissociation constants (K_d values), which are an indicator of the affinity of the receptor for the labeled ligand. Competition assays allow the determination of receptor-binding characteristics of compounds that are not radiolabeled by measuring the relative abilities of the unlabeled ligands to compete with a known, labeled ligand for binding to a receptor. Dissociation constants for these interactions can then be calculated by using the determined IC_{50} values (the concentration of unlabeled compound required to prevent 50% of the radiolabeled ligand from binding to the receptor). A variety of techniques have been developed to separate the receptor-bound ligand from unbound ligand in the sample *(6–8)*, which have been modified and improved over the years for each of the different intracellular receptors.

From: *Methods in Molecular Biology, Vol. 89: Retinoid Protocols*
Edited by: C. P. F. Redfern © Humana Press Inc., Totowa, NJ

With the advent of the production of antibodies against the intracellular receptors, various immunoassays have been developed for use in receptor characterization. The first applications used antireceptor polyclonal antibodies to increase the sedimentation of a radiolabeled ligand–receptor complex in sucrose-density gradients *(9–10)*. As researchers set out to produce antireceptor monoclonal antibodies (MAbs), many used a type of tritiated ligand-binding immunoprecipitation technique as a method to screen fusion clones for the presence of specific secreted antibody *(11–13)*. Radiolabeled ligand was incubated with a crude-receptor preparation followed by the incubation of the ligand–receptor complex with the supernatant of the hybridoma. If a specific antireceptor antibody was present the radiolabeled ligand receptor–antibody complex could be precipitated and radioisotope measured by scintillation counting. Upon the successful development of specific antireceptor MAbs the ligand-binding immunoprecipitation assay has been further adapted for use in detection and characterization of receptors in cells and/or tissues. The ligand-binding immunoprecipitation assay is useful in that it is a highly specific dual-function assay that exploits two characteristics of the receptor: receptor ligand-binding interaction and receptor–antibody interaction. The assay proves especially useful in the measurement of certain receptors or in the use of certain ligands, i.e., upon the use of synthetic ligands that bind to multiple receptors (RU-486 binding to glucocorticoid receptor and progesterone receptor) or in the measurement of one member of a family of receptors that share a common ligand, i.e., the thyroid receptors or the retinoid receptors.

There are six known retinoid receptors: three retinoic acid receptors (RARs α,β,γ) and three retinoid X receptors (RXRs α,β,γ), all of which bind to 9-*cis* retinoic acid (9cRA), and three of which (the RARs) also bind to all-*trans* retinoic acid (tRA) *(14,15)*. Most cell types that have been assayed have been shown to express at least one RAR subtype and one RXR subtype (by RNA techniques) and therefore a conventional ligand-binding assay would not allow the determination of which receptor subtype protein(s) were present. In other words, in order to detect the specific retinoid-receptor subtype proteins endogenously present in cells and tissues a more specific technique is required. To achieve this end there are a number of techniques that can be employed. Ligand-binding assays would be suitable if subtype-selective ligands were available for the receptor subtype of interest. Whereas certain RARα subtype-selective ligands have been reported that would be potentially useful in this type of assay *(16)*, there are not as of yet subtype-selective compounds of sufficient affinity and specificity available for each of the six individual retinoid receptors. However, subfamily-selective ligands that fit these criteria have been developed.

These are ligands that bind to each of the RXRs or to each of the RARs with high affinity and specificity, without binding to the other subfamily's members. These compounds include LG100268 *(17)* and potentially others *(18)* for RXRs and tRA and TTNPB (**ref.** *19*, and our data not shown) and others *(20)* for RARs. These compounds can be used as competitors of [^{3}H]-9cRA or [^{3}H]-tRA in ligand-binding assays to determine the total amount of each receptor subfamily in a given cell or tissue type. As high affinity and high-specificity subtype-selective ligands are developed this assay could easily be extended to a retinoid receptor subtype-selective binding assay.

Other methods that are useful for detection of the retinoid-receptor proteins in cells and/or tissues involve the use of antireceptor antibodies. Direct Western blotting is usually not useful for detection of the low levels of endogenous receptor that are encountered in cells because the signal-to-noise ratio is too low to be useful. Immunocytochemistry and immunohistochemistry are techniques that utilize antibodies (monoclonals are generally required) to visualize receptor through cell or tissue staining. These techniques should ideally utilize more than one antireceptor antibody per receptor in order to increase the believability of the result. Advantages of this assay include the ability to use a small sample of cells or a section of tissue, or samples that are not fresh, such as formalin-fixed samples. Such work has been described using anti-RAR antibodies prior to the discovery of the RXRs *(21,22)*. Ligand-binding immunoprecipitation is also a useful technique that is more specific than most antibody techniques, in that it employs the receptor-ligand binding function as well as receptor-antibody binding in the assay. Advantages of the assay are a high confidence level in the result, an ability to quantify various retinoid-receptor subtypes in a given cell or tissue, and the ability to detect functional receptor (i.e., ligand-binding capable receptor). Another advantage of this assay, unlike the staining techniques, is that polyclonal antibodies can generally be used with satisfactory results. A disadvantage of the assay is that fresh or frozen protein extracts are needed in fairly large amounts (2–10 mg total soluble protein extract, depending on how many of the six receptors will be assayed for). We have generated subtype-selective antibodies against each of the six retinoid receptors, which react only with their respective recombinantly expressed receptors by both Western-blotting and immunoprecipitation techniques *(23)*. These antibodies have proven useful in the detection of endogenous-retinoid receptors in cells and tissues by ligand binding-immunoprecipitation assays. This chapter discusses the use of selective ligand-binding assays and ligand-binding immunoprecipitation techniques for detection and quantification of endogenously expressed retinoid-receptor subtype proteins.

2. Materials

2.1. Ligand-Binding Assay

1. [³H]-9cRA (~28 Ci/mmol *[24]*, Ligand Pharmaceuticals, San Diego, CA) or [³H]-tRA (NEN): ~50 Ci/mmol, store at –80°C in ethanol or toluene/ethanol, shielded from light.
2. Nitrogen gas.
3. Binding buffer (KT0.15C): 0.15 *M* KCl, 10 m*M* Tris-HCl, pH 7.4, 0.5% CHAPS detergent (Boehringer Manheim), 8% glycerol. Store at 4°C shielded from light.
4. Unlabeled competitor ligands: 9cRA, tRA, and/or other known receptor subfamily- or subtype-specific ligands, i.e., LG100268, an RXR-subfamily selective ligand *(17)* (*see* **Notes 1** and **2**). Store in ethanol at –80°C shielded from light.
5. Cell extracts expressing recombinant retinoid receptors or cell or tissue extracts expressing endogenous levels of retinoid receptors (store as aliquots at –80°C).
6. Hydroxylapatite slurry: 50% w/v in KTO.15C buffer.
7. Scintillation fluid and counter.

2.2. Ligand-Binding Immunoprecipitation Assay

1. All reagents listed in **Subheading 2.1.** for ligand-binding assay.
2. Siliconized microcentrifuge tubes (Fisher).
3. Anti-retinoid receptor (RAR and/or RXR) subtype-selective antibodies (polyclonal or monoclonal, purified away from serum components), stored at 4°C or –20°C.
4. Protein A Sepharose (Pharmacia), freshly equilibrated in KT0.15C buffer, and stored at 4°C, shielded from light.

3. Methods

3.1. Receptor Subfamily-Selective Ligand-Binding Assay

We have developed a ligand-binding assay especially suited for retinoids *(14)* utilizing a modified version of a hydroxylapatite method for separation of bound ligand from unbound ligand *(8)*. The availability of RXR subfamily-selective ligands *(17,18)* has allowed the performance of subfamily-specific binding assays which can be performed to quickly determine the total amounts of RARs and/or RXRs in a certain cell type or tissue *(23)*. This assay is easily converted to a retinoid receptor subtype-selective binding assay upon the availability of a subtype-selective ligand (a ligand that binds to one of the retinoid receptor subtypes without binding to the others). We have utilized a highly specific and high affinity RXR-selective ligand, LG100268, that was synthesized at Ligand Pharmaceuticals *(17)* for the purposes of this assay. This compound binds equally well to each RXR subtype with high affinity (K_ds = 1–2 n*M* *[17]*; equal to that of 9cRA *[14]*) without binding to the RARs (K_d >5000 nM *[17]*). Protein extracts are prepared from the cell or tissue of interest and are incubated with [^{3}H]-9cRA at concentrations sufficient to saturate both RARs and RXRs (10–20 nM). Binding is performed in the absence and presence of a 200-fold molar excess of unlabeled competitor ligand. Specific binding is determined by subtracting nonspecific binding (in the presence of competitor) from total binding (in the absence of the competitor). If the competitor ligand used is 9cRA, the resultant specific counts are representative of all RAR and RXR in the sample (as both RARs and RXRs bind to 9cRA). If the competitor used is an RXR-selective (LG100268) or RAR-selective ligand (tRA or TTNPB), then the specific counts derived from the use of that competitor is representative of either the total RXR component or total RAR component of the sample, respectively.

1. Prepare cell or tissue extracts by standard procedures (*see* **Note 3**). Wash cultured cells or tissues well with PBS prior to preparation of extract to remove serum components (albumin binds to retinoic acid). Buffers containing high salt concentrations (0.3–0.6 M KCl) are usually used to ensure removal of receptor from DNA. Determine protein concentrations of all samples, store in aliquots at –80°C, and avoid freeze–thawing more than once.

2. Aliquot radioligand into borosilicate-glass tubes in amounts necessary to achieve final concentrations of 10–20 nM (to ensure saturation of the receptors in the sample) in duplicate or triplicate at each concentration. Displace air in the opened stock vial with a gentle stream of nitrogen gas to help prevent compound oxidation before returning to storage at –80°C. Work under dim light.

3. Use a gentle stream of nitrogen gas to evaporate solvents that are detrimental to receptor integrity (i.e., toluene). Resuspend dried ligand in a suitable volume of ethanol (final concentration in assay to be 4–8% for ligand solubility).

4. Add unlabeled competitor ligands to half of the assay tubes in duplicate or triplicate in ethanol to achieve a final ligand concentration of 200-molar excess of the labeled ligand (take into consideration the receptor-binding affinity of the selective ligand that you are using and adjust the concentration accordingly) (*see* **Notes 1** and **2**).

5. Add protein extracts (for example, 10–50 µg for high levels of recombinantly expressed proteins or ~500 µg per assay tube for measuring endogenous receptor levels in cells or tissue samples). More or less protein may be required depending on the receptor amounts in the sample.

6. Bring up volume accordingly with KT0.15C buffer (usually 100 µL for 50 µg protein extract and 500 µL for 500 µg protein extract).

7. Incubate ligands with proteins overnight at 4°C shielded from the light.

8. Add 50–100 µL of hydroxylapatite slurry. Incubate 30 min on ice or at 4°C. Mix by vortexing of tubes every 10 min. Add 1–2 mL KT0.15C, vortex, and spin at 4°C for 2 min at 2000g. Decant supernatant. Repeat three times.

9. Transfer slurry with water or buffer to a scintillation vial with two aliquots of 500 μL. Add scintillation fluid and count. Determine specific counts for each receptor and convert into fmol/mg protein or other measure.

Figure 1 shows the application of this method to determine the relative amounts of RARs and RXRs in four cell lines. Tritiated 9cRA (open bars) was used in the absence and presence of competitor ligands to assay for the total amounts of each receptor subfamily in the cultured cells. For example, the total amount of RXRs in HeLa cells is represented by ~2000 dpm specific binding (A, lane 6; competitor is LG100268), while the total RAR component is ~3000 dpm (A, lane 4; competitor is tRA). These two numbers added together should represent the total RAR and RXR complement in the cells (~5000 dpm) and is independently confirmed by using 9cRA as a competitor (A, lane 5). This technique gives a quick determination of the retinoid-receptor subfamily protein complement in a particular sample with a high degree of confidence. If tritiated 9cRA is not available, it is possible to use tritiated tRA; however, only the cellular RAR complement can be determined using this ligand, as it binds only to RARs, not to RXRs. **Figure 1** illustrates that using tritiated tRA (solid bars), HeLa cells display ~6500 specific dpm representative of RARs (A, lane 2; 9cRA as competitor) and is approximately equal to the amount of RARs determined by using tritiated 9cRA (~3000 dpm), considering that the specific activity of [^{3}H]-tRA is twice as high as that of [^{3}H]-9cRA used in these experiments. The increased specific binding observed with [^{3}H]-tRA in the presence of unlabeled tRA (A–D, lane 1) as compared with that in the presence of unlabeled 9cRA (A–D, lane 2) is due to the binding of tRA to CRABP in the cell extracts (9cRA does not bind to CRABPs; *see* **ref. 23**).

3.2. Receptor Subtype Ligand-Binding-Immunoprecipitation Assay

This method utilizes the ligand-binding assay for retinoids described previously *(14,23)* and in **Subheading 3.1.**, in combination with immunoprecipitation with anti-retinoid receptor subtype-selective antibodies *(23)*. This assay is therefore a modification of earlier described intracellular receptor ligand-binding immunoprecipitation techniques (**ref. *11–13; see* Note 4**).

1. See protocol for ligand-binding assay. Incubate ligands with protein extracts for 16 h at 4°C.
2. Aliquot ligand-receptor complexes (~500 μg total souble-protein extract in a total volume of ~500 μL binding buffer) into siliconized microcentrifuge tubes. Add purified polyclonal antibodies (500–800 μg per assay tube depending on the titer of the antibody) or MAbs (3–10 μg per assay tube, depending on affinity of antibody) in duplicate tubes. Incubate for 2–4 h at 4°C, shielded from light (*see*

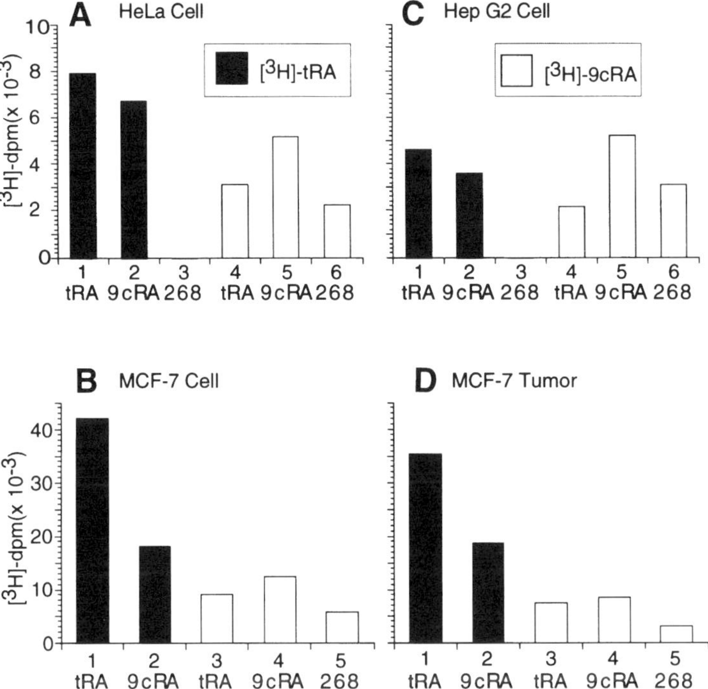

Fig. 1. Determination of retinoid-receptor subfamily complement in various cell extracts using a subfamily-specific ligand binding assay. Extracts (500 µg) from HeLa (**A**), MCF-7 (**B**), or Hep G2 (**C**) cells, or a MCF-7 cell-derived tumor from an estrogen-treated mouse *(23)* (**D**) were incubated in duplicate with 10 n*M* [3H]-tRA (A,C: lanes 1–3; B,D: lanes 1 and 2) or with 10 n*M* [3H]-9cRA (A,C: lanes 4–6; B,D: lanes 3–5) in the presence of a 200-fold molar excess of unlabeled tRA (A,C: lanes 1 and 4; B,D: lanes 1 and 3), unlabeled 9cRA (A,C: lanes 2 and 5; B,D: lanes 2 and 4), or unlabeled LG100268 (A,C: lane 3 and 6; B,D: lane 5). Specific binding was determined by subtracting the counts obtained in the presence of a 200-fold molar excess of competitor ligand (nonspecific binding) from the counts obtained in the absence of competitor (total binding).

Note 5). Add appropriate amounts of species-specific IgG-control antibodies to separate tubes.

3. Add protein-A Sepharose, protein-G Sepharose or other precipitating antibody (1:1 slurry in KT0.15C, 150–500 µL) for 45 min at 4°C, rotating end over end, shielded from light.
4. Spin at 4°C, 1 min at ~15,000 rpm in microcentrifuge and remove supernatants, wash four times, each with 0.8–1.0 mL KT0.15C, to remove unbound counts.
5. Resuspend washed Sepharose pellets in water or buffer (two aliquots of 0.5 mL each) and transfer to scintillation vials to count for tritium dpm.

6. Specific counts are determined by subtracting the counts precipitated with the nonspecific antibodies from the counts precipitated with the antireceptor antibodies. A more specific assay is obtained by including tubes with a 200-fold molar excess of competitor ligand along with the labeled ligand. This adds another degree of specificity to the assay and is especially useful when measuring receptor levels that are very low, near to the limit of detection.

Figure 2 shows that each of the anti-retinoid receptor antibodies that we developed is able to recognize its respective receptor in its native conformation, i.e., the antibodies precipitate the ligand-occupied receptors. Also, each antibody only reacts with the receptor that it was generated against and not to the other receptors; as shown by the fact that each of the recombinantly expressed retinoid receptors *(14,23)* labeled with either [³H]-9cRA or [³H]-tRA are immunoprecipitated only with their respective specific antibodies (**Fig. 2** and data not shown; for additional details, *see* **ref. *23***). **Figure 3** shows a representative experiment in which the subtype-selective antibodies were utilized to detect endogenously expressed retinoid receptors in MCF-7 cells and an MCF-7 cell-derived tumor (for additional details, *see* **ref. *23***). **Table 1** is a compilation of the retinoid–receptor complement of a number of cultured cell, tissues, and tumors that we have analyzed. It is interesting to note that, from the receptor-protein levels that we have observed from a variety of cell types, it is indicated that RARα, RARγ, and RXRα are the most commonly expressed of the retinoid-receptor subtype proteins and are also the most abundant. Analysis of mouse kidney and liver has shown that RXRα and RARγ are the most predominant receptor subtypes, although all six are expressed in those tissues (**Table 1** and data not shown). Also of interest, we have yet to observe any cell or tissue that does not express at least one subtype of each of the retinoid–receptor subfamilies, implicating the retinoid receptors in the basic functions necessary to cells.

4. Notes

1. In using the subfamily-specific (or subtype-selective) ligand-binding assay it is essential that the compounds chosen as specific competitors have a high enough degree of selectivity, i.e., the differential of binding activity between the RXRs and RARs should be at least 1000 times. This is necessary because the ligands are used in a 200-fold molar excess in the competition ligand-binding assay and at these concentrations, they must not bind to the other receptors.

2. In the determination of the concentration of an unlabeled compound to use as a tritiated-ligand competitor, the affinity of the competitor compound should be known for each receptor subtype, and this affinity should be used to calculate the amount that is necessary to completely compete off the tritiated ligand from

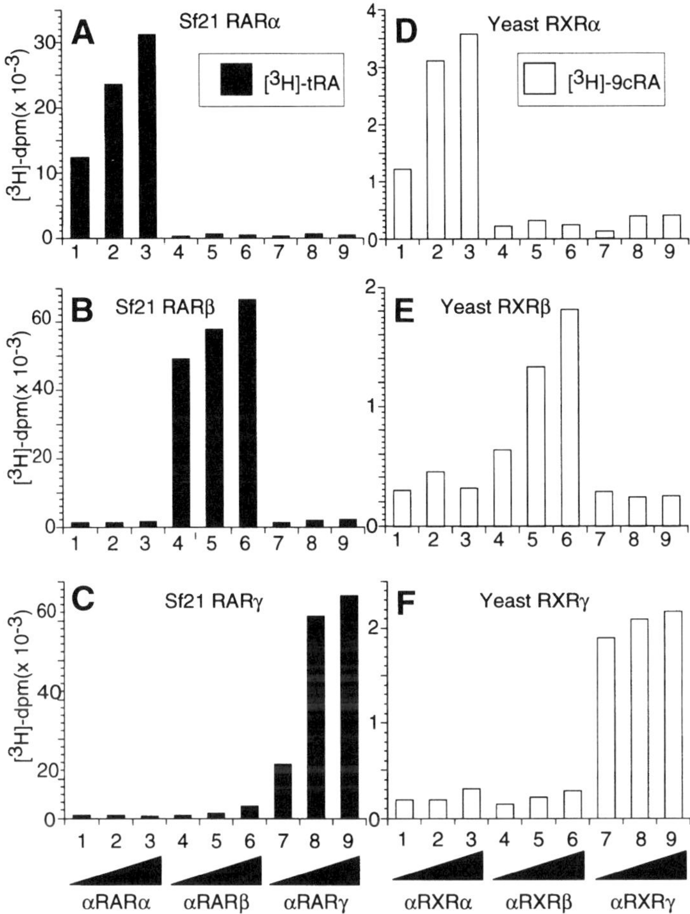

Fig. 2. Ligand-binding immunoprecipitation of ligand-receptor complexes using subtype-specific retinoid–receptor antibodies. Sf21-cell extracts (50 µg total-soluble protein) containing recombinantly expressed hRARα (**A**), hRARβ (**B**), or hRARγ (**C**) were labeled with 10 nM [^{3}H]-tRA; incubated with anti-RARα (A–C: 1 µg, lane 1; 5 µg, lane 2; 25 µg, lane 3), anti-RARβ (A–C: 10 µL, lane 4; 50 µL, lane 5; 100 µL, lane 6), or anti-RARγ (A,B: 10 µL, lane 7; 50 µL, lane 8; 100 µL, lane 9; C: 2 µL, lane 7; 5 µL, lane 8; 20 µL, lane 9) antibodies; and then incubated with protein-A Sepharose (the pellets were washed and then counted for tritium). Yeast extracts (50 µg total-soluble protein) containing recombinantly expressed hRXRα (**D**), mRXRβ (**E**), or mRXRγ (**F**) were labeled with 10 nM [^{3}H]-9-*cis* RA, incubated with anti-RXRα (D–F: 10 µL, lane 1; 50 µL, lane 2; 100 µL, lane 3), anti-RXRβ (D–F: 10 µL, lane 4; 50 µL, lane 5; 100 µL, lane 6), or anti-RXRγ (D–F: 10 µL, lane 7; 50 µL, lane 8; 100 µL, lane 9) antibodies and then precipitated as in A–C.

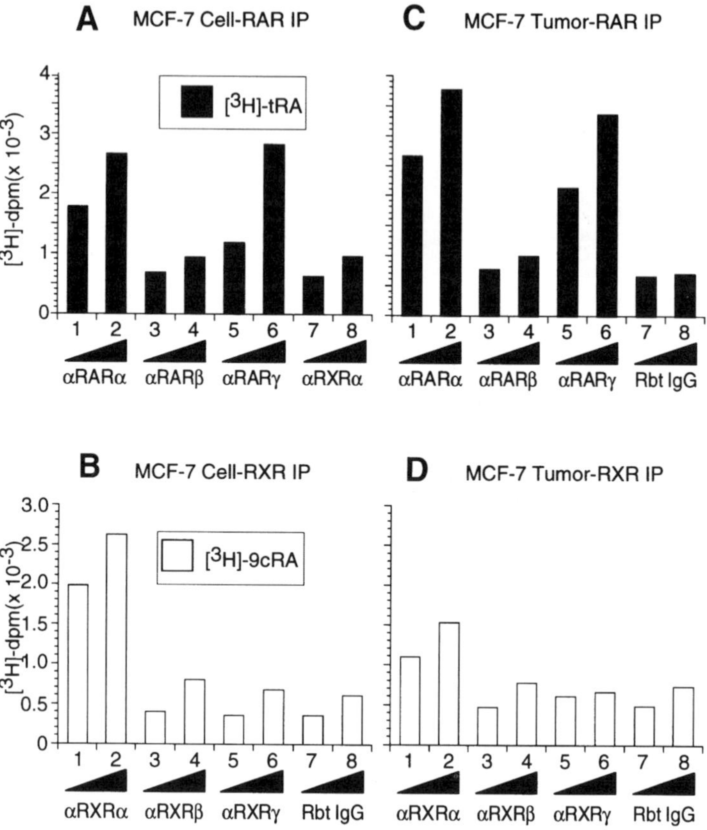

Fig. 3. Ligand-binding immunoprecipitation of endogenously expressed retinoid receptors in MCF-7 cells and in a MCF-7 cell-derived tumor. Extracts (500 μg) from MCF-7 cells or tumor were labeled with 10 n*M* [³H]-tRA (**A,C**) or 10 n*M* [³H]-9cRA (**B,D**), and incubated with primary antibodies (amounts as in **Fig. 2**) generated against RARα (A,C; lanes 1 and 2), RARβ (A,C; lanes 3 and 4), RARγ (A,C; lanes 5 and 6), RXRα (A; lanes 7 and 8; B,D; lanes 1 and 2), RXRβ (B,D; lanes 3 and 4), RXRγ (B,D; lanes 5 and 6), or with nonspecific rabbit IgG (B–D; lanes 7 and 8) and then precipitated as in **Fig. 2**.

the receptor. In other words, because LG100268 and 9cRA have approximately equal affinities for RXRs, a 100- to 200-fold molar excess of either LG100268 or 9cRA is sufficient to compete off [³H]-9cRA from the RXRs in the sample. However, if a compound's affinity for RXRs is 10 times lower than that of 9cRA, then a 1000- to 2000-fold molar excess is required to completely compete off the 9cRA. Make sure that at the concentration used, the compound will not bind to the other receptors (*see* **Note 1**). For example, the concentration of LG100268

Table 1
**Retinoid Receptor Subtype Protein Levels
in Various Cultured Cell, Tissue, and Tumor Extracts**

Source	RARα	RARβ	RARγ	RXRα	RXRβ	RXRγ
HL60 cells	30[a]	ND[b]	ND	60	ND	ND
HeLa cells	28	9[c]	16	50	28	9
Hep G2 cells	20	5	ND	45	ND	ND
CV-1 cells	38	5	46	42	ND	15
MCF-7 cells-E$_2$	32	ND	35	60	ND	ND
MCF-7 cells+E$_2$	80	ND	34	12	ND	ND
MCF-7 tumor+E$_2$	57	ND	49	24	ND	ND
ME-180 tumor	7	ND	28	30	ND	ND
Mouse kidney	35 (total RARs)			52 (total RXRs)		
Mouse liver	16 (total RARs)			67 (total RXRs)		

[a]fmol receptor/mg protein, values were the mean of at least two experiments, variation was within 20%.

[b]ND: not detected.

[c]Values under 10 fmol/mg were confirmed by immunoprecipitation of extracts incubated with tritiated ligand in the absence and presence of a 200-fold excess of unlabeled ligand.

used to compete for 10 nM [^{3}H]-9cRA (200-fold molar excess = 2 μM), does not bind to the RARs *(17)*. tRA is also a good compound to use as [^{3}H]-9cRA competitor to determine the presence of RARs, because its affinity for the RARs is equal to that of 9cRA *(14,15)*.

3. Upon preparation of cell or tissue-protein extracts, take care to use as gentle conditions as possible to lyse the cells. Ligand-binding activity can be partially or totally destroyed during cell lysis. One freeze–thaw cycle followed by swelling of cells in low-salt buffer and dounce homogenization on ice only until cells are lysed is one of the best methods. KCl or NaCl is then added to bring up the salt concentration to 0.4–0.6 M in order to remove receptor from chromatin components. For tissues, a tissue homogenizer such as a Polytron (Brinkmann Instruments) should be used with rinsed, minced tissue on ice, with as little frothing as possible. After extracts are prepared, glycerol can be added up to 20% of the total-extract volume to help stabilize receptor. Aliquot extracts and store at –70°C, avoid more than one freeze/thaw cycle of extracts, as ligand-binding activity will be reduced.

4. Detection limits of the ligand-binding immunoprecipitation assay are about 5 fmol receptor /mg total-soluble protein, using the tritiated ligands now available that have specific activities of ~28–50 Ci/mmol. The production of ligands with higher specific activities will increase the detection limit and/or help to reduce the amount of protein extract that is required per assay point.

5. The maximum efficiency of the ligand-binding immunoprecipitation assay that we have achieved is about 80–90%, as determined by calculation of the amount

of receptor that is precipitated vs the amount of receptor that is present as determined by ligand-binding assay. To ensure that good efficiency of ligand–receptor binding to antibody occurs, do not incubate the liganded receptor with antibody more than 4 h, as loss of liganded receptor occurs. The Sepharose washing procedure should be quick, but efficient. It may help to precoat protein-A Sepharose with a saturating amount of the specific antibody (wash off the excess antibody prior to storage of resin) before incubation with ligand–receptor complexes to increase the efficiency of the interaction. Keep in mind that absolute receptor quantification by ligand binding is usually not possible. A certain loss is always sustained upon cell breakage, and 100% of the liganded receptor will not be precipitated with antibody. Therefore, the assays described in this chapter can be used to quantify relative amounts of the retinoid–receptor subtype proteins in cells or tissues.

Acknowledgments

I thank Dean Edwards, Wes Pike, and Rich Heyman for helpful discussions, and Matthew Titcomb and Ali Haghighi for experimental contributions.

References

1. Toft, D. and Gorski, J. (1966) A receptor molecule for estrogens: isolation from the rat uterus and preliminary characterization. *Proc. Natl. Acad. Sci. USA* **55,** 1574–1581.
2. Jensen, E. V., Hurst, D. J., DeSombre, E. R., and Jungblut, P. W. (1967) Sulfhydryl groups and estradiol-receptor interaction. *Science* **158,** 385–387.
3. Gardner, R. S. and Tomkins, G. M. (1969) Steroid hormone binding to a macromolecule from hepatoma tissue culture cells. *J. Biol. Chem.* **244,** 4761–4767.
4. Spelsberg, T. C., Steggles, A. W., and O'Malley, B. W. (1971) Progesterone-binding components of chick oviduct. 3. Chromatin acceptor sites. *J. Biol. Chem.* **246,** 4188–4197.
5. Brumbaugh, P. F. and Haussler, M. R. (1974) 1 Alpha,25 dihydroxy-cholecalciferol receptors in intestine. II. Temperature-dependent transfer of the hormone to chromatin via a specific cytosol receptor. *J. Biol. Chem.* **249,** 1258–1262.
6. Korenman, S. G. (1968) Radio-ligand binding assay of specific estrogens using a soluble uterine macromolecule. *J. Clin. Endocrinol. Metab.* **28,** 127–130.
7. Santi, D. V., Sibley, C. H., Perriard, E. R., Tomkins, G. M., and Baxter, J. D. (1973) A filter assay for steroid hormone receptors. *Biochemistry* **12,** 2412–2416.
8. Williams, D. and Gorski, J. (1974) Equilibrium binding of estradiol by uterine cell suspensions and whole uteri in vitro. *Biochemistry* **13,** 5537–5542.
9. Greene, G. L., Closs, L. E., Fleming, H., DeSombre, E. R., and Jensen, E. V. (1977) Antibodies to estrogen receptor: immunochemical similarity of estrophilin from various mammalian species. *Proc. Natl. Acad. Sci. USA* **74,** 3681–3685.

10. Logeat, F., Hai, M. T., and Milgrom, E. (1981) Antibodies to rabbit progesterone receptor: crossreaction with human receptor. *Proc. Natl. Acad. Sci. USA* **78,** 1426–1430.

11. Greene, G. L., Fitch, F. W., and Jensen, E. V. (1980) Monoclonal antibodies to estrophilin: Probes for the study of estrogen receptors. *Proc. Natl. Acad. Sci. USA* **77,** 157–161.

12. Pike, J. W., Donaldson, C. A., Marion, S. L., and Haussler, M. R. (1982) Development of hybridomas secreting monoclonal antibodies to the chicken intestinal 1α,25-dihydroxyvitamin D_3 receptor. *Proc. Natl. Acad. Sci. USA* **79,** 7719–7723.

13. Logeat, F., Hai, M. T., Fournier, A., Legrain, P., Buttin, G., and Milgrom, E. (1983) Monoclonal antibodies to rabbit progesterone receptor: crossreaction with other mammalian progesterone receptors. *Proc. Natl. Acad. Sci. USA* **80,** 6456–6459.

14. Allegretto, E. A., McClurg, M. R., Lazarchik, S. B., Clemm, D. L., Kerner, S. A., Elgort, M. G., Boehm, M. F., White, S. K., Pike, J. W., and Heyman, R. A. (1993) Transactivation properties of retinoic acid and retinoid X receptors in mammalian cells and yeast: correlation with hormone binding and effects of metabolism. *J. Biol. Chem.* **268,** 26,625–26,633.

15. Allenby, G., Bocquel, M.-T., Saunders, M., Kazmer, S., Speck, T., Rosenberger, M., Lovey, A., Kastner, P., Grippo, J. F., Chambon, P., and Levin, A. A. (1993) Retinoic acid receptors and retinoid X receptors: interactions with endogenous retinoic acids. *Proc. Natl. Acad. Sci. USA* **90,** 30–34.

16. Fukasawa, H., Iijima, T., Kagechika, H., Hashimoto, Y., and Shudo, K. (1993) Expression of the ligand binding domain-containing region of retinoic acid receptors alpha, beta and gamma in *E. coli* and evaluation of ligand-binding selectivity. *Biol. Pharm. Bull.* **16,** 343–348.

17. Boehm, M. F., Zhang, L., Zhi, L., McClurg, M. R., Berger, E., Wagoner, M., Mais, D. E., Suto, C. M., Davies, P. J. A., Heyman, R. A., and Nadzan, A. M. (1995) Design and synthesis of potent retinoid X receptor selective ligands that induce apoptosis in leukemia cells. *J. Med. Chem.* **38,** 3146–3155.

18. Lehmann, J. M., Jong, L., Fanjul, A., Cameron, J. F., Lu, X. P., Haefner, P., Dawson, M. I., and Pfahl, M. (1992) Retinoids selective for retinoid X receptor response pathways. *Science* **258,** 1944–1946.

19. Crettaz, M., Baron, A., Siegenthaler, G., and Hunziker, W. (1990) Ligand specificities of recombinant retinoic acid receptors RARα and RARβ. *Biochem. J.* **272,** 391–397.

20. Johnson, A. T., Klein, E. S., Gillet, S. J., Wang, L., Song, T. K., Pino, M., and Chandraratna, R. A. S. (1995) Synthesis and characterization of a highly potent and effective antagonist of retinoic acid receptors *J. Med. Chem.* **38,** 4764–4767.

21. Gaub, M. P., Lutz, Y., Ruberte, E., Petkovich, M., Brand, N., and Chambon, P. (1989) Antibodies specific to the retinoic acid human nuclear receptors α and β. *Proc. Natl. Acad. Sci. USA* **86,** 3089–3093.

22. Gaub, M. P., Rochette-Egly, C., Lutz, Y., Ali, S., Matthes, H., Scheuer, I., Chambon, P. (1992) Immunodetection of multiple species of retinoic acid receptor alpha: evidence for phosphorylation. *Exp. Cell. Res.* **201,** 335–346.

23. Titcomb, M. W., Gottardis, M. M., Pike, J. W., and Allegretto, E. A. (1994) Sensitive and specific detection of retinoid receptor subtype proteins in cultured cell and tumor extracts. *Mol. Endocrinol.* **8,** 870–877.
24. Boehm, M. F., McClurg, M. R., Pathirana, C., Mangelsdord, D. J., White, S. K., Hebert, J., Winn, D., Goldman, M. E., and Heyman, R. A. (1994) Synthesis of high specific activity [^{3}H]-9-*cis*-retinoic acid and its application for identifying retinoids with unusual binding properties. *J. Med. Chem.* **37,** 408–414.

18

Nonisotopic *In Situ* Hybridization for the Detection of Nuclear Retinoid Receptor Transcripts in Tissue Sections

Xiaochun Xu and Reuben Lotan

1. Introduction

Retinoids, a group of structural and functional analogs of vitamin A and clinically important agents for chemoprevention, can modulate epithelial differentiation and suppress carcinogenesis in various tissues, including skin, bladder, oral cavity, lung, and mammary gland in experimental animal-model systems by acting primarily as inhibitors of tumor promotion *(1–3)*. More importantly, they are able to reverse premalignant lesions and inhibit the development of second primary tumors in the head and neck area and the skin in xeroderma pigmentosum patients *(4–9)*. They are also useful in the therapy of several types of human cancer, primarily acute promyelocytic leukemia *(9–13)*. It is thought that the ability of retinoids to modulate gene expression enables them to modulate the differentiation and growth of malignant cells, or to suppress the progression of premalignant cells to frank neoplastic lesions by redirecting their differentiation *(14–18)*.

The understanding of the mechanism by which retinoids modulate gene expression implies that retinoids activate a signal-transduction pathway in which nuclear-retinoid receptors, which are members of the steroid hormone-receptor superfamily, play a pivotal role *(14,16,18–22)*. Like other members of this family, the retinoid receptors are ligand-activated, DNA-binding *trans-*acting, transcription-modulating proteins. Two types of receptor have been identified: retinoic acid receptors (RARs) and retinoid X receptors (RXRs); each type includes three subtypes of RAR (α, β, and γ) and of RXR (α, β, and γ) with distinct amino- and carboxy-terminal domains. RXR-RAR

From: *Methods in Molecular Biology, Vol. 89: Retinoid Protocols*
Edited by: C. P. F. Redfern © Humana Press Inc., Totowa, NJ

heterodimers bind to a specific DNA sequence or retinoic acid (RA) response element in the promoter regions of genes that are regulated by retinoids. Because each subtype exhibits specific patterns of expression during embryonal development and different distributions in adult tissues, each is thought to regulate the expression of distinct genes. *In situ* hybridization (ISH) supports the concept that each subtype exhibits different distributions in adult tissues as well as specific patterns of expression in developing mouse embryos *(23–29)*. Several isoforms that result primarily from alternative splicing have been identified for each of these receptors and a distinct tissue distribution of these isoforms had also been reported *(21)*.

Because the RARs appear to be the ultimate mediators of RA action on gene expression, increases or decreases in their expression during normal development, in pathological states, and during cancer development may indicate their involvement in physiological and pathological processes. Furthermore, determination of the pattern of RAR expression in premalignant and malignant tissues relative to normal tissues may provide prognostic value. It may also be important for the rational selection of receptor-specific retinoids *(30)* in prevention or treatment of cancer. Last, because the expression of some of the RARs, notably RAR-β, can be stimulated by retinoids *(31,32)*, their analysis in premalignant lesions before and during chemoprevention with retinoids may serve as an intermediate endpoint for response to retinoids in vivo.

The expression of nuclear-retinoid receptors in different tissues has been analyzed, usually by Northern blotting of RNA extracted from various cells and tissues *(32–39)*. The advantage of this method is that the size and number of transcripts is revealed. Northern-blot analysis requires the availability of fresh or rapidly frozen cells or tissues in sufficient amounts to extract RNA. However, it is often difficult to obtain freshly frozen surgical specimens or to obtain sufficient quantities of tissues such as particular embryo organs. Another inherent disadvantage of Northern-blot analysis is its inability to identify the cellular origin of the mRNA when more than one cell type is being extracted, as is usually the case when tissues that contain mesenchymal and epithelial cells are analyzed or when tumor specimens that may contain infiltrating host-immune cells and stromal cells admixed with the cancer cells are analyzed.

An alternative method that overcomes these limitations is ISH. With this method, one uses tissue sections to detect the expression of RAR mRNAs in individual cells after hybridization with antisense RNA probes. A commercially available digoxigenin-11-dUTP can be incorporated into the antisense RNA during in vitro transcription. The binding of this antisense probe to cellular mRNA after hybridization can be detected by specific alkaline phosphatase-conjugated antidigoxigenin antibodies and a subsequent enzyme-catalyzed color reaction that stains the cells blue *(40)*. This method provides results in about 2 d.

2. Materials

All reagents are purchased from Sigma Chemical Co. (St. Louis, MO) and should be prepared with sterile double-distilled water and stored at room temperature unless stated otherwise (*see* **Note 1**).

1. Diethyl pyrocarbonate (DEPC)-treated water: Add 10 mL of DEPC to 10 L of H_2O, stir overnight, and autoclave.
2. 1 *M* Tris-HCl, pH 8.0: Dissolve 121 g of Tris base (Boeringer Mannheim, Indianapolis, IN) in 800 mL of water and adjust the pH to 8.0 with concentrated HCl, add water to produce a volume of 1 L, and autoclave.
3. Phosphate-buffered saline (PBS) 3X, pH 7.2: Add 22.8 g of NaCl, 2.13 g of Na_2HPO_4, and 1.80 g of NaH_2PO_4 and dissolve in 1 L of water.
4. 4% Paraformaldehyde (PFA): Heat 2/3 vol of 100 mL water to 60°C, add 4 g PFA and 6–7 drops of 4 *N* NaOH, and, after PFA is completely dissolved, add 1/3 vol of 3X PBS and adjust the pH to 7.2 with HCl. Filter it before use.
5. Proteinase-K solution: Add 4 µL of Proteinase K (Boehringer Mannheim) from a 25-mg/mL stock solution to a solution consisting of 40 mL of sterilized water, 400 µL of 1 *M* Tris-HCl buffer, pH 8.0, and 80 µL of 1 *M* $CaCl_2$. Incubate at room temperature for 20 min and at 37°C for 10 min before use.
6. 0.1 *M* Triethanolamine (TEA), pH 8.0: Add 1.4 g of TEA-Cl (Fluka, Ronkonkoma, NY) to DEPC-treated water to make 100 mL solution, adjust the pH to 8.0 by adding NaOH.
7. 1 *M* $CaCl_2$: Dissolve 147 g of $CaCl_2$ in water in a final volume of 1 L, then filter.
8. 5 *M* NaCl: Dissolve 146 g of NaCl in water in a final volume of 500 mL, then autoclave.
9. 1 *M* $MgCl_2$: Dissolve 19.4 g of $MgCl_2$ in water in a final volume of 200 mL, then autoclave.
10. 1 *M* HCl: Mix 86.2 mL of concentrated HCl and 913.8 mL of water.
11. 3 *M* Sodium acetate, pH 5.2: Dissolve 40.8 g of sodium acetate-$3H_2O$ in 80 mL of water, adjust the pH to 5.2 with glacial acetic acid. Add water to 100 mL and autoclave.
12. Saline sodium citrate (SSC) 20X: Dissolve 175.3 g of NaCl and 88.2 g of sodium citrate in 800 mL of water, adjust the pH to 7.0 with a few drops of NaOH. Add water to produce 1 L and autoclave.
13. Tris-acetic acid-EDTA (TAE) electrophoresis buffer 50X: Dissolve 242 g of Tris base and 37.2 g of $Na_2EDTA \cdot 2H_2O$ in 800 mL of water and add 57.1 mL of glacial acetic acid. Adjust the pH to 8.5 and add water to 1 L.
14. 10% Sodium dodecyl sulfate (SDS): Dissolve 100 g of electrophoresis-grade SDS (Boehringer Mannheim) in 900 mL of water heated to 68°C. Adjust the pH to 7.2 with HCl. Add water to 1 L and autoclave.

15. ISH buffer: To 5 mL of 100% deionized formamide (Boehringer Mannheim), 1 mL of 20X SSC; 200 μL of 100X Denhardt's solution, 1 g of dextran sulfate, 4 mg of yeast tRNA, 2.5 mg of salmon-sperm DNA (predenatured), 200 μL of 1 M dithiothreitol (DTT), add DEPC-treated water to a final volume of 10 mL. Stir overnight, aliquot, and store at –80°C.

16. Denhardt solution 100X: Disssolve 10 g of Ficoll 400, 10 g of polyvinylpyrrolidone, 10 g of bovine serum albumin (BSA) in water to make up 500 mL. Filter and store at –20°C.

17. 1 M Dithiothreitol (DTT): Dissolve 15.45 g of DTT in 100 mL of water, then store at –20°C.

18. 0.5 M EDTA: Dissolve 186.1 g of $Na_2EDTA \cdot 2H_2O$ in 800 mL of water by stirring vigorously, adjust the pH to 8.0 with about 20 g of NaOH, then add water to make up 1 L and autoclave.

19. Buffer 1: Dissolve 16 g of maleic acid and 8.77 g of NaCl in 900 mL of DEPC-treated water, adjust the pH to 7.5 with HCl, add DEPC treated-water to a volume of 1 L, autoclave.

20. Buffer 2: Dissolve 10 g of the blocking reagent provided in the Genius 3 detection kit (Boehringer Mannheim) in buffer 1 by heating and store at –20°C (stock solution). Dilute the stock solution 1:5 (v:v) in buffer 1.

21. Buffer 3: Mix 100 mL of 1 M Tris-HCl, 20 mL of 5 M NaCl, and 50 mL of 1 M $MgCl_2$ with 800 mL of DEPC-treated water, adjust the pH to 9.5 and add water to a volume of 1 L. The buffer can be filtered with Whatman No. 1 filter if precipitation occurs.

22. Buffer 4 (Tris-EDTA, TE): Mix 10 mL of 1 M Tris-HCl and 2 mL of 0.5 M EDTA in 1 L of DEPC-treated water, then adjust the pH to 8.0.

23. 4 M LiCl: Dissolve 16.95 g of LiCl in 100 mL of DEPC-treated water, and autoclave.

24. Substrate solution: Mix 45 μL of nitroblue tetrazolium (NBT) solution and 35 μL of X-phosphate solution in 10 mL buffer 3. This solution should be prepared freshly before use.

25. 5% poly-L-lysine (Sigma) prepared in DEPC-treated water.

26. Triton-X-100.

27. Ethanol: absolute and graded solutions of 95, 80, 70, and 50% v/v, prepared with DEPC-treated water.

28. Genius 3 Nucleic-acid detection kit (Boehringer Mannheim).

29. Genius 4 RNA-labeling kit (Boehringer Mannheim).

30. 1:1 mixture of phenol (molecular-biology grade) and chloroform/isoamyl alcohol (49:1).

31. Acetic anhydride.

32. RNase A (Boehringer Mannheim).

33. RNase T1 (Boehringer Mannheim).

34. Aqua-mount medium (Baxter, Houston, TX).

35. Normal sheep serum.

36. Nylon-blotting membrane (Boehringer Mannheim).

3. Methods

3.1. Preparation of Tissue Sections (see Note 1)

3.1.1. Preparation of Glass Slides

1. Treat new glass slides with 70% ethanol/1% HCl for at least 24 h.
2. Wash the glass slides with 70% ethanol once and then with H_2O.
3. Place the glass slides in a slide holder and rinse with H_2O.
4. Bake the glass slides at 180°C for at least 4 h.
5. Cool the glass slides to room temperature and coat them with 5% poly-L-lysine for 5 min. Dry the slides overnight at room temperature or incubate them at 48°C before use.

3.1.2. Preparation of Paraffin Sections from Cultured Cells

1. Grow the cells in 10-cm diameter tissue-culture dishes (2–3 dishes for one block) to 80% confluency.
2. Aspirate the growth medium and wash the cells twice with PBS.
3. Add 5 mL of PFA fixative to the dish, scrape the cells off the dish, transfer the suspension into 15-mL tubes, and fix the cells for 10–30 min at room temperature. Alternatively, remove the attached cells from the dishes by repeated pipetting after a brief exposure to 0.25% trypsin/1% EDTA. Collect the cells by centrifugation, wash them with PBS, and then fix them with 4% PFA for 10–30 min at room temperature or overnight at 4°C.
4. Centrifuge the cells at ca. 250*g* for 10 min to remove the fixative and wash the cells with PBS.
5. Dehydrate the cells by repeated suspension and pelleting with ascending ethanol solutions (50, 70, 80, 95, and 100%).
6. Transfer the cells into microcentrifuge tubes, centrifuge, and then aspirate the 100% ethanol.
7. Add 1 mL of xylene to clear the cells for about 30 min, then centrifuge the cells at 2500*g* for 10 min.
8. Immerse the cells in a solution containing paraffin:xylene (2:1) for 1 h at 59°C.
9. Carefully pour out the paraffin-xylene mixture and embed the cells in a paraffin block.
10. Cut 4- to 6-μm sections using a microtome, float the sections on DEPC-water at 42–45°C, and collect them on glass slides prepared as in **Subheading 3.1.1.**

3.1.3. Growing Cells on Slides

1. Subculture and grow the cells on glass slides or coverslips (that should be precleaned, autoclaved, and placed in a Petri dish using heat-sterilized forceps) for 2–3 d (30–50% confluency).
2. Wash the slides 2–3 times in PBS, label the slides with a pencil, and put in fixative (4% paraformaldehyde or 10% buffered formalin) for 20 min at room temperature or leave overnight at 4°C.

3. Pour out the fixative, wash the slides three times in PBS, and treat them with 0.05–0.3% v/v Triton X-100 for 5–10 min at room temperature.
4. Wash the slides three times in PBS and treat the slides with 0.2 *M* HCl for 10 min. Continue to the ISH protocol (*see* **Subheading 3.3.1., step 4**) or dehydrate the slides through increasing concentrations of ethanol (50–100%) and store the dehydrated specimens in a light-proof box at room temperature before treatment with proteinase K (*see* **Subheading 3.3.1., step 4**).

3.2. Preparation of cRNA Probes

3.2.1. Linearizing Vectors Containing RAR or RXR cDNAs

1. For a total volume of 100 µL, pipet into a clean microcentrifuge tube 5–10 µg of DNA in water or TE buffer and 10 µL 10X restriction buffer; add restriction endonuclease (1–5 U/µg DNA); and make up a total volume of 100 µL with water.
2. Incubate the reaction mixture for 2 h or overnight at 37°C.
3. Stop the reaction by adding 100 µL of 1:1 phenol:chloroform/isoamyl alcohol (49:1).
4. Vortex briefly and centrifuge at maximal speed (ca. 12,000*g*) in a microcentrifuge for 15 min at 4°C.
5. Transfer the supernatant into a new tube.
6. Add 1/10 volume of 3 *M* sodium acetate, pH 5.2, to the solution; mix by vortexing briefly or by flicking the tube several times with a finger.
7. Add 2–2.5 vol of ice-cold 100% ethanol, mix by vortexing, and place in a –70°C freezer for at least 1 h or overnight.
8. Spin for 15 min at ~12,000*g* in a microcentrifuge at 4°C.
9. Pour off the supernatant carefully.
10. Wash with 1 mL of 70% ethanol at room temperature. Invert the tube several times and spin in a microcentrifuge as noted.
11. Aspirate off the supernatant as before.
12. Dry the pellet in a desiccator under vacuum or in a Speedvac evaporator for 10 min.
13. Dissolve the dry pellet in 20–50 µL of DEPC-treated water; the linearized DNA is ready for use in in vitro transcription.

3.2.2. In vitro Transcription of Digoxigenin-Labeled cRNA Probes

The digoxigenin-labeled RAR or RXR probes are synthesized by using Genius 4 RNA-labeling kit.

1. Add the following to a microcentrifuge tube on ice: 2 µL of 10X transcription buffer; 2 µL nucleotide triphosphate (NTP) labeling mixture; 1 µg of DNA; 2 µL of T7, T3, or Sp6 RNA polymerase; 1 µL of RNase inhibitor; and make total volume to 20 µL with water.

2. Centrifuge briefly and incubate for 2 h at 37°C. Longer incubation does not increase the yield of labeled RNA.
3. Add 2 μL of RNase-free DNase I to remove the template DNA and incubate for 15 min at 37°C.
4. Add 1 μL of 0.5 *M* EDTA, pH 8.0, to stop the reaction.
5. Precipitate the labeled RNA with 2.5 μL of 4 *M* LiCl, and 75 μL of prechilled (–20°C) ethanol (75%) and mix well.
6. Leave for at least 2 h or overnight at –70°C.
7. Centrifuge (at 12,000*g*) at 4°C for 15 min, pour off the supernatant, wash the pellet with 100 μL of cold ethanol, 75% (v/v), recentrifuge and aspirate the supernatant, dry the pellet under a vacuum, and dissolve it in 100 μL of TE buffer or in DEPC-treated water for 30 min at 37°C. 1 μL of RNase-inhibitor can be added to inhibit possible contaminating RNase. May be stored at –70°C for long time periods before use or at –20°C for daily use (*see* **Note 2**).

3.2.3. Estimation of Yield of the Probes

Two methods are used to estimate the yield of newly synthesized digoxigenin-labeled cRNA probes. The first is dot blotting and the second measures RNA concentration using a spectrophotometer.

3.2.3.1. ESTIMATION OF THE cRNA PROBE CONCENTRATION BY DOT BLOTTING

1. Take 2 μL of newly synthesized probe and make five serial dilutions of a 10-fold decreased concentration each in DEPC-treated water in microcentrifuge tubes.
2. Take 2 μL of labeled control cRNA (from Genius 4 labeling kit, a stock of 100 ng/μL) and dilute as aforementioned.
3. Spot 1 μL of each dilution, from **steps 1** and **2** on a nylon membrane and fix the RNA to the membrane by UV irradiation or bake for 1 h at 80°C.
4. Incubate the membrane in buffer 2 for 10 min at room temperature then continue incubation in anti-digoxigenin-conjugate solution (diluted 1:5000 in buffer 2) for 10–20 min at room temperature followed by two washes in buffer 1.
5. Incubate the membrane briefly in buffer 3 and change to chromogenic-substrate solution to develop color in the dark. Usually color develops within 20 min. Stop the reaction by adding buffer 4.
6. Compare the highest dilution that still gives a colored spot of newly synthesized probe to the colored spots produced by the standard control.

3.2.3.2. ESTIMATION OF THE CONCENTRATION BY SPECTROPHOTOMETER

Measure the concentration of new probe by using spectrophotometer at 260 nm and compare to a standard curve produced using dilutions of the control RNA (*see* **Note 3**).

3.3. Step by Step In Situ Hybridization

3.3.1. In Situ Hybridization

1. Deparaffinize tissue sections in two changes of xylene, 10 min each.
2. Hydrate the sections by consecutive immersion in descending grades of ethanol solution (100, 95, 80, 70, 50%) 1 min each.
3. Treat sections with 0.2 *N* HCl (10 mL of 1 *M* HCl + 40 mL of DEPC treated-water) for 10 min at room temperature with agitation; then wash the sections three times with PBS.
4. Predigest any contaminating RNase in the proteinase-K solution (*see* **Subheading 2., item 5**). Then treat the slides with proteinase K at 37°C for 10–15 min and wash the sections with PBS three times.
5. Postfix the sections with 4% PFA (freshly made) for 5 min; then wash the sections with PBS three times at room temperature.
6. Transfer the sections to fresh TEA buffer. Add acetic anhydride to a concentration of 0.25% v/v (125 µL of acetic anhydride + 50 mL of 0.1 *M* TEA buffer). Mix quickly and incubate sections for 10 min with agitation at room temperature. Wash the sections with PBS three times.
7. Transfer the sections into the slide holder and dehydrate with graded-alcohol solutions: one min each in 70, 80, 95, and 100%.
8. Air-dry the sections and use them for prehybridization and hybridization. The sections may be stored at this stage in airtight boxes with drying agent at –20°C.
9. Prehybridize the sections in hybridization buffer for 1 h in a closed, humidified box at 48°C (cover the box with plastic wrap and place 2X SSC or DEPC water in bottom of the box).
10. Heat probe + hybridization buffer (20 ng/50 µL) for 5 min at 95–100°C to linearize the probe and then immediately cool on ice.
11. Hybridize the slides with 50–100 µL of digoxigenin-labeled cRNA probe in hybridization buffer for 4 h at 48°C in a closed, humidified box. Cover the tissue sections with parafilm or baked cover slips.
12. After hybridization, remove the parafilm or cover slips in 2X SSC.
13. Digest RNA with RNase A (40 µg/mL) and RNase T1 (2 µg or 10 U/mL) in 10 m*M* Tris-HCl, pH 8.0, 1 m*M* EDTA, 0.5 *M* NaCl for 30 min at 37°C (this step is optional).
14. Wash sections twice with 2X SSC.
15. For the posthybridization washing, incubate the sections with agitation for 2 h with 2X SSC containing 0.05% Triton X-100, and 2% normal sheep serum (NSS) at room tempeature (*see* **Note 4**).

3.3.2. Immunodetection of Positive Signal

The Genius 3 Nucleic-acid detection kit (from Boehringer Mannheim) is used for immunodetection of hybridized digoxigenin-conjugated cRNA. The kit contains antidigoxigenin-alkaline phosphatase conjugate, chromogenic substrate, and blocking reagent. The following protocol is a minor modification of the instructions provided with the kit.

1. Wash slides from **step 15** of the ISH protocol with buffer 1 briefly.
2. Incubate the sections with 2% v/v NSS, 0.3% v/v Triton X-100 in buffer 1 for 30 min at room temperature.
3. Dilute antibody conjugate 1:500–1:1000 with buffer 1 containing 1% v/v NSS and 0.3% v/v Triton X-100.
4. Apply 50–100 μL of the diluted-antibody conjugate to sections, cover the tissue sections with parafilm, and then incubate overnight at 4°C.
5. The next day, put the box containing the sections at room temperature for 10 min, remove the parafilm with buffer 1, and wash the sections twice with buffer 1 for 10 min each at room temperature.
6. Wash slides once with buffer 3 for 1–2 min at room temperature.
7. Develop color by incubating slides with chromogenic-substrate solution (about 300 μL/slide). Place slides in a humidified light-tight box and check occasionally for color development. Color may develop for up to 6 h.
8. Stop color reaction with buffer 4.
9. Cover the tissue sections with cover slips in Aqua mount medium and allow them to dry.
10. Examine the sections using a microscope. The positive signal should be purple or blue and the negative cells or tissue should not show any staining (**Fig. 1**) *(41–43)*.

4. Notes

1. The preparation of tissue sections and all procedures should be carried out in RNase-free conditions, i.e., gloves should be worn during the handling of glassware, reagents, tissue blocks, and equipment related to ISH. DEPC-treated water should be used to wash equipment and staining jars and containers that cannot be heated to high temperatures in the oven. Glassware should be baked (except staining glass containers and holders, which break easily at high temperatures). DEPC-treated water should be used to float tissue sections while cutting paraffin blocks. Equipment used in pre- and posthybridization should be kept separate.
2. It is very important to handle and store the digoxigenin-labeled riboprobes with care to avoid degradation by RNase contamination. The probes should be stored at –80°C with an RNase inhibitor for long term use and can be stored at –20°C for daily use.
3. It is important to estimate the concentration of the probes accurately. The accuracy of the spectrophotometeric measurement depends on whether residual undigested DNA is present in the preparation.
4. There are several choices for posthybridization washing, one of which may be better than others depending on the probe and tissue to be analyzed. The washing conditions are chosen such that the sense probe will give no or minimal staining, whereas the antisense probe will give positive staining. For example:

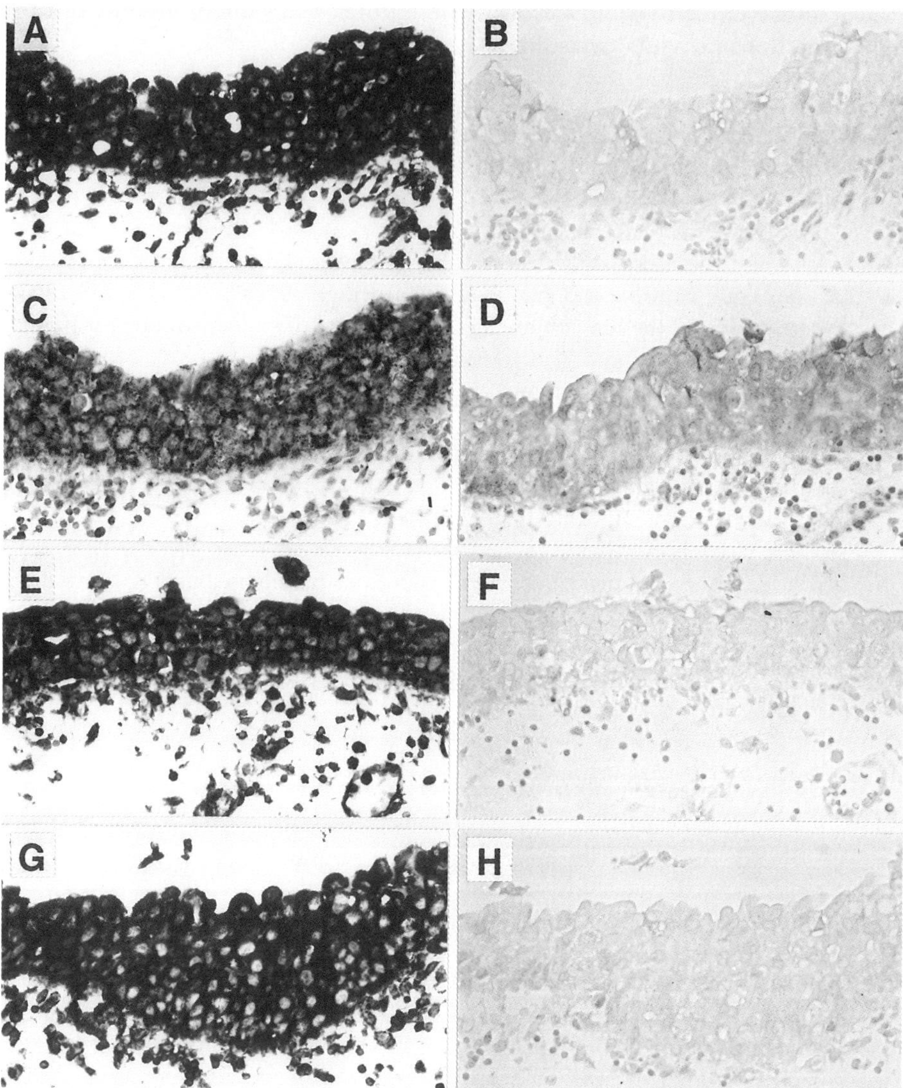

Fig. 1. Localization of nuclear RAR and RXR mRNAs in formalin-fixed and paraffin-embedded surgical specimens by *in situ* hybridization. Sections of normal epithelium adjacent to human-bladder carcinoma were hybridized with RARα antisense (**A**) or sense (**B**); RARβ antisense (**C**) or sense (**D**); RARγ antisense (**E**) or sense (**F**); RXRα antisense (**G**) or sense (**H**).

a. Wash the sections with 2X SSC containing 2% v/v NSS and 0.05% v/v Triton X-100 for 2 h at room temperature;

b. Wash the sections with 2X SSC containing 2% v/v NSS and 0.05% v/vTriton X-100 for 2 h at room temperature and for 20 min at 48°C;

c. Wash the sections with 2X SSC containing 2% v/v NSS and 0.05% v/v Triton X-100 for 2 h at room temperature and 0.1X SSC containing 2% v/v NSS and 0.05% v/v Triton X-100 for 20 min at 48°C;

d. Treat the sections with RNase, and then wash with 2X SSC containing 2% v/v NSS and 0.05% v/v Triton X-100 for 2 h at room temperature.

References

1. Bertram, J. S., Kolonel, L. N., and Meyskens, F. L. Jr. (1987) Rationale and strategies for chemoprevention of cancer in humans. *Cancer Res.* **47,** 3012–3031.

2. Moon, R. C. and Mehta, R. G. (1990) Retinoid inhibition of experimental carcinogenesis, in *Chemistry and Biology of Synthetic Retinoids* (Dawson, M. I. and Okamura, W. H., eds.), CRC, Boca Raton, FL, pp. 501–518.

3. Pollard, M., Luckert, P., and Sporn, M. (1991) Prevention of primary prostate cancer in Loblund-Wistar rats by N-(4-hydroxyphenyl)retinamide. *Cancer Res.* **51,** 3610,3611.

4. Hong, W. K., Endicott, J., Itri, L. M., Doos, W., Batsakis, J. G., Bell, R., Fofonoff, S., Byers, R., Atkinson, E. N., Vaughan, C., Toth, B. B., Kramer, A., Dimery, I. W., Skipper, P., and Strong, S. (1986) 13-Cis retinoic acid in the treatment of oral leukoplakia. *N. Engl. J. Med.* **315,** 1501–1505.

5. Lippman, S., Kessler, J. F., and Meyskens, F., Jr. (1987) Retinoids as preventive and therapeutic anticancer agents. *Cancer Treat Rep.* **71,** 391–405 (Part 1); 493–515 (Part 2).

6. Kraemer, K. H., DiGiovanna, J. J., Moshell, A. N., Tarone, R. E., and Peck, G. L. (1988) Prevention of skin cancer in xeroderma pigmentosum with the use of oral isotretinoin. *N. Engl. J. Med.* **318,** 1633–1637.

7. Lippman, S. M. and Meyskens, F. L. (1989) Results of the use of vitamin A and retinoids in cutaneous malignancies. *Pharmacol. Ther.* **40,** 107–122.

8. Hong, W. K., Lippman, S. M., Itri, L. M., Karp, D. D., Lee, J. S., Byers, R. M., Schantz, S. S., Kramer, A. M., Lotan, R., Peters, L. L., Dimery, I. W., Brown, B. W., and Goepfert, H. (1990) Prevention of second primary tumors with isotretinoin in squamous-cell carcinoma of the head and neck. *N. Engl. J. Med.* **323,** 795–801.

9. Smith, M. A., Parkinson, D. R., Cheson, B. D., and Friedman, M. A. (1992) Retinoids in cancer therapy. *J. Clin. Oncol.* **10,** 839–864.

10. Huang, M. E., Ye, Y. I., Chen, S. R., Chai, J. R., Lu, J. X., Zhoa, L., Gu, L. J., and Wang, Z. Y. (1988) Use of all-trans-retinoic acid in the treatment of acute promyelocytic leukemia. *Blood* **72,** 567–572.

11. Castaigne, S., Chomienne, C., Daniel, M. T., Ballerini, P., Berger, R., Fenaux, P., and Degos, L. (1990) All-trans retinoic acid as a differentiation therapy for acute promyelocytic leukemia. I. Clinical results. *Blood* **76,** 1704–1709.

12. Warrell, R. P., Frankel, S. R., Miller, W. H., Scheinberg, D. A., Itri, L. M., Hittelman, W. N., Vyas, R., Andreeff, M., Tafuri, A., Jakubowski, A., Gabrilove, J., Gordon, M., and Dmitrovsky, E. (1991) Differentiation therapy of acute promyelocytic leukemia with tretinoin (all-trans retinoic acid). *N. Engl. J. Med.* **324,** 1385–1393.

13. Lippman, S. M., Kavanagh, J. J., Paredes-Espinoza, M., Delgadillo-Madrueno, F., Paredes-Casillas, P., Hong, W. K., Holdener, E., and Krakoff, I. H. (1992) 13-cis-retinoic acid plus interferon alpha-2a: highly active systemic therapy for squamous cell carcinoma of the cervix. *J. Natl. Cancer Inst.* **84,** 241–245.

14. Lotan, R. and Clifford, J. L. (1990) Nuclear receptors for retinoids: mediators of retinoid effects on normal and malignant cells. *Biomed. Pharmacother.* **45,** 145–156.

15. Lotan, R., Lotan, D., and Sacks, P. G. (1990) Inhibition of tumor cell growth by retinoids. *Methods Enzymol.* **190,** 100–110.

16. DeLuca, L. M. (1991) Retinoids and their receptors in differentiation, embryogenesis and neoplasia. *FASEB J.* **5,** 2924–2933.

17. Jetten, A. M., Nervi, C., and Vollberg, T. M. (1992) Control of squamous differentiation in tracheobronchial and epidermal epithelial cells: role of retinoids. *J. Natl. Cancer Inst. Monog.* **13,** 93–100.

18. Leroy, P., Krust, A., Kastner, P., Mendelsohn, C., Zelent, A., and Chambon, P. (1992) Retinoic acid receptors, in *Retinoids in Normal Development and Teratogenesis* (Morriss-Kay, G., ed.) Oxford Universty Press, New York, pp. 7–25.

19. Evans, R. (1988) The steroid and thyroid hormone receptor superfamily. *Science* **240,** 889–895.

20. Glass, C. K., DiRenzo, J., Kurokawa, R., and Han, Z. (1991) Regulation of gene expression by retinoic acid receptors. *DNA Cell Biol.* **10,** 623–638.

21. Leid, M., Kastner, P., and Chambon, P. (1992) Multiplicity generates diversity in the retinoic acid signalling pathways. *Trends Biochem. Sci.* **17,** 427–433.

22. Mattei, M. G., Riviere, M., Krust, A., Ingvarsson, S., Vennstrom, B., Islam, M. Q., Levan, G., Kautner, P., Zelent, A., Chambon, P., Szpirer, J., and Szpirer, C. (1991) Chromosomal assignment of retinoic acid receptor (RAR) genes in the human, mouse, and rat genomes. *Genomics* **10,** 1061–1069.

23. Noji, S., Yamaai, T., Koyama, E., Nohno, T., and Taniguchi, S. (1989) Spatial and temporal expression pattern of retinoic acid receptor genes during mouse bone development. *FEBS Lett.* **257,** 93–96.

24. Dolle, P., Ruberte, E., Kastner, P., Petkovich, M., Stoner, C. M., Gudas, L. J., and Chambon, P. (1989) Differential expression of genes encoding alpha, beta and gamma retinoic acid receptors and CRABP in the developing limbs of the mouse. *Nature* **342,** 702–705.

25. Noji, S., Yamaai, T., Koyama, E., Nohno, T., Fujimoto, W., Arata, J., and Taniguchi, S. (1989) Expression of retinoic acid receptor genes in keratinizing front of skin. *FEBS Lett.* **259,** 86–90.

26. Ruberte, E., Dolle, P., Krust, A., Zelent, A., Morriss-Kay, G., and Chambon, P. (1990) Specific spatial and temporal distribution of retinoic acid receptor gamma transcripts during mouse embryogenesis. *Development* **108,** 213–222.

27. Smith, S. M. and Eichele, G. (1991) Temporal and regional differences in the expression pattern of distinct retinoic acid receptor-beta transcripts in the chick embryo. *Development* **111,** 245–252.

28. Ruberte, E., Dolle, P., Chambon, P., and Morriss-Kay, G. (1991) Retinoic acid receptors and cellular retinoid binding proteins. II. Their differential pattern of transcription during early morphogenesis in mouse embryos. *Development* **111,** 45–60.

29. Finzi, E., Blake, M. J., Celano, P., Skouge, J., and Diwan, R. (1992) Cellular localization of retinoic acid receptor-gamma expression in normal and neoplastic skin. *Am. J. Pathol.* **140,** 1463–1471.

30. Lehman, J. M., Dawson, M. I., Hobbs, P. D., Husmann, M., and Pfahl, M. (1991) Identification of retinoids with nuclear receptor subtype-selective activities. *Cancer Res.* **51,** 4804–4809.

31. de The, H., Vivanco-Ruiz, Md. M., Tiollais, P., Stunnenberg, H., and Dejean, A. (1990) Identification of a retinoic acid responsive element in the retinoic acid receptor β gene. *Nature* **343,** 177–180.

32. Clifford, J., Petkovich, M., Chambon, P., and Lotan, R. (1990) Modulation by retinoids of mRNA levels for nuclear retinoic acid receptors in murine melanoma cells. *Mol. Endocrinol.* **4,** 1546–1555.

33. Petkovich, M., Brand, N. J., Krust, A., and Chambon, P. (1987) A human retinoic acid receptor which belongs to the family of nuclear receptors. *Nature* **330,** 444–450.

34. Zelent, A., Krust, A., Petkovich, M., Kastner, P., and Chambon, P. (1989) Cloning of murine α and β retinoic acid receptors and a novel receptor γ predominantly expressed in skin. *Nature* **339,** 714–717.

35. Brand, N., Petkovitch, M., Krust, A., Chambon, P., de The, H., Marchio, A., Tiollas, P., and Dejean, A. (1988) Identification of a second human retinoic acid receptor. *Nature* **332,** 850–853.

36. Cox, K. H., DeLeon, D. V., Angerer, L. M., and Angerer, R. C. (1984) Detection of mRNAs in sea urchin embryos by in situ hybridization using asymmetric RNA probes. *Dev. Biol.* **101,** 485–502.

37. Giguere, V., Ong, E. S., Segui, P., and Evans, R. M. (1987) Identification of a receptor for the morphogen retinoic acid. *Nature* **330,** 624–629.

38. Benbrook, D., Lernhardt, E., and Pfahl, M. (1988) A new retinoic acid receptor identified from a hepatocellular carcinoma. *Nature* **332,** 669–672.

39. Rees, J. L. and Redfern, C. P. F. (1989) Expression of the α and β retinoic acid receptors in skin. *J. Invest. Derm.* **93,** 818–820.

40. Springer, J. E., Robbins, E., Gwag, B. J., Lewis, M. E., and Baldino, F., Jr. (1991) Non-radioactive detection of nerve growth factor receptor (NGFR) mRNA in rat brain using in situ hybridization histochemistry. *J. Histochem. Cytochem.* **39,** 231–234.

41. Xu, X-C., Clifford, J. L., Hong, W. K., and Lotan, R. (1994) Detection of nuclear retinoic acid receptor mRNA in histological tissue sections using nonradioactive in situ hybridization histochemistry. *Diag. Mol. Pathol.* **3,** 122–131.

42. Lotan, R., Xu, X-C., Lippman, S. M., Ro, J. Y., Lee, J. S., Lee, J. J., and Hong, W. K. (1995) Suppression of retinoic acid receptor-β in premalignant oral lesions and its up-regulation by isotretinion. *N. Engl. J. Med.* **332,** 1405–1410.
43. Xu, X-C., Ro, J. Y., Lee, J. S., Shin, D. M., Hong, W. K., and Lotan, R. (1994) Differential expression of nuclear retinoid receptors in normal, premalignant, and malignant head and neck tissues. *Cancer Res.* **54,** 3580–3587.

19

In Situ Hybridization with ^{35}S-Labeled Probes for Retinoid Receptors

Karen Niederreither and Pascal Dollé

1. Introduction

In situ hybridization (ISH) is a technique that enables one to localize an endogenous mRNA in histological and cytological samples to the resolution of nearly single cells. For this reason, ISH is widely used in the fields of genetics, developmental biology, and neurobiology, as well as for diagnostic purposes in medicine. Our laboratory has established conditions for ISH using ^{35}S-labeled antisense-RNA probes (riboprobes), which allow the specific detection of each type of murine retinoic acid receptor (RAR), retinoid X receptor (RXR), cellular retinol-binding protein (CRBP), and cellular retinoic acid binding protein (CRABP) gene transcripts. ISH studies have thus revealed that most of the retinoid receptor genes have complex and distinct expression patterns during mouse development *(1–6)*. The protocol described hereafter has been inspired by the pioneering work of Angerer and colleagues *(7)*. It has been designed for histological sections of mouse embryos or adult tissues, but can be applied to samples from other species, including human. Nevertheless, slightly distinct ISH protocols have been described for the detection of RAR transcripts in chick *(8,9)* and *Xenopus* embryos *(10)*.

We shall describe an ISH procedure in which the same riboprobes can be used either on sections of samples prefixed in paraformaldehyde and embedded in paraffin, or on cryosections of directly frozen samples. Although ISH using paraffin-embedded material has been commonly used to date, we have noticed that ISH on cryosections results in a far higher sensitivity. This is due to both lower "background" grain levels and higher signal intensities. In particular, ISH on cryosections has allowed the detection of specific isoforms of RAR mRNAs (unpublished results from our laboratory), which had been

From: *Methods in Molecular Biology, Vol. 89: Retinoid Protocols*
Edited by: C. P. F. Redfern © Humana Press Inc., Totowa, NJ

unsuccessful on paraffin-embedded tissue sections. A major drawback of cryosections, though, is their poorer histological resolution, especially at early developmental stages. Additionally, it is difficult to orient small specimens precisely, and sectioning itself may require more skill. On the other hand, paraffin sections are easy to perform serially, allow a precise orientation of the tissue sample, and result in high histological resolution. Altogether, we consider that cryosections provide acceptable results in embryos over 12.5 gestational days. The method chosen for the analysis of younger embryos or other tissue types depends on whether priority is given to the sensitivity of signal detection or to histological resolution. Both methods may actually be tested in parallel in a first series of experiments.

More background information can be found in two books dedicated to ISH *(11,12)*. Note that retinoid receptor or binding-protein transcript distribution may also be analyzed by whole-mount ISH of digoxigenin-labeled riboprobes, a method that is particularly informative at early stages of development and is described in Chapter 5 and in **refs.** *12* and *22*. In Chapter 18, Xu and Lotan describe the use of digoxigenin-labeled RAR riboprobes for ISH of paraffin-embedded tissue sections.

2. Materials

2.1. Linearization of Template DNA

1. Purified-plasmid DNA containing the appropriate RAR or RXR sequence (*see* **Note 1**) cloned into a vector having SP6, T7, or T3 RNA-polymerase promoters (e.g., pBluescript plasmids, Stratagene, Palo Alto, CA, or pGem plasmids, Promega, Madison, WI; *see* **Note 2**).
2. Appropriate restriction enzyme for linearizing the plasmid (*see* **Note 2**).
3. Agarose (gel electrophoresis grade).
4. Electrophoresis buffer 1X Tris-acetate-EDTA (TAE): 0.04 M Tris-acetate, 0.001 M EDTA.
5. Phenol/chloroform (1:1) equilibrated in 0.1 M Tris-HCl, pH 8.0, chloroform.
6. Ethanol.

2.2. In Vitro Transcription

1. Diethylpyrocarbonate (Sigma, St. Louis, MO) (*see* **Note 3**).
2. Template DNA (*see* **Subheading 3.1.**).
3. SP6, T7, and/or T3 RNA polymerase and manufacturer's reaction buffers.
4. Adenosine triphosphate (ATP), guanosine triphosphate (GTP), uridine triphosphate (UTP): 5 mM each stock solution.
5. Cytidine triphosphate (CTP): 0.25 mM stock solution.
6. [α^{35}S]-CTP, 800 Ci/mMol (>30 TBq/mmol) (Amersham, Arlington Heights, IL).
7. Dithiothreitol (DTT): 1 M stock solution.
8. RNase-free bovine serum albumin (BSA): 2 mg/mL stock solution.

9. Placental RNase inhibitor (Amersham).
10. RNase-free DNase I (10 U/µL) (Boehringer, Indianapolis, IN).
11. DNase I digestion buffer: 10 mM MgCl$_2$; 20 mM Tris-HCl, pH 7.9.
12. Vanadyl ribonucleoside complex (Gibco-BRL, Gaithersburg, MD).
13. Yeast tRNA (Sigma), 10 µg/µL stock solution (*see* **Note 4**).
14. Sodium dodecyl sulfate (SDS) 0.2% w/v.
15. GF/C glass fiber disks (Whatman, Clifton, NJ), Erlenmeyer filtration flask, 10 and 5% w/v trichloroacetic acid (TCA) diluted in sterile water.
16. Phenol/chloroform (1:1).
17. NaOH 10 N stock solution.
18. Ammonium acetate 3 M (*see* **Note 5**).
19. Ethanol.

2.3. Subbed Slides

1. Large histological staining dishes and metal slide-holding racks (e.g., 50 slides capacity). These can be obtained from any histological material supplier.
2. Precleaned histological slides.
3. Gelatin.
4. Chrome alum (potassium chromium III sulfate).
5. Oven set at 160°C.

2.4. Paraffin Sections

2.4.1. Collection, Fixation, and Embedding of Samples

1. Dissection equipment (forceps, scissors) suitable for the size of the tissues to be collected.
2. Paraformaldehyde.
3. Phosphate-buffered saline (PBS) 10X stock: 80 g NaCl, 2 g KCl, 11.5 g Na$_2$HPO$_4$.7H$_2$O, 2 g KH$_2$PO$_4$ for 1 L.
4. Ethanol.
5. Xylene (this can be replaced by toluene or trichlorethane-containing solvents: e.g., LMR-sol, Labo-Moderne, Paris).
6. Polyethylene or glass test tubes (resistent to organic solvents).
7. Incubating oven set to 58°C.
8. Histological paraffin wax (histological grade, melting point 54–56°C).
9. Plastic embedding moulds: e.g., Peel-a-way™ (Polysciences, Warrington, PA).
10. Whatman No. 1 paper.

2.4.2. Sectioning Paraffin

1. Standard rotary microtome.
2. Subbed slides (*see* **Subheading 3.3.**).
3. Heating plate.
4. Silica gel.
5. Plastic boxes for storage of slides.

2.4.3. Pretreatment of Paraffin Sections

1. Rectangular glass or plastic staining dishes (resistant to steam autoclaving), metallic or plastic racks holding 20 slides.
2. Proteinase-K digestion buffer, 10X stock solution: 0.5 M Tris-HCl, pH 7.5, 50 mM EDTA.
3. Proteinase K 10 mg/mL stock solution.
4. Triethanolamine buffer 0.1 M, pH 8.0: prepared by dilution of a 7.5 M triethanolamine commercial stock solution, adjusted to pH 8.0 by adding 500 µL concentrated HCl per 300 mL buffer.
5. Acetic anhydride.
6. Xylene (or toluene, or trichlorethane-containing solvents: e.g., LMR-sol, Labo-Moderne).
7. Absolute ethanol and a graded series of aqueous ethanol solutions (30, 60, 85, 95%).

2.5. Frozen Sections

2.5.1. Collection and Freezing of Samples

1. Dissection equipment
2. Dry ice.
3. 2-Methylbutane (Aldrich, Milwaukee, WI).
4. Embedding moulds (e.g., Polysciences).
5. Tissue-Tek™ O.C.T. Compound embedding medium (Sakura Finetek, Torrance, CA).

2.5.2. Preparation of Cryosections

1. Cryostat.
2. Subbed slides (*see* **Subheading 3.3.**).
3. Heating plate set at 40°C.
4. Plastic boxes for storage of slides.

2.5.3. Pretreatment of Frozen Sections

1. PBS (*see* **Subheading 2.4.1.**).
2. Standard saline citrate (SSC), 20X stock solution: 3 M NaCl, 0.3 M trisodium citrate.
3. Triethanolamine buffer 0.1 M, pH 8.0, and acetic anhydride (*see* **Subheading 2.4.3.**).
4. Deionized formamide (*see* **Subheading 2.6.**).
5. Formaldehyde diluted to 4% in 1X PBS.
6. Absolute ethanol, 50 and 70% ethanol solutions.

2.6. Hybridization

1. One large water bath (preferably a shaking water bath) set at 52°C and a second water bath set at 75°C.
2. Plastic incubation boxes.

3. Parafilm™ (Greenwich, CT) or glass slide cover slips.
4. Formamide, deionized by stirring Dowex XG8 resin beads (5 g/100 mL formamide) for 1 h, filtered through Whatman No. 1 paper, and stored in aliquots at –20°C.
5. Denhardt's 50X stock solution: 10 g Ficoll, 10 g polyvinylpyrolidone, 10 g BSA (Fraction V) for 1 L of water, stored at –20°C.
6. Salts 10X stock solution: 3 M NaCl, 200 mM Tris-HCl, pH 6.8, 50 mM EDTA, 10 mM sodium phosphate buffer pH 6.8, 10X Denhardt's solution, stored at –20°C.
7. Dextran sulfate: 50% w/v stock solution in water, dissolved at 60°C, cooled and stored at –20°C. This is a very viscous solution and is easier to pipet when prewarmed.
8. DTT: 1 M stock solution, stored at –20°C.
9. Yeast tRNA: 50 µg/µL stock solution, stored at –20°C (*see* **Note 4**).
10. [^{35}S]-labeled riboprobe stock solution (2 ng/µL) prepared as described in **Subheading 3.2.**

2.7. Posthybridization Washes

1. Two large water baths.
2. Staining dishes and racks for slides.
3. Washing solution: 50% formamide, 1X salts, 10 mM DTT. 10X salts stock solution is as described in **Subheading 2.6.** except that Denhardt's solution is omitted. The formamide need not be deionized.
4. RNAase digestion buffer, 10X stock solution: 5 M NaCl, 100 mM Tris-HCl, pH 7.5, 50 mM EDTA.
5. RNAase A: 100 µg/µL stock solution, stored in aliquots at –20°C.
6. SSC: 20X stock solution.
7. 30, 60, 85, 95% Ethanol solutions each containing 0.3 M ammonium acetate.

2.8. Autoradiography

2.8.1. Film Autoradiography

1. Kodak BIOMAX MR film (Eastman Kodak, Rochester, NY).
2. Autoradiography cassettes.
3. Darkroom.

2.8.2. Emulsion Autoradiography

1. Darkroom equipped with a good safe light (e.g., with Kodak Wratten Series II filters) and a water bath.
2. Autoradiography emulsion (Kodak NTB-2), stored in the dark, and away from any source of radiation.
3. Dipping chamber. This can be a made-to-measure glass container with dimensions approx 3 × 6 × 0.5 cm, or simply a plastic slide mailer.
4. Light-tight boxes for storage of slides, containing silica gel in a paper towel.

2.8.3. Developing and Staining

1. Histological staining dishes and slide-holding racks.
2. Developer (Kodak D19), diluted in tap water as indicated by the manufacturer. Once diluted, the developer can be stored for several wk at room temperature.
3. Fixer (Kodak AL4), used as supplied (undiluted).
4. Tap water stored in container within the darkroom (*see* **Note 29**).
5. Toluidine blue diluted to 0.02% w/v in distilled water (*see* **Note 30**).
6. Glass cover slips and mounting medium (e.g., Permount™, Fischer Scientific, Springfield, NY).

2.9. Examination and Photography

1. Dissecting stereomicroscope equiped with dark-field illumination (which makes the silver grains appear white) for low-power examination.
2. Conventional light microscope equipped with dark-field or epiluminescent illumination.
3. Camera adapted to the microscope(s).

3. Methods

3.1. Preparation of Template DNA

^{35}S-labeled riboprobes are prepared according to standard in vitro transcription reactions. The DNA insert that will be used as template for transcription (*see* **Note 1**) is therefore subcloned in a plasmid vector containing promoters for the SP6, T7, or T3 bacteriophage-RNA polymerases. Prior to transcription, the recombinant plasmid has to be linearized using a restriction enzyme to cut at the end of the insert (or at any convenient internal site) opposite the appropriate promoter used to generate an antisense transcript. Template DNA should be linearized to completion because any uncut template will result in read-through transcription leading to high background or poor detection of mRNA signal. Complete plasmid linearization is confirmed by agarose-gel electrophoresis of an aliquot of the digestion reaction.

1. Digest 10–20 µg of purified plasmid DNA (*see* **Note 2**) to completion with the appropriate restriction enzyme.
2. Verify complete linearization of the DNA by running an aliquot of the digest in a 0.8% agarose gel.
3. Add sterile water to the digestion reaction to obtain a total volume of 100 µL. Add 100 µL of phenol/chloroform, vortex, and centrifuge for 2 min at 12,000*g*. Transfer the upper-aqueous phase in a clean tube.
4. Add 100 µL of chloroform and repeat the extraction as in **step 3**.
5. Add 10 µL of 3 *M* sodium acetate, 2.5 vol of absolute ethanol, vortex, and centrifuge for 20 min in a refrigerated centrifuge.
6. Discard the supernatant and add 1 ml of 70% ethanol. Mix and centrifuge for 10 min.

7. Discard the supernatant and dry the DNA pellet.
8. Resuspend the DNA to 1 µg/µL in sterile water and store at –20°C.

3.2. In Vitro Transcription (see Note 3)

1. Prepare the following 30-µL "transcription mix," sufficient for two reactions: 5 µL of 10X transcription buffer, 5 µL of ATP/GTP/UTP mix (5 mM each), 2 µL of 1 M DTT, 2.5 µL of RNAase-free BSA (2 mg/mL), 2 µL of placental RNase inhibitor (50 U/µL), 13.5 µL of water.
2. Set up a 20-µL in vitro transcription reaction, using reagents supplied by the manufacturer of the RNA polymerase, and including: 12 µL of transcription mix, 1 µL of linearized template DNA (1 µg/µL), 1 µL of 0.25 mM CTP, 3 µL of $[\alpha^{35}S]$-CTP (800 Ci/mmol), 1 µL of RNA polymerase (20–25 U/µL), 2 µL of water. Incubate 2 h at 37°C.
3. Prepare the "DNAase I mix," sufficient for four reactions: 20 µL of DNAase I digestion buffer, 1 µL of vanadyl ribonucleoside complex, 1 µL of RNAase-free DNAase I (10 U/µL). Let stand at room temperature for 5 min.
4. To each transcription reaction, add 25 µL of DNAase I digestion buffer and 4.5 µL of DNAase I mix. Incubate for 10 min at 37°C.
5. Add 3 µL of a 10 µg/µL yeast tRNA solution (*see* **Note 4**), 150 µL of water, 200 µL of 0.2% SDS.
6. Take two 2-µL aliquots for scintillation counting.
7. Add an equal volume of phenol/chloroform to the main reaction mixture. Vortex and separate the phases by centrifugation in a microcentrifuge. Transfer the upper (aqueous) phase to a clean tube.
8. Add 0.1 vol of 3 M ammonium acetate (*see* **Note 5**) and 2.5 vol of absolute ethanol to the aqueous phase. Place the mixture at –20°C for at least 30 min. Then pellet the RNA precipitate by centrifugation for 30 min in a refrigerated microcentrifuge.
9. Discard the ethanol, resuspend the RNA pellet (no lyophilization required) in 100 µL water and add 100 µL of 0.2 M NaOH (*see* **Note 6**). Incubate on ice for t minutes as determined by the formula:

$$t = \frac{(Lo - Lf)}{(k \times Lo \times Lf)}$$

where Lo = initial length of the fragment (kb); Lf = final length (kb), i.e., 0.15; and k = hydrolysis rate constant = 0.11.

10. Add DTT to 10 mM, ammonium acetate to 0.3 M, and precipitate the RNA with ethanol as in **step 8**. Wash the RNA precipitate in 70% ethanol and dry the pellet.
11. In parallel, proceed to scintillation counting (*see* **Note 7**) as follows: to one of the 2-µL reaction aliquots (from **step 4**), add 70 µL of water, 25 µL of BSA (2 mg/mL) and 100 µL of 20% trichloroacetic acid (TCA). Allow the RNA to precipitate at 0°C for 10 min then collect the RNA by filtration onto a GF/C glass fiber disk. Wash the filter with 5% TCA. Spot the second 2-µL reaction aliquot (initial

radioactivity) onto another filter. Air-dry and count both disks using a scintilla-tion counter. Calculate the amount of probe synthesised as follows:

$$\text{Probe synthesized (ng)} = \frac{[\text{TCA precipitable radioactivity} \times (0.15 + 0.25) \times 4 \times 330]}{\text{initial radioactivity}}$$

with 0.15 and 0.25 corresponding to the molarities of the $[\alpha^{35}S]$-CTP and CTP used in the transcription reaction (*see* **Note 8**), and 330 being the average molecular weight of one nucleotide.

12. Resuspend the probe (from **step 10**) to 1.2 ng/µL in 10 m*M* DTT (*see* **Note 9**). This represents a 20X stock solution of probe, which can be stored at –20°C for ~1 mo and may be used for several hybridization experiments (*see* **Note 10**).

3.3. Preparation of Subbed Slides

Tissue sections must be mounted on subbed microscope slides in order to avoid detachment of the sections during the hybridization procedure. Subbing in a gelatin/chrome alum solution is an efficient and inexpensive method. Poly-L-lysine may also be employed *(12)*.

1. Prepare a 0.5% gelatin, 0.5% chrome-alum subbing solution as follows. First dissolve the gelatin at 65°C with stirring, then add the chrome alum, filter through Whatman no. 1 filter paper, and allow to cool.
2. Dip precleaned microscope slides (preferably with frosted edges for labeling) for a few seconds in the subbing solution. It is convenient to use large histological staining dishes and metal slide-holding racks (e.g., 50 slides capacity).
3. Drain the slides on a paper towel and allow them to air-dry completely in a dust-free environment (usually overnight).
4. Bake the slides for 4 h at 160°C, then store them in dust-free boxes at room temperature.

3.4. Paraffin Sections

3.4.1. Collection, Fixation, and Embedding of Samples

Tissue fixation accomplishes the following: it prevents the loss of mRNA, it preserves tissue morphology, and it increases the ability of the probe to penetrate target tissue. Crosslinking fixatives such as paraformaldehyde offer the best compromise between the preservation of tissue morphology and the retention of cellular RNA. Formaldehyde is another suitable fixative, allowing ISH to be performed on clinical-tissue specimens fixed with a standard formalin procedure.

1. Dissolve paraformaldehyde to 4% w/v in sterile PBS at 65°C with stirring. Cool and store on ice. Use the dissolved paraformaldehyde within 1 d.

2. Collect the embryos or tissues as quickly as possible and place them in the paraformaldehyde solution (*see* **Note 11**). The volume of fixative solution should be at least 10 times that of the tissue sample(s). Fix the tissues overnight (*see* **Note 12**) at 4°C with gentle agitation (e.g., on a rocking tray).

3. Proceed through the following changes of solution: PBS for 30 min at 4°C; 70% ethanol (*see* **Note 13**), 95% ethanol, and twice 100% ethanol for 30 min each (or 1 h each in the case of large samples >1 cm); twice xylene (*see* **Note 14**) for 30 min each.

4. Transfer the tissue samples into melted paraffin wax (*see* **Note 15**) in a 58°C oven. Incubate in the paraffin for at least 1 h with two changes of paraffin, or overnight for tissues over 5 mm thick. Place the tissues in disposable plastic moulds for the last incubation.

5. Take the moulds out of the oven, orientate the tissue with warm forceps (if necessary, this can be done on a heating plate and/or under a stereomicroscope), and let the paraffin solidify at room temperature. Paraffin blocks can then be stored indefinitely at 4°C.

3.4.2. Sectioning Paraffin

Histological sections are performed using a standard rotary microtome at 5–7 μm thickness after proper orientation and trimming of the paraffin block. The paraffin ribbons are separated for desired placement on slides and can be collected in either of these ways (1. or 2.):

1. Sections are floated on a clean water bath filled with distilled water at a temperature slightly below the melting point of the wax. After the sections spread out, subbed slides are placed beneath the floating ribbons to collect them and dried on a 37°C heating plate for ~1 h, and then overnight in a 37°C oven.

2. Sections are collected on drops of 10% ethanol placed on subbed slides, warmed on a heating plate at ~45°C. When the sections have spread out (~1 min), the excess fluid is drained with paper towels. Sections are then air-dried for ~2 h (protected from dust) and further dried overnight in a sealed box (*see below*) at room temperature. Alternately, sections may be collected on a 37°C heating plate. After draining of excess fluid, the slides are further dried for ~1 h on the heating plate, and then overnight in a 37°C oven.

   Slides are then placed in a sealed box containing silica gel wrapped in a paper towel, and stored at 4°C. Slides can be stored for several mo without any effect on the quality of the ISH.

   Adjacent sections may be placed on several sets of slides in order to perform comparative analyses with various probes, or for control experiments. Sections should not be placed near the edges of the slides, because this may cause hybridization artefacts. It is recommended to wear disposable latex gloves and to use disposable microtome blades. More detailed explanations about these procedures can be found in manuals describing histological methods (*see* **ref. *13***).

3.4.3. Pretreatment of Paraffin Sections

All steps are carried out at room temperature except the proteinase-K treatment.

1. Let the slides warm up to room temperature and place them on slide holding racks (*see* **Note 16**).
2. Immerse the slides in xylene (or trichlorethane-containing solvent) twice for 3 min each time.
3. Rehydrate the sections by passing the slides for 1 min in sequence in 100, 100, 95, 70, and 30% ethanol, while gently moving the rack up and down in each ethanol solution, followed by a few seconds in sterile water (*see* **Note 17**).
4. Place the slides in 1 µg/mL proteinase K (*see* **Note 18**) diluted in 1X proteinase-K digestion buffer (prewarmed in a 37°C water bath) and incubate them for 30 min (*see* **Note 19**).
5. Place the slides in 0.1 *M* triethanolamine buffer (pH 8.0) for at least 30 s.
6. Just before use, dilute acetic anhydride to 0.25% v/v in 0.1 *M* triethanolamine buffer. Mix vigorously and immediately immerse the slides in this solution. Incubate for 10 min while occasionally moving the slide rack up and down.
7. Dip the slides in distilled water, then in 30, 70, 95, and 100% ethanol for about 30 s each. The same ethanol series of **step 3** can be used again, but 100% ethanol must be replaced because it is contaminated with xylene.
8. Allow the slides to air-dry protected from dust. The slides should be hybridized the same d (*see* **Note 20**).

3.5. Frozen Sections

3.5.1. Collection and Freezing of Samples

In contrast to paraffin-embedded material, tissues are frozen and sectioned without previous fixation step. A "postfixation" will be performed on the tissue sections during the treatments prior to the hybridization (*see* **Subheading 3.5.3.**). Small pieces of tissues or embryos are collected and placed in moulds containing a thin layer of OCT-embedding medium. After proper orientation of the specimen, OCT medium is added to fill the mould, which is then frozen on the surface of dry ice. In the case of small embryos collected in PBS, excess fluid has to be drained before placing them in the moulds. Larger samples are frozen by dipping them directly into a beaker containing 2-methylbutane placed in dry ice. Note that 2-methylbutane is stored at 4°C and placed on dry ice just before adding the samples, because too rapid freezing may result in histological artefacts. Frozen samples can then be stored indefinitely at –80°C, but should never be allowed to defrost.

3.5.2. Preparation of Cryosections

Tissue sections are performed on a standard cryostat. The tissue block is first allowed to reach –20°C by placing it in the cryostat chamber for at least

30 min. Samples that had not been frozen in OCT-containing molds are fixed onto the sample holder with water. In order to avoid sample defrosting, cold water and/or aerosol cooling agents can be used. Ten-micrometer-thick sections are made at –20°C. Note that the exact temperature and thickness of sectioning have to be adapted to obtain the best possible sections. Sections are successively collected on subbed slides (*see* **Subheading 3.3.**). Slides are dried on a heating plate at ~40°C for at least 3 min and placed in sealed boxes. Slides can be safely stored at –80°C for several years.

3.5.3. Pretreatment of Frozen Sections

Unless otherwise mentioned, the solutions are at room temperature.

1. Dip the cold slides in acetone at 4°C for 5 min and allow to air-dry.
2. Place the slides in 4% formaldehyde (dissolved in PBS) at 4°C for 15 min.
3. Rinse in PBS twice for 5 min each time.
4. Acetylate the sections by performing **steps 5** and **6** of **Subheading 3.4.3.**
5. Rinse in 2X SSC twice for 2 min each time.
6. Incubate in 50% formamide, 1X SSC at 60°C for 10 min (prewarm the solution).
7. Pass the slides for 1 min each through 50% ethanol, 70% ethanol (each of these cooled to – 20°C), and 100% ethanol.
8. Allow the slides to air-dry and use the same day for hybridization.

3.6. Hybridization

The hybridization and washing conditions are identical for both paraffin sections and cryosections. However, different steps of treatments prior to hybridization have to be performed for each type of sections (described, respectively, in **Subheadings 3.4.3.** and **3.5.3.**). In the case of paraffin sections, the pretreatments consist of an incubation in an organic solvent to remove the paraffin wax, followed by a mild protease treatment to improve the penetration of probe, and an acetylation step that reduces nonspecific electrostatic binding of the negatively charged probe. In the case of cryosections, the tissues are postfixed sequentially in acetone and formaldehyde and acetylated as for paraffin sections.

The entire procedure is performed over 2 d, the slides being incubated overnight for hybridization (*see* **Note 21**), and washed the subsequent day.

1. Prepare the hybridization mix. The volume required depends upon the number of slides to be hybridized. Use 5–10 µL of hybridization mix per cm^2 of tissue section, or 50 µL per slide covered by a 24 × 50-mm cover slip (*see* **Note 22**). For 2 mL hybridization mix, combine:1 mL of deionized formamide, 400 µL of 50% dextran sulfate, 200 µL of 10X salts solution, 20 µL of 1 *M* DTT, 100 µL of 10 µg/µL yeast tRNA, 180 µL of distilled water. Note that this recipe allows for the volume of the probe (1/20th of the final volume of the hybridization mix) to be added in **step 3** (*see* **Note 23**).

2. Aliquot the hybridization mixture according to the number of probes to be used.
3. Thaw the stock riboprobe(s). These should be placed on ice and frozen again as soon as possible if it is planned to reuse them. Add 1/20th of stock probe to each aliquot of hybridization mixture. Vortex and incubate at 75°C for 2 min. Vortex again and place on ice.
4. Place the dry slides horizontally. Apply the hybridization mixture to the dry slides by depositing a drop in the center of the area containing the tissue sections. Cover each slide with a cover slip consisting of piece of Parafilm™ precut in order to cover the sections. The cover slips are handled with a forceps and applied gently to avoid trapping air bubbles.
5. Place the slides in an incubation box containing paper towels or filter paper soaked in 50% formamide, 2X SSC to produce a moist chamber (*see* **Note 24**). Seal the box and incubate overnight in a 52°C water bath. Agitation is not necessary.

3.7. Posthybridization Washes (see Note 25)

1. Prewarm the washing solution and the RNase digestion buffer to 55°C and 37°C, respectively.
2. Incubate the slides at 55°C for ~45 min in the washing solution. This initial step is best performed in staining dishes with slots to hold the slides vertically. This facilitates cover slip removal, because the parafilm cover slips will eventually float off. If necessary, removal of the cover slips can be gently completed with flat forceps. Quickly transfer the slides to holding racks for the remainder of the procedure (*see* **Note 26**).
3. Incubate in washing solution at 55°C for an additional 1 h (*see* **Note 27**).
4. Incubate successively at 37°C in: RNAase digestion buffer for 15 min, RNAase digestion buffer containing 20 µg/mL RNase A for 30 min, RNAase digestion buffer for 15 min.
5. Incubate in washing solution at 55°C twice, for 1 h each time.
6. Incubate successively in: 2X SSC for 15 min at room temperature, 0.1X SSC for 15 min at 50 or 55°C, and 3 L of 0.1X SSC for 30 min at room temperature with very gentle magnetic stirring.
7. Dehydrate the slides by passing them successively for 1 min in 30, 70, and 95% ethanol (each diluted in 0.3 *M* ammonium acetate), followed by absolute ethanol. Dry the slides vertically.
8. Slides can be exposed for autoradiography as soon as they are completely dry.

3.8. Autoradiography

3.8.1. Film Autoradiography

It is useful to first place the hybridized slides in an autoradiography cassette and expose them against a high resolution X-ray film for 1–3 d at room temperature. This type of autoradiography is insufficient for analytical purposes, but is helpful to predict the time required for emulsion autoradiography, which must be adapted for each individual probe and purpose. As a guide, the almost complete absence of background signal after an overnight film exposure will allow the slides to be exposed for at least 2–3 wk under emulsion.

3.8.2. Emulsion Autoradiography

The following steps must be performed in a darkroom equipped with a good safe-light, a water bath and possibly a light-safe door, allowing one to leave the room during the drying of the slides (**step 5**). Temperature in the darkroom should not exceed 20–22°C, as too rapid drying of the slides may result in cracking or peeling of the emulsion.

1. Predilution of the emulsion (*see* **Note 28**). In the dark, dilute a vial of emulsion with an equal volume of distilled water after warming both for ~30 min in a 45°C water bath. Mix gently to avoid the formation of bubbles. The diluted emulsion can then be stored at 4°C in small aliquots (e.g., in photography film containers) protected from light with aluminium foil and from any source of radiation.
2. Melt a vial of diluted emulsion for ~20 min at 45°C and pour it slowly into the dipping chamber placed in the 45°C bath. Any air bubbles in the emulsion will lead to strong labeling artefacts. Thus it is critical to allow the emulsion to sit for ~30 min to let the bubbles float to the surface, and then remove them by dipping a few blank slides.
3. Dip each slide by immersing it slowly for about 3 s in the emulsion.
4. Carefully wipe the emulsion from the back of the slide with paper towels (this step may be omitted since the back of the slides can be cleaned during the staining procedure; *see* **Subheading 3.8.3.**).
5. Place the slides vertically (frosted edges up) on special holders or against a support and allow them to air-dry for at least 1 h. Thus, the emulsion will form an even layer before drying.
6. Place the slides in light-tight boxes containing silica gel, seal the boxes with tape, and wrap them in aluminium foil.
7. Leave them to stand overnight at room temperature for complete desiccation and then store at 4°C for the desired length of time away from sources of radiation.

3.8.3. Developing and Staining

1. Allow the sealed boxes containing the slides to warm up to the darkroom temperature for at least 1 h.
2. In the dark, place the slides in racks and proceed through the following steps: 2 min in developer; 2 min in tap water; 2 min in fixer; twice, for 2 min each time, in tap water (can be extended). Light can be switched on when all the slides are in second tap-water wash. During this procedure, the staining dishes can occasionally be gently rocked back and forth.

3. Without allowing the slides to dry, proceed at room temperature through: 1 min in 0.02% toluidine blue solution, twice for 30 s in distilled water.
4. If necessary, scrape off the layer of emulsion from the back of the slides using a razor blade.
5. Allow the slides to air-dry completely (e.g., overnight), then mount them with glass cover slips, applied slowly and from one side to avoid trapping air bubbles.

3.9. Examination and Photography

The analysis of labeling patterns can be done using any conventional light microscope. Silver grains are clearly visible at high magnifications ($\times$ 200 and over). Low-power examination can be performed with a dissecting stereomicroscope equipped with dark-field illumination (which makes the silver grains appear white), or with a microscope equipped with epiluminescent or dark-field illumination. When using a stereomicroscope, two photographs of the same field under bright or dark-field illumination can be aligned to compare histology and labeling patterns. Alternatively, one can perform double exposure of bright-field and dark-field (using a color filter for dark-field views) of the same section. Computer-aided image-processing methods have also been developed *(14–16)*.

4. Notes

1. ISH studies of retinoid-receptor gene expression have been performed in several vertebrate species. **Table 1** provides the references of various RAR and RXR ISH probes described in the literature. The following comments may be useful in designing novel ISH probes. We have found that probes derived from the entire coding sequence of a given RAR or RXR cDNA give the most satisfying results *(1–3,6)*. Such probes provide high specific-signal intensities, because they hybridize to a long fragment of the target RNA. On the other hand, nonspecific background grain levels are not proportional to the initial length of the probe (note that all probes are hydrolyzed to an average length of ~150 nt; *see* **Note 6**). Probes of similar initial length should be used in comparative studies in order to estimate the relative abundance of various mRNAs, although it should be kept in mind that ISH is not an accurate method for quantitation.

   Importantly, we found no example of crosshybridization despite the fact that the ISH probes contain regions such as those encoding the DNA-binding and ligand-binding domains, which are highly conserved across the nuclear-receptor superfamily. Such lack of crosshybridization in spite of strong sequence conservation at the amino acid level is explained by the fact that sufficient nucleotide mismatches exist between the various mRNA sequences. Thus, imperfect RNA hybrids with transcripts of related receptors are degraded by the RNAase treatment during posthybridization washes. As a corollary, it must be stressed that the use of an ISH probe from a species different than that of the sample is unlikely to

Table 1
In Situ Hybridization Probes for Nuclear Retinoid Receptors

Gene probes	Ref.
Human RARα	23
Human RARγ	24
Human RARα, β, and γ	25,26[a]
Mouse RARγ	3
Mouse RARα, β, and γ	1,2,4,5
Mouse RXRα, β, and γ	6,27
Chick RARβ	9,28
Chick RARβ$_2$	17
Chick RARβ$_1$	18
Chick RARγ	29
Chick RXRγ	30
Xenopus RARγ	10[b]
Xenopus RARγ1 and γ2	19[c]
Zebrafish RARγ	31[c]

[a]Studies with nonradioactively labeled ISH probes (*see also* Chapter 18). Note that even though there has been no report of ISH with human RXR probes, the corresponding cDNAs have been characterized in the laboratory of R. Evans.

[b]Other *Xenopus* RARs and RXRs have been identified (**ref. 32**) but have not been studied by ISH.

[c]Whole-mount ISH studies with digoxigenin-labeled probes.

give satisfactory results. For instance, we found that human RAR probes did not work on mouse tissues. Only ISH on a very related species (e.g., using a mouse probe on rat tissue) might be expected to be successful. It is thus advised to use species-specific probes rather than to adapt the ISH protocol and decrease its stringency for non-species-specific probes.

Most of the published studies have used probes which detect all mRNA isoforms of a given RAR or RXR. It is also possible to generate isoform-specific riboprobes by limiting the template DNA to the specifically spliced region. *In situ* detection of individual RAR (RXR) isoforms poses problems, however, because of limitation in the nucleotide length of isoform-specific probes (e.g., 400–500 nt in the case of RAR isoforms), and in the very low abundance of individual mRNA isoforms. Nevertheless, there have been recent reports of ISH detection of RARβ isoforms in chick embryos *(17,18)* and RARγ isoforms can be specifically detected on cryosections of mouse embryos (unpublished data).

2. Detailed procedures for cloning, amplification, and purification of plasmid DNA, which are beyond the scope of this chapter, can be found in **refs. 20** and **21**.

Plasmid DNA can be purified by the classical alkaline lysis/caesium chloride gradient procedure or through rapid methods involving the use of small anion-exchange resin columns (e.g., Qiagen™ or Nucleobond™ columns).

3. Care should be taken to prevent any contamination by ribonucleases. We prepare the in vitro transcription reagents and solutions with diethylpyrocarbonate (DEPC)-treated water. DEPC inactivates ribonucleases and is used in the following way: dilute 1 mL DEPC per liter milliQ or double distilled water. Mix and let stand overnight. Autoclave the next day to inactivate the DEPC. Note however, that some researchers simply use sterile milliQ or distilled water.

4. The yeast tRNA (Sigma) solution is prepared in the following manner: lyophilized tRNA is dissolved in water at ~30 mg/mL, extracted twice with phenol and twice with phenol/chloroform, precipitated with ethanol, and resuspended in water as a 10 µg/µL (OD measurement) stock solution stored in aliquots at –20°C.

5. Ammonium acetate is preferred to sodium acetate because it is more efficient in preventing free nucleotides to be pelleted down in the precipitate. Some researchers actually add 0.5 vol of a 7.5-*M* ammonium acetate solution.

6. Partial alkaline hydrolysis reduces the riboprobe to smaller molecules, which are believed to reach target RNA more efficiently *(7)*. Some researchers do not perform alkaline hydrolysis of the riboprobes. In particular, this step is omitted in the case of digoxigenin-labeled whole-mount ISH probes, at least up to 1 kb *(12,22)*. We recommend systematically performing the alkaline hydrolysis of ^{35}S-labeled probes.

7. TCA precipitation may be replaced by DE81 filter-binding assay *(20)*. Actually, this entire step may be omitted in routine experiments and replaced by simple counting of an aliquot of the final reaction product (resuspended e.g., in 100 µL). Note, however, that a fraction of the unincorporated radionucleotides may not be eliminated after the two ethanol-precipitation steps. Therefore, a final radioactivity count may overestimate the actual amount of probe.

8. In our conditions, unlabeled CTP is used in slight molar excess compared to ^{35}S-CTP in the transcription reaction. This leads to probes with specific activities of ~5 × 10^8 cpm/µg, which provide adequate signals. Alternative protocols may use only ^{35}S-CTP to yield probes with higher specific activity. In this case, the estimation of probe concentration must be adjusted in consequence.

9. It is possible, in initial experiments or in case of difficulties, to check the integrity of the probe(s) and the extent of hydrolysis by migration of aliquots of the transcription reaction before and after hydrolysis on a 4% polyacrylamide-urea gel followed by drying and autoradiography. The hydrolyzed probe will appear as a large smear centered at its average final length.

10. It is advised, at least in a first series of experiments, to also synthesize a "sense-strand" riboprobe, which will be used as negative control for possible artefactual labeling. Note however, that some sense-strand probes may give higher background levels than that observed with the antisense probe. Another control consists of using the antisense probe on sections incubated with RNase A prior to hybridization, in order to show that the signal emanates from RNA–RNA hybrids.

A frequently encountered problem is to distinguish between low and diffuse (ubiquitous) signal and background labeling. In such cases, competition experiments using at least 10- to 20-fold excess of unlabeled antisense probe may be performed, which should compete only for specific binding.

11. Adult mouse organs can be placed directly into paraformaldehyde, whereas mouse embryos are dissected in sterile PBS (if necessary, a stereomicroscope can be used). Very small embryos can either be left in the uterus or deciduum, or dissected out in PBS. Small embryos or tissues can be safely transferred between solutions using a glass Pasteur pipet that has had its edges smoothed with a flame, or a Pipetman with 1-mL tip. Other specimens can be handled with a spatula or fine-tipped forceps. Sterile polyethylene tubes, which are both resistent to organic solvents and RNAse free, can be used for larger samples. In this case, samples can be left in the same tube through all changes of solutions of **step 3**. Microcentrifuge (Eppendorf) tubes are used for very small embryos, and the solutions are changed by pipeting.

12. Fixation times might be reduced for very small embryos. Young mouse embryos, if left inside the decidua, can be fixed overnight safely. Fixation can be reduced to 2–4 h for embryos that are dissected out, although overnight fixation is not obviously detrimental. Chick embryos below stage 17, however, should be fixed only for 2 h (S. Smith, personal communication).

13. Ethanol is diluted in milliQ or distilled water. The duration of any of the ethanol steps can be safely extended. Samples can be stored in 70% ethanol at –20°C.

14. Trichloroethane-containing solvents, which are widely used for histology, may be used instead of xylene.

15. Paraffin (histological grade, melting point 54–56°C) is melted beforehand in the oven and funnel-filtered through Whatman No. 1 paper. It can then be stored melted in the oven.

16. For both the prehybridization treatments and the washes, slides can be placed on metallic or plastic racks holding 20 or 50 slides. The incubations can be performed in rectangular glass or plastic staining dishes (resistant to steam autoclaving and to organic solvents), which usually require 200–300 mL of solution for slides to be immersed.

17. In order to protect from contaminating RNAases, all solutions used prior to or during the hybridization are prepared with sterile milliQ or distilled water; sterile autoclaved glassware is used throughout. Note that contaminating RNases are not efficiently inactivated using a steam autoclave, but can be destroyed by washing with 0.5 NaOH or by baking at 150°C. Less-stringent conditions can be used during the washes that actually include an RNAase A treatment step. It is thus important to always keep separate the batches of glassware for prehybridization treatments and for washes.

18. We have found that incubation in 1 µg/mL proteinase K works satisfactorily on mouse embryos at any stage as well as on adult tissues. Various proteinase K concentrations might be tested for optimal results with other specimens, but overly high concentrations will lead to tissue destruction.

19. Some authors perform a postfixation of paraffin sections with paraformaldehyde after the proteinase-K step. Such a step did not improve the results in our hands.
20. We found that a prehybridization step was not advantageous. Nevertheless, prehybridization including nonradioactive thio-ATP as a competitor was found to reduce background in the case of chick embryos (S. Smith, personal communication; *see* **refs.** *8* and *9*).
21. Hybridization is performed overnight for convenience, although it has been reported that 5–6 h of incubation may suffice *(12)*. It is carried out in 50% formamide to avoid exposing the slides to elevated temperatures. It is routinely performed at the same temperature whatever the GC content of the probe. This represents rather stringent hybridization conditions, and any potential heterologous hybrids should be eliminated during the washes, which includes RNAase treatment.
22. The volume of hybridization mix placed on each slide should be sufficient to spread over the entire area occupied by the tissue sections without extending outside of the cover slip, because excess amounts of probe lead to background artefacts. Siliconized-glass cover slips may be used instead of Parafilm™ *(12)*.
23. The final dilution of probe in the hybridization mix is 60 ng/mL. We have found that 50–100 ng/mL is a reasonable dilution range. An overly high concentration of probe produces excessive background, which is resistant to high-stringency washes. Instead of calculating standard-probe concentration, one may simply dilute the probe to 20,000–25,000 cpm/μL in the hybridization mixture. However, this approach may introduce variations in probe concentration from experiment to experiment, owing to decay of the radioisotope. Note also that short riboprobes (<500 nt) should be used at lower concentrations in order to remain in the same molar range. In cases of high-background levels, it is advisable to test various dilutions of the probe in the hybridization mix.
24. To provide a moist chamber, one can either place humidified paper towels in the bottom of the box, or stand small containers holding the humidifying solution inside the box. The humidifying solution must be the same ionic strength as the hybridization mixture in order to prevent evaporation. It is essential that the slides are never allowed to dry at any step before completion of the washes, because dried probe binds nonspecifically and cannot be washed off.
25. This relatively stringent washing procedure is used routinely without modification. The most critical step is the RNAase-A treatment, which will degrade all imperfect RNA–RNA hybrids. However, it is important to remove the formamide and DTT prior to RNAase incubation, because these may inhibit the enzyme. There is little that can be done for probes that give excessive background hybridization, since this usually stems from nonspecific "sticking" of the probe to various cellular molecules, this sticking being resistant to any change in washing conditions. Nevertheless, it may be worth testing higher temperatures in the first washing step (prior to RNAase treatment); the temperature can be raised to at least 65°C without detriment. RNAase T1 (1–10 U/mL) may be added to RNAase A. Increasing the concentration of DTT in the washing solution (up to a maximum of 100 m*M*) may also be helpful (D. Bachiller, personal communication).

26. In contrast to all steps prior to hybridization, washes can be performed with nonsterile material, except for the 0.1X SSC final steps, because RNA–RNA hybrids are sensitive to RNases at low-ionic strength.
27. Gentle agitation during washes is not critical but may be performed, for instance, by using a water bath with moving platform.
28. A 1:1 dilution of the Kodak NTB-2 autoradiography emulsion in distilled water is performed to provide a thinner coating of the slides.
29. All developing solutions as well as the tap water used for washing should be at exactly the same temperature. To ensure this, all of these are stored in containers kept in the darkroom. Temperature in the darkroom should not exceed 20–22°C to avoid risks of emulsion cracking or peeling.
30. Several staining protocols are suitable after emulsion autoradiography. If silver grains have to be observed under low-power magnification using dark-field illumination, light nuclear-staining dyes such as toluidine blue are preferred. Staining with Giemsa and hematoxylin is also suitable. H&E staining reveals the cellular morphology better, but eosinophilic structures may produce some reflections in dark-field illumination. Note that overly heavy nuclear staining will quench the overlying silver grains.

Acknowledgments

We are grateful to Prof. Pierre Chambon for his support, and Didier Décimo, Valérie Fraulob, and Brigitte Schuhbaur for their essential contributions in setting up ISH procedures applied to retinoid receptors. We are indebted to Dr. Susan Smith for communicating her ISH conditions on chick embryos and Dr. Simon Ward for useful comments. Our laboratory is supported by funds from the Centre National de la Recherche Scientifique, the Institut National de la Santé et de la Recherche Médicale, The Centre Hospitalier Universitaire Régional, the Association pour la Recherche sur le Cancer, and the Fondation pour la Recherche Médicale.

References

1. Dollé, P., Ruberte, E., Kastner, P., Petkovich, M., Stoner, C. M., Gudas, L. J., and Chambon, P. (1989) Differential expression of genes encoding α, β and γ retinoic acid receptors and CRABP in the developing limbs of the mouse. *Nature* **262,** 702–705.
2. Dollé, P., Ruberte, E., Leroy, P., Morriss-Kay, G., and Chambon, P. (1990) Retinoic acid receptors and cellular retinoid binding proteins. I. A Systematic study of their differential pattern of transcription during mouse organogenesis. *Development* **110,** 1133–1151.
3. Ruberte, E., Dollé, P., Krust, A., Zelent, A., Morriss-Kay, G., and Chambon, P. (1990) Specific spatial and temporal distribution of retinoic acid receptors gamma transcripts during mouse embryogenesis. *Development* **108,** 213–222.

4. Ruberte, E., Dollé, P., Chambon, P., and Morriss-Kay, G. (1991) Retinoic acid receptors and cellular retinoid binding proteins II. Their differential pattern of transcription during early morphogenesis in mouse embryos. *Development* **111,** 45–60.
5. Ruberte, E., Friederich, V., Chambon, P., and Morriss-Kay, G. (1993) Retinoic acid receptors and cellular retinoid binding proteins. III. Their differential transcript distribution during mouse nervous system development. *Development* **118,** 267–282.
6. Dollé, P., Fraulob, V., Kastner, P., and Chambon, P. (1994) Developmental expression of murine retinoid X receptor (RXR) genes. *Mech. Dev.* **45,** 91–104.
7. Cox, K. H., DeLeon, D. V., Angerer, L. M., and Angerer, R. C. (1984) Detection of mRNAs in sea urchin embryos by in situ hybridization using asymmetric RNA probes. *Dev. Biol.* **101,** 485–502.
8. Wedden, S., Pang, K., and Eichele, G. (1989) Expression pattern of homeobox containing genes during chick embryogenesis. *Development* **105,** 639–650.
9. Smith, S. M. and Eichele, G. (1991) Temporal and regional differences in the expression pattern of distinct retinoic acid receptor-beta transcripts in the chick embryo. *Development* **111,** 245–252.
10. Ellinger-Ziegelbauer, H. and Dreyer, C. (1991) A retinoic acid receptor expressed in the early development of Xenopus laevis. *Genes Dev.* **5,** 94–104.
11. Polak, J. M. and McGee, J. O'D. (eds.) (1990) *In Situ Hybridization: Principles and Practice.* Oxford University Press, Oxford, UK.
12. Wilkinson, D. G. (ed.) (1992) *In Situ Hybridization: A Practical Approach.* IRL, Oxford University Press.
13. Prophet, E. B., Mills, B., Arrington, J. B., and Sobin, L. H. S. (eds.) (1992) *Laboratory Methods in Histotechnology.* American Registry of Pathology, Washington DC.
14. Wilkinson, D. G. and Green, J. (1992) In situ hybridization and the three-dimensional reconstruction of serial sections, in *Postimplantation Mammalian Embryos: A Practical Approach* (Copp, A. J. and Cockroft, D. L., eds.). Oxford University Press, pp. 155–171.
15. Monaghan, A. P., Davidson, D. R., Sime, C., Graham, E., Baldock, R., Bhattacharya, S. S., and Hill, R. E. (1991) The Msh-like homeobox gene define domains in the developing vertebrate eye. *Development* **112,** 1053–1061.
16. Vonesch, J. L., Nakshatri, H., Philippe, M., Chambon, P., and Dollé, P. (1994) Stage and tissue-specific expression of the alcohol dehydrogenase 1 (Adh-1) gene during mouse development. *Dev. Dynamics* **199,** 199–213.
17. Smith, S. M. (1994) Retinoic acid receptor isoform $\beta 2$ is an early marker for alimentary tract and central nervous system positional specification in the chicken. *Dev. Dynamics* **200,** 14–25.
18. Smith, S. M., Kirstein, I. J., Wang, Z. S., Fallon, J. F., Kelley, J., and Bradshaw-Rouse, J. (1995) Differential expression of retinoic acid receptor-β isoforms during chick limb ontogeny. *Dev. Dynamics* **202,** 54-66.22.
19. Pfeffer, P. L. and De Robertis, E. M. (1994) Regional specificity of RARγ isoforms in *Xenopus* development. *Mech. Dev.* **45,** 147–153.

20. Sambrook, J., Fritsch, E. F., and Maniatis, T. (1989) *Molecular Cloning—A Laboratory Manual*, 2nd ed., Cold Spring Harbor Laboratory, Cold spring Harbor, NY.
21. Ausubel, F. M., Brent, R., Kingston, R. E., Moore, D. D., Seidman, J. G., Smith, J. A., and Struhl, K. (eds.) (1987) *Current Protocols in Molecular Biology.* Green Publishing Associates and Wiley Interscience, Wiley & Sons.
22. Décimo, D., Georges-Labouesse, E., and Dollé, P. (1995) In situ hybridization of nucleic acid probes to cellular RNA, in *Gene Probes 2*, (Hames, B. D., and Higgins, S., eds.), Oxford University Press, pp. 183–210.
23. Noji, S., Yamaai, T., Koyama, E., Nohno, T., Fujimoto, W., Arata, J., and Taniguchi, S. (1989) Expression of retinoic acid receptor genes in keratinizing front of skin. *FEBS* **259**, 86–90.
24. Finzi, E., Blake, M. J., Celano, P., Skouge, J., and Diwan, R. (1992) Cellular localization of retinoic acid receptor-gamma expression in normal and neoplastic skin. *Am. J. Pathool.* **140**, 1463–1471.
25. Xu, X. C., Clifford, J. L., Hong, W. K., and Lotan, R. (1994) Detection of nuclear retinoic acid receptor mRNAs in histological tissue sections using non-radioactive in situ hybridization histochemistry. *Diagnostic Mol. Pathol.* **3**, 122–131.
26. Xu, X. C., Ro, J. Y., Lee, J. S., Shin, D. M., Hong, W. K., and Lotan, R. (1994) Differential expression of nuclear retinoic acid receptors in normal, premalignant, and malignant head and neck tissues. *Cancer Res.* **54**, 3580–3587.
27. Mangelsdorf, D. J., Borgmeyer, U., Heyman, R. A., Zhou, J. Y., Ong, E. S., Oro, A. E., Kakizuka, A., and Evans, R. (1992) Characterization of three RXR genes that mediate the action of 9-cis retinoic acid. *Genes Dev.* **6**, 329–344.
28. Rowe, A., Richman, J. M., and Brickell, P. M. (1991) Retinoic acid treatment alters the distribution of retinoic acid receptor-β transcripts in the embryonic chick face. *Development* **111**, 1007–1016.
29. Michaille, J. J., Blanchet, S., Kanzler, B., Garnier, J. M., and Dhouailly, D. (1994) Characterization of cDNAs encoding the chick retinoic acid receptor γ2 and preferential distribution of retinic acid receptor γ transcripts during chick skin development. *Dev. Dynamics* **201**, 334–343.
30. Rowe, A., Eager, N. S. C., and Brickell, P. (1991) a member of the RXR nuclear receptor family is expressed in neural-crest-derived cells of the developing chick peripheral nervous system. *Development* **111**, 771–778.
31. White, J. A., Boffa, M. B., Jones, B., and Petkovich, M. (1994) a zebrafish retinoic acid receptor expressed in the regenerating caudal fin. *Development* **120**, 1861–1872.
32. Blumberg, B., Mangelsdorf, D. J., Dyck, J. A., Bittner, D. A., Evans, R. M., and De Robertis, E. M. (1992) Multiple retinoid-responsive receptors in a single cell: families of retinoid "X" receptors and retinoic acid receptors in the *Xenopus* egg. *Proc. Natl. Acad. Sci. USA* **89**, 2321–2325.

20

Isolation of Retinoid Receptors from Mammalian Cells

Ann K. Daly and Christopher P. F. Redfern

1. Introduction

Retinoic acid receptors (RARs) have been detected in nuclear extracts from various cell types on the basis of their ability to bind radiolabeled retinoic acid *(1–3)*. Prior to the isolation of RAR cDNAs in 1987 *(4,5)*, biochemical evidence for the existence of nuclear retinoic-acid binding proteins distinct from cytoplasmic CRABP had been obtained *(1)*. The techniques used in these experiments can be used to study the biochemical properties of the RARs and may also be applicable to the study of the related receptors known as RXRs, which bind 9-*cis* retinoic acid with high affinity *(6)*.

There are a number of technical problems associated with biochemical studies on RARs. These include the fact that these proteins are not abundant in most cells and tissues, that high-specific activity retinoids will be required to detect binding and that high concentrations of salt are required for the solubilization of these proteins from nuclei and this may interfere with certain types of experiment. Higher levels of RAR expression may be obtained by transfection of cells with RAR cDNAs. RARs have been successfully extracted from transfected cells, although the subcellular localization may be different to that generally observed in normal cell lines *(7)*.

To isolate RARs, we routinely incubate cultured cells with radiolabeled retinoic-acid isomers, prepare nuclei, extract the ligand-bound receptors from the purified nuclei and fractionate them using either sucrose density-gradient centrifugation or high-performance size-exclusion chromatography (HPSEC). It is also possible to prepare nuclear extracts from untreated cells and incubate this material with radiolabeled retinoic acid prior to fractionation *(3)*. How-

From: *Methods in Molecular Biology, Vol. 89: Retinoid Protocols*
Edited by: C. P. F. Redfern © Humana Press Inc., Totowa, NJ

ever, for this approach to be successful, relatively large amounts of RARs, which can be obtained by use of large numbers of cells or transfection of cells with RAR-expression vectors, are usually required.

2. Materials
2.1. Purification and Separation of Retinoic Acid Isomers

1. Radiolabeled retinoic acid isomers: [11,12-^{3}H(N)] all-*trans* retinoic acid is available commercially from Dupont NEN (Stevenage, Hertfordshire, UK). This material can also be used for preparation of the 13-*cis* and 9-*cis* isomers. [^{3}H] 9-*cis* retinoic acid is available from Amersham Life Science (Little Chalfont, Buckinghamshire, UK).
2. Unlabeled retinoic acid isomers: all-*trans*, 13-*cis* and 9-*cis* retinoic acid are available from a range of suppliers including Sigma (Poole, UK) and ICN (Thame, UK).
3. Reversed-phase HPLC C_{18} column (e.g., Waters Novapak or Hewlett Packard Spherisorb ODS).
4. 75:25 (v/v) methanol/water buffered with 10 mM sodium acetate, pH 7.4.
5. 90:10 (v/v) methanol/water buffered with 10 mM sodium acetate, pH 7.4.

2.2. Incubation of Retinoic Acid with Cells

1. Dulbecco's modified Eagle's medium (DMEM) or an alternative tissue culture medium. DMEM is available commercially from a number of sources including ICN.

2.3. Preparation of Nuclei

1. Phosphate-buffered saline (PBS): Dulbecco's formula, without calcium and magnesium. Contains 0.2 g/L KCl, 0.2 g/L KH_2PO_4, 8 g/L NaCl, 1.15 g/L Na_2HPO_4 in water (available commercially in the UK from Sigma and ICN).
2. Dounce homogenizer for preparation of nuclei (Kontes, Vineland, NJ).
3. Buffer A for preparation of nuclei: 15 mM KCl, 3.75 mM NaCl, 37.5 mM spermine, 125 mM spermidine, 0.5 mM EDTA, 3.75 mM Tris-HCl, pH 7.4.
4. Buffer B: as for buffer A except that EDTA is omitted.
5. 10% (w/v) Digitonin in water: digitonin (Sigma) is dissolved by heating to 100°C. The solution is allowed to cool and is then filtered through Whatman No. 1 filter paper. Store in aliquots at –20°C.

2.4. Fractionation of Nuclei

1. DNase I (Pharmacia, Herts, UK or Boehringer-Mannheim, Lewes, UK): dissolve in water at a concentration of 10 mg/mL and store in aliquots at –20°C.
2. 4 M NaCl.

2.5. Sucrose Density Gradient Centrifugation

1. 5% w/v Sucrose in 0.6 *M* NaCl, 10 m*M* Tris-HCl, pH 7.4.
2. 20% w/v Sucrose in 0.6 *M* NaCl, 10 m*M* Tris-HCl, pH 7.4.
3. Gradient-forming device such as Bio-Rad model 385 (Bio-Rad, Hemel Hempstead, UK) or Gibco-BRL model 150 (Gibco-BRL, Inchinnan, UK).
4. Swinging-bucket ultracentrifuge rotor such as Sorvall TR-641 (Du Pont NEN, Stevenage, UK) or Beckman SW-41 (Beckman Instruments, High Wycombe, UK).

2.6. Analysis of Nuclear Extracts by HPSEC

1. 0.3 *M* Potassium phosphate buffer, pH 7.8.
2. A column suitable for HPSEC such as a GF-250 column (Dupont NEN).

3. Methods

3.1. Purification and Separation of Retinoic-Acid Isomers

All-*trans* retinoic-acid photo-isomerizes readily to give a mixture of stereo-isomers with the 13-*cis* and all-*trans* forms predominating. Some 9-*cis* retinoic acid is also produced. For this reason, two important precautions are required for work with radioactive retinoic-acid isomers: first, all manipulations should be done under darkroom conditions or in rooms equipped with yellow lights, and, second, the purity of the retinoic acid should be checked by HPLC analysis, and the required isomer repurified if necessary. A number of methods for the HPLC analysis of retinoic-acid isomers are available (**ref. 8**; *see* Chapter 1). We routinely use reversed-phase chromatography with a C_{18} column.

1. Apply the sample to the column (*see* **Note 1**) and elute with a linear gradient of 75:25 methanol/water increasing to 90:10 methanol/water buffered with 10 m*M* sodium acetate, pH 7.4, over 20 min at 2 mL/min. Retinoic-acid isomers are detected by measurement of the absorbance of the column effluent at 343 nm. To avoid reisomerization during the purification of specific isomers, the lamp in the UV detector should be extinguished just before the peak of interest is eluted.

3.2. Incubation of Retinoic Acid with Cells

A variety of cell lines, including murine-F9 embryonal-carcinoma cells, murine-melanoma cells (B16 and S9), HeLa cells and HL-60 promyelocytic leukemia cells, have been successfully used for the extraction of RARs from nuclei *(1,9)*. For cell lines growing as monolayers, we have used three 75-cm^2 flasks at cell densities of 5×10^6 to 10^7 cells per flask for a typical experiment. In the case of HL-60 cells, which grow only in suspension culture, the cells are normally maintained at densities of 0.3×10^6 to 0.9×10^6 cells/mL but may be pelleted and resuspended at a 5- to 10-fold higher density prior to incubation with radiolabeled retinoic acid.

Incubation of cells with radiolabeled retinoic acid is best carried out in serum-free medium (*see* **Note 2**).

1. Remove the medium from each 75-cm^2 flask of cells, and replace with 5 mL of DMEM or an equivalent serum-free culture medium containing radiolabeled retinoic acid (*see* **Note 2**).
2. Incubate the cells with the radiolabeled retinoic acid for 2–4 h at 37°C. It is important that the incubation should not be prolonged, because retinoic acid may be metabolized quite quickly (*see* **Note 3**).

The specificity of binding of radiolabeled retinoic acid to nuclear proteins can be tested by incubating the cells with the radiolabeled retinoic acid in the presence of 100-fold excess of unlabeled retinoic acid.

3.3. Preparation of Nuclei

A wide range of methods for preparation of nuclei have been described *(10)*. For our studies on RARs in cultured cells, we isolate nuclei by homogenization in a hypotonic buffer containing detergent and polyamines.

1. After detaching the cells from the tissue-culture flask, pellet the cells by centrifugation at 2000*g* for 5 min.
2. Wash the cells with PBS and resuspend in 10 mL of buffer A. Centrifuge the suspension at 2000*g* for 5 min and resuspend the cell pellet in 10 mL of buffer A containing 0.1% digitonin which has been precooled to 4°C.
3. Homogenize the cells on ice in a Dounce glass/glass homogenizer (15 strokes) and sediment the released nuclei by centrifugation at 1000*g* for 5 min.
4. Resuspend, homogenize, and recentrifuge the nuclei twice more, wash once in buffer B, and finally resuspend in 0.5 mL of buffer B.
5. Assess the purity of the nuclear preparation by phase-contrast microscopy and, if whole cells are present, repeat the homogenization.

3.4. Fractionation of Nuclei

To isolate RARs, nuclei are digested with DNase I and extracted with a 0.6 *M* NaCl solution (*see* **Note 4**).

1. Digest the nuclei (5–10 A$_{260}$ U) for 1 h with DNase I (100 mg/mL) on ice in 0.5 mL of buffer B containing 5 m*M* MgCl$_2$.
2. After digestion, add NaCl from a 4-*M* stock solution to give a final concentration of 0.6 *M*. The solution should become less turbid as the majority of nuclear proteins are solubilized (*see* **Note 4**).

3. Centrifuge the solution at 15,000g for 5 min in a microcentrifuge and retain the supernatant for further fractionation by sucrose density-gradient centrifugation or HPSEC.
4. If required, the supernatant may be stored at –20°C prior to further fractionation. The extracts are stable for at least 1 wk under these conditions.

3.5. Sucrose Density Gradient Centrifugation

1. Apply up to 0.5 mL of nuclear extract to a 11.5 mL linear density-gradient of 5–20% sucrose using 0.6 M NaCl, 10 mM Tris-HCl, pH 7.4, as buffer.
2. Centrifuge the gradients at 205,000g for 40–60 h at 4°C in a swinging-bucket rotor.
3. Following centrifugation, fractionate the gradient into 0.3- to 0.5-mL fractions.
4. Suitable size standards such as sheep IgG (7 S), hemoglobin (4.5 S) and myoglobin (2 S) should be run in a parallel tube and detected in fractions on the basis of A_{280}.
5. Retinoic-acid receptors can be identified by scintillation counting of all or part of each fraction. They should show a sedimentation coefficient of approx 4 S.
6. Further chromatographic or electrophoretic analysis can be carried out once the RAR-containing fraction has been identified.

3.6. Analysis of Nuclear Extracts by HPSEC

As an alternative to sucrose density-gradient centrifugation, RARs can be separated by HPSEC.

1. Inject the nuclear extract (up to 50 µL) onto a GF-250 column or equivalent and elute with 0.3 M potassium phosphate buffer, pH 7.8, at a flow rate of 1 mL/min.
2. Monitor the effluent for absorbance at 280 nm and for radioactivity.
3. A peak of tritium label representing RAR is eluted at a position corresponding to an M_r value of 45,000 (*see* **Note 6**).

4. Notes

1. To enable injection in a volume of not more than 100 µL, the retinoic acid can be concentrated by blowing off the organic solvent with a stream of nitrogen and resuspending in an appropriate volume of methanol.
2. We recommend that the incubation of cells with radiolabeled retinoic acid is performed in the absence of serum. Retinoic acid binds with low affinity to serum albumin *(11)* and this may reduce the concentration of free retinoic acid available for entry into the cells. For studies with [^{3}H] all-*trans* retinoic acid (15–50 Ci/mmol), we have obtained satisfactory results using a final concentration of 20 nM. In the presence of 10% (v/v) fetal calf serum (FCS), we find that it is necessary to increase the retinoic-acid concentration to 0.2 µM to obtain a similar level of binding of retinoic acid to nuclear receptors.

3. Retinoic acid isomers undergo metabolism to more polar metabolites including 4-hydroxyretinoic acid in cultured cells. The rate of metabolism varies between different cell types (Redfern, CPF, unpublished data). We have obtained satisfactory results using incubation times of 2–4 h. Longer incubation times may present problems because of the absence of FCS and the possibility of retinoic-acid metabolism.

4. For efficient extraction of ligand-bound RARs, it is important that both nuclease digestion and high-salt extraction are combined. However, fragmentation of DNA by sonication on ice using a microprobe and three 10 s bursts at 30-s intervals is an alternative to DNase I digestion. We have found that 0.6 M NaCl is the minimum ionic strength required to solubilize the maximum amount of nuclear-associated tritium label in experiments using all-*trans* retinoic acid *(1)*. The combination of nuclease digestion and 0.6 M NaCl extraction releases approx 20% of nuclear-tritium label, whereas sonication combined with 0.6 M NaCl extraction releases approx 36%. The additional radiolabel released by sonication appears to represent non-RAR-associated ligand, because the RAR peak area from extraction by sonication is no greater than that from nuclease extraction, although general background radioactivity is higher *(1)*. It is likely that the radiolabel remaining in the pellet is also non-RAR associated.

5. The DNase I digestion/0.6 M NaCl extraction procedure results in the solubilization of the majority of nuclear proteins. The residual insoluble material, which is often termed the nuclear matrix, consists mainly of the nuclear-lamin proteins *(12)*. It is important that the nuclei are maintained on ice for the DNase I digestion and salt-extraction procedures to ensure adequate RAR extraction, because the incubation of isolated nuclei at 37°C may result in the formation of insoluble complexes resistant to salt extraction *(13)*.

6. RARs are believed to interact with retinoic-acid response elements as heterodimers with RXRs. However, isolated RAR-retinoic-acid complexes consistently show a molecular weight of 45,000 when fractionated by HPSEC, which suggests that it is the monomeric form that is being extracted *(2,7)*. A similar result has been obtained when a nuclear extract is prepared prior to incubation with retinoic acid *(3)*.

References

1. Daly, A. K. and Redfern, C. P. F. (1987) Characterisation of a retinoic-acid binding component from F9 embryonal carcinoma cell nuclei. *Eur. J. Biochem.* **68,** 133–139.
2. Darmon, M., Rocher, M., Cavey, M.-T., Martin, B., Rabilloud, T., Delescluse, C., and Shroot, B. (1988) Biological activity of retinoids correlates with affinity for nuclear receptors but not for cytosolic binding protein. *Skin Pharmacol.* **1,** 161–175.
3. Nervi, C., Grippo, J. F., Sherman, M. I., George, M. D., and Jetten, A. M. (1989) Identification and characterization of nuclear retinoic acid-binding activity in

human myeloblastic leukemia HL-60 cells. *Proc. Natl. Acad. Sci. USA* **86,** 5854–5858.

4. Petkovich, M., Brand, N. J., Krust, A., and Chambon, P. (1987) A human retinoic acid receptor which belongs to the family of nuclear receptors. *Nature* **330,** 624–629.

5. Brand, N. J., Petkovich, M., Krust, A., Chambon, P., De Thé, H., Marchio, A., Tiollais, P., and Dejean, A. (1988) Identification of a second human retinoic acid receptor. *Nature* **332,** 850–853.

6. Levin, A. A., Sturzenbecker, L. J., Kazmer, S., Bosakowski, T., Huselton, C., Allenby, G., Speck, J., Kratzeisen, C., Rosenberger, M., Lovey, A., and Grippo, J. F. (1992) 9-*cis* retinoic acid stereoisomer binds and activates the nuclear receptor RXR-alpha. *Nature* **355,** 359–361.

7. Cavey, M. T., Martin, B., Carlavan, I., and Shroot, B. (1990) *In vitro* binding of retinoids to the nuclear retinoic acid receptor α. *Anal. Biochem.* **186,** 19–23.

8. McCormick, A. M., Napoli, J. L., and DeLuca, H. F. (1980) High-pressure liquid chromatography of Vitamin A metabolites and analogs. *Methods in Enzymol.* **67,** 220–233.

9. Daly, A. K., Rees, J. R., and Redfern, C. P. F. (1989). Nuclear retinoic-acid-binding proteins and receptors in retinoic-acid-responsive cell lines. *Exp. Cell. Biol.* **57,** 339–345.

10. Muramatsu, M. (1970) Isolation of nuclei and nucleoli, in *Methods in Cell Physiology,* vol. 4 (Prescott, D. M., ed.), Academic, New York, NY, pp. 195–230.

11. Goodman, D. S. (1984) Plasma retinol-binding protein, in *The Retinoids,* vol. 2 (Sporn, M. B., Roberts, A. B., and Goodman, D. S., eds.), Academic, New York, NY, pp. 41–88.

12. Lewis, C. D., Lebkowski, J. S., Daly, A. K., and Laemmli, U. K. (1984) Interphase nuclear matrix and metaphase scaffolding structures. *J. Cell. Sci. Suppl.* **1,** 103–122.

13. Littlewood, T. D., Hancock, D. C., and Evan, G I. (1987) Characterization of a heat shock-induced insoluble complex in the nuclei of cells. *J. Cell. Sci.* **88,** 65–72.

21

Analysis of Retinoid Receptor Phosphorylation

Ali Tahayato, Christophe Rachez, Pierre Formstecher, and Philippe Lefebvre

1. Introduction

Protein kinases (PKs) and protein phosphatases (PPases) are activated in response to extracellular stimuli such as peptidic hormones and growth factors. The resulting alteration of the phosphorylation state of target proteins is a common post-translational modification used by cells to modulate rapidly the activity of enzymes and transcription factors *(1,2)*. Most of the investigations have focused on the role of *O*-phosphorylation on serine, threonine, and tyrosine residues, and consequently, most of the techniques developed for these studies apply to these acid-stable residues. Although we will only consider methods to assess the functional importance of *O*-phosphorylation, it should be kept in mind that *N*-phosphorylation is very likely to be as important as *O*-phosphorylation *(3)*. Indeed, the most common biochemical techniques, from gel fixing to peptide analysis and purification, rely on the use of low pH solutions in which *N*-phosphorylated residues are unstable. These technical aspects have so far limited the study of *N*-phosphorylation processes on the function of nuclear receptors, but appropriate methods from the study of *N*-phosphoproteins have been developed and described in detail by Matthews and coworkers *(3)*. In addition, mammalian N-PKs are yet to be fully characterized, and the cloning of these enzymes catalyzing the transfer of phosphate to lysine, histidine, and also arginine residues will undoubtedly pave the way for a better understanding of the physiological role of *N*-phosphorylation. Considering this, the initial choice of the experimental strategy is very important and should take into account both technical-feasibility parameters and access to biological reagents such as purified enzymes, antibodies, and possibly expression vectors coding for PKs and PPases.

From: *Methods in Molecular Biology, Vol. 89: Retinoid Protocols*
Edited by: C. P. F. Redfern © Humana Press Inc., Totowa, NJ

Transfecting cells containing retinoic acid receptors (RARs) and retinoid X receptors (RXRs) with a reporter gene responsive to retinoic acid (RA), and treating them with various pharmacological activators or inhibitors of PKs and PPases, may be a preliminary step to identify signaling pathways that interfere with the retinoid signal. Again, several limitations of this approach, although used by many investigators in the nuclear-receptors field (**refs. *4–8***; reviewed in **ref. *9***), including ourselves *(10,11)*, are to be kept in mind:

1. These modulators are more or less specific for the targeted enzyme.
2. Their effects and mechanism of action are not fully elucidated.
3. PKs, PPases, and their isoforms are in some instances expressed in a cell-specific manner.
4. PKs and PPases may have substrate specificities that are yet unknown. This is exemplified by the recently reported histidine phosphatase activity of PP1 and PP2A *(12)*, enzymes initially described as serine/threonine phosphatases, which means that most of the results obtained by using okadaic acid as an agent interfering with steroid-signaling pathways must be reconsidered.
5. Their effects can be indirect, i.e., interfere with component(s) of the basal-transcription machinery.

The difficulty of deciphering crosstalk processes modulating the retinoid signal is increased by the multiplicity of RAR and RXR isoforms, as well as their differential transcriptional activities, which are both cell- and promoter-specific *(13)*. Another approach is to identify systematically residues that are phosphorylated in a given cell type in which receptors are overexpressed. This will lead to the identification of long-lived phosphorylated residues, but will not give any information about kinase(s) acting on that particular residue. This problem is alleviated by the description of an increasing number of phosphorylation sites, allowing for the definition of phosphorylation-consensus sequences and thereby providing evidence for the implication of a specific kinase. We have used a blend of these two approaches to study some of the signaling pathways interfering with hRXRα and/or hRARα functions on a DR5-response element. The initial screening, with various pharmacological agents like cAMP, TPA, genistein, and okadaic acid, will not be described here, but examples may be found in reports published by our laboratory *(10,11)* and others *(5,14,15)*. We decided to focus on the interference between retinoid and protein kinases C signaling pathways, and we have devised in vitro methods that allow for the identification of phosphorylation sites of the human RARα by protein-kinase C. These methods can, of course, apply to any other kinases, and we will cover in this chapter the details of the construction and purification of full-length His-tagged receptors, and techniques that are used to identify the amino-acids phosphorylated in this in vitro setting. The ultimate goal is to identify and mutate phosphorylated amino acids to assess the contribution of these posttranslational modifications to the various activities of hRARα such as ligand binding, homodimerization, heterodimerization, and activation of transcription.

2. Materials

2.1. Construction of the His₆-Tag Bacterial Expression Vector

1. Recipient expression vector: the choice of the expression vector is essentially determined by the number of appropriate restriction sites allowing for the insertion of the receptor cDNA in the correct reading frame. Qiagen (Diagen Gmbh, Dusseldorf, Germany) and InVitrogen (InVitrogen BV, Netherlands) supply pQE and pRSET vectors, respectively, with various polylinkers, and pET vectors can be purchased from Novagen or Promega (Madison, WI) (*see* **Note 1**).
2. Plasmid containing the desired cDNA.
3. Primers for PCR amplification of the cDNA.
4. Reagents for PCR amplification:
   a. Vent (New England Biolabs) or Pfu (Stratagene) DNA polymerase.
   b. Mineral oil (Perkin-Elmer).
   c. Reaction buffer (provided by the manufacturer).
   d. 100 mM dNTPs (Promega or Pharmacia).
   e. 100 mM $MgCl_2$ (provided by the manufacturer).
5. Reagents for cloning PCR product:
   a. Restriction enzymes (in this example, *Hin*dIII and *Bam*HI).
   b. Appropriate 10X restriction enzyme buffer.
   c. Calf intestine alkaline phosphatase.
   d. Transfer RNA as carrier for precipitation.
   e. T_4DNA ligase and 10X-ligase buffer.
6. Competent JM109 or DH5α *Escherichia coli* strains.
7. Ampicillin (100 mg/mL in sterile water).
8. Luria broth (with 100 µg/mL Ampicillin): A 5X stock solution is prepared by mixing 50 g bactotryptone (Difco), 25 g bactoyeast (Difco), 50 g NaCl per L. Autoclave for 30 min and add 5 mL of 5 N NaOH.
9. LB agar plates (100 µg/mL Ampicillin).
10. SOC medium: Mix 20 g bactotryptone, 5 g bactoyeast, and 0.5 g NaCl. Add 2.5 mL of 1 M KCl, and complete to 1000 mL with water. Autoclave for 30 min, let cool at 35–40°C, and add 5 mL of a sterile 2 M $MgCl_2$ solution and 20 mL of a sterile 2-M glucose solution (Sterilize by filtration through a 0.22-µm filter cartridge).

2.2. Overexpression and Purification of the Cloned Receptor

Materials detailed here are suitable for the overexpression of the receptor cloned into pQE vectors.

1. IPTG (powder, from Promega, Madison, WI).
2. Competent M15 or SG 13009 *E. coli* strains (Qiagen, Courtaboeuf, France) bearing the Rep4 plasmid. This plasmid confers kanamycin resistance to transformed bacteria and encodes the *lac* repressor.
3. Kanamycin (Gibco-BRL, Gaithersburg, MD; 10 mg/mL).
4. LB agar plates supplemented with ampicillin (100 µg/mL) and kanamycin (25 µg/mL).
5. 1X LB medium supplemented with ampicillin (100 µg/mL) and kanamycin (25 µg/mL).
6. Water bath or orbital shakers at 26 and at 37°C.
7. Low-speed centrifuge and 500-mL bottles, precooled at 4°C.
8. Lysozyme.
9. Antiproteases: leupeptin and aprotinin at 250 µg/mL, 0.5 M benzamidin in H_2O, 0.5 M PMSF in ethanol.
10. Sonicator.
11. Lysis buffer $PNI_0\beta$: 2X PBS, pH 7.4, 0.4 M NaCl, 5 mM 2-mercaptoethanol, supplemented upon use with antiproteases (10 µg/mL final concentration leupeptin and aprotinin, 0.1 mM PMSF).
12. High-speed centrifuge, precooled at 4°C (10- to 15-mL tubes).
13. Nitrilotriacetic acid (NiTA resin, Qiagen; *see* **Note 2**).
14. Chromatography buffers (*see* **Note 3**):
    a. $PNGI_0$ Buffer: 2X PBS, 0.2 M NaCl at 4°C.
    b. $PNGI_{30}$ Buffer: buffer $PNGI_0$ supplemented to 30 mM imidazole at 4°C.
    c. $PNGI_{200}$ Buffer: buffer $PNGI_0$ supplemented to 200 mM imidazole at room temperature.
15. Econo-Pac columns (1.5 × 12 cm, Bio-Rad, Ivry, France).
16. Peristaltic pump.
17. Mini-electrophoresis system (e.g., the Mini-Protean II system from Bio-Rad).
18. Protein-blotting apparatus and blotting buffer (14.6 g Tris base; 69.1 g glycine, 1200 mL methanol, add distilled water to a final volume of 6000 mL, pH 8.3).
19. Anti-receptor (Affinity BioReagents, Neshanic Stn, NJ) and antiMRGS-His_6 (Qiagen) antibodies.
20. Nitrocellulose membrane (Schleicher and Schuell, Paris, France; BA-S 85).

2.3. In Vitro Phosphorylation of the Receptor and Acid Hydrolysis

1. Purified receptor.
2. Purified kinase(s). An increasing number of purified kinases and phosphatases are available from suppliers such as Calbiochem, Promega, UBI, and Sigma.
3. $[^{32}P]\gamma$-ATP (adenosine 5′ triphosphate), 3000-6000Ci/mmol (NEN-Dupont, Le Blanc Mesnil, France).
4. Electrophoresis and blotting equipment.
5. 6 M HCl.

6. Phosphoamino-acids standard mixture (Sigma) made of a mix of O-phospho-DL-serine, O-phospho-DL-threonine and O-phospho-DL-tyrosine (10 mg/mL stock solution in water).
7. Glass-backed cellulose thin-layer chromatography (TLC) plates (20 × 20 cm, Aldrich).
8. Methanol.
9. Ninhydrin spray.
10. Thin-layer horizontal electrophoresis apparatus.
11. Electrophoresis buffer: 5% acetic acid, 0.5% pyridine, pH 3.5.

2.4. Electroelution and Reversed Phase-HPLC

All reagents should be of the highest grade available.

1. BioTrap electroelution apparatus and BT-1 and BT-2 membranes (Schleicher & Schuell).
2. 0.5X TAE: 10 mM Tris-acetate, pH 7.4, 0.5 mM EDTA.
3. 10 mM Ammonium bicarbonate, pH 8.3, 01% sodium dodecyl sulfate (SDS).
4. 50% Trichloroacetic acid.
5. Ethanol.
6. Sequencing-grade endoprotease (Trypsin, endoproteinase C, or other enzymes from Boehringer-Mannheim).
7. High performance liquid chromatography (HPLC)-grade acetonitrile and water (JT Baker).
8. 12 M Guanidinium hydrochloride solution in water.
9. HPLC apparatus, with a gradient controller and a UV detector.
10. C18 reversed-phase HPLC column. We use routinely a DeltaPak 15 µm C18-300A, 7.8 mm × 300 mm from Waters (Millipore, Milford, MA).

3. Methods

3.1. Construction of the His$_6$-Tag Bacterial Expression Vector

The protocol is described here for the cloning of hRARα into pQE9 (*see* **Note 1**).

1. Primers for PCR amplification: The hRARα cDNA is prepared by PCR amplification and inserted into the pQE-9 vector as a *Bam*HI-*Hin*dIII fragment, in order to generate an in-frame fusion protein made of a histidine tag followed by the sequence coding for the receptor:
    Upstream primer: 5'-GCGGATCCGCCAGCAACAGCA-3'
    Downstream primer: 5'-GCAAGCTTCCATGTGGCGTGG-3'
2. PCR reaction:
    a. Mix 100 ng of the plasmid containing the cDNA, 10 µM final of each primer, 5 µL of PCR reaction buffer, and adjust to the appropriate concentration in MgCl$_2$ that should be determined in preliminary experiments.

    b. Add distilled water up to 48 μL final volume.

    c. Heat for 10 min at 95°C in a heating block and let cool to room temperature.

    d. Add the required amount of desoxynucleotide triphosphates (dNTPs), which varies according to the type of heat-stable DNA polymerase you are using.

    e. Add 0.1–1.0 U of DNA polymerase.

    f. Run the amplification reaction with the following parameters, which are given as an indication: 1 min at 94°C, 1 min at 58°C, 2 min at 72°C (30 cycles). The slope was 0.5°C per min, in a MJ-Research MiniCycler programmable thermoblock.

    g. Purify the amplified cDNA by agarose electrophoresis and adsorption on an NA-45 membrane.

3. Digest the DNA with 2.5 U of *Hin*dIII and *Bam*HI overnight.

4. Digest 5 μg of pQE9 plasmid with 2.5 U of *Hin*dIII and *Bam*HI, and 5 U of calf-intestine alkaline phosphatase overnight.

5. Inactivate the calf-intestine alkaline phosphatase by heating the sample for 30 min at 68°C.

6. Mix 50 fmol of *Bam*HI/*Hin*dIII-cut pQE9 with increasing amounts of insert such that the ratio vector:insert varies in the range of 1:1 to 1:10.

7. Precipitate DNAs by 2 volumes of ethanol, using 5 μg of tRNA as a carrier.

8. Resuspend in 17 μL H_2O, add 2 μL of 10X ligase buffer and 1 μL of ligase (ca. 1 U), and incubate 60 min at 25°C (or overnight at 12°C).

9. Transform JM109-competent cells (45 μL) with 2–5 μL of the ligation mix (*see* **Note 4**).

10. Grow colonies and characterize positive clones (*see* **Note 5**).

11. Transform M15-Rep4 competent cells. At this stage, it is useful to test several clones, if possible, for their rate of expression of the receptor.

3.2. Overexpression and Purification of the Cloned Receptor

3.2.1. Preparation of the Cell Bacterial Lysate

1. Grow a colony from freshly transformed M15 cells at 37°C in 50 mL of LB medium supplemented with 100 μg/mL ampicillin and 25 μg/mL kanamycin (LB/AK).

2. Seed a 5-L flask containing 1 L of LB/AK with the 50 mL preculture.

3. Grow bacteria until OD_{600} reaches 0.5–0.7 (it usually takes 2–3 h to reach the appropriate cell density).

4. Add 0.5 g of IPTG (ca. 2 m*M* final) to the medium. At this point, several critical parameters should be tested to ensure for an optimal production of the receptor (*see* **Note 6**).

5. Centrifuge bacteria at 1000*g* for 20 min at 4°C and resuspend in 20 mL lysis buffer.

6. Add 200 μg/mL of lysozyme and incubate for at least 30 min at 4°C. The completion of cell lysis is easily visualized by a strong increase in viscosity.

7. Shear the DNA by a mild-sonication step (five pulses of 5 s at 10% intensity) (*see* **Note 7**).

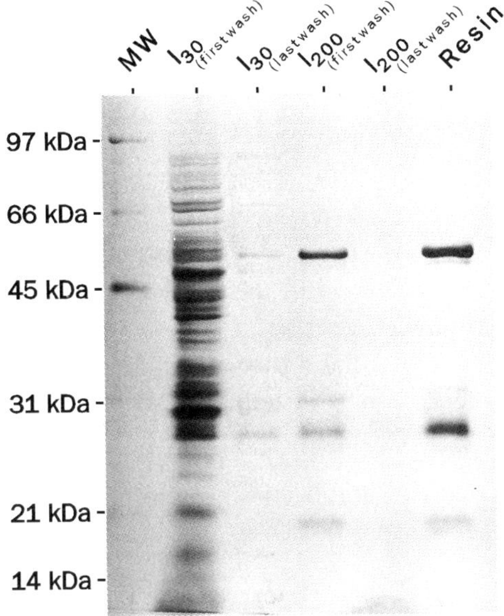

Fig. 1. Purification by NiTA-affinity chromatography of hRARα. The NiTA resin is mixed with the bacterial extract and washed as described in the text. 100-µL aliquots of fractions from the first and last washes with $PNGI_{30}$, as well as fractions from the elution step with $PNGI_{200}$ were resolved on a 8% SDS-polyacrylamide gel, which was silver stained. Molecular masses of standard proteins are indicated on the left.

8. Centrifuge the homogenate for one h at 100,000g at 4°C and adjust the supernatant to 20% v/v glycerol if stored at –80°C.

3.2.2. Receptor Purification

All except the elution step are performed at 4°C.

1. Add 1 mL of resuspended NiTA resin to the EconoColumn and equilibrate with 20 vol (10 mL) of $PNGI_0$ buffer.
2. Transfer to a 50-mL polypropylene tube.
3. Add the cell lysate to the resin and allow batch adsorption to proceed for at least 30 min on a spinning wheel (*see* **Note 8**).
4. Transfer the slurry into a Bio-Rad EconoColumn and allow the gel to settle.
5. Connect the column to a peristaltic pump and wash the resin with 20 vol of $PNGI_0$ buffer, and 100 vol of $PNGI_{30}$ buffer or until OD_{280} reaches zero (*see* **Note 9**).
6. Transfer the column at room temperature, and resuspend it in 400 µL of $PNGI_{200}$. Incubate the slurry for 5 min. Collect the flow-through, and repeat this step five times. Most of the receptor is eluted in the second and third fractions (*see* **Note 10**).
7. Assay the various fractions for protein content, and characterize their receptor content by SDS-PAGE (**Fig. 1**) and Western-blotting using an anti-His tag and/or

an antireceptor antibody (*see* **Note 11**). 10 mL of each fraction are sufficient for these analyses, and the protein content usually ranges from 0.5 to 3 mg/mL (*see* **Note 12**).

3.3. Phosphorylation of His$_6$-hRARα by Purified Kinases and Analysis of the Phosphorylated Polypeptides

In vitro phosphorylation of proteins by purified kinases can be a simple task. More cumbersome is the analysis of the phosphorylated protein, which requires the use of chromatographic procedures, and the overall yield is always a major concern. The rapid decline of the specific radioactivity of the labeled peptide, as well as handling of a fair amount of isotope, necessitates the fine tuning of experimental procedures before proceeding to the actual analysis.

3.3.1. In Vitro Phosphorylation of Receptors

There is no standard set of conditions allowing for the optimal activity of a given kinase on the receptor preparation. One point of importance is to use the optimal ATP concentration for the enzyme activity (*see* **Note 13**). In addition, some kinases may require other cofactors, in addition to bivalent cations (Mg^{2+}, Mn^{2+}, Ca^{2+}), to stimulate their activity.

3.3.2. Analysis of Phosphoamino Acids

1. Resolve the phosphorylation mix by SDS-PAGE under standard conditions.
2. Transfer protocol:
   a. Soak the polyvinylidene fluoride (PVDF) membrane for 10 min in methanol. (Do not use a nitrocellulose membrane. Its degradation causes smearing during electrophoresis.)
   b. Equilibrate the membrane in blotting buffer.
   c. Transfer proteins for 3 h at 60 V.
3. Wash the membrane three times in 250 mL distilled water to remove buffer and detergent.
4. Wrap the wet membrane in Saran wrap, and expose to autoradiographic film. A few hours' exposure is usually sufficient to locate the spot of interest.
5. Excise the piece of membrane containing the labeled receptor. Rewet the membrane if necessary by soaking it in methanol then in water and remove excess water from the membrane with a 3MM Whatman paper.
6. Place the membrane in a screw-capped microcentrifuge tube and submerge the membrane in 6 *M* HCl.
7. Incubate at 110°C for 1 h in a heating block.
8. Let cool and spin down membrane debris.
9. Transfer the hydrolysate into a new microcentrifuge tube and dry down in a SpeedVac evaporator.
10. Resuspend in 10 µL of electrophoresis buffer and mix with 10 µg of each standard phosphoamino acid.

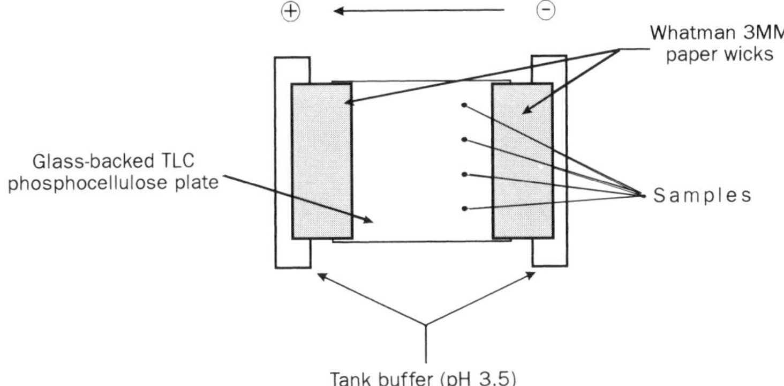

Fig. 2. Schematic representation of the electrophoresis apparatus used for the analysis of phosphoamino acids. The apparatus is connected to a power supply able to deliver a voltage of 1500 V at 50–75 mA.

11. Spot the sample on the TLC plate at 5–8 cm from the edge. Dry with a hair-dryer during application to minimize the area of the spot.
12. Wet the TLC plate by applying a 3MM Whatman sheet damped with electrophoresis buffer. Do not forget to punch a hole at the sample origin. Care should be taken to obtain an even wetting of the plate.
13. Place the TLC plate in the electrophoresis apparatus (*see* **Fig. 2**) and connect tanks to the plate with 3MM Whatman paper wicks, that should cover ca. 2.5 cm of the edge of the TLC plate.
14. Run the electrophoresis for 30 min at 1000 V, 30 mA. Remove the plate and air-dry (*see* **Note 14**).
15. Spray the plate with ninhydrin under a fume hood and heat the plate at 100°C until the coloration appears.
16. Autoradiograph the plate at –70°C with intensifying screens.

3.4. Determination of the Phosphorylated Residues in the Receptor Sequence

The identification of phosphorylated residue(s) requires a highly purified protein in sufficient amounts (1 nmol ≈50 µg) to reach the final microsequencing step of the labeled peptide(s) successfully. Consequently, labeling procedures have to be scaled up accordingly, and it is advisable at this stage to use radioactive phosphate as a tracer.

3.4.1. Purification of the Phosphorylated Receptor

1. Resolve the partially purified and phosphorylated receptor on a preparative 12% SDS-PAGE.

2. Stain the gel with KCl (*see* **Note 15**):
   a. Rinse the gel slab with distilled water
   b. Stain at 4°C in 0.25 *M* KCl, 1 m*M* DTT for 15 min.
   c. Localize the receptor band. It will generate a white band visualized with oblique light when placed on a black background.
3. Excise the desired band.
4. Place the gel slice into a BioTrap apparatus (Schleicher & Schuell) filled with 0.5X TAE and perform the electroelution for 4–5 h at 200 V or overnight at 150 V (*see* **Fig. 3**).
5. Remove the buffer and replace it with 10 m*M* ammonium bicarbonate, 0.1% SDS. This buffer exchange is necessary if you want to cleave the receptor with endoproteases such as trypsin. Repeat this step twice for 2 h.
6. Collect the eluate in the elution chamber and precipitate the protein with TCA (20% w/v final concentration). This will remove traces of SDS.
7. Incubate for 2 h at 4°C and spin down for 15 min at 15,000g.
8. Wash the pellet with cold ethanol.
9. Dry down the pellet in a SpeedVac evaporator.

3.4.2. Proteolytic Cleavage and Purification of Peptides by Reversed-Phase Chromatography

Sequencing grade endoproteinases should be used, as well as HPLC grade solvents.

1. Resuspend the purified protein in 25 m*M* ammonium bicarbonate, pH 8.3. These conditions are suitable for trypsin digestion, and buffer conditions should be adjusted to match with enzyme buffer requirements.
2. Digest the receptor using a 1:50 w:w ratio of trypsin to receptor. To ensure maximal cleavage, a similar amount of trypsin may be added to the sample after several hours of digestion (*see* **Note 16**).
3. Dilute the digestion mix in 100 µL 0.1% trifluoroacetic acid (TFA).
4. Inject the mix on the C18 column. Develop a gradient from 0.1% TFA in water to 70% acetonitrile, 0.085% TFA (*see* **Note 17**).
5. Collect fractions and identify labeled peptide(s) by scintillation counting of 1/10 of each fraction. Monitor OD_{220} to assess the concentration of peptides in each fraction. As a rule of thumb, 1 nmol of a 10-mer peptide yields an OD_{220} of about 0.2. Thus, a peak of 0.002 OD contains 10 pmol of peptide, which should be sufficient for microsequencing. However, it is advisable to work with OD peaks of about 0.1–.01 for a better sensitivity of the Edman-degradation reaction and detection of the phenylthioazolinone (PTH) derivatives of amino acids.
6. Lyophilize the labeled fraction(s) in glass tubes.
7. Resuspend dried material in 1 m*M* DTT for 20 min at 50°C to reduce disulfide bridges.
8. Cool to room temperature and add one volume of 100 m*M* iodoacetic acid or iodoacetamide. Incubate at 20°C for 20 min in the dark, to prevent formation of

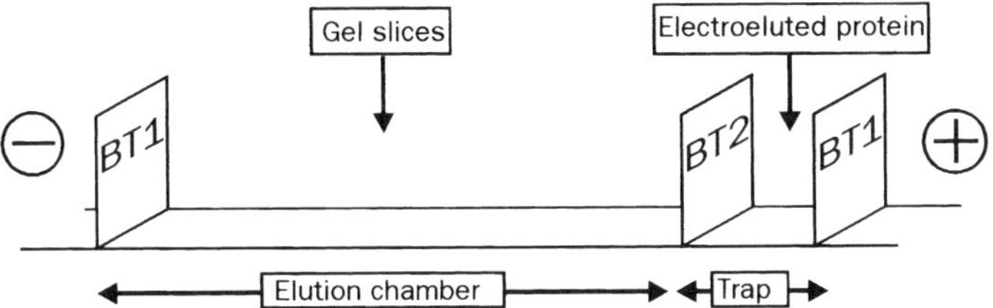

Fig. 3. The electroelution apparatus (BioTrap, Schleicher & Schuell). A schematic representation of the BioTrap is shown here. After KCl staining of the gel, the band containing the receptor is placed in the elution chamber. Do not cut the gel into small pieces, but into large chunks of about 0.3 × 10 mm. The electroeluted protein, which is now purified to homogeneity, will be recovered in about 400 µL buffer. At the end of the electroelution, reverse the polarities for 15–30 s to remove polypeptides adsorbed onto the BT1 membrane.

iodine, which reacts strongly with tyrosine residues. This reduction/alkylation step of cysteines is necessary for their further identification by the Edman degradation procedure.

9. Carry out the microsequencing analysis. This step will not be described here, because it has to be performed according to each specialist and material's requirement(s). You should establish, together with the microsequencing facility adviser, how to prepare the sample appropriately.

4. Notes

Preliminary remark: A wealth of useful technical information complementary to this review can be found in **refs. *16*** and ***17***.

1. If you are reluctant to resort to PCR for the amplification of the cDNA, it may be desirable to modify the polylinker such as to create convenient restriction sites maintaining the correct reading frame for your cDNA. However, cloning a full-length cDNA by classical means is often problematic and it is often easier to resort to PCR. The strategy followed to amplify the desired cDNA for further subcloning is dictated by the cloning sites available in the recipient vector. Whereas standard PCR procedures are followed (*Taq* DNA polymerase should be substitued by other heat-stable DNA polymerases having a proofreading activity such as Vent [New England Biolabs] or Pfu [Stratagene] DNA polymerases), primer lengths should be adjusted to restriction-enzyme requirement with respect to the number of bases flanking the recognition sequence (*see* the New England Biolabs catalog). Another important feature of the commercially available vectors is the development of highly specific monoclonal antibodies (MAbs) directed against the MRGS(His)$_6$ epitope, which are very useful for immunoprecipitation experiments. Care should be taken when choosing the location of the His tag (i.e., N-terminal or C-terminal, *see also* **Note 8**).

2. Other NiTA resins can be purchased from ClonTech (Talon[R]) or Pharmacia (HiTrap[R]). This latter resin is provided as prepacked columns that can be loaded with different metals, and can be adapted to fast performance liquid chromatography (FPLC) systems.

3. Do not substitute phosphate buffers for Tris-buffered solutions.

4. Extraction of pQE plasmids from M15 and SG13009 cells is not recommended, because extremely low yields for plasmid extraction are obtained from these particular strains. Therefore, the expression vector has to be fully characterized before proceeding to the transformation of M15 cells with the expression vector.

5. Sequencing of products is highly recommended in order to avoid any problem owing to mutation(s) generated by DNA polymerases.

6. Two parameters are important at this point: (a) the duration of the induction and (b) the temperature at which it takes place. We use routinely a 3- to 4-h induction at 37°C for the full-length receptor, but we noted that some receptor mutants have a better expression rate if induction takes place at 26°C for 2 h.

7. We have noted an exquisite sensitivity of hRARα to sonication, and especially when testing its ligand-binding activity. Therefore, depending on further assays you want to perform with crude-bacterial extracts (electrophoretic mobility-shift assays, ligand binding, and so on), you can also get rid of DNA by DNAseI digestion or polyethyleneimine precipitation *(18)*, which are techniques less deleterious to His_6-hRARα functions.

8. An overnight incubation is possible at this stage, although it does not increase the yield of the purification.

9. These conditions are appropriate for the purification of His_6-hRARα, but salt and imidazole concentrations should be carefully adjusted for each tagged protein. We observed a biphasic effect of salt concentration on His_6-hRARα purification: high-salt concentrations (above 700 m*M* NaCl) increase the number of contaminant proteins in the eluate, whereas a salt concentration between 200 and 450 m*M* NaCl favors the washing out of these contaminants. Note also that very low salt concentrations (<150 m*M*) used for the adsorption and washing steps decrease the overall yield of the purification procedure. This is owing to the loss of His_6-hRARα during washing of the matrix, and we suggest that this could be the result of the formation of unstable His_6- hRARα homodimers upon adsorption on the NiTA beads, which are readily dissociated at low salt concentrations.

10. A variable, but nonnegligible amount of receptor remains bound to the resin in these conditions. You may want to retrieve this column-bound receptor by bringing the elution buffer to 20 m*M* EDTA. However, we suspect this column-bound fraction to be misfolded and more strongly bound to the matrix by nonspecific hydrophobic interactions.

11. The combined use of both antibodies is useful to assess the occurrence of proteolysis products of the receptor in the preparation, because the anti-His tag antibody recognizes the N-terminal sequence, whereas antibodies against specific RAR subtypes are generally directed against the C-terminal domain F. When proteolysis occurs, a 30-kDa product is obtained, representing a receptor poly-

peptide deleted from its N-terminal portion. We observed also that denaturation in Laemmli buffer prior to SDS-PAGE analysis for times longer than 3–5 min at 100°C induces cleavage of the protein, in the presence of high-imidazole concentrations.

12. At this stage, we routinely obtain preparations containing the receptor and a major contaminant of about 30 kDa. The receptor represents 50% of proteins, whereas this 30-kDa bacterial protein accounts generally for 30–40% of total proteins. Several procedures can be used to obtain a pure preparation. Basically, two strategies should be considered: a preliminary chromatographic step involving anion-exchange chromatography (DEAE-sepharose) or a further purification step, after NiTA chromatography, allowing for the separation of the two polypeptides. We have favored, in our experiments, the latter option since the 30-kDa polypeptide did not interfere with any of the biological properties of the receptor. Preparative SDS-PAGE and subsequent electroelution of the receptor from the gel was used as a less expensive and more convenient alternative to preparative anion-exchange chromatography.

13. This will often require the isotopic dilution of ^{32}P-γ-ATP as purchased from NEN-Dupont or Amersham. 3000 Ci/mmol ATP is generally 3–5 μM, whereas kinases may require 10- to 100-fold higher ATP concentration for a full activity.

14. An optimal resolution of phosphoamino acids can be achieved by two-dimensional electrophoresis (pH 1.9, then pH 3.5). The protocol is similar but in our hands, it did not prove to be required for the identification of the type of phosphorylated amino acid.

15. Never use the Coomassie blue staining of the gel. The fixing step is incompatible with the elution step that follows the staining procedure, and Coomassie blue may block the free N-terminal NH_2 group, thereby preventing a successful Edman degradation of the N-terminal peptide.

16. Polypeptide denaturation is often partial under these conditions. A complete denaturation may be obtained in 6 M guanidium choride, and further reduction with 5 mM DTT. This treatment yields a polypeptide fully accessible to endoproteases.

17. The TFA percentage has to be adjusted in solvent B (acetonitrile 70%) to minimize baseline drift. Percentages vary from 0.09 to 0.07% and should be adjusted for each batch of acetonitrile.

Acknowledgments

We acknowledge the help of M.-H Metz-Boutigue for initial advices on peptide microsequencing, and the skillful technical assistance of J.-M. Wojtasik and Christophe Dessoit. This work was supported by grants from Institut National de la Santé et de la Recherche Médicale (CJF 92-03), Association pour la Recherche sur le Cancer, Fédération Nationale des Centres de Lutte contre le Cancer, Université de Lille II, and C.H.R.U. de Lille. A. Tahayato and C. Rachez were supported by fellowships from the Conseil Régional du

Nord-Pas-de-Calais and from the Association pour la Recherche sur le Cancer (A.R.C.), respectively.

References

1. Edwards, D. R. (1994) Cell signalling and the control of gene transcription. *Trends Pharmacol. Sci.*, **15**, 239–244.
2. Krebs, E. G. (1993) Protein phosphorylation and cellular regulation. 1. (Nobel lecture). *Angew. Chem. Int. Ed.*, **32**, 1122–1129.
3. Matthews, H. R. (1995) Protein kinases and phosphatases that act on histidine, lysine, or arginine residues in eukaryotic proteins: A possible regulator of the mitogen-activated protein kinase cascade. *Pharmacol. Ther.*, **67**, 323–350.
4. Matkovits, T. and Christakos, S. (1995) Ligand occupancy is not required for vitamin D receptor and retinoid receptor-mediated transcriptional activation. *Mol. Endocrinol.*, **9**, 232–242.
5. Zhang, Y. X., Bai, W. L., Allgood, V. E., and Weigel, N. L. (1994) Multiple signaling pathways activate the chicken progesterone receptor. *Mol. Endocrinol.* **8**, 577–584.
6. Beck, C. A., Weigel, N. L., Moyer, M. L., Nordeen S. K., and Edwards, D. P. (1993) The progesterone antagonist RU486 acquires agonist activity upon stimulation of cAMP signaling pathways. *Proc. Natl. Acad. Sci. USA*, **90**, 4441–4445.
7. Power, R. F., Conneely, O. M., and O'Malley, B. W. (1992) New insights into activation of the steroid hormone receptor superfamily. *Trends Pharmacol. Sci.* **13**, 318–323.
8. Power, R. F., Mani, S. K., Codina, J., Conneely, O. M., and O'Malley, B. W. (1991) Dopaminergic and ligand-independent activation of steroid hormone receptors. *Science* **254**, 1636–1639.
9. O'Malley, B. W., Schrader, W. T., Mani, S., Smith, C., Weigel, N. L., Conneely, O. M., and Clark, J. H. (1995) An alternative ligand-independent pathway for activation of steroid receptors. *Recent Prog. Horm. Res.* **50**, 333–347.
10. Tahayato, A., Lefebvre, P., Formstecher, P., and Dautrevaux, M. (1993) A protein kinase C-dependent activity modulates retinoic acid-Induced transcription. *Mol. Endocrinol.* **7**, 1642–1653.
11. Lefebvre, P., Gaub, M. P., Tahayato, A., Rochette-Egly, C., and Formstecher, P. (1995) Protein phosphatases 1 and 2A regulate the transcriptional and DNA binding activities of retinoic acid receptors. *J. Biol. Chem.* **270**, 10,806–10,816.
12. Kim, Y. H., Huang, J. M., Cohen, P., and Matthews, H. R. (1993) Protein phosphatases-1, phosphatases-2A, and phosphatases-2C are protein histidine phosphatases. *J. Biol. Chem.* **268**, 18,513–18,518.
13. Leid, M., Kastner, P., and Chambon, P. (1992) Multiplicity generates diversity in the retinoic acid signalling pathways. *Trends Biochem. Sci.* **17**, 427–433.
14. Moyer, M. L., Borror, K. C., Bona, B. J., DeFranco, D. B., and Nordeen, S. K. (1993) Modulation of cell signaling pathways can enhance or impair glucocorticoid-Induced gene expression without altering the state of receptor phosphorylation. *J. Biol. Chem.* **268**, 22,933–22,940.

15. Lin, K. H., Ashizawa, K., and Cheng, S. Y. (1992) Phosphorylation stimulates the transcriptional activity of the human beta1 thyroid hormone nuclear receptor. *Proc. Natl. Acad. Sci. USA,* **89,** 7737–7741.
16. Hunter, T. and Sefton, B. M. (eds.) (1991) Protein phosphorylation, Part A (vol. 200), in *Methods in Enzymology.* Academic, New York.
17. Hunter, T. and Sefton, B. M. (eds.) (1991) Protein phosphorylation, Part B (vol. 201), in *Methods in Enzymology.* Academic, New York.
18. Lefebvre, B., Rachez, C., Formstecher, P., and Lefebvre, P. (1995) Structural determinants of the ligand-binding site of the human retinoic acid receptor alpha. *Biochemistry* **34,** 5477–5485.

22

Photoaffinity Labeling of RARs and Mapping of Labeled Sites by an Endoproteinase Combination Technique

Yuichi Hashimoto and Toru Sasaki

1. Introduction

Photoaffinity labeling and mapping of the labeled site(s) are a superior and convenient methodology for initial structural investigation of ligand-receptor complexes. The method we describe in this chapter consists of five steps:

1. Design and synthesis of a photoreactive probe for retinoic acid receptor (RAR)-labeling.
2. Photoaffinity labeling of RAR.
3. Sequential digestion of the labeled RAR with endoproteinases.
4. High-performance liquid chromatography (HPLC) analysis of the digests to determine the amino-acid sequence context of the labeled sites (endoproteinase combination technique).
5. Mapping of the labeled site(s) by comparison of the determined context with the known amino-acid sequence of the RAR.

The last step can be applied only when the amino acid sequence of the substrate that is labeled has been established.

There are three requirements for a useful RAR-labeling probe, i.e., a specific-binding activity to the correct ligand-binding pocket of RAR, a high-covalent reactivity in the pocket, and specific detectability of the covalent adduct. Compounds with any structure which satisfies these three requirements, for example, a radiolabeled retinoid analog possessing a photoreactive functional group in its skeleton, can be used in the method described in this chapter. From the standpoint of convenience of preparation, a retinoid combined with a commercially available azidonaphthalene sulfonic acid through a short-spacer moiety (introduction of an azidodansyl group) is suitable. An example of a suc-

From: *Methods in Molecular Biology, Vol. 89: Retinoid Protocols*
Edited by: C. P. F. Redfern © Humana Press Inc., Totowa, NJ

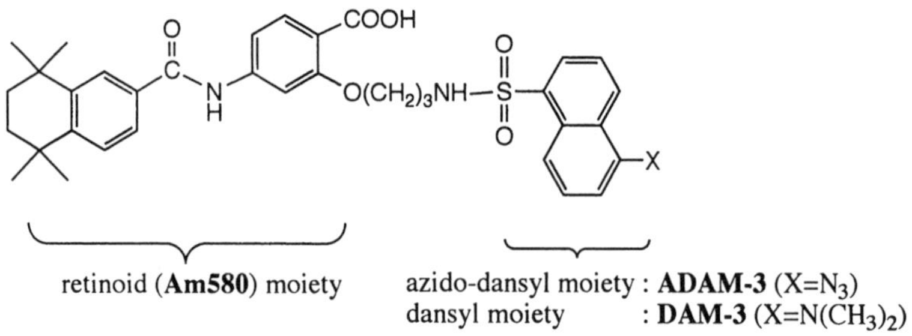

retinoid (**Am580**) moiety   azido-dansyl moiety : **ADAM-3** (X=N₃)
                                            dansyl moiety   : **DAM-3** (X=N(CH₃)₂)

Fig. 1. Structure of a RAR-photoaffinity labeling agent, ADAM-3, and a fluorescent RAR-probe, DAM-3.

cessfully used probe is ADAM-3 (**Fig. 1**) *(1–3)*, which consists of the skeleton of a potent synthetic retinoid (Am580) *(4)* and a fluorescent/photoreactive azidodansyl moiety. Its dimethylamino derivative, DAM-3 (**Fig. 1**), can be used as a fluorescent probe, whose fluorescence intensity is greatly increased by binding to RAR *(3,5)*. ADAM-3 is a superior labeling probe, because it specifically binds RARs α and β (it does not bind RARγ) at the cognate ligand-binding pocket, and it covalently reacts only after irradiation to give fluorescent adducts. The only disadvantage of ADAM-3 as a probe, if any, is that the photoreactive "business end" is situated apart from the retinoid skeleton itself. Of course, it can covalently bind with the receptor only when a reactive amino-acid residue is located nearby, as is the case for every photoaffinity-labeling agent.

For mapping the labeled site(s), direct amino acid sequencing and/or analysis by mass spectroscopy of the fragment containing the labeled site should be tried when the purity and the amount of the labeled fragment are sufficient. Otherwise, mapping by the endoproteinase combination technique, i.e., determination of the amino acid sequence context around the labeled site by the use of several specific endoproteinases and their sets of combinations, can be readily applied. The technique is based on the principle that, if the retention time (t_R) of a fragment that is cut out by one endoproteinase (specific to an amino acid residue X) changes as a result of a second digestion with another endoproteinase (specific to another amino acid residue Z), then the labeled site should be in the fragment having an amino acid sequence context of X-Z-X (**Fig. 2**) *(3,5)*. If the two retention times are the same, the labeled fragment with X's at both ends contains no Z (**Fig. 2**). Suitable commercially available endoproteinases include Lys-C, Arg-C, Asp-N, and Glu-C. The use, according to the procedure described in **Fig. 2**, of these four endoproteinases—i.e., twelve combinations in total (*vide infra*)—would allow us to determine the relationship of the labeled site to the amino-acid residues which should be recognized by the four endoproteinases, i.e., Lys, Arg, Asp, and Glu.

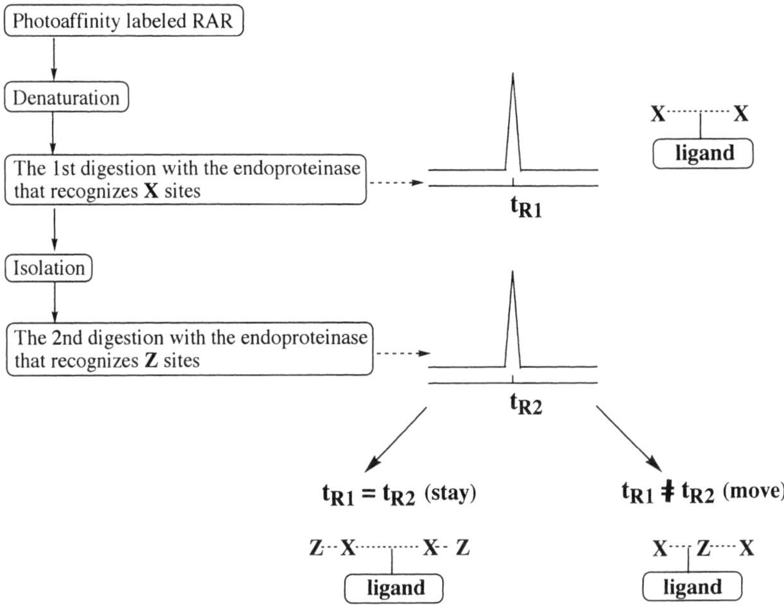

Fig. 2. Principle of the endoproteinase-combination technique.

In the case of an RAR sample that is photoaffinity labeled with ADAM-3, the labeled fragment obtained after enzymatic digestion can be specifically detected as a fluorescent peak on HPLC, and can be easily isolated. Application of the endoproteinase-combination technique to the ADAM-3-bound RAR allows us to determine the amino acid sequence context around the labeled site, and the comparison of the context with the reported amino-acid sequence of the RAR makes it possible to map the labeled site.

This method has several advantages:

1. It can be applied easily and can place the labeled site in a relatively limited sequence without amino-acid sequencing.
2. Purification of the labeled fragment is not crucial because the labeled fragments can be specifically detected (*vide supra*).
3. A small amount of material is enough and high sensitivity of the probe is not crucial, because loss of the labeled fragments during the procedure is slight.

On the other hand, this method has some deficiencies. First, it can not identify a single amino acid as the labeled site in principle. Second, the mapping procedure becomes complicated if more than one labeled site exists between adjacent recognition sites of one proteinase. Third, if the labeled site overlaps with a recognition site of an endoproteinase, the mapping procedures again become complex (*vide infra*). In such a case, further analysis, for example, a third digestion, might be necessary.

2. Materials

2.1. Preparation of Fluorescent Photoaffinity Probe, ADAM-3

The synthetic scheme for ADAM-3 is shown in **Fig. 3**. All the procedures are basic synthetic organic-chemical techniques.

1. Treat 2,5-dimethyl-2,5-hexanediol [1] with HCl gas in conc. HCl at 0°C for 15 min to obtain 2,5-dichloro-2,5-dimethylhexane [2]. Store [2] in a sealed tube because the compound is highly volatile *(4)*.
2. Refluxing of a solution of [2] in dry benzene in the presence of AlCl$_3$ (Friedel-Craft's alkylation) gives tetrahydrotetramethylnaphthalene [3], which can be purified by distillation under vacuum *(4)*.
3. Treatment of [3] with acetylchloride in dichloroethane in the presence of AlCl$_3$ (Friedel-Craft's acylation) gives tetrahydrotetramethylacetylnaphthalene [4], which can be converted to tetrahydrotetramethylnaphthoic acid [5] by treatment with high-test bleaching powder (haloform reaction) *(4,6)*.
4. Treatment of [5] with thionyl chloride gives the acid chloride [6], which can be condensed with methyl aminosalicylate [7] (prepared by esterification of aminosalicylic acid with methanol containing HCl gas) to give [8] *(6)*.
5. Condensation of [8] with Br(CH$_2$)$_3$NHCOOCH$_2$Ph [9] (prepared by benzyloxycarbonylation of bromopropylamine) in dry dimethylformamide in the presence of NaH, followed by sequential deprotection with KOH and CF$_3$COOH gives amino propyloxy-Am580 [10] *(7)*.
6. Convert the commercially available 5-azidonaphthalene sulfonic acid sodium salt [11] to acid chloride by treatment with thionyl chloride, and condense [11] with aminopropyloxy-Am580 [10] to give ADAM-3 *(2)*. Store ADAM-3 in a sealed tube to keep it dry in a refrigerator in the dark. ADAM-3 is stable on storage under these conditions. Usage of 5-dimethylaminonaphthalene sulfonic acid instead of [11] gives DAM-3, which is a nonphotolabile fluorescent analog of ADAM-3 (**Fig. 1**) *(2,3)*.

2.2. Binding Assay and Photoaffinity Labeling

1. Incubation buffer: 20 m*M* Tris-HCl, pH 8.0, 50 μ*M* EDTA, 0.25 *M* NaCl (*see* **Note 1**).
2. Washing solution: 25% v/v ethanol in distilled water.
3. Filtration apparatus and nitrocellulose filter: circular filters with 25 mm diameter and a pore size of 45 μm, which can be set in a filtration apparatus with a glass-mesh support.
4. Dextran-coated charcoal (DCC) suspension: 1% w/v activated charcoal (Norit A), 0.01% w/v dextran, 0.6 *M* NaCl, 20 m*M* Tris-HCl, pH 8.0, 10% v/v dimethyl sulphoxide (DMSO) *(8)*. Store in refrigerator (*see* **Note 2**).

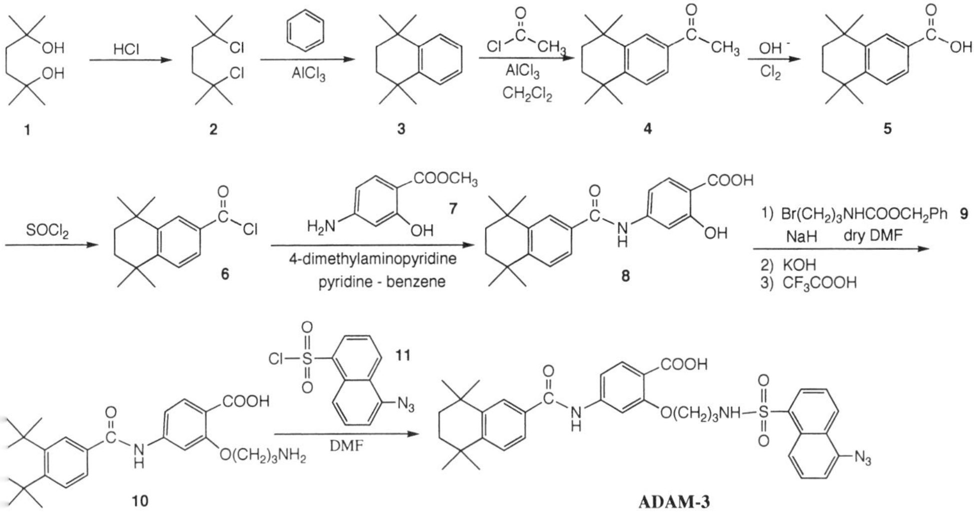

Fig. 3. Preparation of ADAM-3.

5. Ligands: Both radiolabeled and cold ligands should be stored at –20°C in the dark in DMSO or ethanol. [³H]All-*trans*-retinoic acid is commercially available from NEN. [³H]Am80 was prepared at Amersham *(9)*.

6. High-pressure mercury lamp (450 W) with cooling jacket.

7. Dialysis buffer: 20 mM Tris-HCl, pH 8.0, 0.5 mM EDTA, 10 mM 2-mercaptoethanol (2ME).

8. 2% Sodium dodecyl sulfate (SDS).

2.3. Endoproteinase Combination Technique

1. Denaturing buffer: 120 mM 2ME, 8 M urea, 0.1 % EDTA, 0.35 M Tris-HCl, pH 8.8.

2. *S*-Carboxymethylating (S-CM) buffer: 1.1 M ICH₂COONa, 8 M urea.

3. Endoproteinase digestion buffers:

    Lys-C buffer: 25 mM Tris-HCl, pH 8.5, 1 mM EDTA.

    Arg-C buffer: 10 mM Tris-HCl, pH 7.6, 8.5 mM CaCl₂, 5 mM dithiothreitol (DTT), 0.5 mM EDTA.

    Asp-N buffer: 25 mM sodium phosphate buffer, pH 7.8.

    Glu-C buffer: 25 mM ammonium carbonate buffer, pH 7.8.

4. HPLC: ODS column (e.g., µBondasphere 5µm C₁₈ 100Å, 3.9 × 150 mm). Elution buffer A: 0.1% (v/v) trifluoroacetic acid (TFA). Elution buffer B: 29.95% (v/v) acetonitrile, 69.95% (v/v) isopropanol, 0.1% (v/v) TFA. Linear gradient of 0–60% buffer B in buffer A. Detection: OD 215 nm and fluorescence intensity (excitation at 350 nm, emission at 520 nm).

3. Methods

3.1. RAR-Binding Assay

Efficiency and specificity of the prepared probe (e.g., ADAM-3) for the binding to RAR are crucial. These fractions can be evaluated by ligand competition assay using radiolabeled retinoids and RAR samples. As RAR samples, RARs extracted from various cells or tissues, including cells (mammalian, *E. coli*, and so on) transfected with an RAR-expression vector, can be used. *(1–3,5,7,9–11)* (*see* **Note 3**).

3.1.1. Filter-Binding Method **(9,11)**

1. Incubate an RAR sample (500 μL) with 1–4 nM of radiolabeled retinoid in the incubation buffer in the presence or absence of competitors (ADAM-3 and other retinoids, 0.1- to 1000-fold) at 4°C for 9–24 h.
2. Filter the incubation mixture though a presoaked nitrocellulose filter by gentle suction using a filtration apparatus (*see* **Notes 4** and **5**).
3. Wash the filter two times with 1 mL of ice-cold incubation buffer, and then wash with 1 mL of ice-cold washing buffer.
4. Remove the filter from the apparatus, and dry it in air.
5. Put the filter at the bottom of a liquid-scintillation counter vial with the front side (the face to which protein is adhering) of the filter uppermost, and add liquid-scintillation cocktail to measure the radioactivity (*see* **Note 6**).

3.1.2. DCC Method **(8)**

1. Incubation conditions are the same as described in **Subheading 3.1.1.**
2. Add 150 μL of ice-cold DCC suspension to the incubation mixture under ice cooling, and stand on ice for exactly 15 min (*see* **Note 7**).
3. Remove DCC by centrifugation at 10,000g for 10 min.
4. Take 550 μL of the supernatant, and mix it with liquid scintillation cocktail to measure the radioactivity.

3.1.3. Evaluation

Plot the measured radioactivity against the concentration of the added competitors, and draw binding-competition curves to judge the binding affinity and specificity of the competitors.

ADAM-3 competes with [³H]all-*trans*-retinoic acid or [³H]Am80 for binding to RARs α and β with a binding efficiency of 1/10–1/30 of that of all-*trans*-retinoic acid or Am80. This efficiency is sufficient for the photoaffinity labeling of RARs.

3.2. Photoaffinity Labeling of RARs **(1–3)**

1. Incubate RAR samples (5–10 μM for purified RAR proteins, or 1–5 mg/mL for RAR-containing cell extracts) with ADAM-3 (10–20 μM) in the incubation buffer in the presence or absence of competitor (20–100 μM) at 4°C for 6–18 h using a Pyrex tube (*see* **Note 8**).

2. Irradiate the mixture using a high-pressure mercury lamp under ice-cooling for 10 min (*see* **Note 9**).
3. Add 1/20 volume of 2% SDS and boil for 1 min.
4. Add 1/4 volume of DCC suspension to collect unbound and/or photodecomposed ADAM-3, and stand the mixture on ice for 15 min.
5. Remove the DCC by centrifugation and measure the fluorescence intensity of the supernatant.

3.3. Endoproteinase Combination Technique (1,12–14)

3.3.1. Sequential Digestion

1. Conditions of photoaffinity labeling of RARs and the SDS treatment of the labeled samples are the same as described in **Subheading 3.2.**
2. Put the photoaffinity-labeled RAR in a dialysis bag and dialyze against the dialysis buffer at 4°C for 1 h twice to remove the unbound ADAM-3 and/or its decomposed products.
3. Change the dialysis buffer to the denaturing buffer, and allow to stand at room temperature for 12–24 h.
4. Put the denatured sample in a Pyrex tube, and add 1/9 volume of the S-CM buffer. Then stand the mixture in the dark for 1 h at room temperature for *S*-carboxymethylation.
5. Dialyze the mixture against four changes of water at 4°C for more than 2 h each change of water (*see* **Note 10**).
6. Freeze-dry the sample in aliquot portions. The dried samples can be stored at –20°C.
7. Digest the obtained ADAM-3-bound RAR samples with endoproteinases (the first-stage digestion) in an appropriate buffer listed in **Subheading 2.3.** Amounts of endoproteinases added to the substrate should be 1/50–1/100 w/w for Lys-C, Arg-C and Glu-C, or 1/200 for Asp-N. Incubation conditions are 36°C/18 h for Lys-C and Arg-C, 25°C/18 h for Glu-C, and 36°C/6 h for Asp-N. The digested mixtures can be stored at 4°C.
8. Adjust the concentration of the peptides and TFA to 0.1–0.5 mg/mL and 0.1% v/v, respectively. Subject 400–450 μL of the sample solution to HPLC. Examples of HPLC profiles for the digests of recombinant hRARα labeled with ADAM-3 are shown in **Fig. 4**.
9. Determine the retention times of the fluorescent peaks (t_{R1}'s), estimate the amount of the peaks from the UV-monitored chart, and collect the fluorescent peaks (*see* **Notes 11–13**). Whether the fluorescence of the peak is owing to the specific binding of ADAM-3 can be judged by disappearance (or decrease) of the fluorescence intensity of the peak when the photolabeling reaction is conducted in the presence of a retinoid competitor (all-*trans*-retinoic acid, Am80, and so on).
10. Freeze-dry the collected fractions and digest the residue (the second-stage digestion) with an endoproteinase other than the one employed in the first-stage digestion. The conditions of enzymatic digestion are the same as for **step 7** in this section.

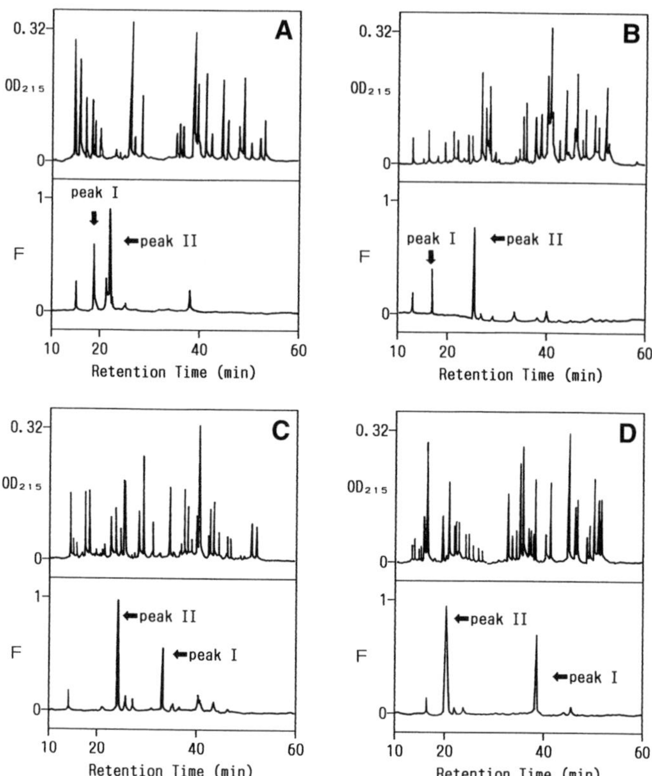

Fig. 4. HPLC Profiles of the first-stage digests of recombinant hRARα labeled with ADAM-3. Ligand-binding domain of hRARα fused with maltose-binding protein (MBP), expressed in *E. coli*, was photoaffinity labeled by ADAM-3 *(1,3,10)*. The labeled recombinant hRARα was digested with Arg-C (**A**), Asp-N (**B**), Glu-C (**C**) and Lys-C (**D**). Conditions for HPLC analysis are described in the text. The top profiles were monitored by measuring OD_{215}. The bottom profiles were monitored by measuring fluorescence intensity. Peaks I and II are specific (they disappeared when photoaffinity labeling was performed in the presence of retinoid competitors). Other fluorescent peaks are nonspecific.

11. Analyze the second stage digest under the same conditions as those for the first stage digests, and determine the retention times of the fluorescent peak (t_{R2}'s) (*see* **Note 11**).

3.3.2. Analysis of Amino Acid Sequence Context (3,12–14)

The method for the amino acid sequence context determination depends on the HPLC t_R's (**steps 9** and **11** in **Subheading 3.3.1.**). Therefore, reproducibility of the HPLC is critical. To obtain the t_R's to be compared, HPLC conditions should be exactly the same, the reproducibility should be checked, and com-

Table 1
t_{R1}-t_{R2} Correlation[a] for Peak II in Fig. 4.

		t_{R1}			
		Arg-C	Asp-N	Glu-C	Lys-C
t_{R2}	Arg-C	—	Move (1)	Move (2)	Move (3)
	Asp-N	Move (4)	—	Move (5)	Move (6)
	Glu-C	Move (7)	Stay (8)	—	Move (9)
	Lys-C	Stay (10)	Stay (11)	Stay (12)	—

[a]The requirements deduced from the combination technique results (1)–(12) are:

[1] Asp—Arg—Asp	[2] Glu—Arg—Glu
[3] Lys—Arg—Lys	[4] Arg—Asp—Arg
[5] Glu—Asp—Glu	[6] Lys—Asp—Lys
[7] Arg—Glu—Arg	[8] Glu—Asp—Asp—Glu
[9] Lys—Glu—Lys	[10] Lys—Arg—Arg—Lys
[11] Lys—Asp—Asp—Lys	[12] Lys—Glu—Glu—Lys

parison of the t_R's should be done with t_R's obtained in one set of HPLC analyses performed on the same day.

1. **Steps 1–11** are the same as those described in **Subheading 3.3.1.**
2. Compare the t_{R1} with t_{R2}, judge whether they are the same (stay) or different (move), and prepare t_{R1}-t_{R2} correlation tables. Examples for recombinant hRARα labeled with ADAM-3 are shown in **Tables 1** (for the peak II in **Fig. 4**) and **2** (for the peak I in **Fig. 4**).
3. Determine the amino acid sequence context (*see* **Tables 1** and **2**).
4. Search for the sequences in the known RAR sequences that satisfy all of the twelve amino acid sequence contexts determined at **step 13**. In **Table 2**, two sequences satisfy all of the twelve requirements: Asp288 - Arg294 and Thr295-Leu306 (**Fig. 5**). The results indicate that the labeled site in the fragment corresponding the peak I in **Fig. 4** exists in the sequence Asp288-Leu 306.
5. If sequences that satisfy all of the determined contexts cannot be found (this is the case for the contexts shown in **Table 1**), reanalyze the t_{R1}-t_{R2} correlation tables on the hypothesis that one amino acid residue which should be recognized by one of the employed endoproteinases is blocked (not recognized) by ADAM-3-labeling. In the case of **Table 1**, a sequence which satisfies all twelve requirements can be found only when Arg385 is regarded as blocked: the sequence Ile381-Lys390. The results suggest that the peak II in **Fig. 4** includes Arg385, and labeling by ADAM-3 occurred at this residue or in its vicinity (**Fig. 5**) (*see* **Note 14**).

Recently, the structure of a dimeric apo-ligand-binding domain of hRXRα was determined by X-ray crystallography *(15)*. In spite of the difference between RXRα and RARα in their ligand-binding selectivity, the determined

Table 2
t_{R1}-t_{R2} Correlation[a] for Peak I in Fig. 4.

		t_{R1}			
		Arg-C	Asp-N	Glu-C	Lys-C
t_{R2}	Arg-C	—	Move (1)	Move (2)	Stay (3)
	Asp-N	Move (4)	—	Move (5)	Move (6)
	Glu-C	Move (7)	Move (8)	—	Stay (9)
	Lys-C	Move (10)	Move (11)	Move (12)	—

[a]The requirements deduced from the combination technique results (1)–(12) are:

[1] Asp—Arg—Asp   [2] Glu—Arg—Glu
[3] Arg—Lys—Lys—Arg   [4] Arg—Asp—Arg
[5] Glu—Asp—Glu   [6] Lys—Asp—Lys
[7] Arg—Glu—Arg   [8] Asp—Glu—Asp
[9] Glu—Lys—Lys—Glu   [10] Arg—Lys—Arg
[11] Asp—Lys—Asp   [12] Glu—Lys—Glu

structure is thought to be similar to that of hRARα, because the amino acid sequence of the eleven α-helices that compose the basic structure of the ligand-binding domain of hRXRα is highly homologous to that of hRARα. On the basis that the overall structure of the ligand-binding domain is highly conserved between hRXRα and hRARα, the labeled sites determined by the endoproteinase-combination technique, i.e., Asp288-Leu306 and Ile381-Lys390 in hRARα, correspond to the region located around the bottom of one putative binding-pocket, which is designated pocket B *(15)* and one of the α-helices composing the pocket B, respectively.

More recently, the structure of a holo-ligand-binding domain of hRARγ was reported *(16)*.

4. Notes

1. Incubation in a high-salt buffer results in a stronger binding of retinoids to RARs. In our experiments, the highest association constants were obtained by the use of a buffer of 20 mM Tris-HCl, pH 8.0, 0.6 M KCl. For the incubation, the content of organic solvent (EtOH and/or DMSO) should not exceed 5% v/v.

2. The quality of the DCC suspension is critical. In our experiments, charcoal (C-5260, Sigma) and dextran (D-9260, Sigma) were used. On preparation, the DCC suspension should be sonicated for more than 2 h. Then stand the suspension in refrigerator-without agitation overnight. Take off the floating materials, which cause deviation of the binding-assay results, with a paper towel.

3. For extraction of RARs from mammalian cells, we recommend cell disruption in 20 mM Tris-HCl, pH 8.0, 0.6 M KCl. Addition of proteinase inhibitors and/or reducing agents including 2ME and DTT is not crucial if the operations are per-

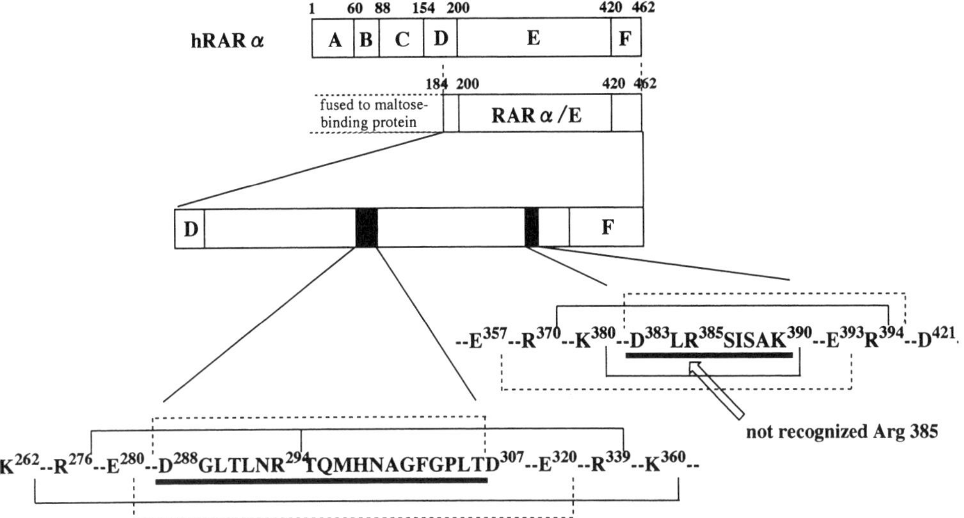

Fig. 5. Schematic illustration of the ADAM-3 binding site on the recombinant hRARα. The maltose-binding protein (MBP) moiety is omitted in the illustration. There was no sequence in the MBP moiety that satisfies the amino acid contexts shown in **Tables 1** and **2**.

formed quickly at a temperature below 4°C. As recombinant RARs, the ligand-binding domains fused to maltose-binding protein and expressed in *E. coli* were used in our experiments *(3,10,12)*.

4. Nitrocellulose filter should be presoaked by floating it on an incubation buffer. The down side of the filter from where the soaking starts should be the face on which protein samples adhere. The filter should not be dried until **step 4**.

5. Temperature is critical for this assay. Therefore, the filter support and the washing buffers should be precooled.

6. After addition of the liquid-scintillation cocktail, the vial should be allowed to stand for 20 min. During this time, the cocktail completely soaks into the filter. Small bubbles on the filter, if any, should be removed by gentle agitation.

7. DCC suspension should be shaken vigorously before use. Stirring of the DCC suspension should be continued during pipetting of the suspension. For example, a stirring bar is put inside the bottle of DCC suspension, and stirring is continued on a magnetic stirrer.

8. Thiols including 2ME should not be added, because thiols are known to reduce the photoreactive-azido groups.

9. Irradiation for 10 min is enough. Longer irradiation time causes damage to the proteins. Degassing is not necessary. The temperature of the sample solution should be kept below 4°C. In our experiments, the lamp was equipped with cold-water cooling jacket and the system was held in an ice-water bath. The Pyrex sample tube was attached to the surface of the lamp system for irradiation.

10. If aggregation of proteins occurs, change the dialysis buffer to 2 m*M* Tris-HCl. Chelator molecules such as EDTA should be completely removed, because endoproteinase Asp-N is a Zn-proteinase. The dialysate can be stored at 4°C. Add 2ME for long-term storage.

11. Complete digestion is a critical requirement for the endoproteinase combination technique. To confirm the completeness of the digestion, checking the time course of the enzymatic reaction by HPLC is useful. When there are two or more peaks with comparable fluorescence intensity, assignment of each peak is crucial. For this purpose, comparison of t_{R2}'s (retention times of the second digests) of the same combination of two endoproteinases with the reversed order is useful. For example, if there are two major fluorescent peaks in the HPLC profile of the sample digested with endoproteinases A (retention times of $t_{R1}A1$ and $t_{R1}A2$) and B ($t_{R1}B1$ and $t_{R1}B2$), isolate each fluorescent peak and digest each peak with endoproteinases B (for $t_{R1}A$'s) and A (for $t_{R1}B$'s). The second digests will give $t_{R2}B1$ from the peak of $t_{R1}A1$, $t_{R2}B2$ from $t_{R1}A2$, $t_{R2}A1$ from $t_{R1}B1$, and $t_{R2}A2$ from $t_{R1}B2$, respectively. The retention time $t_{R2}B1$ should be exactly the same as one of the $t_{R2}A$'s. Similar investigation for every two endoproteinase combination sets allows the assignment of the fluorescent peaks.

12. When two t_R's that should be compared are very similar, it is recommended to coinject the two samples to check the identity of the t_R's.

13. Purity of the isolated peak is not crucial for this method if the isolated sample contains only one fluorescent peak. Therefore, isolation of the peak should be done so as to recover the fluorescence in a good yield.

14. In the case of **Table 2**, first use combinations of three endoproteinases to identify the labeled site as one that should be recognized by the omitted endoproteinase. For example, on the assumption that one of the recognition sites of Lys-C is blocked, search for sequences that satisfy the requirements obtained with the combination of Arg-C/Asp-N/Glu-C, and then examine whether these sequences contain Lys. These procedures give the following sequences: 245–262 (Asp256 as the site not recognized) and 381–390 (Arg385 not recognized). Further examination of these candidate sequences to see whether they are consistent with all the results in **Table 1** leads to the conclusion mentioned in this step.

References

1. Sasaki, T., Shimazawa, R., Sawada, T., Iijima, T., Fukasawa, H., Shudo, K, Hashimoto, Y., and Iwasaki, S. (1995) Determination of the photoaffinity-labeled site on the ligand-binding domain of retinoic acid receptor α. *Biochem. Biophys. Res. Comm.* **207,** 444–451.

2. Shimazawa, R., Sanda, R., Mizoguchi, H., Hashimoto, Y., and Iwasaki, S. (1991) Fluorescent and photoaffinity labeling probes for retinoic acid receptors. *Biochem. Biophys. Res. Comm.* **179,** 236–265.

3. Hashimoto, Y. and Shimazawa, R. (1993) Fluorescent and photoaffinity labeling probes for retinoic acid receptors. *Clin. Dermatol.* **5,** 585–598.

4. Kagechika, H., Kawachi, E., Hashimoto, Y., Himi, T., and Shudo, K. (1988) Retinobenzoic acids. 1. Structure-activity relationships of aromatic amides with retinoidal activity. *J. Med. Chem.* **31,** 2182–2192.

5. Shimazawa, R., Hibino, S., Mizoguchi, H., Hashimoto, Y., Iwasaki, S., Kagechika, H., and Shudo, K. (1991) Fluorescent probes for retinoic acid receptors: molecular measures for the ligand binding pocket. *Biochem. Biophys. Res. Comm.* **180,** 249–254.

6. Newman, M. S. and Holmes, H. L. (1943) β-Naphthoic acid. *Org. Syn. Col.* **2,** 428–430.

7. Kagechika, H., Hashimoto, Y., Kawachi, E., and Shudo, K. (1988) Affinity gels for purification of retinoid-specific binding protein (RSBP). *Biochem. Biophys. Res. Comm.* **155,** 503–508.

8. Hashimoto, Y. and Shudo, K. (1991) Cytosolic-nuclear tumor promoter-specific binding protein: association with the 90kDa heat shock protein and translocation into nuclei by treatment with 12-O-tetradecanoylphorbol 13-acetate. *Jpn. J. Cancer Res.* **82,** 665–675.

9. Hashimoto, Y., Kagechika, H., Kawachi, E., and Shudo, K. (1988) Specific uptake of retinoids into human promyelocytic leukemia cells HL-60 by retinoid-specific binding protein: Possibly the true retinoid receptor. *Jpn. J. Cancer Res.* **79,** 473–484.

10. Fukasawa, H., Iijima, T., Kagechika, H., Hashimoto, Y., and Shudo, K. (1993) Expression of the ligand-binding domain-containing region of retinoic acid receptors α, β and γ in *Escherichia coli* and evaluation of ligand-binding selectivity. *Biol. Pharm. Bull.* **16,** 343–348.

11. Hashimoto, Y., Petkovich, M., Gaub, M. P., Kagechika, H., Shudo, K., and Chambon, P. (1989) The retinoic acid receptors α and β are expressed in the human promyelocytic leukemia cell line HL-60. *Mol. Endocrinol.* **3,** 1046–1052.

12. Eyrolles, L., Kagechika, H., Kawachi, E., Fukasawa, H., Iijima, T., Matsushima, Y., Hashimoto, Y., and Shudo, K. (1994) Retinobenzoic acids 6. Retinoid antagonists with a heterocyclic ring. *J. Med. Chem.* **37,** 1508–1517.

13. Sasaki, T., Shimazawa, R., Sawada, T., Iijima T., Fukasawa, H., Shudo, K., Hashimoto, Y., and Iwasaki, S. (1996) Location of two photoaffinity-labeled sites on the ligand-binding domain of retinoic acid receptor α. *Biol. Pharm. Bull.* **19,** 659–664.

14. Sawada, T., Kobayashi, H., Hashimoto, Y., and Iwasaki, S. (1993) Identification of the fragment photoaffinity-labeled with azidodansyl-rhizoxin as Met363-Lys379 on β-tubulin. *Biochem. Pharm.* **45,** 1387–1394.

15. Bourguet, W., Ruff, M., Chambon, P., Gronemeyer, H., and Moras, D. (1995) Crystal structure of the ligand-binding domain of the human nuclear receptor RXR-α. *Nature* **375,** 377–382.

16. Renaud, J.-P., Rochel, N., Rutt, M., Vivat, V., Chambon, P., Gronemeyer, H., and Moras, D. (1995) Crystal structure of the RAR-γ ligand-binding domain bound to all-trans retinoic acid. *Nature* **378,** 681–689.

23

PCR Cloning of N-Terminal RAR Isoforms and APL-Associated PLZF-RARα Fusion Proteins

Arthur Zelent

1. Introduction

Vitamin A (retinol), a simple fat-soluble molecule discovered in 1913 by McCollum and Davis *(1)*, is essential for proper development, growth, and maintenance of a normal adult vertebrate organism *(2–4)*. Dietary deficiency of vitamin A results in structural and functional abnormalities of a multitude of organs and organ systems *(5–7)*. Likewise, excess of vitamin A is deleterious to life, particularly in its early development where hypervitaminosis A leads to profound embryonic malformations *(6,8,9)*.

Over the years, considerable effort has been devoted to studying the cellular metabolism of vitamin A and characterizing its biologically active derivatives *(10,11)*. The term retinoids has been used to encompass a continual growing family of synthetic and natural compounds that are both structurally and functionally related to vitamin A and its physiologically active metabolites, such as all-*trans* retinoic acid (RA). The association of physiological effects of RA with dramatic changes in gene expression *(4)* suggested that it may directly participate in regulating gene transcription. This hypothesis proved to be correct by the discovery of a nuclear retinoic acid receptor (RAR), a member of a superfamily of steroid/thyroid hormone nuclear receptors *(12,13)*.

Regulation of gene expression at the transcriptional level is an essential component of important cellular and developmental processes such as growth, differentiation, and lineage determination *(14–17)*. In this respect, the concerted action of cell-type specific transcription factors, which bind to the DNA elements (response elements) located in the regulatory regions (such as promoters and/or enhancers) of specific genes and either inhibit or stimulate the rate of

From: *Methods in Molecular Biology, Vol. 89: Retinoid Protocols*
Edited by: C. P. F. Redfern © Humana Press Inc., Totowa, NJ

transcription initiation by RNA polymerase II *(18–20)*, is of extreme importance. Nuclear receptors are soluble proteins that bind as dimers to specific hormone-response elements and act as cell-type and promoter-specific transcription factors. In contrast to other transcription factors, however, their activities can be modulated through binding of the corresponding hydrophobic ligands, such as steroid hormones or retinoids *(21–23)*.

The pleiotropic nature of RA effects suggested an inherent complexity of retinoid signaling pathway(s) that may involve multiple retinoid receptors and ligands. Subsequent discoveries of two other *RAR* genes (*RARβ* and *γ*) and their isoforms *(24–31)*, as well as three different *retinoid-X receptor (RXR)* genes (*α*, *β*, and *γ*)—which encode proteins that selectively bind 9-*cis*-retinoic acid *(32,33)* and heterodimerize with RARs and some nonretinoid nuclear receptors *(34–36)*—proved this hypothesis to be correct.

To date, it has been shown that in the mouse and human, and probably in all vertebrates, three *RAR* and three *RXR* genes exist, each encoding multiple isoforms generated by alternative-promoter usage and exon splicing *(28–30,34,37–41)*. The murine and human *RAR* genes are very similar in their genomic organization and structure. Each gene possesses two promoters (P1 and P2) directing expression of at least two major isoforms (α1 and α2, β1 and β2, γ1 and γ2) that differ in the sequences encoding their 5'-untranslated regions (5'-UTR) and N-terminal A regions, i.e., upstream of the 5' border of the exon encoding the B region (*see* **Fig. 1A–C**). Two additional major RARβ isoforms (RARβ3 and RARβ4) can be generated from the *RARβ* gene by the alternative splicing of transcripts originating from P1 and P2 promoters. The RARβ3 differs from RARβ1 by an additional 27 amino acids encoded in an exon that is spliced between the exons encoding the A and B regions of RARβ1 (*see* **ref.** *30* and **Fig. 1B**). RARβ4 on the other hand is generated from an alter-

---

Fig. 1. *(see facing page)* Schematic representation of the three RAR genes and their major isoforms. Exons are represented by shaded boxes and are numbered consecutively. Regions which are not shaded represent 5' and 3' untranslated sequences. Exon/intron structures of *RARα* (**A**), *RARβ* (**B**), and *RARγ* (**C**) genes are as previously described (*see* **Subheading 1.**). Note that only exons encoding the major isoforms of the *RAR* genes are indicated. Two promoters, P1 and P2, in each *RAR* gene are indicated by broken arrows. Various RARα (A), RARβ (B), and RARγ (C) isoforms are indicated by hatched rectangles subdivided into their conserved (B-F) and divergent (A regions) functional domains. Different patterns are used to represent RAR isotype and isoform-specific sequences. Regions encoded by different exons are indicated by arrowheads underneath each diagram. DNA (DBD) and ligand-binding domains (LBD) (regions C and E, respectively) are indicated above the scheme for each isoform. Note that this diagram is not drawn to scale.

A
P1
P2
1 2 3 4 5 6 7 8 9 10
RARα
DBD
LBD
A1 B C D E F
RARα1
DBD
LBD
A2 B C D E F
RARα2
B
P1
P2
1 2 3 4 5 6 7 8 9 10 11
RARβ
DBD
LBD
A1 B C D E F
RARβ1
DBD
LBD
A2 B C D E F
RARβ2
DBD
LBD
A3 B C D E F
RARβ3
CUG
DBD
LBD
B C D E F
RARβ4
C
P1
P2
1 2 3 4 5 6 7 8 9 10 11
RARγ
DBD
LBD
A1 B C D E F
RARγ1
DBD
LBD
A2 B C D E F
RARγ2

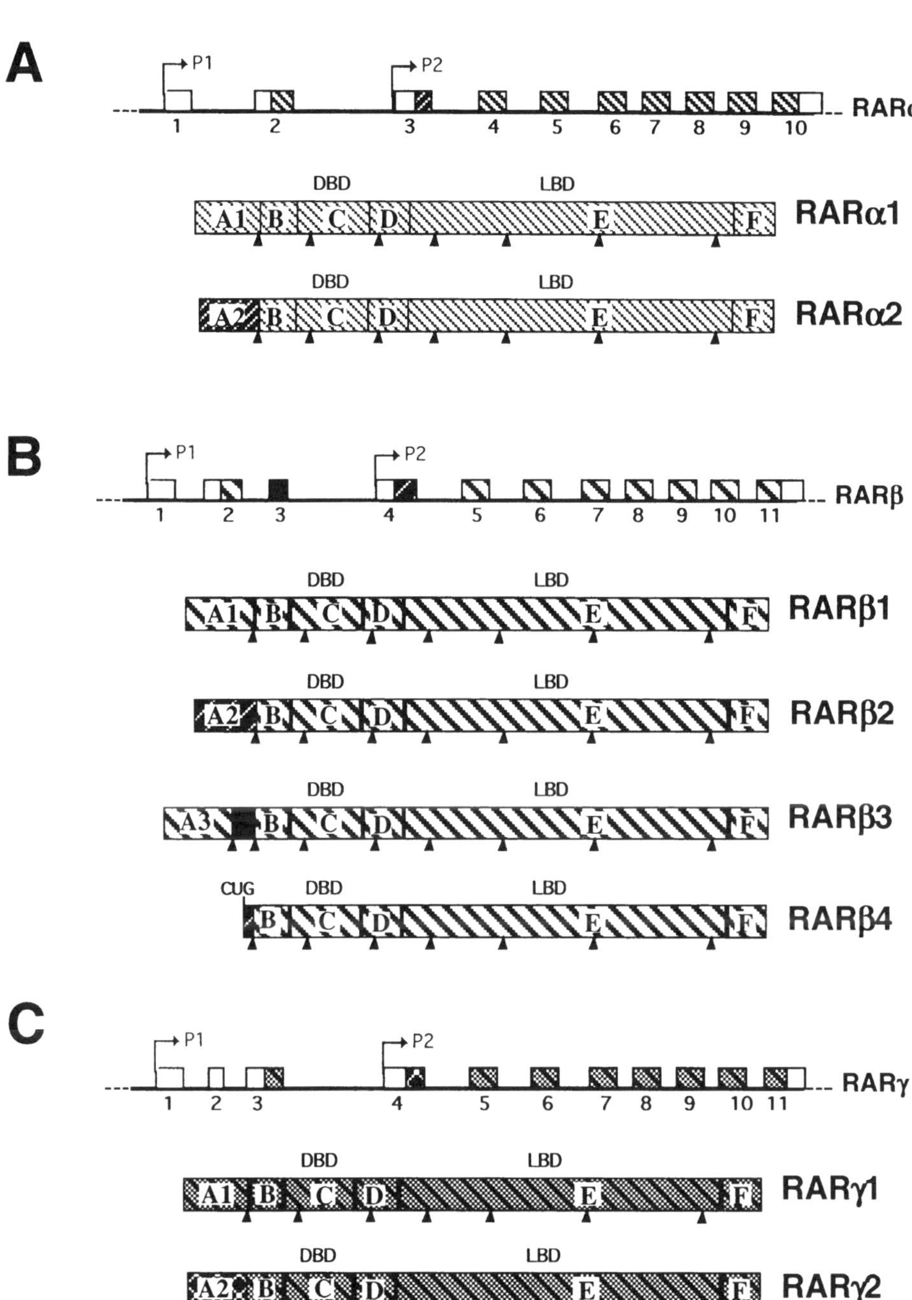

Fig. 1.

natively spliced RARβ2 transcript, which lacks a part of the 5'-UTR and nearly the entire A-region coding sequence, and its translation is initiated at a non-AUG codon *(31)*. Furthermore, the activity of the P2 promoter in each gene is retinoid inducible, and therefore, the RARα2, β2 and β4, and γ2 expression is regulated by retinoids. All the P2 promoters have been cloned and their corresponding retinoic-acid response elements (RAREs) characterized *(42–44)*.

Given the widespread role of RA in regulation of cell growth and differentiation, it is perhaps not surprising that somatic mutations in *RAR* genes have been associated with carcinogenesis *(24,45–48)*. It is, however, remarkable that in one malignant disease, RA as a single therapeutic agent is able to induce a complete remission *(49,50)*. Acute promyelotic leukemia (APL), which represents a block in granulocytic differentiation and the only human cancer so far to be successfully treated by differentiation therapy using RA, is associated with three different chromosomal translocations—t(5;17), t(11;17), and t(15;17)—which consistently involve the *RARα* gene *(51–54)*. The *RARα* fusion oncogene products, NPM-RARα, PLZF-RARα, or PML-RARα structurally appear as variant RARα isoforms, i.e., they all possess non-RARα amino acid sequences upstream of the B-region (*see* **Fig. 2** for their schematic representation).

The aim of this chapter is to describe the experimental approach and methodology which was utilized to isolate a large number of N-terminal isoforms of *RARα, β*, and *γ* as well as to clone the PLZF-RARα cDNA from APL cells with t(11;17) chromosomal translocation. It is worth pointing out that the same approach, with minor modifications, can be used to isolate N-terminal, or even C-terminal isoforms *(55,56)* of any gene for which some of the sequence is known. Therefore, this chapter could serve as a good reference for planning and executing such experiments.

At approximately the same time as the first RARs were being discovered, a new technique was developed that revolutionized molecular biology as well as diagnostic and forensic medicine. This technique, called the polymerase chain reaction (PCR), involves the use of a thermostable DNA polymerase, usually that of *Thermus aquaticus (Taq)*, to amplify specific DNA sequences in vitro *(57)*. As a result, a given DNA sequence from a single cell can be quickly and reliably amplified to quantities visualized in an ethidium bromide-stained agarose gel, using a pair of specific oligonucleotide primers flanking the region of interest and a large number of PCR cycles (involving repeating denaturation, annealing, and extension steps).

The ease with which known, or even unknown genes, can be cloned and sequenced using PCR has introduced a new element of competition into the scientific community. This remarkable technique is not, however, without flaws. Because of its high sensitivity, extreme care is required to avoid con-

RARα FUSION PROTEINS IN ACUTE PROMYELOCYTIC LEUKAEMIA

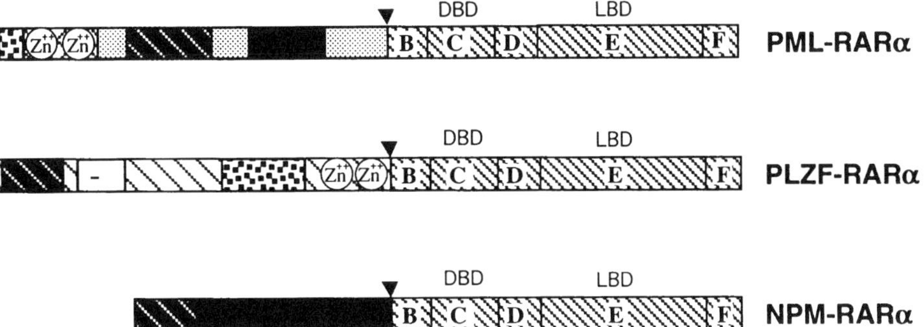

Fig. 2. RARα fusion proteins. The sequences encoding B–F regions of RARα, which are present in all the fusion proteins, are indicated as in **Fig. 1**. Different patterns are used to represent various putative functional regions of the promyelocytic leukemia (PML), promyelocytic leukemia zinc finger (PLZF), and nucleophosmin (NPM) proteins. Black rectangle with white stripes represents protein–protein interaction motifs present in N-termini of all the RARα chimeras. Circled Zn^{2+} symbols represent the Ring-finger and nine krüppel-like zinc-finger motifs of the PML and PLZF proteins, respectively. Filled triangles denote boundaries between sequences encoded in separate exons lying on the opposite sides of a given translocation breakpoint.

tamination with traces of DNAs that are identical to those being studied. When working to isolate unknown sequences, the lack of 3'-5' proof reading ability of *Taq* DNA polymerase causes errors requiring sequencing of multiple-independent clones of the same gene. The error rate for *Taq* DNA polymerase is approx 10- and 100-fold greater than for modified T7 and T4 DNA polymerases, respectively *(58–60)*. In my own experience, the abundance of errors appeared to vary among different sequences of the same length, suggesting some relationship between the sequence and the error rate. Recently, other thermostable-DNA polymerases, which possess 3'-5' exonuclease proofreading activities and hence generate fewer errors than *Taq*, became available. Vent™ DNA polymerase (New England Biolabs, Hertfordshire, UK) from *Thermococcus litoralis (61,62)* and *Pfu* DNA polymerase (Stratagene, Cambridge, UK) from *Pyrococcus furiosus (63)* have approx 10-fold lower error rate than *Taq*. In my experience, *Pfu* DNA polymerase performs with a better fidelity and reproducibility than *Vent™*. Nevertheless, Taq DNA polymerase still remains my enzyme of choice for most applications, including those described here. In my hands, this enzyme is more robust than the other two DNA polymerases, generates more reproducible results and can amplify longer fragments of DNA and with greater sensitivity.

The anchored-PCR approach (*see* **Fig. 3A** for schematic representation) used to isolate RAR isoforms was originally described in 1991 by Loh et al. *(64)*. In addition to the anchored-PCR method, I also describe in this chapter the use of semiquantitative-reverse transcriptase (RT)-PCR approach to assay the relative levels of different RAR isoforms in various tissues and cell lines. I use the term semiquantitative, as opposed to quantitative, to denote the fact that although this approach can be used with great reproducibility and sensitivity to compare relative levels of expression for a given isoform in different samples, it cannot be used accurately to assign values to the differences. In addition, this analysis may not always reflect accurately the relative levels of different isoforms in the same sample, especially if the differences in their levels of expression are small. Although I will not discuss this separately, it should be noted that all RARα chimeric transcripts associated with APL can be isolated by anchored PCR or detected by RT/PCR using the methods described here and an appropriate set of oligonucleotide primers (*see* **Table 1**).

2. Materials

Unless otherwise specified all reagents are prepared in double-distilled water (ddH$_2$O) and stored at room temperature (rm temp). Suppliers are indicated only when reagents are not readily available or a specific source is recommended over others.

2.1. Total RNA Isolation

1. Solution I: 4.5 *M* Guanidinium thiocyanate (GnSCN), 0.5% sodium *N*-lauroylsarcosine, 25 m*M* sodium citrate, pH 7.0, 0.1 *M* 2-mercaptoethanol, 25 m*M* EDTA, pH 7.0, and 0.1% Antifoam A. 100 mL of above solution is made up by combining 56 g GnSCN, (Fluka, Dorset, UK), 0.5 g of sodium

---

Fig. 3. *(see facing page)* Schematic representation of anchored PCR **(A)** and RT/PCR **(B)** procedures. Top of each diagram indicates receptor encoding mRNA with a 5'-cap (^{7}MeG) and a polyA tail (A$_n$). Protein initiation (ATG) and termination (TGA) codons are indicated and regions encoding various domains are identified with their corresponding letters. Note that RXRs do not possess the F regions. Succession of steps, as described in the text, and resulting products are indicated consecutively. Different oligonucleotide primers and probes are indicated by arrows and brackets. Note that RT could represent general (RT) or any receptor specific RT primer (RTα-γ, for mRARs and 35-37, for mRXRs) shown in **Table 1**. Likewise, letters and numbers refer not to individual primers but rather to their positions along the mRNA. In anchored PCR schematic, primers 1, 2, and B represent primers 1α-γ, 2α-γ, and Bα-γ in **Table 1**, respectively. In a diagram for RT/PCR, roman numerals I and II, as well as letter A, represent any of the receptor/isoform specific 3'–5' oligonucleotide primer pairs, as well as probes, indicated in **Tables 1** and **2**.

A
ANCHORED PCR FOR RAR ISOFORMS
A B C D E F
5'UTR ATG
TGA 3'UTR
7MeG
An
A B 2 1 RT
RT, dNTPs
Sephadex G50 column to remove unincorporated dNTPs
5' end ssDNA
TdT, dGTP
Ethanol ppt
dCn
3
dGn
5' end ssDNA
4 1
PCR 25-30 cycles
Gel purification of longer fragments
dCn
3' end ssDNA
4 2
PCR 25-30 cycles
dCn
dsDNA
dGn
Subcloning, screening and sequencing

B
RT/PCR FOR RARs/RXRs
A B C D E F
5'UTR ATG
TGA 3'UTR
7MeG
An
II A B I RT
RT, dNTPs
5' end ssDNA
II I
PCR 25-30 cycles
II
dsDNA
I
electrophoresis / hybridization

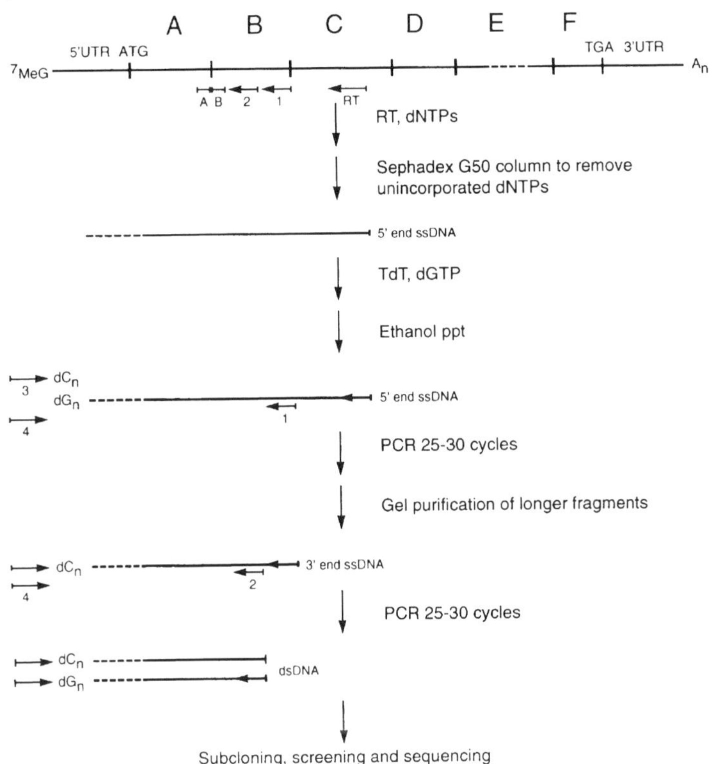

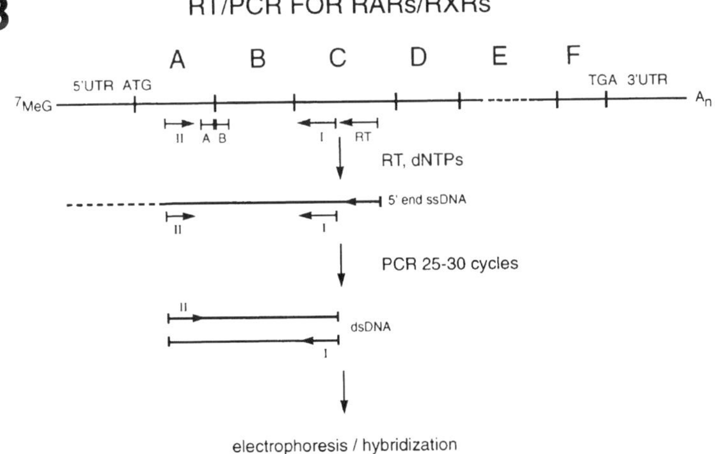

Fig. 3.

Table 1
Oligonucleotide Probes and Primers

No.	Oligonucleotide sequence	Polarity (relative to mRNA)	Specificity	Application
RT	C T T C T G G A T G C T T C G T C G G A A G A A G C C C T T	anti-sense	mRXR/RARs/C-r	primer/RT
RTα	T G G A T G C T T C G T C G G A A G A A G C C C T T A C A G	anti-sense	mRARα/C-r	primer/RT
hRTα	T G G A T G C T G C G G C G G A A G A A G C C C T T G C A G	anti-sense	hRARα/C-r	primer/RT
RTβ	T G A A T A C T T C T G C G G A A A A A G C C C T T G C A C	anti-sense	mRARβ/C-r	primer/RT
RTγ	T G A A T G C T G C G T C T G A A G A A G C C C T T G C A G	anti-sense	mRARγ/C-r	primer/RT
1α	A T G G A T C C A G G C T T G T A G A T G C G G G G C A G G G G C	anti-sense	mRARα/B-r	primer/PCR
1β	A T G G A T C C C C G A G G A G G A G G A A G T G G A G A T G G T	anti-sense	mRARβ/B-r	primer/PCR
1γ	A T G G A T C C T G G C T T A T A G A C C C G A G G A G G T G G T	anti-sense	mRARγ/B-r	primer/PCR
2α	A T G G A T C C T G G T G A G G G A G G G C T G G G T A	anti-sense	mRARα/B-r	primer/PCR
2β	A T G G A T C C T G G G C T C G G G A C G A G C T C C T	anti-sense	mRARβ/B-r	primer/PCR
2γ	A T G G A T C C T G A G G G A G A G C T G G G T A C C A	anti-sense	mRARγ/B-r	primer/PCR
Bα	C T G C T C T G G G T C T C G A T	anti-sense	mRARα/B-r	probe
Bβ	G T A C T C T G T G T C T C G A T	anti-sense	mRARβ/B-r	probe
Bγ	G T G C T C T G T G T C T C C A C	anti-sense	mRARγ/B-r	probe
3	G C A T G C G C G C G G C C G C G G A G G C C C C C C C C C C C C C C	-	anchor	primer/PCR
4	G C A T G C G C G C G G C C G C G G A G G C C	-	anchor	primer/PCR
5	G C T T C C A G T T A G T G G A T A T A	sense	hRARα1/A-r	probe
6	G C T T C C A G T C A G T G G T T A C A	sense	mRARα1/A-r	probe
7	T T G G A A T G G C T C A A A C C A C T	sense	m/hRARα2/A-r	probe
8	G T G C G T G G A C A C A T G A C	sense	mRARβ1/β3/A-r	probe
9	A G A T C C T G G A T T T C T A C	sense	mRARβ2/A-r	probe
10	G T G G A C A C C A A G C T T G A	sense	mRARβ3/A-r	probe
11	G G A T C G G C C G A G C G A G C	sense	mRARβ2/β4/A-r	probe
12	G A G A T G C T G A G C C C T A G	sense	mRARγ1/A-r	probe
13	T T C G C C G G A C T T G A G T C	sense	mRARγ2/A-r	probe
14	A A C T C T C C A A C G G G T C G	sense	mRXRα/A-r	probe
15	A G G A T T C T C C G G G C C T G	sense	mRXRβ/A-r	probe
16	T G G C T C G A C G T C C A T G A	sense	mRXRγ/A-r	probe
17	C T C A C A G G C G C T G A C C C C A T	anti-sense	m/hRARα/C-r	primer/PCR
18	C T C G C A G G C A C T G A C G C C A T	anti-sense	mRARβ/C-r	primer/PCR
19	T T C A C A G G A G C T G A C C C C A T	anti-sense	mRARγ/C-r	primer/PCR
20	T C A C A A C T G T A T A C C C C A T A	anti-sense	mRXRα/C-r	primer/PCR
21	T C G C A G C T G T A A A C C C C A T A	anti-sense	mRXRβ/C-r	primer/PCR
22	T C A C A G C T G T A C A C A C C G T A	anti-sense	mRXRγ/C-r	primer/PCR
23	G C G G G C A C C T C A A T G G G T A C	sense	m/hRARα1/A-r	primer/PCR
24	G A A C C G G G C C T G T T T G C T C C	sense	m/hRARα2/A-r	primer/PCR
25	G C A T G A G C A C C A G C A G C C A C	sense	mRARβ1/β3/A-r	primer/PCR
26	G T T T G A C T G T A T G G A T G T T C	sense	mRARβ2/A-r	primer/PCR
27	C A G G C A T G T C A G A G G A C A A C	sense	mRARβ3/A-r	primer/PCR
28	G G A G A A C T T G G G A T C G G T G C	sense	mRARβ2/β4/A-r	primer/PCR
29	T G G G G C C T G G A T C T G G T T A C	sense	mRARγ1/A-r	primer/PCR
30	G C C G G G T C G C G A T G T A C G A C	sense	mRARγ2/A-r	primer/PCR
31	A C C C A G G T G A A C T C T T C G T C	sense	mRXRα/A-r	primer/PCR
32	T T C C C A G T C A T C A G T T C T T C	sense	mRXRβ/A-r	primer/PCR
33	A C A T G T A T G G A A A T T A T T C C	sense	mRXRγ/A-r	primer/PCR
34	G A T G A A G A C G T A C G G G T	sense	hPLZF	probe
35	T T T G C G T A C T G T C C T C T T G A A G A A G C C C T T	anti-sense	mRXRα/C-r	primer/RT
36	C T T C C G A A T G G T G C G C T T G A A G A A A C C C T T	anti-sense	mRXRβ/C-r	primer/RT
37	T T T C C T G A T G G T C C T T T T G A A G A A G C C T T T	anti-sense	mRXRγ/C-r	primer/RT
38	G G A G C A G C A C A G G A A G C T G C	sense	hPLZF	primer/PCR

N-laurylsarcosine (or 1.67 mL of Sarcosyl NL-30), 2.5 mL of 1 *M* sodium citrate, pH 7.0, 12.5 mL of 0.2 *M* EDTA, pH 7.0, 0.33 mL of Antifoam A (30% solution, Sigma, St. Louis, MO) and ddH$_2$O to a final volume of 100 mL. Dissolve solids by stirring (warm if necessary), filter through a 0.45-μm cellulose nitrate filter (Nalge, Rochester, NY) and add 0.7 mL of 2-mercaptoethanol (14.3 *M* stock solution). This solution can be stored in the dark for up to 1 mo.

2. Solution II: 7.5 *M* Guanidine hydrochloride (GnHCl), 25 m*M* sodium citrate, pH 7.0, and 5 m*M* dithiotriethol (DTT). To make 100 mL of this solution, dissolve 71.7 g of GnHCl (Fluka, Dorset, UK) in ddH$_2$O, adjust the final volume to 100 mL and filter as aforementioned. Neutralize to pH 7.0 with a few drops of 50 m*M* NaOH (*see* **Note 1**). Add 2.5 mL of 1 *M* sodium citrate, pH 7.0, and 255 μL of 2 *M* DTT. Store for 1 mo.

3. 5.7 *M* CsCl solution: dissolve 479.85 g of CsCl (optical grade, Gibco-Brl, Life Technologies, Paisley, UK) in ddH$_2$O. Add 100 mL of 0.5 *M* EDTA, pH 8.0 and adjust pH to 7.0. Adjust the volume to 500 mL with ddH$_2$O and filter as noted previously. Treat this solution with diethylpyrocarbonate (DEPC) (0.1% final concentration) overnight (O/N) and autoclave (*see* **Note 2**).

4. 5 *M* potassium acetate: this solution is identical to that used in the alkaline-lysis method for plasmid preparation and can be made up by combining 60 mL of 5 *M* potassium acetate, 11.5 mL glacial acetic acid, and 28.5 mL ddH$_2$O.

2.2. Isolation of Polyadenylated (pA⁺) RNA

1. Solutions: ETS (10 m*M* Tris-HCl, pH 7.5, 1 m*M* EDTA, 0.5% SDS), ETS/0.1 *N* NaOH, ETS/0.5 *M* NaCl and ET/0.5S (10 m*M* Tris-HCl, pH 7.5, 1 m*M* EDTA, 0.05% SDS).

2. Oligo(dT)-cellulose, Type III (Collaborative Biomedica Products, Becton Dickinson Labware, Bedford, MA). 1 g of the oligo(dT)-cellulose is sufficient to bind approx 3 mg of poly(A)⁺ RNA. It is usually sufficient to use 100 mg of oligo(dT)-cellulose in ETS for one chromatography column.

3. 0.8 × 4 cm, 12-mL Poly-Prep chromatography columns (Bio-Rad, Hercules, CA).

2.3. Reverse Transcription, G-Tailing and PCR

1. Oligonucleotides: synthetic oligonucleotides should be diluted in ddH$_2$O to 100 ng/μL, 1 μg/μL, and 25 pmol/μL for reverse transcription, PCR and

---

Table 1 *(see opposite page)* Except for the human RARa2, all oligonucleotides in **Table 1** have been derived from published sequences (*see* **Subheading 1.**). Their orientations are indicated as either sense (for 5'-oligonucleotide primers) or antisense (for 3'-oligonucleotide primers). All oligonucleotides used in the anchored-PCR procedure are identified with numbers or letters corresponding to those in the **Fig. 3A**; and when appropriate, each number or letter is followed by a symbol to indicate specificity (see also the fourth column in the table). BamHI and NotI restriction enzyme recognition sequences are underlined. RAR and RXR oligonucleotides used in RT/PCR are numbered.

sequencing/hybridization, respectively (an average molecular weight of 1 base is 330 g/mol). **Table 1** lists the sequences of all the oligonucleotides required for the procedures described in this chapter (*see* **Note 3**). Oligonucleotides should be stored in 50- to 100-µL aliquots at –20°C.

2. Reverse transcriptase (RT): Moloney murine leukemia virus (M-MLV) RT (200 U/µL) should be purchased from Gibco-BRL. The enzyme is supplied with a vial of 5X reaction buffer (250 mM Tris-HCl, pH 8.3, 375 mM with KCl, and 15 mM MgCl$_2$) and a vial of 100 mM DTT. It is advisable to aliquot the enzyme into 3–4 smaller portions and use them in succession. Store at –20°C.

3. RNasin (20–40 U/µL), recombinant or purified from human placenta, should be purchased from Promega (Madison, WI).

4. dNTP mix: 400 µL of 10 mM dNTPs should be made in ddH$_2$O using commercially available 100 mM stock solutions of each dNTP (Pharmacia Biotech, Uppsala, Sweden). Store at –20°C.

5. *Taq* DNA polymerase (5 U/µL) should be purchased from Perkin Elmer (Norwalk, CT).

6. 10X PCR buffer: 100 mM Tris-HCl, pH 8.7 (*see* **Note 4**), 500 mM KCl, 15 mM MgCl$_2$, 200 µg/mL bovine serum albumin (BSA). Commercially available buffer from Perkin Elmer is identical to the above except it contains gelatin instead of BSA (*see* **Note 5**). Keep in 1-mL aliquots at –20°C.

7. Terminal deoxynucleotidyl transferase (TdT, 10–20 U/µL) can be purchased from Gibco-BRL. The enzyme is supplied with a vial of 5X reaction buffer (500 mM potassium cacodylate, pH 7.2, 10 mM CoCl$_2$, 1 mM DTT). Store at –20°C. Note that potassium cacodylate and CoCl$_2$ are highly toxic and should be handled with care.

8. 10 mM dGTP solution diluted from 100 mM dGTP stock (Pharmacia Biotech). Store at –20°C.

9. Sephadex G50 (*see* **Note 6**) or Bio-Spin30 chromatography column (Bio-Rad).

10. 6X DNA loading buffer: 30% glycerol, 0.25% bromophenol blue, 0.25% xylene cyanol FF in ddH$_2$O. Aliquot into 1-mL samples and store at 4°C.

11. Agarose gels: 1.4% agarose made up and run in 1X TAE buffer (50X TAE: 242 g Tris, 57.1 mL glacial acetic acid, 100 mL 0.5 M EDTA, pH 8.0, per liter of ddH$_2$O). Gels should be run at 80–100 V until bromophenol blue dye (lower) has gone through 2/3–3/4 of the gels length.

12. 0.2 µm, 30 × 60-cm sheets, reinforced nitrocellulose membranes (Schleicher & Schuell, Keene, NH).

13. Denaturing (1.5 M NaCl, 0.5 M NaOH) and neutralizing (1.5 M NaCl, 0.5 M Tris-HCl, pH 7.4) solutions.

14. 20X SSC (3.0 M NaCl, 0.3 M sodium citrate, pH 7.0) and 20X SSPE (3.6 M NaCl, 0.2 M sodium phosphate, pH 7.7 [Na$_2$HPO$_4$ is added to NaH$_2$PO$_4$ to bring pH of the solution to 7.7], 20 mM EDTA).

15. 10 mg/mL solution of salmon-sperm DNA (ssDNA): dissolve 1 g of ssDNA Type III, sodium salt (Sigma) in 100 mL of ddH$_2$O and autoclave. Store at –20°C.

16. 50X Denhardts reagent: 5 g Ficoll Type 400 (Pharmacia Biotech), 5 g BSA Fraction V (Sigma, St. Louis, MO), 5 g polyvinylpyrrolidone and ddH$_2$O to 500 mL.
17. T4-polynucleotide kinase (New England BioLabs, Beverley, MA); 10X kination buffer: 500 mM Tris-HCl, pH 7.6, 1 mM spermidine, 1 mM EDTA, 100 mM MgCl$_2$, 50 mM DTT.
18. Prehybridization solution: 5X SSPE, 5X Denhardts, 1 mg/mL ssDNA, 0.1% SDS.
19. Hybridization solution: 5X SSPE, 1X Denhardt's, 1 mg/mL ssDNA.
20. Hybridization wash solution: 2X SSPE/0.1% SDS.
21. GeneClean II kit (Bio 101, Vista, CA).

2.4. Ligation of PCR Products, Bacteria Transformation, and Screening

1. 0.45-μm, 87-mm circles, reinforced nitrocellulose membranes (Schleicher & Schuell).
2. T4-DNA ligase, as well as *Bam*HI and *Not*I restriction endonucleases, supplied with their respective 10X reaction buffers, should be purchased from New England BioLabs.
3. L-Broth (LB): 10 g Bacto-tryptone, 5 g yeast extract, 10 g NaCl, 250 μL of 10 N NaOH per liter of ddH$_2$O; LB agar: L-broth with 1.5% bacto-agar.
4. pBluescript SK (II)+ plasmid DNA.
5. Competent *Escherichia coli* XL1-Blue (Stratagene, La Jolla, CA).
6. Elutip-d columns (Schleicher & Schuell).
7. Low salt buffer: 0.2 M NaCl, 10 mM Tris-HCl, pH 7.5, 1 mM EDTA.
8. High salt buffer: 1 M NaCl, 10 mM Tris-HCl, ph 7.5, 1 mM EDTA.

3. Methods

3.1. Total RNA Isolation

Good-quality RNA serves as the foundation for a successful outcome of anchored PCR and/or RT/PCR experiments. I recommend using total and poly(A)$^+$ RNAs for RT/PCR and anchored PCR, respectively. Anchored PCR can also be performed using total RNA but this will increase the background and reduce the probability of finding rare clones. The procedures described here, as well as in the next section, are based on the original papers by Chirgwin et al. *(65)* and Aviv and Leder *(66)*, and yield highly pure and intact RNAs. I recommend using them rather than the variety of commercially available kits, especially those for preparation of poly(A)$^+$ RNA. When working with a large number of samples, or with samples containing less than 0.5 g of tissue (or 5 × 10^6 cells), I recommend (*see* **Note 7**) using the one-step acid GnSCN-phenol/chloroform method of Chomczynski and Sacchi *(67)*. It is worth noting that RNAs from various species and tissues are available commercially (Clontech, Palo Alto, CA), albeit at a relatively high price.

1. Freeze 1 g of tissue in liquid nitrogen and grind it to a very fine powder with a mortar and pestle (*see* **Note 8**).
2. Suspend the frozen powder in 15–20 mL of solution I (in a 50-mL polypropylene tube) and homogenize with an ultra-turrax (or a polytron) for 1–2 min at maximum speed. Likewise, a cell pellet can be disrupted with a glass or Teflon homogenizer (*see* **Note 9**).
3. Transfer the homogenate to a 30 mL-Corex tube and spin for 10 min at room temperature and 8000 rpm (rcf ~10,000g) in a Sorvall HB-4 rotor.
4. Save the supernatant and resuspend the pellet in additional 5–10 mL of solution I. Homogenize again as noted previously. Remove any remaining insoluble material by additional centrifugation (as noted) and combine the supernatants.
5. To the total supernatant, add CsCl to a final concentration of 0.2 g/mL, and layer it over a 5.7 M CsCl cushion occupying 1/6–1/4 of the total centrifuge tube volume.
6. Centrifuge in a Beckman SW28 rotor at 25,000 rpm (~120,000g) and 20°C for 16–20 h; use slow deceleration (*see* **Note 10**).
7. After centrifugation carefully remove the supernatant and resuspend the RNA pellet in 2–5 mL of solution II. To ensure complete resuspension of the pellet, heat the solution at 60°C for 30 s and subsequently vortex vigorously. This last step may be repeated several times. If any insoluble material remains, centrifuge the sample at rm temp and 8,000 rpm (~10,000g) in a Sorvall HB-4 rotor; save the supernatant (*see* **Note 11**).
8. Precipitate the RNA from solution II by adding 0.025 vol of 1 M acetic acid and 0.5 vol of absolute ethanol (with gentle agitation by vortexing), and incubating O/N at –20°C.
9. Sediment the RNA by 5 min centrifugation in a Sorvall HB-4 rotor at 6000 rpm (~6000g) and –10°C. To remove any remaining GnHCl, wash the RNA pellet twice with 80% ethanol (in ddH$_2$O), centrifuging the RNA at 8000 rpm (~10,000g) in a HB-4 rotor after each wash, and dry it under vacuum.
10. Dissolve the RNA pellet in DEPC-treated ddH$_2$O. Store at –20°C as an ethanol precipitate (*see* **Note 12**), or at –80°C in ddH$_2$O.

3.2. Isolation of Poly(A)$^+$ RNA

1. Dissolve RNA in ETS at ~0.65 mg/mL.
2. Heat the RNA solution at 85°C for 5 min and cool rapidly to room temperature.
3. Add NaCl (from 5 M stock) to 0.5 M final concentration and pass over an oligo(dT)-chromatography column at ~0.25 mL/min (*see* **Note 13**).
4. Wash the column with 15 vol of ETS/0.5 M NaCl and elute the poly(A)$^+$ RNA with 4 vol of ET/0.5S into a 15-mL siliconized Corex centrifuge tube.
5. Adjust the NaCl concentration to 1 M and add 3 vol of absolute ethanol. Mix and precipitate the poly(A)$^+$ RNA O/N at –20°C.

6. Sediment the poly(A)$^+$ RNA by 15 min centrifugation in a Sorvall HB-4 rotor at 4°C and 8000 rpm (~10,000g), wash with 80% ethanol, recentrifuge, and dry under vacuum. Resuspend following the same recommendations as those given in **Subheading 3.1., step 10** for total RNA.

3.3. Reverse Transcription, G-Tailing, and PCR

1. In a 1.5-mL microcentrifuge tube, combine 1 μg of poly(A)$^+$ RNA (or 2.5 μg of total), 100 ng of a given antisense RT oligonucleotide primer (RTα, β, or γ, *see* **Table 1**), and ddH$_2$O to the final volume of 27.6 μL. Using higher or lower amounts of RNA in this step reduces the efficiency of subsequent PCR.
2. Incubate the tube at 70°C for 5 min and then transfer to ice.
3. Add 8 μL of 5X RT buffer, 2 μL of 10 mM dNTPs, 0.4 μL of 0.1 M DTT, 1 μL of RNasin (1 U/μL final concentration), and 1 μL of M-MLV RT (5 U/μL final concentration).
4. Mix by tapping the bottom of a tube with a finger and incubate at 37°C for 45 min (*see* **Note 14**). In the RT-PCR protocol, 2.5 μL of this reaction can be used directly for PCR (*see* **Subheading 3.5.**).
5. To remove excess dNTPs from the newly synthesized cDNA, bring the volume of reaction to 100 μL with TE (10 mM Tris-HCl, 1 mM EDTA), pH 8.0, and apply it onto either a Sephadex G50 or Bio-Spin30 chromatography column. Centrifuge 2–5 min at 3000–4000 rpm in a table-top centrifuge (>1100g) collecting the liquid in a 1.5-mL microcentrifuge tube.
6. Precipitate the synthesized cDNAs by adding 0.1 vol of 3 M sodium acetate and 3 vol of 100% ethanol, and placing the tube at –20°C O/N or 15 min at –80°C (1 μg of tRNA can be added as carrier for precipitation).
7. Centrifuge for 15 min at 4°C in a refrigerated microcentrifuge (full speed, ~16,000g).
8. Wash the precipitate with 80% ethanol, recentrifuge for 5 min as noted earlier, decant the supernatant, and dry the pellet under vacuum in a desiccator.
9. Resuspend the pellet in 12 μL of ddH$_2$O. Add 4 μL of 5X TdT-reaction buffer, 2.0 μL of 10 mM dGTP, and 2 μL of TdT (final concentration 1–2 U/μL). Incubate the reaction for 1 h at 37°C.
10. Precipitate, centrifuge, and wash as in **steps 6–8**. Dry under vacuum and resuspend in 20 μL of ddH$_2$O; use 1 μL for PCR reaction. Store at –20°C.
11. Set up the PCR reaction as follows: in a 0.5-mL microcentrifuge tube, combine 83.6 μL of ddH$_2$O, 10 μL of 10X PCR buffer, 2 μL of 10 mM dNTPs, 1 μL (1 μg) of RAR specific 3' primer (*see* **Table 1**), 0.1 μL of dC-tailed anchored PCR oligonucleotide primer (100 ng), and 0.9 μL (900 ng) of anchored PCR primer lacking the dC-tail (primers 3 and 4 in **Table 1**, using this primer combination prevents progressive elongation of dC-tail). Mix by vortexing. Add 0.4 μL of *Taq* DNA polymerase and 1 μL of dG-tailed cDNA. If required, add two drops of mineral oil (Perkin-Elmer 2400 or 9600 PCR machines, for example, do not require mineral oil) and transfer immediately to a thermal cycler (*see* **Note 15**).

12. For amplification, use 25 cycles of denaturation at 95°C for 30 s, annealing at 60°C (50°C for the first 4–5 cycles if using 3' primers with heterologous restriction enzyme recognition sequences at their 5' ends) for 1.5 min and extension at 72°C for 3 min. The last cycle should be followed by a final extension at 72°C for 15 min. After completion, the reaction can be stored frozen or, for a short time, at 4°C.

13. Run 20–40 µL of the aforementioned reaction, together with a size marker loaded in adjacent lane, on a 1.4% agarose gel in 1X TAE. Stain the gel with 1 µg/mL ethidium bromide and visualize the DNA under long-wave UV light. One should expect to see a smear whose intensity decreases with increasing molecular weight.

14. Cut out a part of the gel that contains the amplified DNAs ranging in size from 300 to 1500 bp as compared with the adjacent size markers. There may be very little or no ethidium bromide staining in that region as the amplification of long DNA fragments at this stage is very inefficient. If strong staining is observed above 300 bp after the first round of anchored PCR, the second round (**steps 15–18**) can be omitted.

15. Purify nucleic acid from the gel slice using GeneClean II kit (Bio 101, Vista, CA). Store the DNA in TE, pH 7.5, at –20°C.

16. As in **step 11**, set up a new PCR reaction, this time using only 1 µg of the 5' anchor oligonucleotide primer, which lacks the dC-tail (primer 4 in **Table 1** and **Fig. 3**) and a receptor specific 3' oligonucleotide primer nested with the primer used for the first round of PCR (Primer 2 in **Fig. 3**). For example, when performing PCR for RARα isoforms, the 3' oligonucleotide primers for the first and second rounds of PCR would be those designated as 1α and 2α in **Table 1**, respectively.

17. Set the thermal cycler at the same conditions as for the first round.

18. After completion of the second round of anchored PCR, run 20 µL of the reaction on an agarose gel as in **step 13**. Visualize the amplified DNA under a short-wave UV light.

19. Denature the DNA in the gel by two treatments at rm temp with denaturing solution (2.5 mL of solution/1 mL of gel volume), 15 min each time. Rinse the gel in ddH$_2$O and then soak it twice at rm temp in neutralizing solution, 15 min each time.

20. Using 20X SSC transfer the DNA in the gel O/N onto a nitrocellulose filter.

21. Bake the nitrocellulose filter at 80°C under vacuum for 1 h to fix the transferred DNA to the membrane.

22. To ^{32}P-label an oligonucleotide probe, combine in a 1.5-mL microcentrifuge tube 5 pmol of a given oligonucleotide (B in **Fig. 3** and Bα-γ in **Table 1**), 2.5 µL ddH$_2$O, 1 µL of 10X kination buffer, 5 µL (50 µCi) of ^{32}P-γ-ATP (specific activity >5000 Ci/mmol, Amersham, Arlington Heights, IL), and 1 µL T4-polynucleotide kinase. Incubate 30 min at 37°C. Stop the reaction by heating at 65°C for 5 min and store it at –20°C in a shielded container. Aliquot of this reaction can be used as a probe (5 µL/20 µL of hybridization solution) without any further purification.

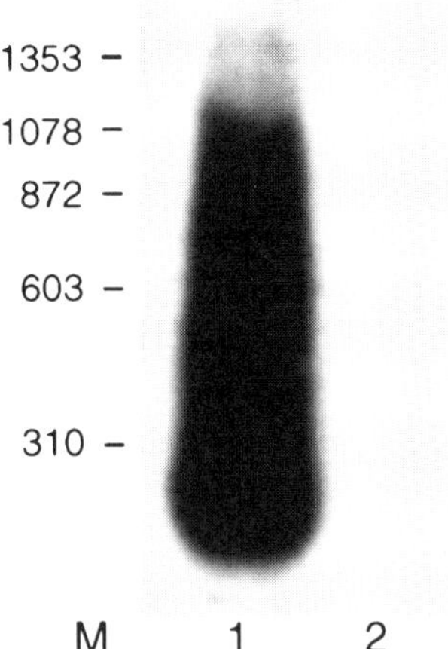

Fig. 4. High-molecular-weight DNA smear obtained following two rounds of anchored PCR using nested RARα oligonucleotide primers and poly(A)$^+$ RNA from APL cells with t(11;17). Lane 2, negative control corresponding to the same PCR reaction, but where the reverse transcription step was omitted.

23. Prehybridize for 2 h and hybridize with a ^{32}P-labeled oligonucleotide probe O/N at a temperature (T_H) equal to the T_m of the probe –5°C.
24. Wash the hybridized filters twice for 15 min with agitation in 2X SSPE/0.1% SDS, first at rm temp and then at T_m –5°C.
25. Expose for 1 h at –80°C to a Kodak XAR-5 film using intensifying screens. Successful outcome of the procedure is confirmed by the presence of a high molecular-weight smear (*see* **Fig. 4**, for example). The same ^{32}P-labeled oligonucleotide probe can be used for screening of bacterial colonies after transformation (*see* **Subheading 3.4.**).

3.4. Ligation of PCR Products, Bacterial Transformation, and Screening

1. Digest 10 µg of Bluescript SKII(+) with *Not*I and *Bam*HI endonucleases for 4 h at 37°C (these two enzymes can be used simultaneously using 10X *Bam*HI buffer).
2. Run the digest on a 1% agarose gel in 1X TAE. Cut out the band corresponding to the linear DNA and purify the vector using GeneClean II kit. After purification, run the vector DNA on a gel again to verify its integrity and purity. Dilute to 30–50 ng/µL in TE, pH 7.5, and store at –20°C.
3. Test the vector for self-ligation and the presence of contaminating uncut vector DNA by transforming supercompenent XL1-Blue *E. coli* (Stratagene) with 30 ng unligated and ligated vector. If the background is high (>15 colonies) recut and repurify the vector.

4. To remove traces of the mineral oil (if necessary), extract the PCR reaction with ddH$_2$O-saturated ether and precipitate with 0.1 vol of sodium acetate and 3 vol of ethanol. Centrifuge at full speed (~16,000g) in a bench-top microcentrifuge for 10 min, wash with 80% ethanol, recentrifuge, decant supernatant, and dry under vacuum.

5. Resuspend the dried pellet in 75 µL of ddH$_2$O, add 10 µL of 10X *Bam*HI buffer, and mix. Add 7.5 µL each of *Bam*HI and *Not*I. Digest at 37°C for 4 h.

6. Dilute the reaction with 1 mL of low-salt buffer (0.2 *M* NaCl, 10 m*M* Tris-HCl, pH 7.5, 1 m*M* EDTA) and pass slowly through an Elutip-d column (Scleicher & Schuell) previously hydrated with a high salt (1 *M* NaCl, 10 m*M* Tris-HCl, pH 7.5, 1 m*M* EDTA) and equilibrated with the low-salt buffers, respectively.

7. Wash the column with several mL of low-salt buffer, elute with 0.4 mL of high-salt buffer into a 1.5-mL microcentrifuge tube, and precipitate at –20°C O/N after adding 1 mL of 100% ethanol to the eluted sample.

8. Centrifuge at maximum speed in a microcentrifuge, wash, and resuspend the sample in 20–50 µL of ddH$_2$O and store at –20°C.

9. For ligation, combine 30 ng of vector DNA and 30–50 ng of purified PCR-amplified DNA with 2 µL of 10X ligase buffer and 0.5 µL of T4-DNA ligase in a total of 20 µL of reaction mixture (*see* **Note 16**). Carry out ligation either for 3–4 h at rm temp (~20°C) or O/N at 14°C.

10. Transform 100 µL of supercompetent XL-1 Blue *E. coli* with 1–5 µL of the ligation reaction as directed by the supplier (Stratagene). It is essential to obtain high-transfection efficiency, especially when looking for rare clones. I found it very reliable and convenient to purchase supercompetent cells for this purpose. If this is not possible, I recommend using electrocompetent cells and electroporation to transform bacteria *(68)*.

11. Plate out all the bacteria on a LB agar plate with 50 µg/mL ampicillin and grow O/N at 37°C.

12. Pick 100 clones with autoclaved toothpicks and using a grid, transfer them, in duplicate, to corresponding positions on two different plates. If necessary, more colonies can be analyzed in this way. Grow colonies O/N at 37°C and transfer to nitrocellulose membranes.

13. In order to lyse the bacteria and denature the DNA float the nitrocellulose membrane, with transferred bacterial colonies facing up, for 5 min on top of the denaturation solution. Neutralize the DNA on the filter by floating it twice (5 min each time) on top of neutralization solution.

14. Fix the nucleic acid onto nitrocellulose by baking it under vacuum at 80°C for 1 h.

15. Prehybridize the filters for 2 h and hybridize O/N (as in **Subheading 3.3., step 23**) one filter with a RAR specific B-region ^{32}P-labeled oligonucleotide probe and its duplicate with a mixture of A-region ^{32}P-labeled oligonucletide probes corresponding to previously isolated sequences. For example, when cloning PLZF-RARα cDNA we hybridized one filter with the ^{32}P-labeled RARα

B-region oligonucleotide probe (Bα in **Table 1**), and a duplicate filter with a mixture of RARα1 and RARα2 specific probes (5 and 7 in **Table 1**).

16. Wash the hybridized filters as in **Subheading 3.3., step 24** and expose them to a KODAK XAR-5 film (usually 1–2 h at –80°C with two intensifying screens is sufficient).

17. Pick colonies which hybridize with the B-region probe (common to all isoforms) and not with the mixture of unique probes corresponding to divergent sequences of known isoforms. Grow the bacteria in 5 mL of LB with 50 µg/mL ampicillin.

18. Miniprep the plasmid DNA from 1.5 mL of bacterial culture using a standard alkaline-lysis protocol (*see* **Note 17**).

19. Digest 10 µL of each isolate with *Not*I and *Bam*HI for 2 h. Run on 1% agarose gel in TAE to check the insert size. Sequence each unique clone to determine its identity.

3.5. Semiquantitative RT/PCR

For sake of simplicity this procedure will describe RT/PCR (*see* **Fig. 3B** for a schematic) for human RARα isoforms. **Figure 5** compares the results of such analysis to those obtained with Northern blot using the same panel of different cell lines. **Table 1** lists all RT and PCR primers required to carry out a comparable RT/PCR analyses on all major murine RAR isoforms as well as mouse RXRα, β and γ. To avoid confounding results due to amplification of RAR/RXR sequences that may be present in small amounts of contaminating-genomic DNA, the 5' and 3' oligonucleotide primers for PCR amplification were derived from regions encoded in separate exons.

1. Using RT primer common to all RARs/RXRs, or one of the specific RT primers (*see* **Table 1**), and 2.5 µg of total RNA, set up and perform reverse transcription as in **Subheading 3.3., steps 1–4**. After completion, store the reactions at –20°C.

2. Using isoform specific 5' and 3' oligonucleotide primers (*see* **Table 2**), assemble the PCR reaction as described in **Subheading 3.3., step 11**. This time use 0.8 µg of each primer per reaction (*see* **Note 18**).

3. Immediately after **step 2**, add 2 drops of mineral oil to each tube, transfer to a thermal cycler, set the annealing temperature as recommended for a given oligonucleotide-primer pair used in the reaction (*see* **Table 2**) and start the program. For example, the following conditions should be used to amplify RARα1 and/or α2 sequences: denaturation for 30 s at 95°C, annealing for 1.5 min at 60°C and extension for 3 min at 72°C. These steps should be repeated 25–30 times (*see* **Note 19**). The last PCR cycle is followed by a 15 min extension at 72°C. After PCR is completed, tubes can be stored for a short time at 4°C (for longer times store at –20°C).

4. Aliquot 20 µL of each reaction to a separate microcentrifuge tube, add 4 µL of 6X gel loading buffer and run on a 1.4% agarose gel in 1X TAE.

5. Stain the gel in 10 µg/mL ethidium bromide and visualize under short-wave UV light. Transfer to a nitrocellulose filter and hybridize with an appropriate isoform-

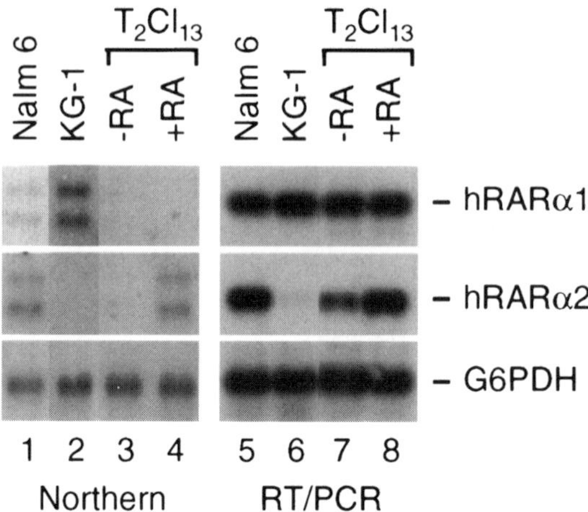

Fig. 5. Comparison of Northern blot (lanes 1–4) and semiquantitative RT/PCR (lanes 5–8) results. Both methods were used to analyze expression of RARα1 and RARα2 isoforms in different human cell lines. Note that as expected the expression of RARα2 is induced with RA in T_2Cl_{13} embryonal-carcinoma cells. In this experiment, 30 cycles of PCR have been carried out. Note that RT/PCR is considerably more sensitive (compare lines 1–4 to 5–8) and generates results which are in agreement with the Northern blot.

specific ^{32}P-labeled oligonucleotide probe (*see* **Table 2** for list of probes and temperatures of hybridization) as described in **Subheading 3.3., steps 19–23**.

6. Wash as described in **Subheading 3.3., step 24**, using temperatures recommended in **Table 2** for the highest stringency wash. Expose for 1 h to a Kodak XAR-5 film at –80°C with two intensifying screens.

4. Notes

1. Because of the high-salt content of this solution it is difficult to obtain an accurate reading using a pH meter. Its initial pH is close to 7.0 and requires only a small adjustment. After the pH adjustment, it is worth recallibrating the pH meter and repeating the measurement. Further adjustments can be made if necessary. Also verify the pH with a pH paper.

2. All solutions that are used for RNA work, except Tris buffers and solutions I and II, should be treated O/N with DEPC and then autoclaved. Solution II can be made up using DEPC-treated ddH$_2$O. All glassware should also be treated with 0.1% DEPC and autoclaved. Tris-containing buffers should be made up from components which themselves have been treated with DEPC and then reautoclaved.

3. The anchored-PCR procedure described here utilizes specific RAR 3' oligonucleotide primers with heterologous sequences encoding unique restriction-enzyme

Table 2
Recommended Annealing Temperatures

Receptor isoform	5' + 3' oligonucleotide pair	Annealing temperatures	Oligonucleotide probe	Temperature of hybridisation	Product size in bp
h/mRARα1	23+17	60°C	5/6	51/55°C	283
h/mRARα2	24+17	60°C	7	53°C	266
mRARβ1/β3	25+18	60°C	8	49°C	320/401
mRARβ2	26+18	50°C	9	43°C	295
mRARβ3	27+18	55°C	10	47°C	213
mRARβ2/β4	28+18	59°C	11	55°C	560/206
mRARγ1	29+19	57°C	12	49°C	281
mRARγ2	30+19	57°C	13	49°C	301
mRXRα	31+20	50°C	14	49°C	437
mRXRβ	32+21	53°C	15	51°C	272
mRXRγ	33+22	47°C	16	49°C	470
PLZF-RARα	38+17	60°C	34	47°C	251

Recommended annealing temperatures for PCR and hybridization (in 5X SSPE) for given 3'–5' oligonucleotide primer pairs (I and II in **Fig. 3B**) and a ^{32}P-labeled oligonucleotide probe (A in **Fig. 3B**), respectively. Numbers correspond to oligonucleotides in **Table 1**. The most stringent wash should be carried out in 2 x SSPE/0.1% SDS at the temperature of hybridization.

sites at their 5' ends (*see* oligonucleotides 1α-γ and 2α-γ in **Table 1**). However, one can also use primers devoid of these heterologous sequences in conjunction with a commercially available T/A-cloning vector (InVitrogen, San Diego, CA), which exploits the fact that *Taq* DNA polymerase adds a single deoxyadenosine to the 3' ends of all duplex molecules provided by PCR. Using these vectors may be advantageous because it simplifies cloning and requires shorter oligonucleotide primers with higher specificity. To obtain the highest cloning efficiency (>80%) one should not use more than 25 PCR cycles per given round of amplification; extension step at the end of a PCR round should be 30 min and the products should be ligated immediately after the end of the PCR run. Most of the T/A cloning vectors provide white-blue selection, however, my recommendation is not to use it as one may miss positive clones which ligate in frame with the *LacZ* gene.

As a rule of thumb, the oligonucleotides should be approx 20 bases (except 30 bases for RT primers) in length and should have similar T_m. One should avoid runs of the same nucleotide, repetitive sequences and homologies between 3' and 5' oligonucleotide primers which could facilitate basepairing between them ("primer dimers"), especially at their 3' ends. The T_m for oligonucleotides up to 20 bases can be accurately determined using $T_m = 2(A+T) + 4(G+C)$ formula. For longer nucleotides, especially those with engineered restriction-enzyme sites at their 5' ends, T_m is not easily predicted (*see* **refs. *69,70***). Nevertheless, it is usually possible to obtain good results by just making an educated guess (using aforementioned formula), and performing several experiments over a small range of temperatures around the estimated T_m. With oligonucleotide primers possessing restriction enzyme sites at their 5' ends, to allow more efficient hybridization to the target sequence the first four cycles should be done at 5–10°C below the final annealing temperature. The subsequent rise in the annealing temperature for the remaining cycles is necessary to reduce background.

4. pH of the PCR buffer is very important. Most protocols give it as 8.3 at room temperature, but if you check the 10X PCR buffer from Perkin Elmer with a pH paper, you'll see that it appears higher. In my hands, the best results were obtained using 1 *M* Tris-HCl, pH 8.7 (at rm temp) to make up the 10X PCR buffer. When in doubt, I recommend using a range of Tris-HCl stock solutions pH (8.3, 8.5, 8.7) and performing test reactions with some standard template. 1.5 m*M* MgCl$_2$ concentration works well for most applications. If there are problems, however, several buffers with a whole range of MgCl$_2$ concentrations (between 0.5 and 2.5 m*M*) should be tested.

5. I have been able to obtain somewhat better results using BSA instead of gelatin.

6. If not available commercially, Sephadex G50 column can be prepared in a 1- to 2-mL syringe. Sephadex G50 (Pharmacia Biotech) should be suspended in TE pH 8.0 (10 m*M* Tris-HCl, pH 8.0, m*M* EDTA) and soaked at least O/N or autoclaved at 15 lb/in. for 15 min on liquid cycle. The tip of the syringe can be plugged with glass wool.

7. Acid GnSCN-Phenol-chloroform procedure:
   a. Resuspend a small amount of tissue or a cell pellet in solution I and homogenize with a glass or Teflon homogenizer. If tissue is difficult to disrupt in this way, use ultra-turrax.
   b. Transfer the homogenate to a polypropylene tube (Falcon); centrifuge briefly to remove debris if necessary.
   c. Add sequentially 0.1 mL 2 *M* sodium acetate, pH 4.0, 1 mL of ddH$_2$O saturated phenol, and 0.2 mL of chloroform/isoamyl alcohol (49:1). Mix vigorously by inversion after adding each component. Shake the final suspension vigorously for 10 s and incubate on ice for 15 min.
   d. Centrifuge at 4°C and 10,000*g* for 20 min.
   e. Transfer the aqueous phase to a fresh tube, add 1 mL of isopropanol, and incubate at –20°C for a minimum of 1 h. Centrifuge as before, resuspend the pellet in 0.1–0.5 mL of solution II and transfer to a 1.5-mL microcentrifuge tube.

    f. Precipitate with 0.025 vol of 1 *M* acetic acid and 0.5 vol 100% Ethanol at –20°C O/N.

    g. Centrifuge (at ~6000*g*) in a refrigerated microcentrifuge at –10°C for 5 min. Wash with 80% ethanol, recentrifuge, decant supernatant, dry the pellet, and resuspend the RNA in a small vol of ddH$_2$O. Store at –80°C.

8. Either freshly removed tissue (or harvested cells) or tissues (or cells) previously snap frozen in liquid nitrogen and stored at –80°C can be used. Unless tissue is rich in RNases, or is very difficult to disrupt, pulverization in liquid nitrogen is not essential.

9. During homogenization the tube can be covered with parafilm in order to prevent any loss of the material. Once the foam generated during homogenization settles, the sample can be homogenized once more to increase yield. If homogenate is too viscous, dilute it with more solution I.

10. Other Beckman rotors and centrifugation conditions that can be used (~150,000*g*): SW-55 at 35,000 rpm for 16–20 h; SW-41 at 32,000 rpm O/N; SW-50 at 39,000 rpm O/N; SW50.1 at 36,000 rpm O/N.

11. If a large amount of insoluble material remains after centrifugation, re-extract it further with a small vol of solution I. Any remaining insoluble material can be removed by additional centrifugation at 10,000*g* and discarded. Supernantants should be combined for further treatment in either 15 or 30 mL Corex centrifuge tube. Falcon 17 by 100 mm round-bottom polypropylene tubes can also be used.

12. Large quantities (>100 µg of total and >25 µg of poly[A]$^+$) of RNA should be precipitated with 1 vol of potassium acetate and 3 vol of 100% ethanol and stored as suspension at –20°C. RNA forms a very fine precipitate and its concentration in the ethanol suspension is a good estimate of an actual amount.

13. Pour the oligo(dT)-cellulose suspended in ETS into a disposable Poly-Prep chromatography column (Bio-Rad). Allow for the oligo(dT)-cellulose to settle and wash it with 10 volumes of ETS/0.1 NaOH. Equilibrate with 10 vol of ETS/0.5 *M* NaCl before loading the sample.

14. If working with many samples, appropriate volumes of 5X RT buffer, dNTPs, DTT, RNasin, and RT can be combined in one tube. Use enough of each reagents to suffice all the samples, plus one extra. Mix and aliquot 12.4 µL of this solution into each sample.

15. This reaction, as much as possible, should be assembled on ice using ice-cold reagents. Immediately after the addition of mineral oil, transfer all the samples into a thermal cycler and start a program.

16. A convenient way of estimating concentrations of small amounts of nucleic acids is to spot, along with some dilutions of a standard DNA solution, 2 µL of a given sample onto a plate containing 1% agarose in 1X TAE, and 5 µg/mL ethidium bromide. After illumination with a short-wave UV light, concentration of a given sample can be estimated by comparing its fluorescence to the standard.

17. Centrifuge 1.5 mL of bacterial culture for 15–30 s (from pressing the start button) in a bench-top microcentrifuge. Remove the medium and resuspend well the bacterial pellet in 100 µL of cold 50 m*M* glucose, 10 m*M* EDTA, 25 m*M* Tris-HCl, pH 8.0.

Incubate at rm temp for 5 min and then add 200 µL of 0.2 *N* NaOH, 1% SDS solution. Mix thoroughly by inverting (do not vortex) and incubate on ice for a further 5 min. Add 150 µL of ice-cold potassium acetate and incubate on ice for 5 min. Centrifuge for 5 min at 4°C (centrifugation at rm temp also works well) at full speed (~16,000*g*) in a bench-top microcentrifuge. Transfer supernatant to a new tube and extract with an equal volume of phenol/chloroform/isoamyl alochol (25:24:1 ratio). Precipitate with 1 mL of 100% ethanol. Centrifuge for 10 min in a microcentrifuge (at ~16,000*g*), wash with 80% ethanol, and dry the pellet under vacuum. Resuspend in 100 µL of TE containing 10 µg/mL RNaseA.

18. When working with a large number of samples, all reagents can be combined on ice in one tube (add the enzyme after all other components have been added and thoroughly mixed). If preparing x PCR reactions, prepare enough stock mix to suffice x + 1 samples. Add 97.5 µL of the reaction mix to each tube containing 2.5 µL of a given RT reaction.
19. I recommend starting with 25 cycles and increase gradually to 30 if the levels of expression remain below detection.

Acknowledgments

I am grateful to Nigel Brand and Leanne Wiedemann for critical reading of this manuscript and to the Leukaemia Research Fund of Great Britain for support.

References

1. McCollum, E. V. and Davis, M. (1913) The necessity of certain lipins in the diet during growth. *J. Biol. Chem.* **15,** 167–175.
2. Bloch, C. E. (1924) Blindness and other diseases in children arising from deficient nutrition (lack of soluble A factor). *Am. J. Dis. Child.* **27,** 139–148.
3. Wolbach, S. B. and Howe, P. R. (1925) Tissue changes following deprivation of fat-soluble A vitamin. *J. Exp. Med.* **42,** 753–777.
4. Sporn, M. B., Roberts, A. B., and Goodman, D. S. (ed.). (1984) *Retinoids*, vols. 1 and 2. Academic, Orlando, FL.
5. Wilson, J. G., Roth, C. B., and Warkany, J. (1953) An analysis of the syndrome of malformations induced by maternal vitamin A deficiency. Effects of restoration of vitamin A at various times during gestation. *Am. J. Anat.* **92,** 189–217.
6. Moore, T. (1967) Effects of vitamin A deficiency in animals: pharmacology and toxicology of vitamin A, in *The Vitamins*, 2nd ed., vol. 1, (Sebrell, W. H. and Harris, R. S., eds.), Academic, New York, pp. 245–266.
7. Livrea, M. A. and Packer, L. (ed.) (1993) *Retinoids*. Marcel Dekker, New York.
8. Kalter, H. and Warkany, J. (1959) Experimental production of congenital malformations in mammals by metabolic procedure. *Physiol. Rev.* **39,** 69–115.
9. Kochhar, D. M. (1961) Teratogenic activity of retinoic acid. *Acta. Pathol. Microbiol. Scand.* **70,** 398–404.
10. Ross, A. C. (1993) Cellular metabolism and activation of retinoids: roles of cellular retinoid-binding proteins. *FASEB J.* **7,** 317–327.

11. Napoli, J. L. (1996) Retinoic acid biosynthesis and metabolism. *FASEB J.* **10,** 993–1001.
12. Giguere, V., Ong, E. S., Segui, P., and Evans, R. M. (1987) Identification of a receptor for the morphogen retinoic acid. *Nature* **330,** 624–629.
13. Petkovich, M., Brand, N. J., Krust, A., and Chambon, P. (1987) A human retinoic acid receptor which belongs to the family of nuclear receptors. *Nature* **330,** 444–450.
14. Karin, M. (1990) Transcriptional control and the integration of cell autonomous and environmental cues during development. *Curr. Opin. Cell Biol.* **2,** 996–1002.
15. Weintraub, H. (1993) The MyoD family and myogenesis: redundancy, networks, and thresholds. *Cell* **75,** 1241–1244.
16. La Thangue, N. B. (1994) DRTF1/E2F: an expanding family of heterodimeric transcription factors implicated in cell-cycle control. *Trends Biochem. Sci.* **19,** 108–114.
17. Shivdasani, R. A. and Orkin, S. H. (1996) The transcriptional control of hematopoiesis. *Blood* **87,** 4025–4039.
18. Tjian, R. and Maniatis, T. (1994) Transcriptional activation: a complex puzzle with few easy pieces. *Cell* **77,** 5–8.
19. Zawel, L. and Reinberg, D. (1995) Common themes in assembly and function of eukaryotic transcription complexes. *Ann. Rev. Biochem.* **64,** 533–561.
20. Struhl, K. and Stargell, L. A. (1996) Mechanism of transcriptional activation in vivo: two steps forward. *Trends Genet.* **12,** 311–315.
21. Beato, M., Herrlich, P., and Schutz, G. (1995) Steroid-hormone receptors—many actors in search of a plot. *Cell* **83,** 851–857.
22. Mangelsdorf, D. J., Thummel, C., Beato, M., Herrlich, P., Schutz, G., Umesono, K., Blumberg, B., Kastner, P., Mark, M., Chambon, P., and Evans, R. M. (1995) The nuclear receptor superfamily—the 2nd decade. *Cell* **83,** 835–839.
23. Chambon, P. (1996) A decade of molecular biology of retinoic acid receptors. *FASEB J.* **10,** 940–954.
24. Brand, N., Petkovich, M., Krust, A., Chambon, P., de The, H., Marchio, A., Tiollais, P., and Dejean, A. (1988) Identification of a second human retinoic acid receptor. *Nature* **332,** 850–853.
25. Krust, A., Kastner, P., Petkovich, M., Zelent, A., and Chambon, P. (1989) A third human retinoic acid receptor, hRAR-γ. *Proc. Natl. Acad. Sci. USA* **86,** 5310–5314.
26. Zelent, A., Krust, A., Petkovich, M., Kastner, P., and Chambon, P. (1989) Cloning of murine α and β retinoic acid receptors and a novel receptor γ predominantly expressed in skin. *Nature* **339,** 714–717.
27. Giguere, V., Shago, M., Zirnbigl, R., Tate, P., Rossant, J., and Varmuza, S. (1990) Identification of a novel isoform of the retinoic acid receptor γ expressed in the mouse embryo. *Mol. Cell. Biol.* **10,** 2335–2340.
28. Kastner, P., Krust, A., Mendelsohn, C., Garnier, J.-M., Zelent, A., Leroy, P., Staub, A., and Chambon, P. (1990) Murine isoforms of retinoic acid receptor γ with specific patterns of expression. *Proc. Natl. Acad. Sci. USA* **87,** 2700–2704.
29. Leroy, P., Krust, A., Zelent, A., Mendelsohn, C., Garnier, J.-M., Kastner, P., Dierich, A., and Chambon, P. (1991) Multiple isoforms of the mouse retinoic acid

receptor-α are generated by alternative splicing and differential induction by retinoic acid. *EMBO J.* **10,** 59–69.

30. Zelent, A., Mendelsohn, C., Kastner, P., Krust, A., Garnier, J. M., Ruffenach, F., Leroy, P., and Chambon, P. (1991) Differentially expressed isoforms of the mouse retinoic acid receptor β are generated by usage of two promoters and alternative splicing. *EMBO J.* **10,** 71–81.

31. Nagpal, S., Zelent, A., and Chambon, P. (1992) RAR-β4, a retinoic acid receptor isoform, is generated from RAR-β2 by alternative splicing and usage of a CUG initiator codon. *Proc. Natl. Acad. Sci. USA* **89,** 2718–2722.

32. Heyman, R. A., Mangelsdorf, D. J., Dyck, J. A., Stein, R. B., Eichele, G., Evans, R. M., and Thaller, C. (1992) 9-*cis* retinoic acid is a high affinity ligand for the retinoid X receptor. *Cell* **68,** 397–406.

33. Levin, A. A., Sturzenbecker, L. J., Kazmer, S., Bosakowski, T., Huselton, C., Allenby, G., Speck, J., Kratzeisen, C., Rosenberger, M., Lovey, A., and Grippo, J. F. (1992) 9-*cis* retinoic acid stereoisomer binds and activates the nuclear receptor RXR-α. *Nature* **355,** 359–361.

34. Leid, M., Kastner, P., and Chambon, P. (1992) Multiplicity generates diversity in the retinoic acid signalling pathways. *Trends Biochem. Sci.* **17,** 427–433.

35. Yu, V., Naar, A. M., and Rosenfeld, M. G. (1992) Transcriptional regulation by the nuclear receptor superfamily. *Curr. Opin. Biotech.* **3,** 597–602.

36. Mangelsdorf, D. J. and Evans, R. M. (1995) The RXR heterodimers and orphan receptors. *Cell* **83,** 841–850.

37. Brand, N. J., Petkovich, M., and Chambon, P. (1990) Characterization of a functional promoter for the human retinoic acid receptor-alpha (hRAR-α). *Nucleic Acids Res.* **18,** 6799–6806.

38. Chambon, P., Zelent, A., Petkovich, M., Mendelsohn, C., Leroy, P., Krust, A., Kastner, P., and Brand, N. (1991) The family of retinoic acid nuclear receptors, in *Retinoids: 10 years on* (Saurat, J.-H., ed.), Karger, Basel, Switzerland, pp. 10–27.

39. Fleischhauer, K., Park, J. H., DiSanto, J. P., Marks, M., Ozato, K., and Yang, S. Y. (1992) Isolation of a full-length cDNA clone encoding a N-terminally variant form of the human retinoid X receptor β. *Nucleic Acids Res.* **20,** 1801.

40. Mangelsdorf, D. J., Borgmeyer, U., Heyman, R. A., Zhou, J. Y., Ong, E. S., Oro, A. E., Kakizuka, A., and Evans, R. M. (1992) Characterization of three RXR genes that mediate the action of 9-*cis* retinoic acid. *Genes Dev.* **6,** 329–344.

41. Liu, Q. and Linney, E. (1993) The mouse retinoid-X receptor-γ gene: genomic organization and evidence for functional isoforms. *Mol. Endocrin.* **7,** 651–658.

42. de The, H., del Mar Vivanco-Ruiz, M., Tiollais, P., Stunnenberg, H., and Dejean, A. (1990) Identification of a retinoic acid responsive element in the retinoic acid receptor β gene. *Nature* **343,** 177–180.

43. Leroy, A., Nakshatri, H., and Chambon, P. (1991) Mouse retinoic acid receptor-α2 isoform is transcribed from a promoter that contains a retinoic acid response element. *Proc. Natl. Acad. Sci. USA* **88,** 10,138–10,142.

44. Lehmann, J. M., Zhang, X.-K., and Pfahl, M. (1992) RARγ2 expression is regulated through a retinoic acid response element embedded in sp1 sites. *Mol. Cell Biol.* **12,** 2976–2985.

45. de The, H., Marchio, A., Tiollais, P., and Dejean, A. (1987) A novel steroid thyroid hormone receptor-related gene inappropriately expressed in human hepatocellular carcinoma. *Nature* **330,** 667–670.

46. Borrow, J., Goddard, A. D., Sheer, D., and Solomon, E. (1990) Molecular analysis of APL breakpoint cluster region on chromosome 17. *Science* **249,** 1577–1580.

47. de The, H., Chomienne, C., Lanotte, M., Degos, L., and Dejean, A. (1990) The t(15;17) translocation of acute promyelocytic leukemia fuses the retinoic acid receptor α gene to a novel transcribed locus. *Nature* **347,** 558–561.

48. Alcalay, M., Zangilli, D., Pandolfi, P. P., Longo, L., Mencarelli, A., Giacomucci, A., Rocchi, A., Biondi, A., Rambaldi, A., Lo Coco, F., Diverio, D., Donti, E., Grignani, F., and Pelicci, P. G. (1991) Translocation breakpoint of acute promyelocytic leukemia lies within retinoic acid receptor α locus. *Proc. Natl. Acad. Sci. USA* **88,** 1977–1981.

49. Warrell, R. P., de The, H., Wang, Z.-Y., and Degos, L. (1993) Acute promyelocytic leukemia. *N. Engl. J. Med.* **329,** 177–189.

50. Chomienne, C., Fenaux, P., and Degos, L. (1996) Retinoid differentiation therapy in promyelocytic leukemia. *FASEB J.* **10,** 1025–1030.

51. Grignani, F., Fagioli, M., Alcalay, M., Longo, L., Pandolfi, P. P., Donti, E., Biondi, A., Lo Coco, F., Grignani, F., and Pelicci, P. G. (1994) Acute promyelocytic leukemia: from genetics to treatment. *Blood* **83,** 10–25.

52. Zelent, A. (1994) Translocation of the RARα locus to the PML or PLZF gene in acute promyelocytic leukaemia. *Br. J. Haem* **86,** 451–460.

53. de The, H. (1996) Altered retinoic acid receptors. *FASEB J.* **10,** 955–960.

54. Redner, R. L., Rush, E. A., Faas, S., Rudert, W. A., and Corey, S. J. (1996) The t(5;17) variant of acute promyelocytic leukemia expresses a nucleophosmin-retinoic acid receptor fusion. *Blood* **87,** 882–886.

55. Frohman, M. A., Dush, M. K., and Martin, G. R. (1988) Rapid production of full-length cDNAs from rare transcripts: Amplification using a single gene-specific oligonucleotide primer. *Proc. Natl. Acad. Sci. USA* **85,** 8998–9002.

56. Ohara, O., Dorit, R. L., and Gilbert, W. (1989) One-sided polymerase chain reaction: The amplification of cDNA. *Proc. Natl. Acad. Sci. USA* **86,** 5673–5677.

57. Saiki, R. K., Gelfand, D. H., Stoffel, S., Scharf, S. J., Higuchi, R., Horn, G. T., Mullis, K. B., and Erlich, H. A. (1988) Primer-directed enzymatic amplification of DNA with a thermostable DNA polymerase. *Science* **239,** 487–491.

58. Keohavong, P. and Thilly, W. G. (1989) Fidelity of DNA polymerases in DNA amplification. *Proc. Natl. Acad. Sci. USA* **86,** 9253–9257.

59. Krawczak, M., Reiss, J., Schmidtke, J., and Rösler, U. (1989) Polymerase chain reaction: replication errors and reliability of gene diagnosis. *Nucleic Acids Res.* **17,** 2197–2201.

60. Ennis, P. D., Zemmour, J., Salter, R. D., and Parham, P. (1990) Rapid cloning of HLA-A, B cDNA by using the polymerase chain reaction: Frequency and nature of errors produced in amplification. *Proc. Natl. Acad. Sci. USA* **87,** 2833–2837.

61. Cariello, N. F., Swenberg, J. A., and Skopek, T. R. (1991) Fidelity of *Thermococcus littoralis* DNA polymerase (Vent™) in PCR determined by denaturing gradient gel electrophoresis. *Nucleic Acids Res.* **19,** 4193–4198.

62. Kong, H., Kucera, R. B., and Jack, W. E. (1993) Characterization of a DNA polymerase from the Hyperthermophile Archaea *Thermococcus litoralis*. *J. Biol. Chem.* **268,** 1965–1975.

63. Lundberg, K. S., Shoemaker, D. D., Adams, M. W. W., Short, J. M., Sorge, J. A., and Mathur, E. J. (1991) High-fidelity amplification using a thermostable DNA polymerase isolated from *Pyrococcus furiosus*. *Gene* **108,** 1–6.

64. Loh, E. Y., Elliot, J. F., Cwirla, S., Lanier, L. L., and Davis, M. M. (1989) Polymerase chain reaction with single-sided specificity: analysis of T-cell receptor delta chain. *Science* **243,** 217–220.

65. Chirgwin, J. M., Przybyla, A. E., MacDonald, J., and Rutter, W. J. (1979) Isolation of biologically active ribonucleic acid from sources enriched in ribonuclease. *Biochem.* **18,** 5294–5299.

66. Aviv, H. and Leder, P. (1972) Purification of biologically active globin mRNA by chromatography on oligothymidylic acid-cellulose. *Proc. Natl. Acad. Sci. USA* **69,** 1408–1412.

67. Chomczynski, P. and Sacchi, N. (1987) Single-step method of RNA isolation by acid guanidinium thiocyanate-phenol-chloroform extraction. *Anal. Biochem.* **162,** 156–159.

68. Calvin, N. M. and Hanawalt, P. C. (1988) High-efficiency transformation of bacterial cells by electroporation. *J. Bacteriol* **170,** 2796–2801.

69. Lathe, R. (1985) Synthetic oligonucleotide pobes deduced from amino acid sequence data. *J. Mol. Biol.* **183,** 1–12.

70. Rychlik, W., Spencer, W. J., and Rhoads, R. E. (1990) Optimization of the annealing temperature for DNA amplification *in vitro*. *Nucleic Acids Res.* **18,** 6409–6412.

24

RT-PCR in Diagnosis and Disease Monitoring of Acute Promyelocytic Leukemia (APL)

David Grimwade, Stephen Langabeer, Kathy Howe, and Ellen Solomon

1. Introduction

1.1. Molecular Characterization and Pathogenesis of APL

Acute promyelocytic leukemia (APL) is one of the most common subtypes of acute myeloid leukemia (AML), accounting for 10–15% of *de novo* cases and typically presenting in early middle age *(1)*. The disease is characterized by a potentially devastating coagulopathy that can lead to rapid demise, particularly owing to cerebral hemorrhage *(1)*, a balanced chromosomal translocation, t(15;17), that is present in virtually all cases of morphological APL *(2–5)* and a unique treatment response to retinoids *(6–9)*. Development of APL reflects two critical processes: leukemic transformation coupled with a block in myeloid differentiation causing the bone-marrow to become replaced by abnormal promyelocytes *(10)*. Retinoids, for example all-*trans* retinoic acid (ATRA) or 9-*cis* retinoic acid (9-*cis* RA), release this block at the promyelocyte stage, such that complete remission is achieved by terminal differentiation of the leukemic clone rather than by a cytotoxic effect *(7–9,11–12)*. Clinical trials have demonstrated that retinoids can achieve remission rates in APL that match those of conventional chemotherapy; indeed remission has even been induced in patients that were previously resistant to chemotherapeutic agents *(7–9)*.

Thirteen years following the original description of the t(15;17) chromosomal rearrangement that has become the diagnostic hallmark of APL, the translocation breakpoint region was ultimately characterized and cloned. This was achieved by two distinct strategies: Borrow et al. employed a physical-

From: *Methods in Molecular Biology, Vol. 89: Retinoid Protocols*
Edited by: C. P. F. Redfern © Humana Press Inc., Totowa, NJ

mapping approach utilizing a *Not*I linking library, identifying clones that spanned the breakpoint region on chromosome 17 *(13)*; whereas other groups screened *RARα* as a potential-candidate gene, knowing its proximity to the breakpoint region on chromosome 17 and taking into account the unique sensitivity of APL to ATRA *(14–16)*. These studies established that the molecular consequence of t(15;17) is a rearrangement between retinoic acid receptor alpha *(RARα)* and the previously unknown promyelocytic leukemia *(PML)* genes. The reciprocal translocation leads potentially to the formation of two fusion-gene products: PML-RARα transcribed from add(15q) is believed to mediate leukemogenesis, retaining virtually all the domains considered to be of functional importance to both PML and RARα, whereas RARα-PML, is transcribed from del(17q) in only 80% cases and its significance remains unclear *(5,13–24)*. The breakpoint on chromosome 17 invariably occurs within the 2nd intron of *RARα (13,16,24)*; whereas 5' *(bcr 3)* and 3' *(bcr 1/2)* breakpoint regions have been delineated within *PML (24)*. 5' *PML* rearrangements, which are found in approx 33% cases of APL, typically occur within intron 3 *(bcr 3)*. 3' breakpoints most commonly occur within intron 6 (approx 60% APL: *bcr 1*); the remainder are generally situated at variable positions within exon 6 *(bcr 2) (5,24–27)* (*see* **Fig. 1**).

In order to determine how the t(15;17) promotes leukemogenesis, much effort has been devoted to the study of the physiological role of PML and RARα. PML is widely expressed and is predominantly localized to the nucleus, incorporated with several other proteins into structures known as PML nuclear bodies (also known as ND10, PODS) *(28–31)*. A variety of experimental approaches have implicated PML in immunological responses. Mice with homozygous deletions of PML are immune compromised with a propensity to bacterial infections *(32)*, and infections induced in cell lines by agents such as

---

Fig. 1. *(see facing page)* Schematic representation of the genomic organization at the chromosome 15 and 17 breakpoint regions in a case of APL associated with a 3' *(bcr 1) PML* breakpoint. The limits of other potential *PML* breakpoints *(bcr 2* and *3)* as defined by Southern blotting are denoted by double-ended horizontal arrows. Although the region encompassed by 5' and 3' breakpoint regions is wide, *bcr 1* is restricted to intron 6 and *PML* breakpoints of virtually all *bcr 2* and *bcr 3* cases fall within exon 6 or intron 3 respectively as indicated by the vertical arrows. Therefore *bcr3* breakpoints generally lead to the translocation of *PML* exons 4–9 to chromosome 17. In patients with a *bcr 2* breakpoint the *RARα-PML* reciprocal fusion gene typically includes variable portions of *PML* exon 6 in addition to exons 7–9. Exons 4–9 of *PML* are subject to alternative splicing, generating multiple PML-RARα and RARα-PML fusion products and up to 13 PML isoforms. (Adapted from **refs.** *1,18,19,21,24*; reproduced courtesy of Springer-Verlag, Heidelberg, Germany.)

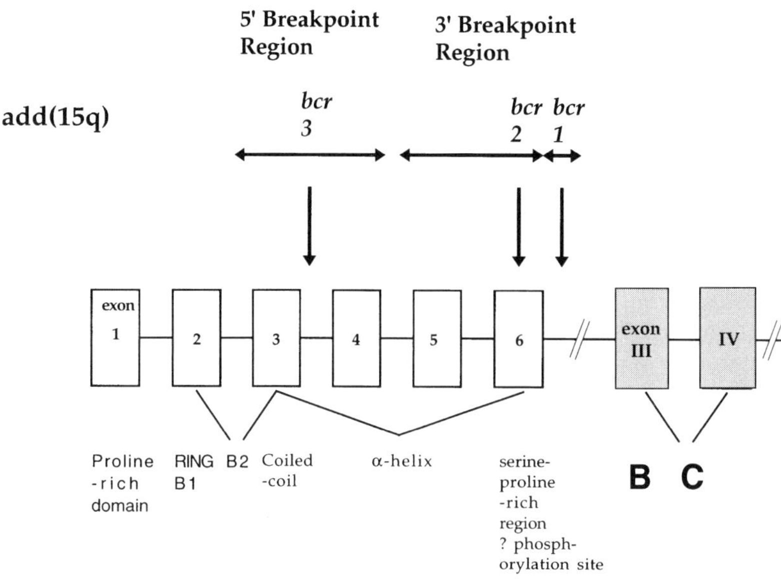

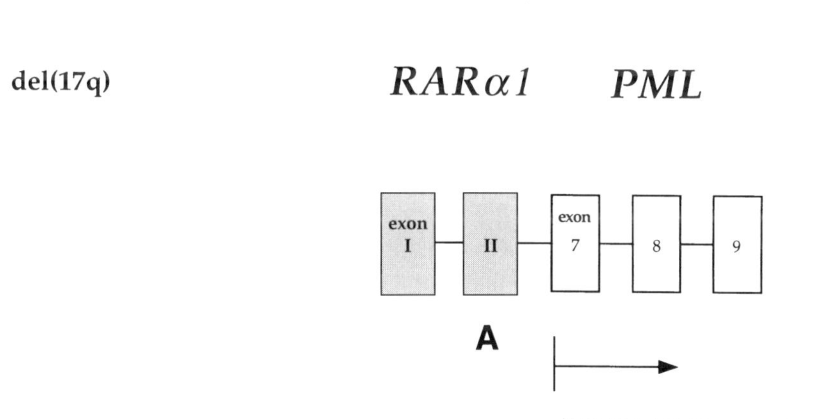

Fig. 1.

herpes viruses precipitate disruption of PML nuclear bodies with subsequent relocation of PML to the cytoplasm *(33)*. Furthermore, histological sections of tissues associated with chronic-inflammatory processes reveal more prominent PML-nuclear staining, which may be related to increased expression of PML in response to interferons *(34–38)*. In addition to its role in the immune system, there is some evidence to suggest that PML itself, or components of the

PML-nuclear bodies, are cell-cycle regulated and can mediate growth-suppressor activity *(39,40–42)*.

PML, in common with RFP and TIF 1 that also form oncogenic-fusion proteins as a result of chromosomal translocation, is characterized by a "tripartite motif" comprising a "RING finger" and coiled-coil, separated by two "B-Boxes" *(22,31,43,44)* (*see* **Fig. 1**). The B-Box and RING finger domains are cysteine histidine-rich regions capable of zinc binding *(43,44)*, which initially raised the possibility of PML being a transcription factor *(17,21,22)*. However, there has been no evidence for direct DNA binding by PML, and components of the tripartite motif are more likely to represent sites of protein-protein interaction *(43)*. In support of this, it has already been demonstrated that integrity of the RING finger, B-boxes, and coiled-coil is required for PML nuclear-body formation and that PML-homodimer interaction is mediated through the coiled-coil domain *(22,31,44,45)*.

In contrast to PML, RARα has been relatively well characterized; it is essentially a transcription factor mediating the effect of RA at specific DNA-retinoid-response elements *(46,47)*. Integrity of the retinoid-signaling pathway has been found to be critical for normal embryogenesis *(48–52)* and postnatal myeloid differentiation *(53)*. For high-affinity DNA binding, RARs must bind with a member of the retinoid X receptor (RXR) family of nuclear receptors, which are also required for high-affinity DNA binding of thyroid and vitamin D receptors (TR and VDR, respectively) *(54,55)*. Specificity of the response is conferred by features of the response element and pattern of response by the configuration of heterodimer binding *(47,56)*.

Hence, PML-RARα is currently considered to promote leukemogenesis by interacting with wild-type PML, leading to the disruption of PML-nuclear bodies *(28–30)*, thus abrogating any growth-suppressor function *(40–42)*. This is coupled with an abnormal pattern of retinoid responses, also mediated by the fusion protein *(17,20,22,23)* associated with sequestration of RXR leading to the block in myeloid differentiation that characterizes the disease *(29)*. Recent transgenic experiments suggest, however, that expression of PML-RARα is not in itself sufficient to generate APL, and further mutational events are required for leukemogenesis in this model *(57)*.

1.2. Importance of Establishing a Molecular Diagnosis of APL

APL is characterized by the t(15;17); yet, in a recent study of 100 patients entered into the UK Medical Research Council (MRC) ATRA trial for APL, conventional cytogenetics failed to identify this translocation in 13% patients in whom its molecular consequence, the *PML-RARα* rearrangement, was ultimately successfully found by reverse transcriptase-polymerase chain reaction (RT-PCR) *(5)*. Inability to detect t(15;17) on karyotype assessment most commonly reflects poor-quality metaphase preparations that are a frequent problem in APL; but some diagnostic failures result from omission to culture cells prior to cytogenetic analysis, such that only normal marrow elements are effectively studied. Despite prior culture, a few patients are still found with normal karyotype, but with evidence of *PML-RARα* rearrangement identifiable by fluorescent *in situ* hybridization (FISH) or RT-PCR, suggesting the occurrence of interstitial insertions or small translocations below the resolution of conventional cytogenetic techniques. RT-PCR or FISH have also proved invaluable in establishing the presence of *PML-RARα* rearrangements in cases of APL where cytogenetics reveals changes other than t(15;17), including three- or four-way translocations involving chromosome 15 and/or 17 *(5,58–61)*.

Establishing the presence of t(15;17) or its molecular consequences in patients with suspected APL is critical, because the existence of a *PML-RARα* rearrangement is a prerequisite for a favorable response to ATRA *(4)*. For example, patients with APL associated with the rare cytogenetic variant: t(11;17) leading to a promyelocytic zinc finger *(PLZF)-RARα* rearrangement have been found to exhibit a poor differentiation response to ATRA and have an adverse prognosis *(62)*. Recent trials in patients with t(15;17)-associated APL have demonstrated that combined therapy with ATRA and chemotherapy confers significant improvements in disease-free survival (DFS) as compared to treatment with chemotherapy alone *(63,64)*. The most efficacious means of combining ATRA and chemotherapy remains to be determined, and indeed preliminary data relating to the intergroup study of APL has suggested that exposure to ATRA at any time during the treatment course entails advantage in terms of DFS *(64)*. ATRA commenced from the time of diagnosis leads to a rapid amelioration of the coagulopathy, and has even been advocated as a means to achieve outpatient-induction therapy of APL. However, the benefits of rapid coagulopathy control may be offset by morbidity or even mortality secondary to a constellation of clinical features that comprise the "ATRA syndrome," which may develop in up to 30% patients, although this can be effectively treated by early employment of high-dose steroids *(65,66)*. The ATRA syndrome is probably related to the differentiation response, and hence is likely to be confined to cases possessing t(15;17); but until this is established it would be wise to restrict ATRA therapy to patients with evidence of *PML-RARα* rearrangement.

Molecular techniques are essential to identify all patients who might benefit from ATRA, and are critical for appropriate analysis of clinical trials of APL that involve retinoids. Studies that even predated the clinical use of ATRA found that the presence of t(15;17) identifies a subgroup of AML with relatively good prognosis *(67)*. This information has been incorporated into current AML trials, e.g., UK AML MRC12 , whereby such patients, in addition to those demonstrating inv(16) or t(8;21), are spared routine use of transplantation in first complete remission (CR) that might confer more risk than benefit. More widespread use of RT-PCR to identify these subtypes of AML would be expected to achieve more accurate allocation of patients to stratified-treatment groups that might ultimately yield improvements in DFS.

A final reason for performing RT-PCR in all patients with APL at diagnosis is to determine whether *PML*-breakpoint patterns confer independent-prognostic information that might be used to influence the design of future trials that could lead to further improvements in cure rates. RT-PCR is the only technique suitable for this purpose, because cytogenetics, FISH, and determination of PML nuclear-staining patterns serve only to demonstrate the presence or absence of the t(15;17), and Southern blotting is rather cumbersome. In this respect, there is some evidence to suggest that *bcr 2-PML* breakpoints are associated with a reduced ATRA response in vitro *(25)* and in some studies the presence of a *bcr 3* breakpoint has been associated with reduced DFS *(19,66,68)*; although this latter finding remains contentious *(69)*.

1.3. Rationale for Disease Monitoring

RT-PCR assays, as currently employed, are capable of detecting a single APL cell among 10^4 to 10^5 nonleukemic cells *(19,26,27)*. This technique, therefore, affords the opportunity for more objective assessment of remission status in the presence of apparently normal bone-marrow morphology and cytogenetics. It also potentially provides an assay capable of evaluating the relative efficacy of different treatment protocols, which can be used in conjunction with clinical parameters of outcome. However, the major aspiration driving investigations of RT-PCR in hematological malignancies has been the expectation that disease monitoring might enable one to tailor therapy more precisely to the requirements of individual patients. Those in whom PCR profiles suggest that they are likely to be cured of their disease could be spared excessive therapy with an inherent risk of excess morbidity and mortality; whereas those predicted highly likely to relapse could be targeted for dose intensification in first CR. A number of early studies claimed that RT-PCR could be reliably employed to predict treatment outcome in APL *(4,70)*; however, such studies presented a somewhat distorted view because they included significant numbers of patients treated in relapse or *de novo* patients treated with ATRA alone, i.e., without subsequent chemotherapy consolidation. In this setting, it was claimed that persistent PCR positivity throughout the treatment course, or recurrence of PCR positivity while in remission, predicted relapse; whereas achievement of PCR negativity was associated with prolonged DFS *(70)* and indeed patients in long-term remission were found to be negative for the putative leukemogenic transcript *PML-RARα (71)*. More recent studies of *de novo*

APL receiving combination therapy with ATRA and chemotherapy, reveal that virtually all patients so treated ultimately become PCR negative *(72,73)*. The few patients that remain PCR positive following consolidation have a high rate of relapse *(74)*. Furthermore, in a study of APL patients who ultimately relapsed despite first-line combination treatment, we found that 11/12 had been PCR negative following chemotherapy consolidation *(75)*. Therefore, it is clear that in *de novo* patients receiving aggressive therapy with ATRA and chemotherapy, assessment of PCR status solely postconsolidation cannot be used to identify all patients who ultimately relapse, highlighting the relative insensitivity of current PCR protocols. Whether the rapidity of achievement of PCR negativity is a better prognostic indicator will become apparent from large ongoing clinical trials.

2. Materials for RT-PCR

2.1. Sample Preparation

1. Bone marrow/peripheral blood sample (*see* **Note 1**).
2. 15-mL plastic centrifuge tubes (e.g., Falcon/Sarstedt).
3. Sterile pastets.
4. Dulbecco's phosphate-buffered saline (PBS), without calcium chloride or magnesium chloride.
5. GTC-ME: 4 M guanidine thiocyanate, 0.025 M sodium citrate, 0.1 M 2-mercaptoethanol, 0.5% *N*-lauryl sarcosine.

2.2. RNA Extraction

1. Sterile 1.5-mL capped tubes.
2. 2 M sodium acetate (pH 4.0).
3. Phenol.
4. Chloroform/isoamyl alcohol (49:1).
5. Isopropanol.
6. 75% ethanol.
7. GTC-ME.
8. DEPC-treated sterile water (0.1% v/v diethyl pyrocarbonate).
9. Agarose.

2.3. RT-PCR

1. Sterile 0.5-mL capped tubes.
2. 18-mer oligo-dT (50 µg in 50 µL sterile water, store at –20°C).
3. Sterile water.
4. Avian mycloblastosis virus (AMV) reverse transcriptase, with the manufacturer's buffer (NBL [Cramlington, UK]; store at –70°C).
5. RNasin (Promega [Southampton, UK]; store at –20°C).
6. Deoxynucleotide triphosphates (dNTPs) (Pharmacia Biotech [St. Albans, UK]; store at –20°C).

7. TE: 10 mM Tris-HCl, pH 8.0, 1 mM EDTA.
8. 10X PCR buffer containing 15 mM MgCl$_2$ (Promega, store at –20°C).
9. *Taq* polymerase (Promega, store at –20°C).
10. PCR primers (stock solution at 1 µg/µL in sterile water, store at –20°C).
11. Liquid paraffin.
12. Techne Programmable Dri-block (PHC-1).
13. 5X Ficoll blue loading buffer: 15% w/v Ficoll (type 400, Pharmacia), 0.25% w/v bromophenol blue, 0.25% w/v xylene cyanol FF.
14. Electrophoresis tanks (Bio-Rad, Hemel Hempstead, UK).
15. Ethidium bromide.
16. 100-bp ladder (Pharmacia, store at –20°C).

3. Methods for RT-PCR

3.1. Sample Preparation

1. For the purposes of molecular detection of t(15;17) in cases of suspected APL, study of presentation bone marrow is preferable, although evidence for a *PML-RARα* rearrangement may also be successfully obtained from peripheral blood samples taken shortly after diagnosis (*see* **Note 1**). For residual-disease assessment, only bone-marrow specimens are considered suitable for RT-PCR assessment.
2. Dispense bone marrow/peripheral blood samples into 15-mL tubes and spin at 1600g for 5 min. Remove the plasma/supernatant with a sterile pastet and discard. Gently remove the buffy-coat layer with a further sterile pastet and transfer to a new 15-mL tube (*see* **Note 2**). Resuspend the buffy coat to a volume of 10 mL in Dulbecco's PBS. Take a small aliquot to assess the total number of cells available for RNA extraction (STKS, Coulter) (*see* **Note 1**).
3. Spin the buffy coat suspended in PBS at 1600g for 5 min and remove the supernatant with a sterile pastet. Resuspend the cells in GTC-ME, which causes proteins to dissolve, nuclear proteins to dissociate from nucleic acids owing to loss of secondary structure, and inactivation of RNAases. The volume of GTC-ME is dependent on the cell-count of PBS-suspended cells: for 1–2 × 10^7, add GTC-ME to a final volume of 1 mL, for >2–4 × 10^7 add GTC-ME to a volume of 2 mL and for >4–6 × 10^7 cells add GTC-ME to a volume of 3 mL, and so on. Cells in GTC-ME may then be stored at 4°C until RNA extraction.

3.2. RNA Extraction (see Note 3)

1. RNA is prepared from 1-mL cells in GTC-ME (i.e., 1–2 × 10^7 cells); rapidly add to this 100 µL 2 M sodium acetate, 1 mL phenol, and 200 µL chloroform/isoamyl alcohol in order to achieve DNA precipitation and disruption of nucleoprotein complexes. Mix and cool on ice for 15 min, then spin the samples at 11,600g for 20 min at 4°C. Separate off the upper-aqueous phase, to which add 600 µL isopropanol to precipitate the RNA, and leave at –20°C overnight (or for at least 1 h).

2. Spin samples at 11,600*g* for 20 min at 4°C, discard the supernatant, and wash the RNA pellets in 500 μL 75% ethanol. Add 500 μL GTC-ME to redissolve, then add 600 μL isopropanol to reprecipitate the RNA and leave the samples at –20°C overnight (or for at least 1 h).

3. Spin the samples and wash the RNA pellets in 75% ethanol as before. Then dissolve the RNA pellets in 50 μL DEPC-treated sterile water and store at –70°C.

4. Prior to storage, assess RNA concentration and quality by optical density (OD) measurements and gel electrophoresis (3 μL RNA run in 0.5% w/v agarose). For quantification purposes, assess the absorbance of RNA diluted in sterile water by UV spectrophotometry at 260 nm. A ratio of readings taken at 260 and 280 nm provides an indication of RNA purity (260/280 ratio values within the range 1.80–2.00 are satisfactory). Gel electrophoresis serves to validate concentration determined by OD, determine the degree of DNA contamination and assess RNA quality by presence of 18S and 28S ribosomal bands. Minimal residual disease (MRD) studies using RNA lacking ribosomal bands should be treated with suspicion.

3.3. Reverse Transcriptase PCR (see Notes 4 and 5)

1. The scheme for reverse transcription and nested PCR is outlined in **Fig. 2**.

2. For reverse transcription, use 1 μg total RNA, with 0.5 μg 18-mer oligo-dT as primer in a 20-μL reaction. In addition to patient samples, a series of control samples should be run in parallel: including RNA derived from a 1 in 1000 dilution of APL cell-line NB4 *(76)* in *PML-RARα* negative-filler cells, serving as positive control in the subsequent PCR reaction, RNA from a non-APL patient/cell-line, e.g., HL60 (-ve control) and a control lacking RNA (RT water control). We have found oligo-dT to be the most reliable primer for reverse transcription; although other groups have used random hexamers *(26,27,70)* or gene-specific primers *(4)* with success. Heat RNA, oligo-dT, and sterile water at 65°C for 5 min, spin samples briefly at 11,600*g* and place on ice prior to addition of the reaction mixture. This comprises 10 U of AMV reverse transcriptase in the manufacturers buffer, 10 U of RNasin, and 1 mmol of each dNTP. Incubate samples for 1 h at 42°C, at which stage stop the reaction by addition of 100–300 μL TE, and transfer the tubes to ice as the PCR stage is set up.

3. For analysis of all patient samples and controls, perform four separate PCR amplifications in order to detect *PML, RARα1, PML-RARα,* and *RARα1-PML* transcripts using two rounds of PCR with nested primers, as indicated in **Figs 2–6**. Primer sets to detect *PML* and *RARα* are used both as a control for the success of reverse transcription and RNA integrity. Typically, we perform PCR reactions on 8–9 patient samples, with the 1 in 1000 dilution of NB4 described earlier serving as a positive and sensitivity control; a non-APL sample and water controls are also run in parallel to exclude contamination at all stages of the RT-PCR procedure. All PCR experiments are performed at least twice from the

NESTED RT-PCR IN APL

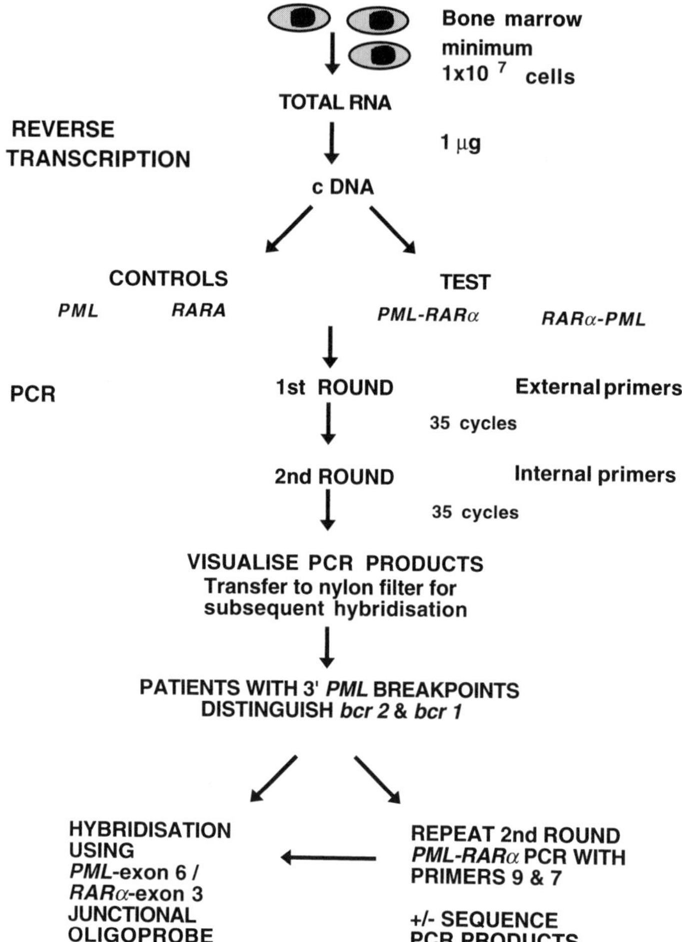

Fig. 2. Flow diagram demonstrating protocol for nested RT-PCR in APL.

reverse-transcription stage and results of patient samples only considered reliable if both normal *PML* and *RARα* coamplify successfully and all other controls are satisfactory.

4. The first-round reaction mixture comprises 10 µL cDNA (store at –40°C), 5 µL PCR buffer, each dNTP at 100 µM, 20 pmol of each external primer, 1 U of *Taq* polymerase, and sterile distilled water to make the total volume up to 50 µL. Prepare reaction mixtures for each of the four separate amplifications (*PML*, *RARα*, *PML-RARα*, and *RARα-PML*) sufficient for all patient samples

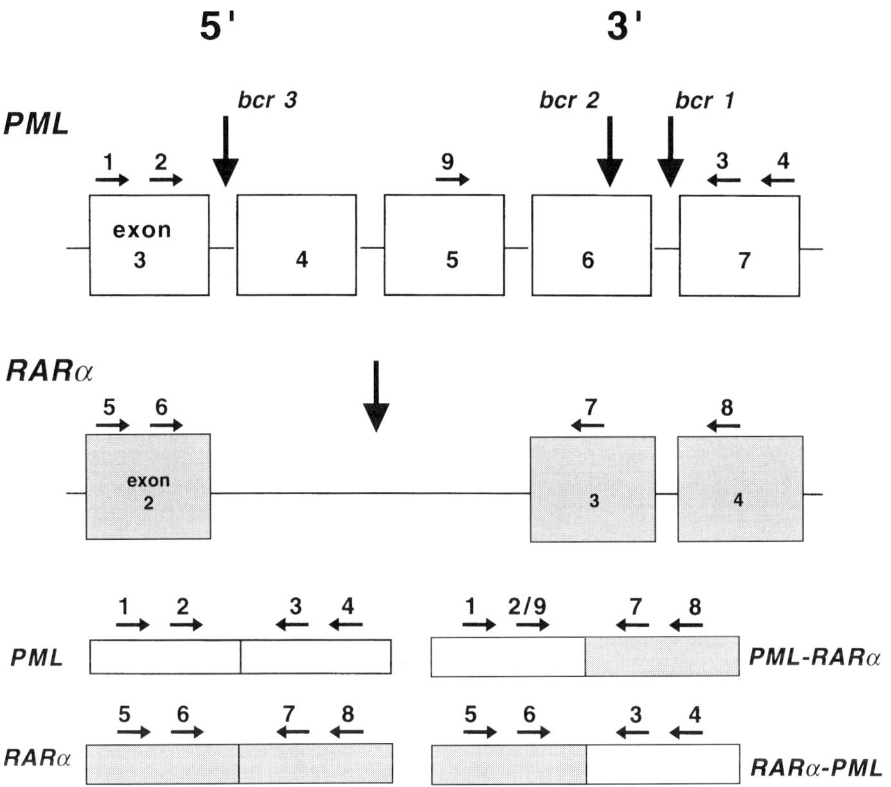

PRIMERS :

PML

1. 5'-AGCTGCTGGAGGCTGTGGAC-3'
2. 5'-TGTGCTGCAGCGCATCCGCA-3'
3. 5'-CTGCTGATCACCACAACGCG-3'
4. 5'-CGGCATCTGAGTCTTCCGAG-3'
9. 5'-AGTGTACGCCTTCTCCATCA-3'

RARα

5. 5'-GGCCAGCAACAGCAGCTCCT-3'
6. 5'-GGTGCCTCCCTACGCCTTCT-3'
7. 5'-GGCGCTGACCCCATAGTGGT-3'
8. 5'-TCTTCTGGATGCTGCGGCGG-3'

Fig. 3. Schematic representation of primer positions employed for nested RT-PCR relative to genomic structure of *PML* and *RARα* and breakpoints generated by the t(15;17) translocation. Positions of the most frequent breakpoints within *PML* are indicated by the vertical arrows. Disruption of *RARα* invariably occurs within intron 2. (Figure adapted from **ref. 5**, reproduced courtesy of Blackwell Scientific Publications, Oxford, UK.)

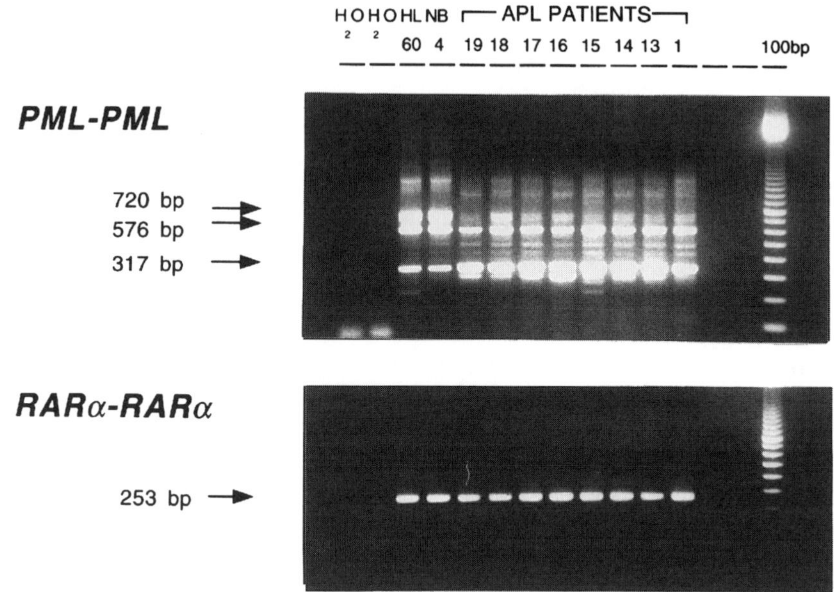

Fig. 4. Controls for RT-PCR and RNA integrity. Pattern of PCR products generated by use of primer sets to amplify wild-type *PML* and *RARα* cDNA serving as controls for RNA integrity and the success of reverse transcription. *PML* amplification involves use of primer set 1 and 4 for first round PCR, followed by primers 2 and 3 for the second round, *see* **Fig. 3**. For *RARα* amplification primers 5 and 8 serve as external primers, with 6 and 7 used for subsequent nested PCR. Multiple *PML* products are indicative of alternative splicing between central exons.

   and controls, and add to PCR tubes in the laminar flow hood, prior to addition of 10 µL cDNA and subsequent overlay with 30 µL of liquid paraffin in a separate laboratory.

5. Transfer the tubes to the PCR machine. The reaction cycle comprises 1 min denaturing at 95°C, 1 min annealing at 55°C, and 1 min extension at 72°C repeated for 35 cycles, with a subsequent 10-min extension period at 72°C. PCR products may be left at room temperature overnight prior to proceeding to the second round of PCR.

6. Second-round PCR-reaction mixture includes 1 µL first-round product and 20 pmol of each internal primer; the concentrations of PCR buffer, dNTPs, and *Taq* polymerase are the same as in the first-round PCR, and again sterile distilled water is added to bring the total volume to 50 µL. As for the first round of PCR, we aliquot the reaction mixture in the laminar-flow hood, prior to addition of the first-round products and 30 µL paraffin-oil overlay in the separate laboratory. PCR conditions are identical to the first round and PCR products may be left at room temperature overnight prior to gel electrophoresis.

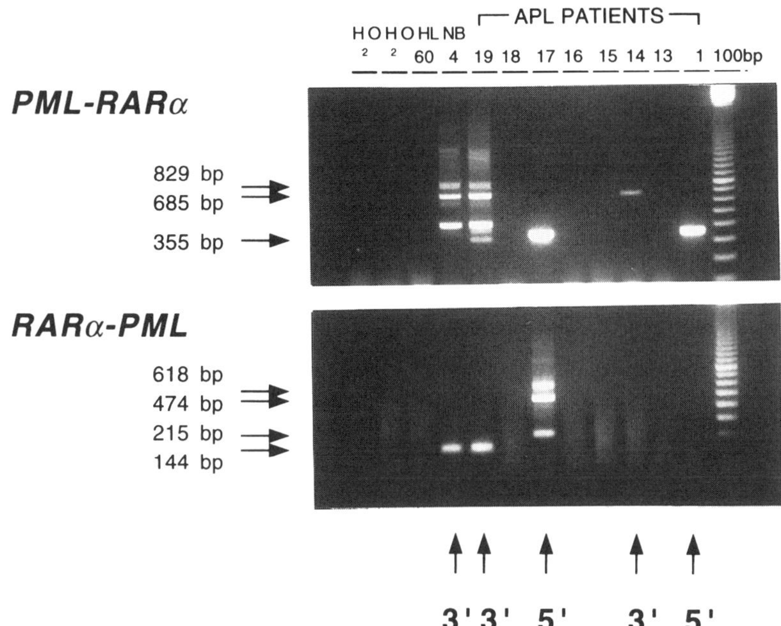

Fig. 5. RT-PCR demonstrates 5' and 3' breakpoints in *PML* gene in APL. Distinction between APL cases with 5' (*bcr 3*) and 3' (*bcr 1/2*) *PML* breakpoints using RT-PCR for amplification of add(15q)-derived *PML-RARα* and del(17q)-associated *RARα-PML* transcripts. *PML-RARα* products have been generated by use of primers 1 and 8 as external primers, followed by primers 2 and 7 for the second round, *see* **Fig. 3**. *RARα-PML* products are generated by primers 5 and 4 followed by 3 and 6. In APL cases with a 3' *PML* breakpoint a series of *PML-RARα* transcripts are generated reflecting alternative splicing between *PML* central exons. These are represented by PCR products of 426 bp (*PML* exon 3-4-*RARα*), 685 bp (*PML* exon 3-4-6-*RARα*), and 829 bp (*PML* exon 3-4-5-6-*RARα*), as demonstrated by APL patients 14, 19 and APL cell-line NB4. In cases with 3' *PML* breakpoints that express *RARα-PML*, a single 144-bp product is detected as demonstrated by APL patient 19 and NB4. Cases with a 5' *PML*, breakpoint (*bcr 3*) exhibit a single 355-bp *PML-RARα* product (*PML* exon 3-*RARα*); however, if expressed, multiple *RARα-PML* transcripts are detected owing to alternative splicing, as demonstrated by patient 17. In such a situation, *RARα-PML* products of 215 bp (RARα-PML exon 4-7), 474 bp (*RARα-PML* exon 4-6-7), and 618 bp (*RARα-PML* exon 4-5-6-7) may be generated (**ref. *19***). APL patients 13,15,16, and 18 were studied in remission for residual-disease assessment; in each case disease-related transcripts were not detected.

7. Mix 20 µL of second-round PCR products with 5 µL 5X Ficoll-blue loading buffer, and size separate by electrophoresis through 1.5% agarose gels (3.75 g agarose in 250 mL Tris-acetate/EDTA electrophoresis buffer (TAE) using TAE as buffer in the electrophoresis tank. Visualize PCR products by incorporation of

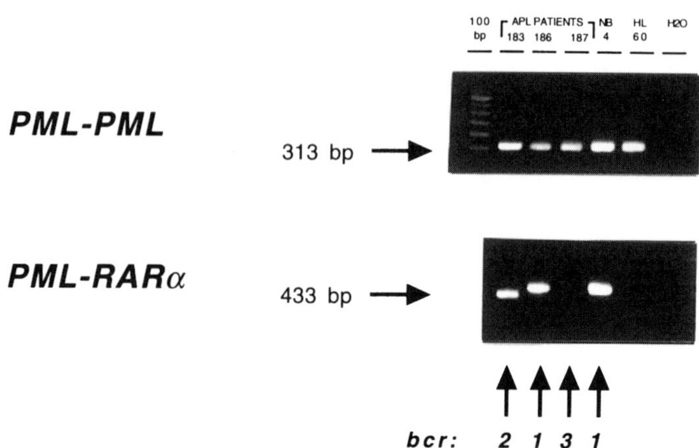

Fig. 6. Distinction between intron 6 (*bcr 1*) and exon 6 (*bcr 2*) 3' *PML* breakpoints by RT-PCR. Employment of primers 9 and 7 for the second round of *PML-RARα* PCR permits distinction of *bcr 2* (exon 6) from *bcr 1* (intron 6) *PML* breakpoints. *Bcr 2* breakpoints are indicated by the generation of product differing in size from that associated with NB4 (*bcr 1*), which is used as a control. The *PML-RARα* fusion gene associated with the *bcr 3*-breakpoint pattern lacks *PML* exon 5, explaining the absence of amplified product in APL patient 187. Amplification of wild-type *PML* using primers 9 and 3 for the second round of PCR demonstrated in the top panel is included as a control.

ethidium bromide (1 μg/mL) into the gel, and photograph under UV light. Assess PCR product sizes by use of a suitable marker, e.g., 100-bp ladder.

8. After electrophoresis, we routinely transfer DNA to nylon filters (Hybond N+) by standard techniques, for subsequent hybridization using a 221bp *RARα* probe (H7b), an *Xho*I, *Kpn*I cDNA fragment that traverses the *RARα*-breakpoint site *(13)*. This serves to confirm the specificity of PCR-reaction products identifying *PML-RARα* and *RARα-PML* transcripts. For further clarification, *PML* probes, which hybridize to 5' and 3' regions of *PML*, have also been used *(21)*. Probes are labeled with α^{32}P-dCTP using standard random-priming techniques, hybridization is performed overnight, and filters washed to a final stringency of 0.1X SSC, 0.1% SDS at 65°C. They are exposed to Kodak XAR or Fuji RX film for 1–12 h, between intensifying screens at –70°C.

9. For those patients in whom RT-PCR at the time of diagnosis demonstrates a 3' *PML* breakpoint pattern, *bcr 1* (intron 6) and *bcr 2* (exon 6) *PML* breakpoints can be distinguished by using primers 9 *(27)* and 7, instead of 2 and 7 for the second-round *PML-RARα* PCR (*see* **Figs. 3** and **6**). A recent study, however, has demonstrated that these primer sets do not reliably identify all patients with *bcr 2* breakpoints, and has suggested that hybridization techniques employing oligoprobes spanning the *PML* exon 6/*RARα* exon 3 border are more suitable to

distinguish *bcr 1* and *bcr 2* cases *(25)*. In practice, with access to an automated sequencer (ABI), we have found it simpler and more reliable to sequence the *PML/RARα* junction in all cases with 3' *PML* breakpoints; but whether this has any clinical relevance remains to be determined. For patients with a 3' *PML* breakpoint being monitored for residual disease, negative *PML-RARα* PCR using primers 2 and 7 may be confirmed by repeating the second round with primers 9 and 7.

3.4. Data Interpretation

The PCR products resulting from *PML, RARα, PML-RARα,* and *RARα-PML* amplifications are demonstrated in **Figs. 4–6**. APL patients with 5' and 3' *PML* breakpoints may be readily distinguished by the pattern of PCR products generated. Alternative splicing between the central exons of *PML* accounts for the multiple bands seen in *PML* RT-PCR and in *PML-RARα* PCR assays for patients with 3' *PML* breakpoints *(19,24,77)*, as demonstrated in **Figs. 4** and **5**. *Bcr 1* and *bcr 2* breakpoints cannot be reliably distinguished by sizing of these alternatively spliced transcripts, and therefore junctional oligoprobes or internal *PML* primers as described previously must be used to achieve this classification. Patients with a 5' breakpoint *(bcr 3)* exhibit a single 355-bp product by *PML-RARα* PCR using nested primers 2 and 7, reflecting fusion of *PML* exon3 to *RARα* exon 3 *(19)*; whereas if the del(17q)-derived product is expressed in such cases, multiple bands are seen in *RARα-PML* assay, again reflecting alternative splicing among *PML* central exons *(18,19)* (*see* **Figs. 3** and **5**). In a recent study of 93 APL patients with a *PML/RARα* rearrangement, we found that 80% demonstrated evidence of *RARα-PML* expression *(5)*. Whether this is of any clinical significance is unclear at present; nevertheless *RARα-PML* PCR has proved a valuable adjunct to the more conventional *PML-RARα* assay, because it provided the sole molecular evidence of the t(15;17) in 7/93 patients with suspected APL. In six patients, the pattern was consistent with a 3' *PML* breakpoint, in accordance with the improved sensitivity of the *RARα-PML* assay for this breakpoint pattern noted in sensitivity studies using mixtures of NB4 and HL60 cells *(75,78)* (*see* **Note 5**). Failure to detect *PML-RARα* transcripts in these patients was probably a reflection of sample quality rather than implying its absence; and indeed this has been confirmed in one patient who has subsequently relapsed with both *PML-RARα* and *RARα-PML* transcripts detectable.

Despite use of nested primers, which serve to improve both PCR sensitivity and specificity, occasional aberrant bands reflecting amplification errors have been reported. In view of this phenomenon, quality control for minimal residual disease (MRD) studies is improved by only reporting samples derived from patients in whom *PML* breakpoint pattern at diagnosis or relapse has been clearly established. Furthermore, where presentation breakpoint data is unknown, it serves to underline the importance of subsequent hybridization techniques to prevent reporting of such bands as false-positives.

4. Notes

1. Presentation peripheral blood or bone-marrow samples are both suitable for RT-PCR assessment of patients with a suspected diagnosis of APL. Furthermore, should cytogenetic analysis of diagnostic bone marrow have failed, we have successfully detected PCR evidence of the *PML/RARα* rearrangement in peripheral-blood samples up to 15 d post-diagnosis, and with WBC as low as $0.3 \times 10^9/l$ (RNA extracted from 3×10^7 cells): in such cases at least 30 mL blood should be requested. PCR evidence of the *PML/RARα* rearrangement may also be found in the majority of bone-marrow samples taken following the first course of chemotherapy to assess remission status. For the purposes of residual-disease assessment, only bone-marrow samples with RNA extracted from at least 1×10^7 cells are considered suitable for analysis *(26)*; for marrow aspirates performed following consolidation courses up to 5 mL of marrow may be required to achieve suitable cell counts. Quality control assessment of RNA evaluated by optical density, gel electrophoresis, and success of RT-PCR has established that tubes/containers using lithium heparin, EDTA, or heparinized tissue-culture medium (e.g., RPMI-1640) as anticoagulants are all suitable for transport of specimens to the laboratory. RNA of sufficient quality to demonstrate *PML-RARα* transcripts has even been extracted from diagnostic specimens that have been in transit for at least 6 d at room temperature. We have also detected *PML-RARα* transcripts in a remission sample in transit for at least 4 d. The majority of our samples, however, are received within 0–2 d, and ideally for MRD purposes samples should be received within 24 h for meaningful analysis.
2. Many groups prefer to isolate mononuclear cells for RT-PCR analysis by centrifugation using a Ficoll gradient (Ficoll-Paque, Pharmacia); however we have found that this is less satisfactory for samples of low cell count. We routinely prepare 3 slides from the buffy coat: 1 for MGG staining to assess cell morphology and to evaluate the degree of leukemic infiltration and 2 slides are stored at –20°C when thoroughly air-dried. The latter slides are suitable for PML-antisera staining to determine the presence of the *PML/RARα* rearrangement *(5,79)*.
3. RNA is extracted by a method based on that described by Chomczynski and Sacchi *(80)*. We have found this to be more satisfactory than the CsCl extraction method, which was associated with problems of contamination between samples. RNA is extracted in a laminar-flow hood, using sterile 1.5-mL tubes and plugged pipet tips; gloves are worn at all times to avoid RNA degradation owing to RNases. RNA is extracted in a separate laboratory from that in which RT-PCR is performed to avoid risks of contamination of RNA by amplified product. RNA should be extracted in parallel from a non-APL patient sample or cell-line (e.g., HL60) to exclude contamination of patient RNA samples with the leukemic clone.

4. RT-PCR is performed using a method based on that developed by Borrow et al. *(19)*. Rigorous steps are taken to avoid risk of contamination to or between samples at all stages of the RT-PCR protocol *(81)*. Reaction mixtures for the reverse transcription and PCR stages are always prepared in a laminar-flow hood; whereas addition of cDNA and mineral oil to first-round PCR mixture, addition of amplified first-round PCR products and minerai oil to second-round PCR mixture, and subsequent gel electrophoresis of amplified second-round products are all performed on a designated bench in a separate laboratory from the laminar-flow hoods where RNA is extracted or those where RT-PCR reaction mixtures are set up. Clones of *PML, RARα ,PML-RARα*, or *RARα-PML* are never permitted in the hood, nor in the vicinity of laboratory bench areas where RT-PCR reactions are set up. Reverse transcription and PCR reactions are performed in 0.5-mL sterile Eppendorf tubes. At all stages of the procedure sterile plugged pipet tips are employed and gloves are worn. Both gloves are changed each time a sample from a different patient is handled; PCR tubes are opened by hand, with the tube-opening glove being changed as each new tube is manipulated (we have found this technique to be more satisfactory than using tube-opening devices, which are associated with a greater risk of contamination).

5. To establish the sensitivity of the assay, RT-PCR should be performed on total RNA derived from a series of 10-fold dilutions of the APL cell-line NB4 (*bcr 1* breakpoint, *RARα-PML* positive), in filler cells that are negative for the *PML/RARα* rearrangement, e.g., HL60 *(75)*. In all cases, RNA is made from a total of 1×10^7 cells; in such circumstances the technique can detect 1 APL cell in 10^4 HL60 cells using primer sets to detect *PML-RARα*, whereas *RARα-PML* can be detected at a sensitivity of 1 in 10^5.

5. Evaluation of the Role of RT-PCR in Diagnosis and Disease Monitoring in APL

5.1. Diagnosis of APL

Over the last few years, it has become clear that for patients with APL, the presence of the t(15;17) or its molecular consequence the *PML-RARα* rearrangement, is a prerequisite for a favorable differentiation response to retinoids *(4,62)*, and that combination therapy with ATRA and chemotherapy confers significant improvements in DFS *(63,64)*. We have demonstrated that cytogenetics fails to identify the t(15;17) in 13% patients with molecular evidence of a *PML-RARα* rearrangement *(5)*. In most cases, this reflects poor-quality metaphase spreads, or failure to culture cells such that only normal marrow elements are effectively analyzed, but in a subgroup is a consequence of cryptic rearrangements below the resolution of conventional cytogenetics. It is critical that in all patients with suspected APL lacking the t(15;17), a *PML-RARα* rearrangement is excluded by molecular techniques, such that suitable patients are not denied retinoid therapy. In such situations, RT-PCR is the method of choice; it can be successfully employed using peripheral-blood samples taken several d after initial diagnosis, and in most situations *PML-RARα* and/or *RARα-PML* transcripts are still detectable in bone-marrow aspirates taken following the first course of treatment. RT-PCR should not, however, be considered a substitute for cytogenetic analysis, which may offer independent prognostic information.

In the context of clinical trials of APL, RT-PCR to detect *PML-RARα* and RARα-PML transcripts should be routinely performed in all cases to determine whether there is any correlation between *PML*-breakpoint pattern or *RARα-PML* expression and various disease characteristics, including response to ATRA/induction chemotherapy and long-term prognosis. Furthermore such analysis will also establish whether ATRA therapy has any beneficial or detrimental effects in patients presumed to have APL on morphological grounds, but lacking cytogenetic or RT-PCR evidence of a *PML-RARα* rearrangement.

5.2. RT-PCR for Disease Monitoring

In contrast to the undeniable role of RT-PCR in confirming a morphological diagnosis of APL and predicting a response to retinoids, the suitability of such techniques to disease monitoring and in particular determining the most appropriate consolidation therapy for individual patients is far less certain. The ultimate aim of RT-PCR monitoring is to predict reliably which patients are cured of their disease, such that they may be spared unnecessary additional therapy that might entail higher-mortality rates, and hence conversely to identify those patients with a high risk of relapse and therefore most likely to benefit from dose intensification in first CR. At current levels of sensitivity, it is clear that patients with PCR-detectable disease after completion of consolidation therapy have a high risk of relapse *(74)*; these patients should be targeted for further treatment to increase their chances of cure; the optimum therapy for such patients is currently being addressed by the Italian AIDA study *(73,82)*. However, the vast majority of ATRA and chemotherapy-treated patients have no evidence of PCR-detectable disease following completion of therapy *(72,73)*. Furthermore, in recent studies of a group of 10 patients who relapsed despite first-line therapy with ATRA and the intensive AML-10 protocol, we established that none had evidence of PCR-detectable disease following chemotherapy consolidation *(75* and unpublished data). Therefore, assessment of "PCR status" at the end of treatment will fail to predict relapse in all patients who ultimately do so, and studies which suggest that achievement of PCR negativity following all therapy is associated with prolonged DFS are merely a reflection of the relatively good outlook of APL patients with t(15;17) once CR is achieved, rather than predictive prowess on the part of current RT-PCR protocols. Ongoing trials will determine whether rapidity of achievement of

PCR negativity is a more useful predictor of DFS. Future trials are also likely to address whether posttreatment surveillance to detect evidence for impending clinical relapse is worthwhile; in particular whether there is any survival advantage inherent in retreating patients with PCR-detectable disease as opposed to awaiting frank clinical relapse.

The limitations of MRD monitoring as an accurate predictor of prognosis, highlighted by a number of studies, has prompted the creation of an International Workshop for RT-PCR in APL by Prof. Christine Chomienne *(26)* in order to improve reliability and sensitivity of the technique with a view to greater standardization of methods employed, hence enabling more meaningful comparisons of different treatment protocols. The workshop aims to address all aspects of the RT-PCR procedure; potential improvements may be evaluated by semi-quantitative PCR techniques *(83)* and ultimately judged by the predictive power of PCR in MRD studies.

In conclusion, RT-PCR analysis has already become established as a prerequisite for meaningful analysis of clinical trials of APL, over the next few years it will become apparent as to whether *PML*-breakpoint pattern assessment and disease monitoring can provide further independent prognostic information that will shape the design of future studies.

Acknowledgments

David Grimwade was supported by an MRC clinical training fellowship and Ellen Solomon and Kathy Howe by EEC grant : BIOMED-CT92-0755. Stephen Langabeer and DNA/RNA banking facilities at University College Hospital, London are supported by the Kay Kendall Leukaemia Fund and molecular studies for the MRC ATRA Trial by the ICRF. We would like to thank David Linch and Katherine Borden for critical reading of the manuscript. We are also grateful to Iain Goldsmith and the oligonucleotide synthesis service at ICRF, Clare Hall; and the photography department at ICRF, Lincoln's Inn Fields.

References

1. Grimwade, D. and Solomon, E. (1997) Characterisation of the *PML/RARα* rearrangement: associated with t(15;17) acute promyelocytic leukaemia. *Curr. Top. Microbiol. Immunol.* **220,** 81–112.
2. Rowley, J. D., Golomb, H. M., and Dougherty, C. (1977) 15/17 translocation, a consistent chromosomal change in acute promyelocytic leukaemia. *Lancet* **1,** 549–550.
3. Larson, R. A., Kondo, K., Vardiman, J. W., Butler, A. E., Golomb, H. M., and Rowley J. D. (1984) Evidence for a 15;17 translocation in every patient with acute promyelocytic leukemia. *Amer. J. Med.* **76,** 827–841.

4. Miller, W. H., Kakizuka, A., Frankel, S. R., Warrell, R. P., DeBlasio, A., Levine, K., Evans, R., and Dmitrovsky, E. (1992) Reverse transcription polymerase chain reaction for the rearranged retinoic acid receptor α clarifies diagnosis and detects minimal residual disease in acute promyelocytic leukemia. *Proc. Nat. Acad. Sci. USA* **89,** 2694–2698.

5. Grimwade, D., Howe, K., Langabeer, S., Davies, L., Oliver, F., Walker, H., Swirsky, D., Wheatley, K., Goldstone, A., Burnett, A., and Solomon, E. (1996) Establishing the presence of the t(15;17) in suspected acute pronyelocytic leukaemia: cytogenetic, molecular and PML immunofluorescence assessment of patients entered into the MRC ATRA Trial. *Brit. J. Haematol.* **94,** 557–573.

6. Breitman, T. R., Collins, S. J., and Keene, B. R. (1981) Terminal differentiation of human promyelocytic leukaemic cells in primary culture in response to retinoic acid. *Blood* **57,** 1000–1004.

7. Huang, M-E., Ye, Y-C., Chen, S-R., Chai, J-R., Lu, J-X., Zhoa, L., Gu, L-J., and Wang, Z-Y. (1988) Use of all-trans retinoic acid in the treatment of acute promyelocytic leukemia. *Blood* **72,** 567–572.

8. Castaigne, S., Chomienne, C., Daniel, M. T., Ballerini, P., Berger, R., Fenaux, P., and Degos, L. (1990) All-trans retinoic acid as a differentiation therapy for acute promyelocytic leukemia. I. Clinical results. *Blood* **76,** 1704–1709.

9. Chomienne, C., Ballerini, P., Balitrand, N., Daniel, M. T., Fenaux, P., Castaigne, S., and Degos, L. (1990) All-trans retinoic acid in acute promyelocytic leukemias. II. In vitro studies: structure-function relationship. *Blood* **76,** 1710–1717.

10. Grignani, F., Ferrucci, P. F., Testa, U., Talamo, G., Fagioli, M., Alcalay, M., Mencarelli, A., Peschle, C., Nicoletti, I., and Pelicci, P. G. (1993) The acute promyelocytic leukemia-specific PML-RARα fusion protein inhibits differentiation and promotes survival of myeloid precursor cells. *Cell* **74,** 423–431.

11. Elliott, S., Taylor, K., White, S., Rodwell, R., Marlton, P., Meagher, D., Wiley, J., Taylor, D., Wright, S., and Timms, P. (1992) Proof of differentiative mode of action of all-*trans* retinoic acid in acute promyelocytic leukemia using X-linked clonal analysis. *Blood* **79,** 1916–1919.

12. Sakashita, A., Kizaki, M., Pakkala, S., Schiller, G., Tsuruoka, N., Tomosaki, R., Cameron, J. F., Dawson, M. I., and Koeffler, H. P. (1993) 9-*cis*-retinoic acid: effects on normal and leukemic hematopoiesis in vitro. *Blood* **81,** 1009–1016.

13. Borrow, J., Goddard, A. D., Sheer, D., and Solomon, E. (1990) Molecular analysis of acute promyelocytic leukemia breakpoint cluster region on chromosome 17. *Science* **249,** 1577–1580.

14. de Thé, H., Chomienne, C., Lanotte, M., Degos, L., and Dejean, A. (1990) The t(15;17) translocation of acute promyelocytic leukaemia fuses the retinoic acid receptor α gene to a novel transcribed locus. *Nature* **347,** 558–561.

15. Longo, L., Pandolfi, P. P., Biondi, A., Rambaldi, A., Mencarelli, A., Lo Coco, F., Diverio, D., Pegoraro, L., Avanzi, G., Tabilio, A., Zangrilli, D., Alcalay, M., Donti, E., Grignani, F., and Pelicci, P. G. (1990) Rearrangements and aberrant

expression of the retinoic acid α gene in acute promyelocytic leukemias. *J. Exp. Med.* **172,** 1571–1575.

16. Alcalay, M., Zangrilli, D., Pandolfi, P. P., Longo, L., Mencarelli, A., Giacomucci, A., Rocchi, M., Biondi, A., Rambaldi, A., Lo Coco, F., Diverio, D., Donti, E., Grignani, F., and Pelicci, P. G. (1991) Translocation breakpoint of acute promyelocytic leukemia lies within the retinoic acid receptor α locus. *Proc. Nat. Acad. Sci. USA* **88,** 1977-1981.

17. Kakizuka, A., Miller, W. H., Umesono, K., Warrell, R. P., Frankel, S. R., Murty, V. V. V. S., Dmitovsky, E., and Evans, R. M. (1991) Chromosomal translocation t(15;17) in human acute promyelocytic leukemia fuses RARα with a novel putative transcription factor, PML. *Cell* **66,** 663–674.

18. Alcalay, M., Zangrilli, D., Fagioli, M., Pandolfi, P. P., Mencarelli, A., Lo Coco, F., Biondi, A., Grignani, F., and Pelicci, P. G. (1992) Expression pattern of the RARα-PML fusion gene in acute promyelocytic leukemia. *Proc. Nat. Acad. Sci. USA* **89,** 4840–4844.

19. Borrow, J., Goddard, A. D., Gibbons, B., Katz, F., Swirsky, D., Fioretos, T., Dube, I., Winfield, D. A., Kingston, J., Hagemeijer, A., Rees, J. K. H., Lister, T. A., and Solomon, E. (1992) Diagnosis of acute promyelocytic leukaemia by RT-PCR: detection of PML-RARA and RARA-PML fusion transcripts. *Brit. J. Haematol.* **82,** 529–540.

20. de Thé, H., Lavau, C., Marchio, A., Chomienne, C., Degos, L., and Dejean, A. (1991) The PML-RARα fusion mRNA generated by the t(15;17) translocation in acute promyelocytic leukemia encodes a functionally altered RAR. *Cell* **66,** 675–684.

21. Goddard, A. D., Borrow, J., Freemont, P. S., and Solomon, E. (1991) Characterization of a zinc finger gene disrupted by the t(15;17) in acute promyelocytic leukemia. *Science* **254,** 1371–1374.

22. Kastner, P., Perez, A., Lutz, Y., Rochette-Egly, C., Gaub, M-P., Durand, B., Lanotte, M., Berger, R., and Chambon, P. (1992) Structure, localisation and transcriptional properties of two classes of retinoic acid receptor α fusion proteins in acute promyelocytic leukemia (APL): structural similarities with a new family of oncoproteins. *EMBO J.* **11,** 629–642.

23. Pandolfi, P. P., Grignani, F., Alcalay, M., Mencarelli, A., Biondi, A., Lo Coco, F., Grignani, F., and Pelicci, P. G. (1991) Structure and origin of the acute promyelocytic leukemia myl/RARα cDNA and characterization of its retinoid-binding and transactivation properties. *Oncogene* **6,** 1285–1292.

24. Pandolfi, P. P., Alcalay, M., Fagioli, M., Zangrilli, D., Mencarelli, A., Diverio, D., Biondi, A., Lo Coco, F., Rambaldi, A., Grignani, F., Rochette-Egly, C., Gaube, M,-P., Chambon, P., and Pelicci, P. G. (1992) Genomic variability and alternative splicing generate multiple PML/RARα transcripts that encode aberrant PML proteins and PML/RARα isoforms in acute promyelocytic leukaemia. *EMBO J.* **11,** 1397–1407.

25. Gallagher, R. E., Li, Y-P., Rao, S., Paietta, E., Andersen, J., Etkind, P., Bennett, J. M., Tallman, M. S., and Wiernik, P. H. (1995) Characterization of acute promyelocytic leukemia cases with PML-RARα break/fusion sites in PML exon

6: Identification of a subgroup with decreased in vitro responsiveness to all-*trans* retinoic acid. *Blood* **86,** 1540–1547.

26. Alcalay, M., Balitrand, N., Barragan, E., Biondi, A., Cambier, N., Chomienne, C., Degos, L., Diverio, D., Fabry, U., Fenaux, P., Grimwade, D., Guidez, F., Haskovec, C., Lo Coco, F., Loukopoulos, D., Luciano, A., McIntyre, E., Miller, W., Momparler, R., Naoe, T., Saglio, G., Seale, J., Stamatopoulos, K., Valent, A., and Zorreguieta, A. (1996) RT-PCR in acute promyelocytic leukemia: second workshop of the European Retinoic Group. *Leukemia* **10,** 368–371.

27. Biondi, A., Rambaldi, A., Pandolfi, P. P., Rossi, V., Giudici, G., Alcalay, M., Lo Coco, F., Diverio, D., Pogliani, E. M., Lanzi, E. M., Mandelli, F., Masera, G., Barbui, T., and Pelicci, P. G. (1992) Molecular monitoring of the myl/retinoic acid receptor-α fusion gene in acute promyelocytic leukemia by polymerase chain reaction. *Blood* **80,** 492–497.

28. Dyck, J. A., Maul, G. G., Miller, W. H., Chen, J. D., Kakizuka, A., and Evans, R. M. (1994) A novel macromolecular structure is a target of the promyelocyte-retinoic acid receptor oncoprotein. *Cell* **76,** 333–343.

29. Weis, K., Rambaud, S., Lavau, C., Jansen, J., Carvalho, T., Carmo-Fonseca, M., Lamond, A., and Dejean, A. (1994) Retinoic acid regulates aberrant nuclear localization of PML-RARα in acute promyelocytic leukaemia cells. *Cell* **76,** 345–356.

30. Koken, M. H. M., Puvion-Dutilleul, F., Guillemin, M. C., Viron, A., Linares-Cruz, G., Stuurman, N., de Jong, L., Szostecki, C., Calvo, F., Chomienne, C., Degos, L., Puvion, E., and de Thé, H. (1994) The t(15;17) translocation alters a nuclear body in a retinoic acid-reversible fashion. *EMBO J.* **13,** 1073–1083.

31. Borden, K. L. B., Boddy, M. N., Lally, J., O'Reilly, N. J., Martin, S., Howe, K., Solomon, E., and Freemont, P. S. (1995) The solution structure of the RING finger domain from the acute promyelocytic leukaemia proto-oncoprotein PML. *EMBO J.* **14,** 1532–1541.

32. Pandolfi, P. P., Rivi, R., Gaboli, M., Giorgio, M., Antoniou, M., Bygrave, A., Fagioli, M., Cordon-Cardo, C., and Pelicci, P. G. (1995) Targeted disruption of the PML gene of acute promyelocytic leukemia. *Blood* **86, Suppl 1,** 261a.

33. Everett, R. D. and Maul, G. G. (1994) HSV-1 IE protein Vmw110 causes redistribution of PML. *EMBO J.* **13,** 5062–5069.

34. Flenghi, L., Fagioli, M., Tomassoni, L., Pileri, S., Gambacorta, M., Pacini, R., Grignani, F., Casini, T., Ferrucci, P. F., Martelli, M. F., Pelicci, P-G., and Falini, B. (1995) Characterization of a new monoclonal antibody (PG-M3) directed against the aminoterminal portion of the PML gene product : Immunocytochemical evidence for high expression of PML proteins on activated macrophages, endothelial cells, and epithelia. *Blood* **85,** 1871–1880.

35. Terris, B., Baldin, V., Dubois, S., Degott, C., Flejou, J-F., Henin, D., and Dejean, A. (1995) PML nuclear bodies are general targets for inflammation and cell proliferation. *Cancer Res.* **55,** 1590–1597.

36. Lavau, C., Marchio, A., Fagioli, M., Jansen, J., Falini, B., Lebon, P., Grosveld, F., Pandolfi, P. P., Pelicci, P. G., and Dejean, A. (1995) The acute promyelocytic leukaemia-associated PML gene is induced by interferon. *Oncogene* **11,** 871–876.

37. Chelbi-Alix, M. K., Pelicano, L., Quignon, F., Koken, M. H. M., Venturini, L., Stadler, M., Pavlovic, J., Degos, L. and de Thé, H. (1995) Induction of the PML protein by interferons in normal and APL cells. *Leukemia* **9,** 2027–2033.

38. Stadler, M., Chelbi-Alix, M. K., Koken, M. H. M., Venturini, L., Lee, C., Säib, A., Quignon, F., Pelicano, L., Guillemin, M-C., Schindler, C., and de Thé, H. (1995) Transcriptional induction of the PML growth suppressor gene by interferons is mediated through an ISRE and a GAS element. *Oncogene* **11,** 2565–2573.

39. Chang, K-S., Fan, Y-H., Andreeff, M., Liu, J., and Mu, Z-M. (1995) The *PML* gene encodes a phosphoprotein associated with the nuclear matrix. *Blood* **85,** 3646–3653.

40. Mu, Z-M., Chin, K-V., Liu, J-H., Lozano, G., and Chang, K-S. (1994) PML, a growth suppressor disrupted in acute promyelocytic leukemia. *Mol. Cell. Biol.* **14,** 6858–6867.

41. Koken, M. H. M., Linares-Cruz, G., Quignon, F., Viron, A., Chelbi-Alix, M. K., Sobczak-Thépot, J., Juhlin, L., Degos, L., Calvo, F., and de Thé, H. (1995) The PML growth-suppressor has an altered expression in human oncogenesis. *Oncogene* **10,** 1315–1324.

42. Liu, J-H., Mu, Z-M., and Chang, K-S. (1995) PML suppresses oncogenic transformation of NIH/3T3 cells by activated *neu. J. Exp. Med.* **181,** 1965–1973.

43. Freemont, P. S. (1993) The RING finger. A novel protein sequence motif related to the zinc finger. *Ann. NY Acad. Sci.* **684,** 174–192.

44. Borden, K. L. B., Lally, J. M., Martin, S. R., O'Reilly, N. J., Solomon, E., and Freemont, P. S. (1996) *In vivo* and *in vitro* characterization of the B1 and B2 zinc-binding domains from the acute promyelocytic leukemia protooncoprotein PML. *Proc. Nat. Acad. Sci.* **93,** 1601–1606.

45. Perez, A., Kastner, P., Sethi, S., Lutz, Y., Reibel, C., and Chambon, P. (1993) PML/RAR homodimers: distinct DNA binding properties and heteromeric interaction with RXR. *EMBO J.* **12,** 3171–3182.

46. Beato, M. (1989) Gene regulation by steroid hormones. *Cell* **56,** 335–344.

47. Stunnenberg, H. G. (1993) Mechanisms of transactivation by retinoic acid receptors. *Bioessays* **15,** 309–315.

48. Damm, K., Heyman, R. A., Umesono, K., and Evans, R. M. (1993) Functional inhibition of retinoic acid response by dominant negative retinoic acid receptor mutants. *Proc. Nat. Acad. Sci. USA* **90,** 2989–2993

49. Kastner, P., Grondona, J. M., Mark, M., Gansmuller, A., LeMeur, M., Decimo, D., Vonesch, J-L., Dolle, P., and Chambon, P. (1994) Genetic analysis of RXRα developmental function: convergence of RXR and RAR signalling pathways in heart and eye morphogenesis. *Cell* **78,** 987–1003.

50. Andersen, B. and Rosenfeld, M. G. (1995) New wrinkles in retinoids. *Nature* **374,** 118–119.

51. Saitou, M., Sugai, S., Tanaka, T., Shimouchi, K., Fuchs, E., Narumiya, S., and Kakizuka, A. (1995) Inhibition of skin development by targeted expression of a dominant-negative retinoic acid receptor. *Nature* **374,** 159–162.

52. Sucov, H. M., Dyson, E., Gumeringer, C. L., Price, J., Chien, K. R., and Evans, R. M. (1994) RXRα mutant mice establish a genetic basis for vitamin A signalling in heart morphogenesis. *Genes Dev.* **8,** 1007–1018.

53. Tsai, S. and Collins S. J. (1993) A dominant negative retinoic acid receptor blocks neutrophil differentiation at the promyelocyte stage. *Proc. Nat. Acad. Sci. USA.* **90,** 7153–7157.

54. Kliewer, S. A., Umesono, K., Mangelsdorf, D. J., and Evans, R. M. (1992) Retinoid X receptor interacts with nuclear receptors in retinoic acid, thyroid hormone and vitamin D_3 signalling. *Nature* **355,** 446–449.

55. Zhang, X-K., Hoffmann, B., Tran, P B-V., Graupner, G., and Pfahl, M. (1992) Retinoid X receptor is an auxiliary protein for thyroid hormone and retinoic acid receptors. *Nature* **355,** 441–446.

56. Kurokawa, R., DiRenzo, J., Boehm, M., Sugarman, J., Gloss, B., Rosenfeld, M. G., Heyman, R. A., and Glass, C. K. (1994) Regulation of retinoid signalling by receptor polarity and allosteric control of ligand binding. *Nature* **371,** 528–531.

57. Grisolano, J. L., Wesselschmidt, R. L., Pelicci, P. G., and Ley, T. J. (1997) Altered myeloid development and acute leukemia in transgenic mice expressing PML-RARα under control of cathepsia G regulatory sequences. *Blood* **89,** 376–387.

58. Borrow, J., Shipley, J., Howe, K., Kiely, F., Goddard, A., Sheer, D., Srivastava, A., Antony, A. C., Fioretos, T., Mitelman, F., and Solomon, E. (1994) Molecular analysis of simple variant translocations in acute promyelocytic leukaemia. *Genes, Chromosomes & Cancer* **9,** 234–243.

59. Chen, Z., Morgan, R., Stone, J. F., and Sandberg, A. A. (1994) Identification of complex t(15;17) in APL by FISH. *Cancer Genet. Cytogen.* **72,** 73,74.

60. Hiorns, L. R., Min, T., Swansbury, G. J., Zelent, A., Dyer, M. J. S., and Catovsky, D. (1994) Interstitial insertion of retinoic acid receptor-α gene in acute promyelocytic leukaemia with normal chromosomes 15 and 17. *Blood* **83,** 2946–2951.

61. McKinney, C. D., Golden, W. L., Gemma, N. W., Swerdlow, S. H., and Williams, M. E. (1994) RARA and PML gene rearrangements in acute promyelocytic leukaemia with complex translocations and atypical features. *Genes, Chromosomes & Cancer* **9,** 49–56.

62. Licht, J. D., Chomienne, C., Goy, A., Chen, A., Scott, A. A., Head, D. R., Michaux, J. L., Wu, Y., DeBlasio, A., Miller, W. H., Zelenetz, A. D., Willman, C. L., Chen, Z., Chen, S-J., Zelent, A., Macintyre, E., Veil, A., Cortes, J., Kantarjian, H., and Waxman, S. (1995) Clinical and molecular characterization of a rare syndrome of acute promyelocytic leukemia associated with translocation (11;17). *Blood* **85,** 1083–1094.

63. Fenaux, P., Chastang, C., Castaigne, S., Archimbaud, E., Sanz, M., Link, H., Guerci, A., Fegueux, N., Zittoun, R., Stoppa, A. M., Travade, P., Lamy, T., Maloisel, F., Sadoun, A., San Miguel, J., Veil, A., Rayon, C., Conde, E., Fey, M., Bordessoule, D., Ganser, A., Bowen, D., Dreyfus, F., Huguet, F., Tilly, H., Guy, H., Auzanneau, G., Chomienne, C., and Degos, L. (1994) Treatment of newly

diagnosed acute promyelocytic leukemia (APL) with all-transretinoic acid (ATRA) followed by intensive chemotherapy (CT). Updated results of the European group. *Blood* **84, Suppl 1,** 379a.

64. Tallman, M. S., Andersen, J., Schiffer, C. A., Appelbaum, F. R., Feusner, J. E., Woods, W. G., Ogden, A., Weinstein, H., Shepherd, L., Rowe, J. M., and Wiernik, P. H. (1995) Phase III randomized study of all-trans retinoic acid (ATRA) vs daunorubicin (D) and cytosine arabinoside (A) as induction therapy and ATRA vs observation as maintenance therapy for patients with previously untreated acute promyelocytic leukemia (APL). *Blood* **86, Suppl 1,** 125a.

65. Frankel, S. R., Eardley, A., Lauwers, G., Weiss, M., and Warrell, R. P. (1992) The "retinoic acid syndrome" in acute promyelocytic leukemia. *Ann. Intern. Medicine* **117,** 292–296.

66. Vahdat, L., Maslak, P., Miller, W. H., Eardley, A., Heller, G., Scheinberg, D. A., and Warrell, R. P. (1994) Early mortality and the retinoic acid syndrome in acute promyelocytic leukemia: Impact of leukocytosis, low-dose chemotherapy, PML-RAR-α isoform, and CD13 expression in patients treated with All-*Trans* retinoic acid. *Blood* **84,** 3843–3849.

67. Burnett A. K. (1994) Karyotypically defined risk groups in acute myeloid leukaemia. *Leukaemia Res.* **18,** 889–890.

68. Huang, W., Sun, G-L., Li, X-S., Cao, Q., Lu, Y., Jang, G-S., Zhang, F-Q., Chai, J-R., Wang, Z-Y., Waxman, S., Chen, Z., and Chen, S-J. (1993) Acute promyelocytic leukemia: Clinical relevance of two major PML/RARα isoforms and detection of minimal residual disease by retro-transcriptase polymerase chain reaction to detect relapse. *Blood* **82,** 1264–1269.

69. Fukutani, H., Naoe, T., Ohno, R., Yoshida, H., Miyawaki, S., Shimazaki, C., Miyake, T., Nakayama, Y., Kobayashi, H., Goto, S., Takeshita, A., Kobayashi, S., Kato, Y., Shiraishi, K., Sasada, M., Ohtake, S., Murakami, H., Kobayashi, M., Endo, N., Shindo, H., Matsushita, K., Hasegawa, S., Tsuji, K., Ueda, Y., Tominaga, N., Furuya, H., Inoue, Y., Takeuchi, J., Morishita, H., and Iida, H. (1995) Isoforms of PML-*retinoic acid receptor alpha* fused transcripts affect neither clinical features of acute promyelocytic leukemia nor prognosis after treatment with all-trans retinoic acid. *Leukemia* **9,** 1478–1482.

70. Lo Coco, F., Diverio, D., Pandolfi, P. P., Biondi, A., Rossi, V., Avvisati, G., Rambaldi, A., Arcese, W., Petti, M. C., Meloni, G., Mandelli, F., Grignani, F., Masera, G., Barbui, T., and Pelicci, P. G. (1992) Molecular evaluation of residual disease as a predictor of relapse in acute promyelocytic leukemia. *Lancet* **340,** 1437–1438.

71. Diverio, D., Pandolfi, P. P., Biondi, A., Avvisati, G., Petti, M. C., Mandelli, F., Pelicci, P. G., and Lo Coco, F. (1993) Absence of reverse transcription-polymerase chain reaction detectable residual disease in patients with acute promyelocytic leukemia in long-term remission. *Blood* **82,** 3556–3559.

72. Miller, W. H., Levine, K., DeBlasio, A., Frankel, S. R., Dmitrovsky, E., and Warrell, R. P. (1993) Detection of minimal residual disease in acute promyelocytic

leukemia by a reverse transcription polymerase chain reaction assay for the PML/RAR-α fusion mRNA. *Blood* **82,** 1689–1694.

73. Lo Coco, F., Diverio, D., Avvisati, G., Luciano, A., Biondi, A., and Mandelli, F. (1995) High frequency of early molecular remission in acute promyelocytic leukemia by combined all-trans retinoic acid and idarubicin (Italian GIMEMA-AIEOP "AIDA" trial). *Blood* **86, Suppl 1,** 265a.

74. Diverio, D. (1994) Monitoring of treatment outcome in acute promyelocytic leukemia by RT-PCR. *Leukemia* **8,** 1105–1107.

75. Grimwade, D., Howe, K., Langabeer, S., Burnett, A., Goldstone, A., and Solomon, E. (1996) Minimal residual disease detection in acute promyelocytic leukemia by reverse-transcriptase PCR: evaluation of *PML-RARα* and *RARα-PML* assessment in patients who ultimately relapse. *Leukemia* **10,** 61–66.

76. Lanotte, M., Martin, V., Najman, S., Ballerini, P., Valensi, S., and Berger, R. (1991) NB4, a maturation inducible cell-line with t(15;17) marker isolated from a human promyelocytic leukemia (M3). *Blood* **77,** 1080–1086.

77. Fagioli, M., Alcalay, M., Pandolfi, P. P., Venturini, L., Mencarelli, A., Simeone, A., Acampora, D., Grignani, F., and Pelicci, P. G. (1992) Alternative splicing of PML transcripts predicts coexpression of several carboxy-terminally different protein isoforms. *Oncogene* **7,** 1083–1091.

78. Tobal, K., Saunders, M. J., Grey, M. R., and Yin, J. A. L. (1995) Persistence of RARα-PML fusion mRNA detected by reverse-transcriptase polymerase chain reaction in patients in long term remission of acute promyelocytic leukemia. *Brit. J. Haematol.* **90,** 615–618.

79. Dyck, J., Warrell, R. P., Evans, R. M., and Miller, W. H. (1995) Rapid diagnosis of acute promyelocytic leukemia by immunohistochemical localization of PML/RAR-α protein. *Blood* **86,** 862–867.

80. Chomczynski, P. and Sacchi, N. (1987) Single-step method of RNA isolation by acid guanidinium thiocyanate-phenol-chloroform extraction. *Anal. Biochem* **162,** 156–159.

81. Kwok, S. and Higuchi, R. (1989) Avoiding false positives with PCR. *Nature* **339,** 237–238.

82. Avvisati, G., Lo Coco, F., Diverio, D., Falda, M., Ferrara, F., Lazzarino, M., Russo, D., Petti, M. C., and Mandelli, F. (1996) AIDA (all-*trans* retinoic acid + idarubicin) in newly-diagnosed acute pronyelocytic leukemia–a Gruppo Italiano Malaltie Ematologiche Maligne dell Adulto (GIMEMA) pilot study. *Blood* **88,** 1390–1398.

83. Seale, J. R. C., Varma, S., Swirsky, D. M., Pandolji, P. P., Goldman, J. M., and Cross, N. C. P. (1996) Quantification of PML-RARα transcripts in acute promyelocytic leukaemia–explanation for the lack of sensitivity of RT-PCR for the detection of minimal residual disease and induction of the leukaemic specific messenger RNA by alpha-interferon *Brit. J. Haematol.* **95,** 95–101.

25

A Two-Hybrid Protein Interaction System to Identify Factors That Interact with Retinoid and Vitamin D Receptors

Paul N. MacDonald

1. Introduction

Steroid hormone receptors activate and repress gene transcription through an integrated series of protein–protein interactions. One principle contact is the interaction of the nuclear receptors with one another to form both homodimeric and heterodimeric complexes that bind with high affinity to their cognate DNA-responsive elements. Following receptor–response element binding, the DNA-bound homodimers and heterodimers must ultimately communicate with the preinitiation complex to influence RNA polymerase II-mediated transcription. This communication process presumably involves additional protein–protein contacts. Identifying these contacts, the proteins that are involved, and the precise nature of these interactions are central to our understanding of this fundamental mechanism.

The two-hybrid system is a yeast-based genetic screen that is used to identify and characterize protein–protein interactions (1). In this system, two fusion proteins are expressed in yeast and their interaction is monitored in vivo using appropriate reporter-gene systems. One hybrid protein, termed the "bait," contains the DNA-binding domain (DBD) of a transactivator fused to protein X. The second fusion protein, termed the prey or "target," contains an activation domain (ACT) of the transactivator fused to protein Y. If DBD-X and ACT-Y are co-expressed in yeast, and if X and Y interact with one another, then the transactivator is reconstituted in a functionally relevant manner and expression of the reporter gene is enhanced. The power of the system resides in the ability to screen cDNA libraries constructed in the ACT vector to conduct global searches for novel proteins that interact with a protein of interest.

From: *Methods in Molecular Biology, Vol. 89: Retinoid Protocols*
Edited by: C. P. F. Redfern © Humana Press Inc., Totowa, NJ

Recently, this system was used to identify several factors putatively involved in the mechanism of steroid hormone-mediated gene transcription. In a two-hybrid screen to identify factors that interact with the vitamin D receptor (VDR), we isolated transcription factor IIB (TFIIB) as a VDR-interactive clone *(2)*. The interaction of VDR, retinoic-acid receptors and other steroid-hormone receptors with TFIIB may represent a fundamental step in the mechanism of transcription mediated by the nuclear-receptor family *(3–5)*. The two-hybrid system has also identified several putative coactivator and corepressor proteins that contact retinoid receptors, thyroid receptors, vitamin D receptors, and other members of the nuclear-receptor family *(6–9)*. Thus, the two-hybrid system is playing an instrumental role in the identification of factors involved in nuclear receptor-mediated gene expression. This chapter discusses several procedures and strategies used to establish a two-hybrid system to examine proteins that interact with retinoid receptors, with the VDR, or with nuclear receptors in general.

2. Materials

2.1. Plasmids and Yeast Strains

1. The DNA-binding domain vector (the "bait"): These vectors consist most commonly of the DBD of either Gal4 ($Gal4_{DBD}$) or LexA ($LexA_{DBD}$) followed by a multiple-cloning site in which a cDNA of interest is introduced. Often, cDNAs are subcloned into these constructs using convenient restriction-enzyme sites. Alternatively, restriction-enzyme sites can be introduced into a nuclear-receptor cDNA using appropriately designed primers in a PCR reaction or via site-directed mutagenesis. The fusion constructs must be engineered so as to preserve the reading frame of the insert cDNA. Most two-hybrid vectors also contain nuclear-localization sequences to target the expressed-fusion proteins to the nucleus. Our laboratory uses the $Gal4_{DBD}$-containing vectors pMA424 *(10)* and pAS1 *(11)*, which contain *HIS3* and *TRP1* selection markers, respectively. The initial bait constructs that were used to identify proteins that interact with the VDR were constructed by subcloning VDR cDNA corresponding to amino acids 93-427 into the *Eco*R1 sites pMA424 and pAS1 (*see* **ref. 2** and **Fig. 1**).
2. The activation domain vector (the prey or target): These eukaryotic expression vectors generally contain the activation domains of Gal4 ($Gal4_{ACT}$) or *Herpes virus* VP16 ($VP16_{ACT}$). Fusion proteins containing these activation domains are generated by subcloning appropriate cDNAs in frame into the multiple-cloning banks of these vectors. Our laboratory uses pGAD.GH *(12)*, which contains the $Gal4_{ACT}$ from amino acid 786 to 881, a nuclear-localization sequence, and a *LEU2* selectable marker. The initial activation domain prey that was engineered to test the AS1-VDR bait construct in the two-hybrid system was pGAD-RXR (**Fig. 1**). pGAD-RXR contains the cDNA corresponding to amino acids 223-462 of human RXRα subcloned 3' to the Gal4 activation domain in pGAD.GH.

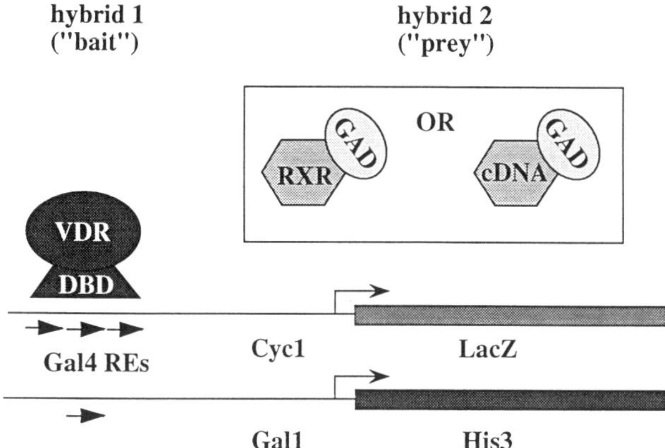

Fig. 1. Schematic illustration of a two-hybrid system used to identify and characterize proteins that interact specifically with the vitamin D receptor.

Because VDR and RXR are authentic heterodimeric partners that interact
through their C-terminal domains *(13–15)*, the AS1-VDR and GAD-RXR constructs proved useful for the initial development of two-hybrid system in our
laboratory *(2)*. These two fusion proteins exhibited strong interaction in this system and this indicated that the AS1-VDR bait was an effective tool to use in a
library screen for other VDR-interactive proteins. Similarly, retinoid receptors
are known to interact with each other and with a variety of other nuclear receptors. Thus, it is straightforward to test the integrity of each retinoid-receptor bait
construct using the cDNAs of several known interacting partners engineered in
the ACT vectors. For example, in specificity studies *(2)*, an AS1-RXR fusion
interacted with GAD-VDR and with GAD-RXR fusions, thus showing both RXR
heterodimerization with VDR and homodimerization of RXR with itself. Moreover, an AS1-RAR fusion interacted with RXR-GAD, but not with VDR-GAD.
This specificity is similar to that observed in vitro and it indicates that both the
AS1-RXR and AS1-RAR constructs are useful baits for a two-hybrid screen.

3. The activation-domain cDNA library: The fusion library consists of cDNA sequences
   derived from a tissue or cell line that are fused to the sequence encoding the activation domain in the prey plasmid. This library is prepared by the individual researcher
   using an appropriate tissue or cell line as a source of mRNA. Alternatively, a variety
   of two-hybrid cDNA-activation domain libraries are now commercially available.
   These libraries are generally random-primed and/or oligo dT primed and they consist
   of $>1 \times 10^6$ inserts. It is important to note that, even in directional libraries, only
   one-third of the inserts are in the appropriate reading frame.
4. Unrelated DBD-fusion constructs: False positives arise frequently in the two-
   hybrid system and several recent reviews have discussed a number of these arti-

factual phenomena in depth *(1,16)*. A common false positive in a cDNA-library screen is that target plasmid which activates reporter gene expression independent of the nature of the bait construct. Thus, an important control in the initial characterization of isolated clones is to examine the specificity of the interaction. For this purpose, a battery of unrelated gal4$_{DBD}$-fusion proteins, including AS1-laminin, AS1-SNF, and AS1-p53, is used to test whether the interaction is selective for a particular nuclear-receptor bait in the two-hybrid assay (*see* **Note 1**).

5. Yeast strains: The most commonly used strains of yeast for two-hybrid screens contain a dual reporter-gene system with the *lacZ* and *HIS3* genes controlled by two distinct GAL4-responsive promoters. These reporter-gene systems are integrated into the yeast genome. The Hf7c strain of yeast is used most often in our laboratory and it has the following genetic composition: *ura3-52 his3-200 ade2-101 lys2-801 trp1-901 leu2-3,112 gal80-538 gal4-542 LYS2::GAL1$_{UAS}$-GAL1$_{TATA}$-HIS3 URA3::GAL4$_{17mers(3X)}$-CYC1$_{TATA}$-lacZ*. The *leu2* and *trp1* auxotrophic markers select for yeast that have been transformed with the GAD.GH activation-domain plasmid and the AS1-DBD plasmid, respectively. Importantly, the *his3* marker is driven by a single *GAL1* upstream-activator sequence (UAS) fused to the basal TATA promoter of the *GAL1* gene. Thus, interaction between a particular GAL4$_{DBD}$ bait and GAL4$_{ACT}$ prey results in Gal4-dependent transcription of the *HIS3* reporter and this interaction is monitored by plating the transformed yeast on histidine-deficient media. A second reporter, the *lacZ* gene, is controlled by a simple basal TATA-containing element from the CYC1 promoter and 3 tandem copies of the GAL4 17-mer UAS. Thus, interaction is also monitored with a β-galactosidase assay. This dual-selection scheme is powerful. In theory, only those clones that interact with a particular bait construct will grow out of the *HIS3* selection, and one can readily eliminate histidine revertants with the *lacZ* screen. Moreover, because the two reporters are driven by distinct GAL4-responsive promoters, one class of false positives (i.e., those GAL4$_{ACT}$ clones that activate on their own because they bind to other DNA sequences in one of the promoters) is dramatically reduced.

2.2. Growth and Maintenance of Saccharomyces cerevisiae

High-quality reagents are essential for the preparation of media to support the growth and maintenance of yeast strains. Bacto-brand agar (0140-01), peptone (0118-01-8), yeast extract (0127-17-9), dextrose (0155-17-4), and yeast-nitrogen base without amino acids (0919-15-3) are obtained from Difco (Detroit, MI). Amino acid supplement mixtures (complete and dropout mixtures) are obtained from Bio-101 (Vista, CA). Adenine hemisulfate salt (A-3159) and 3-amino-1,2,4-triazole (A-8056) are obtained from Sigma (St. Louis, MO). All media are prepared with reagent grade, distilled and deionized water, and are autoclaved at 15 lb/in^2 at 140°C (15 min for 1 L media).

1. YPAD medium: This is a general purpose, nutrient-rich medium for the routine propagation of yeast strains when specific selection conditions are not required. YPAD is prepared by combining 10 g yeast extract, 20 g peptone, 20 g dextrose, and 40 mg adenine sulfate in 1 L of distilled water, and autoclaving for 15 min at 15 lb/in^2. Adenine is included in the culturing medium for certain strains to inhibit reversion of *ade1* and *ade2* mutations.

2. Complete minimal (CM) medium: CM is a defined, minimal medium consisting of 6.7 g of yeast nitrogen base without amino acids and 20 g of dextrose per liter. Amino acid supplement mixtures (either complete or dropout mixtures lacking one or more specific amino acids) are added and the solution is autoclaved as described. In this system, a dropout medium is used to select for yeast that have acquired a particular plasmid in a transformation experiment. For example, CM plates lacking leucine [CM(-leu)] are used to select for those yeast harboring the activation-domain plasmid (GAD.GH derivatives), which carries a *leu2*-selectable marker gene.

3. Solid media: When preparing solid media for yeast work, the same basic recipes described are used and 20 g of Bacto-agar per liter media are added. Generally, 500 mL of medium are autoclaved in a 1-L bottle with a magnetic stir bar. The medium is cooled to approx 50°C in a water bath. Additional reagents such as sterile 3-amino-1,2,4-triazole (**item 4**) are added after cooling. The molten-agar medium is stirred and then is dispensed into 100- or 150-mm culture plates using sterile technique. The plates are allowed to solidify and air-dry for 2–3 d at room temperature. They are stored covered at 4°C.

4. 3-amino-1,2,4-triazole: Most strains of yeast exhibit low-level expression of the *HIS3* marker in the absence of gal4-activated transcription. This "leaky" expression of the reporter gene leads to background growth under histidine-selection conditions, and this background can be effectively eliminated with 3-amino-1,2,4-triazole (3-AT). 3-AT is a chemical inhibitor of imidazole glycerol phosphate dehydratase, the product of the *HIS3* gene. A 2.5-*M* solution of 3-AT is prepared in distilled water and is sterilized by filtration. The sterile-stock solution is stored at –20°C. 3-AT is used at concentrations of 5–50 m*M*, depending on the strain of yeast and the particular vectors used (*see* **Note 2**).

2.3. Introduction of Plasmid DNA into Yeast

1. 10X Lithium Acetate (10X LiAc): 10X LiAc is 1 *M* LiAc, pH 7.5, adjusted with diluted acetic acid and it is sterilized by filtration through a 0.2-μm filter.

2. 10X TE: 10X TE is 0.1 *M* Tris-HCl, pH 7.5, 0.01 *M* EDTA, which is sterilized by autoclaving.

3. 1X LiAc/TE: This solution is prepared from sterile-stock solutions of 10X LiAc and 10X TE in sterile water.

4. 50% polyethylene glycol (PEG): 50 g of PEG 4000 (Sigma P-3640) is dissolved in distilled water and is diluted to a total volume of 100 mL. This solution is sterilized by autoclaving.

5. 40% PEG/LiAc/TE: This solution contains 1X TE, 1X LiAc, and 40% w/v PEG 4000. This is made fresh from the stock solutions of 10X TE, 10X LiAc, and 50% w/v PEG 4000 described previously.

6. Single-stranded carrier DNA: Salmon-testes DNA (Sigma D1626) is cut into small pieces with sterile scissors and is dissolved overnight at 4°C in TE at a concentration of 10 mg/mL. The preparation is sheared using 8–10 passes through an 18-gage needle followed by 3–4 passes through a 22-gage needle. A small aliquot (500 ng) is analyzed by agarose-gel electrophoresis to estimate the average size of the preparation, which should be between 4 and 8 kb. The sheared DNA is extracted once with TE-saturated phenol, once with phenol:chloroform:isoamyl alcohol (25:24:1), and once with chloroform. It is then precipitated twice with ethanol, and the final pellet is resuspended in TE at a concentration of 5 mg/mL by overnight incubation at 4°C. The final preparation is dispensed in 1-mL aliquots into microcentrifuge tubes. These are incubated at 100°C for 20 min, they are immersed in an ice water bath, and are then snap-frozen in a dry ice/ethanol bath for storage at –20°C.

2.4. Analysis of the Interaction

1. 10X Z-buffer: This stock buffer is 0.6 M Na_2HPO_4, 0.4 M NaH_2PO_4, 0.1 M KCl, 0.01 M $MgSO_4$, adjusted to a final pH of 7.0.
2. 1X Z-buffer is prepared fresh for each experiment using the 10X stock, sterile water, and 2-mercaptoethanol to a final concentration of 50 mM.
3. Whatman 50 filter papers (1450), 90 or 145 mm in diameter.
4. Fisherbrand P5 filter paper (09-801), 90 or 145 mm in diameter.
5. For the β-galactosidase filter assay, a stock solution of 5-bromo-4-chloro-3-indolyl-β-D-galactopyranoside (X-gal) is prepared fresh in dimethylformamide at a concentration of 40 mg/mL. This stock solution is added with constant stirring to the 1X Z-buffer to yield a final concentration of 0.33 mg/mL.
6. For liquid β-galactosidase assays, 2-nitrophenyl-β-D-galactopyranoside (ONPG) is dissolved in 0.1 M KH_2PO_4, pH 7.0, at a concentration of 4 mg/mL.
7. Lysis Buffer: This buffer is used to disrupt the yeast cells in the liquid β-galactosidase assay described in **Subheading 3.3.3.** It is 0.1 M Tris-HCl, pH 7.6, containing 0.05% Triton X-100.

2.5. Screening a cDNA-GAD Fusion Library

In addition to several reagents listed in **Subheadings 2.1.–2.4.**, the following items are required:

1. Acid-washed glass beads 425–600 μ in diameter (Sigma, G-8772).
2. STET Buffer: 8% v/v sucrose, 50 mM Tris-HCl, pH 8.0, 50 mM EDTA, 5% Triton X-100.
3. *Escherichia coli* made competent for plasmid transformation by chemical means. Efficiencies on the order of 1×10^8 colonies/μg of DNA are generally required.

Alternatively, electrocompetent bacteria and a suitable electroporation device are required.
4. A replica-plating apparatus and velvet squares are also necessary.

3. Methods

3.1. Growth and Maintenance of Saccharomyces cerevisiae *Strains*

General protocols for routine culturing and maintaining yeast strains are similar to those used for bacteria. Several basic methodologies involved in growing yeast have been described *(17,18)*. Yeast are grown in either liquid media or on the surface of solid-agar plates. Most strains are grown at 30°C and have a doubling time of approx 2 h during exponential growth in rich media. Although their doubling time increases, yeast also grow well in minimal media containing nitrogen, phosphorus, and trace metal salts, defined amino acids, and glucose as a carbon source. As mentioned in **Subheading 2.2.**, complete minimal (CM) dropout media, in which specific amino acids are omitted from this minimal media, are routinely used for the selection of transformants.

3.2. Introduction of Plasmid DNA into Yeast

3.2.1. Preparation of Competent Yeast

The main premise of the two-hybrid system is to express the bait and prey fusions from eukaryotic-expression plasmids in yeast and then monitor in vivo interactions between the two proteins. The first step in this process is to introduce the expression plasmids into the yeast. A straightforward and relatively efficient method is to use alkali cations to make the yeast competent to take up DNA. Lithium acetate (LiAc) and polyethylene glycol (PEG) are two compounds of choice. We routinely prepare competent yeast using protocols developed by Schiestl and Gietz *(19)* with additional modifications *(20)*. This procedure yields competent cells sufficient for 10–20 individual transformations and it is readily scaled up or down to fit the needs of a particular experiment.

1. Pick a single colony of a particular yeast strain from a YPAD plate and grow overnight at 30°C with vigorous shaking in 2 mL of YPAD in a sterile 16 × 150-mm glass-culture tube.
2. Check the OD_{600} of a 1:10 dilution of the overnight culture and expand this culture in 50 mL of YPAD in a 250-mL sterile Erlenmeyer flask at a starting density of 0.2 OD_{600} U. Grow at 30°C with vigorous shaking for 4–5 h to a final OD_{600} of approx 0.7–0.8 U.
3. Harvest the culture in a sterile 50-mL conical-bottom tube by centrifugation approx 1200g for 5 min in a swinging-bucket rotor.

4. Pour off the supernatant and resuspend the cell pellet in 5 mL of sterile water. Spin at 1500*g* for 5 min to harvest the cells.
5. Resuspend the pellet in 1 mL of sterile water and transfer the cells to a sterile 1.7-mL microcentrifuge tube.
6. Spin at maximum speed in a microcentrifuge for 5 s. Aspirate the supernatant and resuspend the pellet in 1.0 mL of LiAc/TE. Harvest cells with a 5-s spin in a microcentrifuge and resuspend the pellet in 200 μL of LiAc/TE. Finally, add enough LiAc/TE so that the final cell concentration is approx 2×10^9 cells/mL (*see* **Note 3**). Store on ice until needed.

3.2.2. Transformation Protocol

Plasmid DNAs are introduced into the yeast along with high-molecular-weight single-stranded carrier DNA. The single-stranded character and the size of the carrier DNA are vital for efficient transformations by this protocol and, thus, the preparation of the carrier DNA is very important. The reader is referred to Schiestl and Gietz *(19)* for a more detailed discussion of the carrier DNA.

1. Plasmid DNAs (500 ng of the bait and prey plasmids each) and carrier DNA (50 μg) are aliquotted into sterile microtubes and placed on ice (*see* **Note 4**).
2. Add 50 μL of the competent-yeast preparation to the DNAs in each of the transformation tubes.
3. Add 300 μL of 40% PEG/LiAc/TE to each tube and mix thoroughly by gently pipetting up and down with a P-1000. Vortex mix each tube for approx 5 s and incubate the yeast/DNA/PEG preparations at 30°C for 30 min.
4. Heat shock all tubes at 42°C for 15 min in a water bath.
5. Spin for 5 s at maximum speed in microcentrifuge to harvest cells.
6. Aspirate the supernatant with a sterile P-200 tip attached to a Pasteur pipet on the end of the vacuum-aspirator device. Place the cell pellets on ice and gently resuspend the pellets in 500 μL of sterile water. Keep the cells on ice until all the reactions have been processed (*see* **Note 5**).
7. Spread 50–100 μL of each transformation on the appropriate selection plate.
8. Grow at 30°C for 3–4 d.

3.3. Analysis of the Interaction

3.3.1. Growth in Histidine-Deficient Media

If the yeast strain has an integrated *HIS3* reporter construct under the control of a GAL4-responsive promoter (**Fig. 1**), then a direct, initial assessment of whether interaction occurs is to plate the transformation on histidine-deficient medium.

Generally, 50–100 μL of each transformation reaction are plated on both CM(-leu-trp) and CM(-leu-trp-his+3AT) plates. The leu/trp selects for those yeast that acquire both plasmids in the transformation (the AS1 and GAD.GH derivatives) and provides an indication of the efficiency of the transformation procedure. The leu/trp/his selection reveals those yeast that acquire both plasmids and further indicates whether the two fusion proteins expressed from each plasmid interact with each other to result in GAL4-dependent transcription of the *HIS3* marker gene (*see* **Note 6**).

After 3–4 d at 30°C, the plates are scored for growth in the absence or presence of histidine selection.

3.3.2. β-Galactosidase Filter Assay

Expression of the second reporter sequence, *lacZ*, may be monitored in one of two ways: by a qualitative colony-lift filter or by a semiquantitative liquid β-galactosidase assay. In the first case, colonies from a transformation are transferred to filter paper and analyzed for β-galactosidase activity directly on the filter. This is especially useful in library screening and in colony purification when one wants to know only whether the *lacZ* gene is expressed or not.

1. Transformed yeast are grown for 3–4 d at 30°C on the appropriate selection medium.
2. Colony lifts are performed using Whatman 50 filter circles. Carefully lay the filters over the colonies on a plate and wet the entire paper by gently tapping the dry areas with forceps. The filters are applied to all of the transformations first and then they are removed in order by carefully lifting the filter with forceps from one edge of the plate. Nearly all of the colony will adhere to the paper. Lay filters colony side up on clean paper towels (*see* **Note 7**).
3. Dispense 3 mL of Z-buffer containing X-gal into the lid of a 100-mm culture dish. Carefully lay filter paper (Fisherbrand, Qualitative P5) onto the Z-buffer avoiding bubbles, waves, or wrinkles. This provides a smooth, evenly wet surface of substrate on top of which the Whatman 50 filters that contain the yeast colonies will be positioned.
4. Freeze the colonies that adhere to the Whatman 50 filter by immersing the filters colony-side down in liquid nitrogen for 10–15 s. Remove the filter and place it colony side up on paper towel. Allow approx 3–5 min for the filter to thaw (*see* **Note 8**).
5. Place the thawed filter (colony side up) onto the Z-buffer impregnated filter in the lid of the culture dish. Position the filter carefully to avoid trapping air bubbles. Use the bottom of the Petri dish as a lid to cover these filters and prevent them from drying out. Incubate at 30°C until the blue color develops (*see* **Note 9**).

3.3.3. Quantitative Liquid β-Galactosidase Assays

A second method to assay *lacZ* expression involves growing individual colonies in liquid-selection medium and then assaying cellular extracts derived from these cultures for β-galactosidase activity. Here, one obtains a measure of the specific activity of β-galactosidase in each culture. This analysis is useful in quantitating the effects of ligands or characterizing the effects of mutations on the interaction of the receptor bait with a target protein. The following is based on protocols described previously *(21,22)*.

1. Transform the yeast with the bait and prey plasmids and grow on selection medium for 3–4 d at 30°C.
2. Pick triplicate colonies from each plate and grow in sterile, 16 × 150-mm culture tubes in 2.5 mL of liquid-selection medium overnight at 30°C with vigorous shaking (*see* **Note 10**).
3. Harvest the individual cultures in 1.7-mL microtubes with multiple 5 s spins.
4. Wash each pellet with 1 mL of sterile water.
5. Resuspend each pellet in 0.25 mL of lysis buffer (*see* **Note 11**).
6. Freeze the cells in a dry ice–ethanol bath and thaw on ice.
7. While the cells are thawing, add 800 µL of Z-buffer to individual microtubes. Add 200 µL of 4 mg/mL ONPG to each tube, mix, and equilibrate at 30°C for 3 min.
8. Initiate the reactions by adding 200 µL of the thawed-cell suspension.
9. After a sufficient yellow color has developed, the reaction is stopped by the addition of 0.5 mL 1 *M* Na_2CO_3. The OD_{420} and the OD_{550} are determined for each reaction (*see* **Note 12**).
10. Dilute 50 µL of the remaining cell suspension to 1 mL with water and determine the OD_{600} of each individual culture. The following formula is used to calculate the specific activity of β-galactosidase for each culture.

$$\frac{\text{Units of }\beta\text{-gal}}{\text{activity}} = \frac{1000 \times [(OD_{420}) - (1.75 \times OD_{550})]}{(t) \times (v) \times (OD_{600})}$$

where t is the time of reaction (min), v is the volume of culture used in the assay (mL), OD_{600} is the cell density at the start of the assay, OD_{420} is the combination of *o*-nitrophenol absorbance and light-scattering debris, OD_{550} is the light scattering of cell debris.

3.3.4. Testing the Effects of Ligands on the Interaction

The ligands for several of the nuclear receptors, including VDR and retinoid receptors, are known to play a role in heterodimer interactions *(2,23)* as well as in the interaction of these receptors with other transcription factors *(6,7,24)*. The two-hybrid system is well-suited to test the ligand dependence of nuclear receptor interactions with other proteins.

1. Triplicate colonies are isolated from a fresh transformation plate and are grown overnight in 2.5 mL of selection medium.
2. The OD_{600} of each culture is determined and duplicate 2.5-mL cultures are started from each original-colony expansion at an initial OD_{600} = 0.02. One culture receives ligand at the desired concentration and the other culture receives an equivalent volume of the solvent vehicle.

3. The cultures are grown for 24 h at 30°C with vigorous shaking.

4. Cells are harvested and β-galactosidase activity is quantitated as described in **Subheading 3.3.3.** Thus, the three original colonies will yield triplicate values for a vehicle control and triplicate values for ligand treatment.

3.4. Screening a cDNA–GAD Fusion Library

3.4.1. Large-Scale Competent Cell Preparation and Transformation

1. Use a single colony to begin a 5-mL culture in YPAD medium and grow overnight at 30°C with vigorous shaking.

2. Dilute the overnight culture to 250 mL with YPAD (30°C) to obtain an OD_{600} of 0.1 U. Incubate this 250-mL culture at 30°C with vigorous shaking until the OD_{600} is approx 1.0 U.

3. Harvest the cells by centrifugation and wash once with 25 mL of sterile water. Resuspend the pellet and wash once in 5 mL LiOAc/TE. Resuspend the final-cell pellet in 1.0 mL of LiAc/TE solution and store on ice. This will yield a final competent-cell preparation of approx 1.5 mL.

4. Add 10 µg of bait plasmid, 10 µg of the cDNA library in the prey plasmid, and 150 µg of carrier DNA to each of seven individual, sterile microcentrifuge tubes (*see* **Note 13**).

5. Add 200 µL of the competent-yeast preparation to each tube. Add 1 mL of 40% PEG solution to each tube and mix gently by pipetting up and down. Incubate for 30 min at 30°C.

6. Heat shock at 42°C for 15 min.

7. Spin to pellet the cells, resuspend each pellet in 200 µL of sterile water, and combine the transformations into a single tube.

8. Spread the entire transformation over 12 × 150-mm plates of CM(-leu-his-trp+3AT). To determine the number of clones screened, dilute a small aliquot of the transformation 1:10 and 1:100 and plate 50 µL of each dilution on duplicate 100-mm plates of CM(-leu-trp). This protocol generally is sufficient to screen 250,000–300,000 individual clones. One can readily scale up the procedure to screen more of the library in a single experiment.

9. After 4 d of growth at 30°C, the colonies that grow under histidine selection are assayed for *lacZ* expression using the β-galactosidase filter assay described in **Subheading 3.3.2.** This protocol uses filter-paper circles that are 145 mm in diameter and these are incubated on 9 mL of Z-buffer containing X-gal dispensed in the lids of 150-mm culture dishes.

10. Colonies that grow well under histidine selection and that express comparatively high β-galactosidase activity are then purified to a single colony. Generally, it is possible to remove the blue colony directly from the β-galactosidase filter assay. This colony is streaked onto a CM(-leu-his-trp+3AT) plate, grown for 4 d at 30°C, and then re-analyzed in a β-galactosidase filter assay.

11. A well-isolated colony that expresses high β-galactosidase activity is chosen for plasmid segregation and plasmid rescue as described in **Subheadings 3.4.2.** and **3.4.3.**

3.4.2. Plasmid Segregation of Positive Clones

Colonies that grow on the CM(-leu-his-trp) plates and that express high levels of β-galactosidase are picked and grown overnight in CM(-leu) medium. This maintains selection for the library plasmid and removes the selection for the bait plasmid. Therefore, a small percentage of yeast will lose the bait plasmid, but retain the prey plasmid. These are readily identified by the following protocol.

1. Pick individual, well isolated colonies and grow in CM(-leu) medium overnight at 30°C with vigorous shaking.
2. Dilute an aliquot of this overnight culture to an OD_{600} of 2×10^{-4} U in CM(-leu) medium and spread 50–100 µL of this dilution onto a CM(-leu) plate. This will yield approx 50–100 colonies per 100-mm plate after 3–4 d at 30°C.
3. Replica plate each dish to CM(-leu), CM(-trp), and CM(-leu,-trp) plates. Grow overnight at 30°C to identify colonies that grow with leu selection, but not with trp selection. Generally, greater than 10% of the colonies lose the bait plasmid (trp selection) under these conditions. These colonies are used as a starting source to:
   a. Reintroduce various unrelated bait constructs to examine specificity;
   b. Mate with an appropriate strain-harboring unrelated baits in a mating assay to examine specificity;
   c. Rescue the library plasmid from the yeast for further analysis.

3.4.3. Plasmid Rescue from Yeast

This procedure is based on the work of Robzyk and Kassir *(25)*.

1. Grow colonies that retain the prey plasmid (leu selection), but that have lost the bait plasmid, overnight at 30°C in 2.5 mL of CM(-leu) medium.
2. Harvest cells in a 1.7-mL microcentrifuge tube by multiple 5-s spins.
3. Resuspend cell pellets in 100 µL of STET buffer, add a 150 µL vol of acid-washed glass beads, and disrupt the cells by vigorous vortexing for 5 min.
4. Add an additional 100 µL of STET buffer and place the cells in a boiling water bath for 3 min.
5. Cool the extracts briefly on ice and then clear by centrifugation in a microcentrifuge at maximal speed for 10 min at 4°C.
6. Remove 100 µL of the supernatant to a fresh tube containing 50 µL of 7.5 *M* ammonium acetate. Mix the tubes and incubate at –20°C for 1 h.
7. Following a 10-min centrifugation, remove the supernatant to a fresh tube and precipitate with 2 vol of ethanol. Wash the pellet once with 70% ethanol, dry, and resuspend in 20 µL of TE.
8. Use 10 µL of the final preparation to transform 100 µL of competent DH5α bacteria. Incubate the freshly thawed bacteria with the DNA for 30 min on ice, heat shock for 45 s at 42°C, add 900 µL of LB medium (no ampicillin), and allow the cells to recover at 37°C with vigorous shaking for 1 h. Harvest the bacteria by centrifugation, resuspend in 200 µL of LB medium containing 100 µg/mL

ampicillin and spread the entire transformation reaction on two 100-mm plates of LB/Amp (*see* **Note 14**).

9. Expand isolated bacterial colonies and store as glycerol stocks for long-term storage. Isolate plasmid DNAs by standard-miniprep protocols for sequencing and for reintroduction into yeast to examine the specificity of the interaction with various unrelated or related AS1 bait constructs (*see* **Note 15**).

4. Notes

1. Generally, the more specificity controls one can examine, the better. We have also constructed AS1-RXR, AS1-RAR, and AS1-ER constructs to examine specificity of GAD-fusions for interaction with AS1-VDR. Moreover, if the search involves identifying proteins that interact with a specific region of your bait, then a strong control is a bait construct in which that region is deleted or mutated. We routinely use C-terminal truncations of our VDR bait construct [AS1-VDR (1-387)] or point mutants in this region to screen for factors that interact selectively with the AF-2 region of VDR or to identify other RXR-related factors that interact with VDR. This extreme C-terminus of VDR (amino acids 387-427) has been shown by biochemical and molecular approaches to contain a transcriptional-activation domain (AF-2) and a distinct domain that is essential for RXR interaction *(15)*. Thus, such a panel of $GAL4_{DBD}$ fusion constructs should be strongly considered early in the development of a particular two-hybrid screening strategy.

2. The level of background *HIS3* expression and the amount of 3-AT required to suppress this residual growth is strain dependent. In our hands, the Y190 strain requires 50 mM 3-AT to reduce this background to acceptable levels after 4 d at 30°C. However, 5 mM 3-AT is all that is required to suppress this background in the Hf7c strain. One should empirically establish the minimal level of 3-AT that is required to reduce residual *HIS3* expression. This is accomplished by introducing an AS1 fusion and an unrelated GAD.GH fusion into the strain and plating the transformation on CM(-leu-trp-his) containing increasing concentrations of 3-AT. Background colony growth is examined after 4 d at 30°C.

3. The density of yeast cells in culture can be estimated spectrophotometrically by measuring the optical density at 600 nm. The culture should be diluted to an OD_{600} < 1.0 U and then, each 0.1 OD_{600} U is roughly equivalent to 3×10^6 cells/mL. For example, an initial OD_{600} of 0.7 units for the 50-mL culture at the start of this procedure would be equivalent to 2.1×10^7 cells/mL or a total of 1×10^9 cells. Thus, the final competent cell preparation is resuspended to a final volume of 0.5 mL in 1X LiAc/TE.

4. Generally, the volume of DNA should not exceed 10% of the volume of competent cells added. If it does, then 10X LiAc/TE should be added so that the transformation reaction is maintained in a 1X LiAc/TE buffer. We have found that for most routine experiments, crude miniprep-DNA prepared with standard alkaline lysis protocols works well in this transformation procedure. However, if the highest efficiencies are required, one should use more purified plasmid DNAs preparations (e.g., cesium banded or Qiagen columns).

5. To save time and effort, we often plate 1/10 of the transformation reaction directly from the heat-shocked cells that are in the 40% PEG solution. Although this tends to decrease the transformation efficiency somewhat, it is generally not a problem in routine assays. However, in cases where the highest transformation efficiencies are desired (i.e., library screening), we routinely remove the PEG before plating as described.

6. It is important in the initial experiments to rule out the possibility that the receptor–bait construct activates reporter-gene expression on its own. The nuclear receptors contain activation domains and some function as transactivators in yeast. If the bait construct is a weak activator by itself, it can lead to a number of false positive interactions and should not be used as a two-hybrid bait.

7. A direct comparison of Whatman 50 filter papers v nitrocellulose filters demonstrated that the filter paper was far superior in this colony-lift filter assay. The signal was more intense and developed more rapidly for colonies on the filter paper compared to colonies on the nitrocellulose filter. It is possible that substrate access may partially explain this difference. Moreover, the colonies appear to adhere better to the Whatman 50 filter paper and it is considerably less expensive than its nitrocellulose counterpart.

8. The yeast colonies adhere well to the filter paper after the freeze–thaw cycle only if sufficient time is given for the cells to thaw. If the filter is applied to the substrate dish too early, then the colonies tend to diffuse and spread out. We have found that somewhere between 3 and 5 min of thawing works well. One indication that the cells have thawed sufficiently is that the colonies develop a depression in their centers taking on the shape of a doughnut or red-blood cell.

9. For strong interactions, the color development may be obvious after 30–60 min of incubation time. Other weaker interactions may require several hours. Generally, we do not observe additional color development beyond 8 h of incubation time.

10. When picking yeast colonies to streak or to begin liquid cultures, an inoculating loop is flamed for several seconds and the hot loop should be quenched completely before picking the colony. Although this may seem intuitively obvious, failure to completely dissipate all of the heat will kill most of the yeast colony and result in little or no growth of the culture. Quenching the tip of the loop in the solid medium of a plate generally is not sufficient. We routinely quench the entire loop in sterile water or sterile-liquid medium.

11. This represents a 10-fold concentration of the original Hf7c culture, which may not be necessary in all cases. *LacZ* expression is strain-dependent, reflecting both the specific-promoter sequence driving the *lacZ* reporter and the strength of the interaction being examined.

12. To ensure linearity of the assay, the final OD_{420} should be less than 1.0 absorbance units. After several trials, one develops a feel for the intensity of the yellow-reaction product that will keep the assay in the linear range.

13. We routinely cotransform the bait and the library in this screen. Alternatively, one may prepare competent cells from yeast that already harbor the bait plasmid.

This would yield much higher transformation efficiencies and the ability to screen more of the library in a single experiment. On the other hand, one may encounter difficulties in the library screen if expression of the bait construct is in any way toxic to the yeast.

14. Electroporation is the method of choice because of the high efficiency of this transformation process. However, we routinely use chemical means to prepare competent bacteria with good results. It is essential that highly competent bacteria be used for this rescue procedure. These are available commercially, but we routinely prepare our own using standard protocols *(26)*. The DH5α strain of bacteria is preferred because, in our hands, this strain yields efficiencies of 5×10^7 to 5×10^8 colonies/µg DNA by this procedure.

15. An alternative method to examine specificity is in a mating assay. Here, a strain of yeast of the opposite mating type, which harbors the individual unrelated bait constructs is crossed with the strain carrying the isolated library plasmid. This is a more efficient method to rapidly screen large numbers of potential clones.

References

1. Fields, S. and Sternglanz, R. (1994) The two-hybrid system: an assay for protein-protein interactions. *Trends Genet.* **10,** 286–292.
2. MacDonald, P. N., Sherman, D. R., Dowd, D. R., Jefcoat, S. C., and DeLisle, R. K. (1995) The vitamin D receptor interacts with general transcription factor IIB. *J. Biol. Chem.* **270,** 4748–4752.
3. Ing, N. H., Beekman, J. M., Tsai, S. Y., Tsai, M.-J., and O'Malley, B. W. (1992) Members of the steroid hormone receptor superfamily interact with TFIIB (S300-II). *J. Biol. Chem.* **267,** 17,617–17,623.
4. Baniahmad, A., Ha, I., Reinberg, D., Tsai, S., Tsai, M.-J., and O'Malley, B. W. (1993) Interaction of human thyroid hormone receptor β with transcription factor TFIIB may mediate target gene derepression and activation by thyroid hormone. *Proc. Natl. Acad. Sci. USA* **90,** 8832–8836.
5. Blanco, J. C. G., Wang, I.-M., Tsai, S. Y., Tsai, M.-J., O'Malley, B. W., Jurutka, P. W., Haussler, M. R., and Ozato, K. (1995) Transcription factor TFIIB and the vitamin D receptor cooperatively activate ligand-dependent transcription. *Proc. Natl. Acad. Sci. USA* **92,** 1535–1539.
6. Onate, S. A., Tsai, S. Y., Tsai, M.-J., and O'Malley, B. W. (1995) Sequence and characterization of a coactivator for the steroid hormone receptor superfamily. *Science* **270,** 1354–1357.
7. Horlein, A. J., Naar, A. M., Heinzel, T., Torchia, J., Gloss, B., Kurokawa, R., Ryan, A., Kamei, Y., Soderstrom, M., Glass, C. K., and Rosenfeld, M. G. (1995) Ligand-independent repression by the thyroid hormone receptor mediated by a nuclear receptor co-repressor. *Nature* **377,** 397–404.
8. Chen, J. D. and Evans, R. M. (1995) A transcriptional corepressor that interacts with nuclear hormone receptors. *Nature* **377,** 454–457.

9. Lee, J. W., Ryan, F., Swaffield, J. C., Johnston, S. A., and Moore, D. D. (1995) Interaction of thyroid-hormone receptor with a conserved transcriptional mediator. *Nature* **374,** 91–94.

10. Ma, J. and Ptashne, M. (1987) A new class of yeast transcriptional activators. *Cell* **51,** 113–119.

11. Durfee, T., Becherer, K., Chen, P.-L., Yeh, S.-H., Yang, Y., Kilburn, A. E., Lee, W.-H., and Elledge, S. J. (1993) The retinoblastoma protein associates with the protein phosphatase type I catalytic subunit. *Genes Dev.* **7,** 555–569.

12. Hannon, G. J., Demetrick, D., and Beach, D. (1993) Isolation of the Rb-related p130 through its interaction with CDK2 and cyclins. *Genes Dev.* **7,** 2378–2391.

13. Yu, V. C., Delsert, C., Andersen, B., Holloway, J. M., Devary, O. V., Naar, A. M., Kim, S. Y., Boutin, J.-M., Glass, C. K., and Rosenfeld, M. G. (1991) RXRβ: a coregulator that enhances binding of retinoic acid, thyroid hormone, and vitamin D receptors to their cognate response elements. *Cell* **67,** 1251–1266.

14. Kliewer, S. A., Umesono, K., Mangelsdorf, D. J., and Evans, R. M. (1992) Retinoid X receptor interacts with nuclear receptors in retinoic acid, thyroid hormone, and vitamin D_3 signalling. *Nature* **355,** 446–449.

15. Nakajima, S., Hsieh, J.-C., MacDonald, P. N., Galligan, M. A., Haussler, C. A., Whitfield, G. K., and Haussler, M. R. (1994) The C-terminal region of the vitamin D receptor is essential to form a complex with a receptor auxiliary factor required for high affinity binding to the vitamin D-responsive element. *Mol. Endocrinol.* **8,** 159–172.

16. Bartel, P., Chien, C., Sternglanz, R., and Fields, S. (1993) Elimination of false positives that arise in using the two-hybrid system. *Biotechniques* **14,** 920–924.

17. Sherman, F. (1991) Getting started with yeast. *Meth. Enzymol.* **194,** 3–21.

18. Treco, D. A. (1989) Basic techniques of yeast genetics, in *Current Protocols in Molecular Biology*, Vol. 2 (Ausubel, F. M., Brent, R., Kingston, R. E., Moore, D. D., Seidman, J. G., Smith, J. A., and Struhl, K., eds.), Wiley, New York, pp. 13.1.1–13.2.11.

19. Schiestl, R. H. and Gietz, R. D. (1989) High efficiency transformation of intact yeast cells using single stranded nucleic acids as a carrier. *Curr. Genet.* **16,** 339–346.

20. Gietz, D., Jean, A. S., Woods, R. A., and Schiestl, R. H. (1991) An improved method for high efficiency transformation of intact yeast cells. *Nucl. Acids Res.* **20,** 1425.

21. Fagan, R., Flint, K. J., and Jones, N. (1994) Phosphorylation of E2F-1 modulates its interaction with the retinoblastoma gene product and the adenoviral E4 19 kDa protein. *Cell* **78,** 799–811.

22. Reynolds, A. and Lundblad, V. (1989) Yeast vectors and assays for expression of cloned genes, in *Current Protocols in Molecular Biology*, vol. 2 (Ausubel, F. M., Brent, R., Kingston, R. E., Moore, D. D., Seidman, J. G., Smith, J. A., and Struhl, K., eds.), Wiley, New York, pp. 13.6.1–13.6.4.

23. MacDonald, P. N., Dowd, D. R., Nakajima, S., Galligan, M. A., Reeder, M. C., Haussler, C. A., Ozato, K., and Haussler, M. R. (1993) Retinoid X receptors stimu-

late and 9-*cis* retinoic acid inhibits 1,25-dihydroxyvitamin D_3-activated expression of the rat osteocalcin gene. *Mol. Cell. Biol.* **13,** 5907–5917.

24. Le Douarin, B., Zechel, C., Garnier, J.-M., Lutz, Y., Tora, L., Pierrat, B., Heery, D., Gronemeyer, H., Chambon, P., and Losson, R. (1995) The N-terminal part of TIF1, a putative mediator of the ligand-dependent activation function (AF-2) of nuclear receptors, is fused to B-raf in the oncogenic protein T18. *EMBO J.* **14,** 2020–2033.

25. Robzyk, K. and Kassir, Y. (1992) A simple and highly efficient procedure for rescuing autonomous plasmids from yeast. *Nucl. Acids Res.* **20,** 3790.

26. Seidman, C. E. (1989) Introduction of plasmid DNA into cells, in *Current Protocols in Molecular Biology*, vol. 1 (Ausubel, F. M., Brent, R., Kingston, R. E., Moore, D. D., Seidman, J. G., Smith, J. A., and Struhl, K., eds.), Wiley, New York, pp. 1.8.1–1.8.3.

26

Gel-Shift Analysis and Identification of RXREs and RAREs by PCR-Based Selection

Myriam I. Baes and Peter E. Declercq

1. Introduction

The electrophoretic mobility-shift assay is one of an array of techniques that are used to identify and characterize protein–DNA interactions. This method is based on the retardation of a labeled-DNA fragment in a nondenaturing gel by bound proteins *(1)*. The binding specificity and affinity can easily be determined by competition analysis with an excess of specific or nonspecific DNA. The retinoid receptors can interact with their target DNA either in a heterodimeric (retinoic acid receptor/retinoic X receptor [RAR/RXR]) or in a homodimeric (RXR/RXR) configuration *(2)*. If the protein source is a complex-cellular extract, one can distinguish between these possibilities by coincubation with specific antibodies to the receptors. When antibodies are added to the binding reaction a DNA/receptor/antibody complex will be formed that is further retarded ("supershift"). Alternatively, the antibodies may impair the receptors ability to bind the DNA or to dimerize, resulting in the disappearance of the shifted band. The mobility-shift assay can easily be acquired, is fast and very sensitive.

The in vitro binding of retinoid receptors to oligonucleotides can also be exploited to identify novel retinoid-responsive elements or to determine the requirements for a DNA sequence to function as an optimal retinoid-responsive element (retinoic X receptor responsive element [RXRE] or retinoic acid receptor responsive element [RARE]). Starting from a randomized pool of oligonucleotides, those with the highest-binding affinities can be selected by consecutive cycles of binding to the receptor, isolation of the bound oligonucleotides and amplification of these oligonucleotides by polymerase chain reaction (PCR). After cloning, the isolated elements are sequenced and

From: *Methods in Molecular Biology, Vol. 89: Retinoid Protocols*
Edited by: C. P. F. Redfern © Humana Press Inc., Totowa, NJ

the composition of the hexamers and flanking bases can be analyzed. This binding-site selection procedure has been applied to define the optimal-binding sites of different members of the nuclear hormone receptor superfamily *(3–8)*.

2. Materials

2.1. Gel-Shift Analysis

1. Protein: The retinoic acid receptors, RAR and RXR, that will be used in the binding reactions can be obtained from several sources. Nuclear extracts can be prepared from tissue or from cell cultures (*see* Chapter 20). Because RARs and RXRs are co-expressed in most cells, heterodimers as well as homodimers will be present in these extracts. On the other hand, these proteins can be overexpressed individually by a number of systems including in vitro transcription and translation, expression in bacteria or yeast, and expression in insect cells using baculovirus and in Hela cells using vacciniavirus vectors (*see* **Note 1**). The proteins do not need to be purified from these extracts to obtain good binding results.
2. Oligonucleotide: a double-stranded (ds) oligonucleotide varying in size between 15 and 200 bases is used as the target in the binding reactions (*see* **Note 2**). This can be obtained by annealing two complementary synthetic oligonucleotides encompassing the binding site. After carefully determining the concentration of each oligonucleotide, 5 μg of each is combined in 100 μL of a medium-salt buffer, e.g., 10 mM Tris-HCl, pH 7.5, 50 mM NaCl, 10 mM MgCl$_2$, 1 mM dithiothreitol (DTT). After heating to 65°C for 2 min, the mixture is allowed to cool slowly to room temperature. The ds oligonucleotide can be stored at –20°C. Alternatively, restriction digest of a plasmid and purification on agarose gels can provide the desired DNA fragment.
3. Binding buffer (5X stock): 100 mM Tris-HCl, pH 8.0, 2.5 mM EDTA, 5 mM DTT, 400 mM KCl, 60% w/v glycerol. A separate 5X stock solution of bovine serum albumin (BSA) (concentration 1 μg/μL) is used. This allows each concentration to vary independently from the other components of the binding reaction.
4. [α^{32}-P]dCTP and [γ^{32}-P]ATP: can be obtained fresh each week from several companies at a specific activity of >3000 Ci/mmol.
5. dNTP: Stock solutions are available from Pharmacia (Uppsala,Sweden) and Boehringer Mannheim (Mannheim, Germany).
6. poly(dI-dC) (Pharmacia): a 20X stock solution (0.5 mg/mL) in TE buffer (10 mM Tris-HCl, pH 8.0, 1 mM EDTA) serves as bulk carrier DNA that will prevent nonspecific interactions of DNA binding proteins with the probe.
7. Gel: A 4 or 5% nondenaturing polyacrylamide gel with an acrylamide to bisacrylamide ratio of 80:1 is composed in 0.5X TBE buffer (10X TBE:121 g Tris, 55 g orthoboric acid, 7.4 g EDTA/L). Combine 9.4 mL of 40% acrylamide, 2.35 mL of 2% *bis*-acrylamide, 3.75 mL of glycerol (50% v/v), 3.75 mL of 10X TBE and bring up to 74 mL with water. Add 1mL of ammonium persulfate (10%) and 50 μL of TEMED and pour a gel with a minimum length of 15 cm and with

a thickness of 1.5 mm. The Bio-Rad Protean II Cell or an analogous instrument is a convenient apparatus. For optimal results the teeth of the comb should be at least 5 mm wide. One of the glass plates is siliconized to facilitate the removal of the gel.

2.2. Binding-Site Selection

1. Randomer: a single stranded (ss) oligonucleotide is synthesized consisting of a stretch of completely randomized G, A, C, and T, flanked by two invariant parts that will serve as anchors for PCR primers and that encompass a restriction site (*see* **Notes 3** and **4**). The randomer should be gel purified in order to select the full-length oligonucleotide: pour an 8% denaturing polyacrylamide gel between two glass plates spaced by 1.5 mm: combine 25.2 g of urea, 6 mL of 10X TBE, 12 mL of 40% acrylamide/2% *bis*-acrylamide and water in a total volume of 60 mL. Add 200 µL of ammonium persulfate (10%) and 30 µL of TEMED. Mix 600 µg of randomer oligonucleotide with loading buffer (9 parts deionized formamide, 1 part 1X TBE, and 0.5% bromophenol blue). Run the gel until the bromophenol blue tracking dye reaches the bottom. Remove the gel from the plates, wrap in plastic foil, and place on a thin-layer chromatography (TLC) plate with fluorescent indicator. Visualize the bands by briefly shadowing with a shortwave UV lamp. Cut out the slowest moving oligonucleotides, crush the gel and elute the oligonucleotides in a rotary shaker overnight in 1–2 mL 0.3 *M* sodium acetate, 0.1% sodium dodecyl sulfate (SDS). Take off the supernatant, extract with phenol/chloroform, and precipitate with ethanol.
2. PCR primers: the sequence of the forward and reverse primers are respectively identical to the 5' invariant part of the randomer and complementary to the 3' invariant part of the randomer. Store the primers as a 5-µ*M* stock solution in TE at 4°C or at –20°C.
3. Protein: a crude bacterial extract in which the receptor is overexpressed is appropriate. This can be obtained by the procedure of Pognonec et al. *(9)*. Sonicate the bacterial pellet from a 250-mL culture in 5 mL of lysis buffer (500 m*M* NaCl, 10% w/v glycerol, 1 m*M* EDTA, 0.1% Nonidet P-40, 5 mg/mL leupeptin, 1 m*M* PMSF, 20 U/mL aprotinin, 10 m*M* Tris-HCl, pH 7.5) for 5 min (15 s on/15 s off) at maximum-energy output. Subject to two freeze–thaw cycles. Centrifuge for 10 min at 10,000*g* at 4°C. Transfer the supernatant to a 50-mL tube, add an equal volume of ice-cold water and add dropwise the same volume of saturated ammonium sulfate while swirling. Keep on ice for 15 min. Centrifuge 15 min at 10,000*g* at 4°C. Suspend the pellet in 1 mL of lysis buffer, and spin for 5 min at maximum speed in a microcentrifuge in the cold. Dialyze the supernatant against 100 m*M* KCl, 10% w/v glycerol, 1 m*M* EDTA, 20 m*M* Tris-HCl, pH 7.5. Store in aliquots at –70°C. Each binding reaction requires approx 100–250 ng of receptor. The receptor concentration can be estimated from stained gels.
4. Antibody-coated magnetic beads: several companies supply magnetic beads coated with sheep antimouse or sheep antirabbit IgG. Antibodies directed to a receptor that have been raised in mice or rabbits can then be bound to these coated

beads, e.g., 1 mg of sheep antimouse-coated beads (Dynabeads M-280 Sheep anti-Mouse IgG, Dynal, Skoyen, Norway) are incubated overnight at 4°C with 10 µg of monoclonal mouse anti-*myc* antibody (Cambridge Research Biochemicals, Cheshire, UK) followed by four washes for 30 min at 4°C with PBS containing 0.1% BSA. These beads are stored at 4°C (*see* **Note 5**).

5. Cloning vector: a plasmid with a convenient reporter (chloramfenicol acetyl transferase, luciferase, alkaline phosphatase) that is useful in cotransfection experiments, e.g., pUTKAT *(10)*, or the pGL2-Promoter and pCAT-Promoter vectors (Promega, Madison, WI).

6. Ampli *Taq*-DNA polymerase and 10X PCR buffer (Perkin-Elmer, Norwalk, CT).

3. Methods

3.1. Gel-Shift of Retinoid Responsive Elements

3.1.1. Labeling of Oligonucleotide

A ds oligonucleotide (with 5' overhang or with blunt end) can be endlabeled efficiently using T4 polynucleotide kinase. Oligonucleotides with 5' overhang can also be labeled by fill-in reaction with Klenow enzyme.

Endlabeling: In a total volume of 10 µL combine 50–100 ng of ds oligonucleotide, 1 µL of 10X kinase buffer (supplied by the manufacturer), 5 µL of $[\gamma\text{-}^{32}P]$ ATP (3000 Ci/mmol) and 1 µL of polynucleotide kinase. Leave at 37°C for 30–60 min. Dilute to 100 µL with TE and spin through Sephadex G-50 (Pharmacia) for 2 min at 1000*g* in a table top centrifuge (use either a commercial column or a homemade column in a 1-mL syringe with a glass bead to hold the resin). This will capture the nonreacted $[\gamma\text{-}^{32}P]ATP$. Typical yields are 20–70X 10^6 cpm/100 ng oligonucleotide (*see* **Note 6**).

Klenow fill in: To 2 µg of ds oligonucleotide (at 500 ng/µL) add 2.5 µL of 10X Klenow buffer (supplied by the manufacturer), 1.5 µL of 5 m*M* dNTP (a mix of three nucleotides other than the labeled nucleotide), 2.5 µL of BSA (0.5 mg/mL), 2.5 µL (25 µCi) of $[\alpha\text{-}^{32}P]$ dCTP (3000 Ci/mmol), 1.5 µL (3 U) of Klenow enzyme in a total reaction volume of 25 µL (*see* **Note 7**). Incubate for 30 min at room temperature. Remove unincorporated nucleotides by centrifugation through Sephadex G-50 as mentioned previously. This technique has the advantage that only ds oligonucleotides can participate in the fill-in reaction yielding radioactive probes that are only ds.

Labeled probes can be kept at –20°C for up to 4 wk.

3.1.2. Binding Reaction

1. Simple binding reaction: Combine 4 µL of 5X binding buffer, 4 µL of the BSA-stock solution, 1 µL (0.5 µg) of poly (dI-dC), and 2 µL of the probe (50,000 cpm equivalent to 40 fmol). Add 2 to 4 µL of receptor extract (in this case a crude-bacterial extract in which the RAR and/or RXR are overexpressed), appropriately diluted in 1X-binding buffer, and adjust the volume to 20 µL with water

(*see* **Notes 8** and **9**). This mixture is incubated for 30 min at room temperature before loading on the gel. In case the binding of heterodimeric receptors is studied, the proteins are preincubated for 20 min at room temperature before addition to the other components of the binding reaction.

2. To further characterize the complexes formed the following steps can be undertaken:

   a. The specificity of the retarded bands can be evaluated by comparing the effect of coincubating with an excess of specific primer vs the same excess of an unrelated oligonucleotide. The unrelated oligonucleotide should not affect the formation of the complex. In contrast, in the presence of the excess-cold probe the complex should disappear. To the binding mixture, add 1 µL of a ds oligonucleotide at a concentration of 4–40 pmol/µL, i.e., a molar excess of 100- to 1000-fold compared to the labeled probe (*see* **Note 10**).

   b. Binding-competition experiments can also be used to compare the affinity of different retinoid-responsive elements. By titrating the amount of competitor DNA added to the binding reaction and quantification of the retarded complexes, an IC50 value can be deduced. To the binding reaction, add 1 µL of a ds oligonucleotide with a concentration ranging from 0.4 pmol/µL to 40 pmol/µL, i.e., a molar excess of 10- to 1000-fold. Quantification of the autoradiograms can be done by cutting out the dried gel followed by liquid-scintillation counting, or by scanning with a Phosphor Imager (Molecular Dynamics) or a similar apparatus.

   c. The importance of certain bases in a retinoid-responsive element can also be assessed in gelshift experiments. Oligonucleotides with defined-base mutations can be used either as labeled probes in binding experiments or as competitors with bonafide retinoid elements.

   d. To unequivocally demonstrate the presence of RAR or RXR in the retarded complex, antibodies directed to these receptors can be used. Monoclonal (purified or ascites fluid) as well as polyclonal antibodies are suitable. The specificity of the antibodies for the receptor should first be evaluated and cross reactivity with other nuclear-hormone receptors should be excluded. The antibodies are preincubated with the protein extract for 30 min at room temperature before addition of the other components of the binding reaction.

3.1.3. Electrophoresis

After polymerization, remove the comb, set the plates up in the apparatus, and flush out the wells. Recirculate the 0.5X TBE electrophoresis buffer and prerun the gel at 100 V for 1 h at room temperature or at 200–300 V in the cold.

The whole binding-reaction mixture is loaded on the gel. In a separate lane, tracking dye can be loaded to visualize the migration of the DNA. The gel is run till the bromophenol blue has moved approx 9 cm, after which it is dried under vacuum at 80°C and exposed overnight without intensifying screens. A typical gel-shift result is shown in **Fig. 1**.

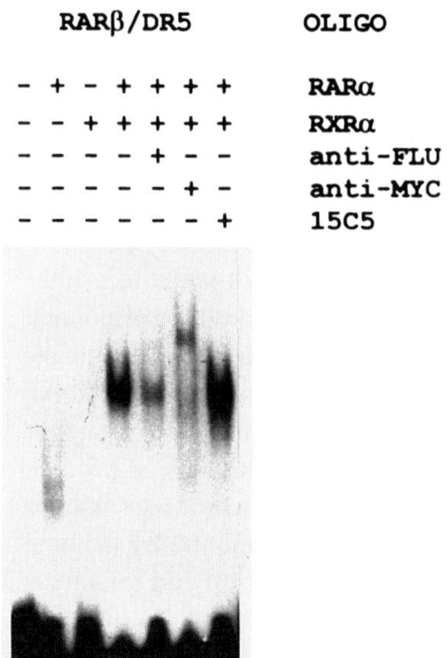

Fig. 1. Gel-shift analysis of a retinoid responsive element of the β-RAR promoter with DR5 configuration that binds RAR and RXR synergistically. The probes were endlabeled with Klenow enzyme and incubated with the bacterially expressed fusion proteins fluRARα [containing the hemaglutinin epitope *(14)*] and *myc*RXRα [containing the c-*myc* 9E10 epitope *(12)*]. Coincubation with anti-*flu* and anti-*myc* antibodies respectively results in a disappearance of the heterodimeric complex and a supershift. In contrast, 15C5—a nonrelated monoclonal antibody—does not affect the complex.

3.2. Identification of Optimal RXREs by Binding-Site Selection

The binding-site selection technique consists of consecutive cycles of binding of a receptor to a pool of oligonucleotides followed by the isolation and amplification of the bound oligonucleotides (**Fig. 2**). A critical step is the separation of the receptor-bound oligonucleotides from the free oligonucleotides. This can be achieved with several methods. Receptor-bound oligonucleotides will be retarded as compared with the unbound material in a 4 or 5% nondenaturing PAGE as described in the gelshift protocol *(3,4,8)*. However, because monomer-bound DNA will have a different retardation than dimer-bound DNA, a broad band should be cut out and eluted if one is interested in all possible binding configurations. To this end, the gel should only be run for a short period of time, and most of the area between the wells and the free probe should be excised. In order to visualize the migration of the free probe in the gel the randomer has to be radiolabeled. Elution of the receptor/DNA complex from a gelslice is not very efficient and needs to be done overnight.

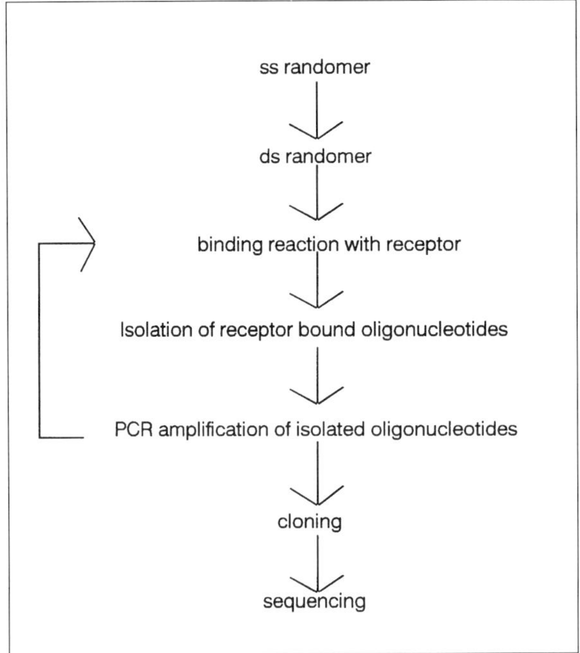

Fig. 2. Outline of the binding-site selection procedure.

We describe here an alternative, more elegant procedure based on the method developed by Wright et al. *(11)* that involves immunoprecipitation of the receptor with antibody-coated magnetic beads. This method has the following advantages:

1. The receptor protein does not need to be purified;
2. The optimal-binding sites of a heterodimer can be studied without interference of homodimers that can also be formed;
3. The isolation of receptor-bound oligonucleotides is much faster; and
4. No radioactivity is needed.

The availability of specific antibodies to the receptor is a requirement although this can be circumvented by producing a fusion protein of the receptor with an epitope to which commercial antibodies are available, e.g., a *myc* tag *(12)*. A procedure for the isolation of high-affinity binding sites for RXRα homodimers is described.

3.2.1. Second-Strand Synthesis of the Randomer

1. Take 300 ng of ss, gel-purified oligonucleotide and incubate with 300 ng of reverse primer in a 20 µL reaction containing 10 m*M* Tris-HCl, pH 7.5, 50 m*M*

NaCl, 10 m*M* MgCl$_2$, 1 m*M* DTT, 200 µ*M* each of dCTP, dGTP, dTTP, and dATP and 1 U of Klenow fragment. Incubate 2–4 h at room temperature.

2. Purify the reaction mix on a Sephadex G-50 spin column to remove unincorporated nucleotides.

3.2.2. Binding Reaction and Isolation of Receptor-Bound Oligonucleotides

1. Combine in a microcentrifuge tube: 100–250 ng of bacterially expressed, partially purified *myc*RXRα receptor protein, 25 ng of ds randomer, 4 µL of 5X binding buffer, 4 µL of BSA 1mg/mL, 1 µL of poly (dI-dC) (0.5 µg/µL), and incubate for 20 min at room temperature.

2. Add 20 µL of anti-*myc* antibody-coated magnetic beads (10 mg of beads per mL) to the binding reaction and agitate gently for 1 h at room temperature. Add 500 µL of PBS containing 0.1% BSA and hold the tube for 1 min against a magnet and withdraw the supernatant. Wash twice with 500 µL of the PBS-0.1% BSA solution. Resuspend the beads in 40 µL of 1X PCR buffer, boil for 5 min in order to disrupt protein–DNA interactions, and spin briefly.

3.2.3. Amplification of the Selected Sites

1. Take 10 µL of the bead supernatant and combine in a PCR-reaction tube with 100 ng of forward primer, 100 ng of reverse primer, 250 µ*M* each of dATP, dCTP, dGTP, and dTTP, and 2.5 U of Ampli*Taq*-DNA polymerase.

2. Initiate the following program for 15 cycles: denaturation at 94°C for 1 min; annealing at 65°C for 1 min; extension at 72°C for 1 min.

3. Run 10 µL of the PCR mixture on a 2% agarose gel to confirm the amplification of the oligonucleotides. Repeat **Subheadings 3.2.2.** and **3.2.3.** five times (*see* **Note 11**).

4. Use each time one-fourth of the PCR mixture from the previous cycle for the next round of binding, selection, and amplification. To avoid formation of artifactual DNA, purify the PCR products obtained after every other round on an 8% nondenaturing PAGE (essentially as described for the randomer purification but in the absence of urea).

3.2.4. Monitoring the Enrichment of Binding Sites

To test whether the oligonucleotides obtained after each round of binding, selection, and amplification are capable of binding the RXRα homodimers, set up a gel-shift experiment.

1. Radiolabel the original randomer and the material obtained after each selection round by PCR: use 50 ng of randomer or one-fourth of the bead supernatant after each selection round and combine each with 100 ng of forward primer, 100 ng of reverse primer, 50 µ*M* each of dATP, dGTP, dTTP, 20 µ*M* of dCTP and 10 µCi of [α-^{32}P] dCTP, 2.5 U of Ampli*Taq* polymerase in 1X PCR buffer in a total volume of 50 µL.

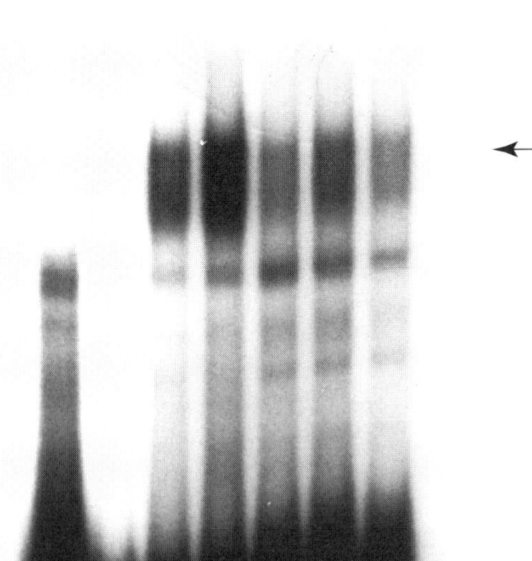

Fig. 3. Enrichment for RXRα homodimer-binding sites during the selection procedure. Bacterially expressed RXRα was incubated with the radiolabeled product obtained after each selection cycle and separated on a 5% nondenaturing polyacrylamide gel. The cycle numbers are indicated with 0 standing for the original randomer. The arrow indicates the position of the RXRα-retarded complex.

2. Subject to 15 cycles of PCR with the same program as described above.
3. Remove the unincorporated label by Sephadex-G50 chromatography.
4. Use the same number of counts (20,000 to 50,000 cpm) of each labeled probe in a gel-shift assay. As can be seen in **Fig. 3**, a signal of RXRα binding appeared starting from the second round of selection.

3.2.5. Cloning of Selected Oligonucleotides in a Reporter Plasmid

1. After the sixth round, precipitate one-half of the PCR products with 2.5 vol of ethanol in the presence of 0.3 *M* sodium acetate.
2. Dissolve in 10 µL and digest with the two restriction enzymes whose sites were incorporated in the invariable part of the randomer. At the same time, digest a reporter vector with the same enzymes.
3. Run the digests on a 2% gel and recover the DNA fragments by any conventional technique, e.g., extraction with glass beads, or, if a low-melt agarose gel was used, an agarase digestion can be done.
4. Ligate the oligonucleotides with the vector in a molar ratio of 3:1 after estimating the concentration of these components on an agarose gel.

5. Transform the ligation reaction in competent cells.
6. Pick at least 50 of the transformed colonies, inoculate in 3 mL Luria Broth base and grow overnight.
7. Isolate the plasmid DNA by any miniprep procedure and identify the plasmids with inserted oligonucleotides by restriction-digest analysis. The insert positive-reporter plasmids can be sequenced according to any of the usual protocols. The selected oligonucleotides can then be functionally tested using either gel-shift analysis or cotransfection experiments.

4. Notes

1. Hela cells contain endogenous RARs and RXRs that will contaminate the receptor protein that is overexpressed with the vaccinia virus.
2. Because the synthetic oligonucleotides that are used in binding reactions may also need to be tested in co-transfection experiments after insertion in a reporter vector, it is advisable to design these oligonucleotides such that the termini contain a cloning site e.g., a GATC overhang can be cloned in either a *Bam*HI or *Bgl*II site.
3. The design of the randomer will depend on the questions to be answered: if one wants to detect novel-hexamer arrangements the stretch of randomized nucleotides should be long enough, e.g., 20–25 bases. This would allow to identify two hexamers separated by 8–13 bases. This randomer could also be used to examine the optimal bases in a responsive element in all positions of the hexamers and of the flanking bases. However, the longer the randomer, the greater the complexity. Because only a limited amount of DNA can be brought in the first binding reaction, not all possible sequences may be represented if the randomer is more than 20 bases long. Alternatively, one can start from the assumption that the recognition sequence of one receptor is AGGTCA and evaluate the optimal spacing, orientation, and actual sequence of the second hexamer and of the intervening bases. In the latter case, the randomized part can be kept much shorter, e.g., 11 bases *(7)*.
4. The invariant part of the randomer oligonucleotide should not include combinations of bases that resemble the consensus sequence AGGTCA and must be 15–25 bases long. The choice of restriction sites will depend on those available in the reporter vector in which the selected oligonucleotides will be cloned.
5. The antibody-coated magnetic beads to which the primary antibody was bound can be kept for at least 2 wk at 4°C. Prepare enough beads to carry out six cycles of binding, isolation, and amplification.
6. Oligonucleotides with 5' protruding ends are phosphorylated more efficiently than blunt ends, whereas labeling of 3' overhanging oligonucleotides with T4-polynucleotide kinase is rather inefficient. The enzyme is inhibited by low levels of phosphate buffer or ammonium salts. The latter may contaminate the oligonucleotide as a result of the synthesis procedure. The ammonium can be eliminated using Sephadex G-50 spin columns followed by ethanol precipitation in the presence of sodium salts.

7. Which of the four [α-^{32}P] dNTPs is added to the reaction depends on the sequence of the protruding 5' termini at the ends of the DNA. It is preferable that the radioactive nucleotide to be inserted is not the most 3' nucleotide. By choosing appropriate ends, the radioactive nucleotide can be incorporated several times, resulting in a probe with higher-specific activity.

8. The concentration of bulk carrier DNA and of BSA that needs to be added in the binding reaction, in order to prevent nonspecific interactions with the probe, greatly depends on the purity of the retinoid-receptor preparation. Titrations of these components need to be carried out to determine the optimal conditions for complex formation.

9. The required concentration of the receptor needs to be established. A wide range, e.g., 10-fold serial dilutions can be tested at first. To visualize synergistic binding, the concentrations of the RAR and RXR proteins should be kept low, such that the individual proteins do not produce visible complexes.

10. In competition experiments with excess specific and nonrelated oligonucleotides, the protein should be added last so that it can interact with all oligonucleotides simultaneously.

11. The number of selection rounds can be modified depending on the specific goal of the experiment. With each additional round, a further enrichment for the highest-affinity interactions is achieved. Thus, in order to observe a variety of binding sites it is necessary to limit the number of cycles. Usually four to eight rounds of selection are performed. After six rounds of selection with RXRα, all sequenced oligonucleotides contained two or three copies of the consensus (or variant) binding-site AGGTCA present in different configurations (direct repeats with 1, 2, and 6 interspacing basepairs and inverted repeats without interspacing basepairs) *(13)*.

12. When evaluating the sequences of the isolated clones it is import to carefully check whether no tandem repeats of the oligonucleotides were inserted. These would certainly be more active than the single-binding sites.

References

1. Carey, J. (1991) Gel retardation. *Methods Enzymol.* **208,** 103–117.
2. Glass, C. K. (1994) Differential recognition of target genes by nuclear receptor monomers, dimers, and heterodimers. *Endocr. Rev.* **15,** 391–407.
3. Roch, P. J., Hoare, S. A., and Parker, M. G. (1992) A consensus DNA-binding site for the androgen receptor. *Mol. Endocrinol.* **6,** 2229–2235.
4. Harding, H. P. and Lazar, M. A. (1993) The orphan receptor Rev-ErbAα activates transcription via a novel response element. *Mol. Cell. Biol.* **13,** 3113–3121.
5. Katz, R. W. and Koenig, R. J. (1993) Nonbiased identification of DNA sequences that bind thyroid hormone receptor α1 with high affinity. *J. Biol. Chem.* **268,** 19,392–19,397.
6. Dowhan, D. H., Downes, M., Sturm, R. A., and Muscat, G. E. O. (1994) Identification of deoxyribonucleic acid sequences that bind Retinoid-X Receptor-γ with high affinity. *Endocrinology* **135,** 2595–2607.

7. Kurokawa, R., Yu, V. C., Näär, A., Kyakumoto, S., Han, Z., Silverman, S., Rosenfeld, M. G., and Glass, C. K. (1993) Differential orientations of the DNA-binding domain and carboxy-terminal dimerization interface regulate binding site selection by nuclear receptor heterodimers. *Genes Dev.* **7,** 1423–1435.

8. Subauste, J. S., Katz, R. W., and Koenig, R. J. (1994) DNA binding specificity and function of retinoid X receptor α. *J. Biol. Chem.* **269,** 30,232–30,237.

9. Pognonec, P., Kato, H., Sumimoto, H., Kretzschmar, M., and Roeder, R. G. (1991) A quick procedure for purification of functional recombinant proteins over-expressed in E.Coli. *Nucl. Acid. Res.* **19,** 6650.

10. Prost, E. and Moore, D. D. (1986) CAT vectors for analysis of eukaryotic promoters and enhancers. *Gene* **45,** 107–111.

11. Wright, W. E., Binder, M., and Funk, W. (1991) Cyclic amplification and selection of targets (CASTing) for the myogenin consensus binding site. *Mol. Cell. Biol.* **11,** 4104–4110.

12. Evan, G. I., Lewis, G. K., Ramsay, G., and Bishop, J. M. (1985) Isolation of monoclonal antibodies specific for human c-myc proto-oncogene product. *Mol. Cell. Biol.* **5,** 3610–3616.

13. Castelein, H., Jansen, A., Declercq, P. E., and Baes, M. (1996) Sequence requirements for high affinity retinoid X receptor-a homodimer binding. *Mol. Cell. Endocr.* **119,** 11–20.

14. Carter, M. E., Gulick, T., Moore, D. D., and Kelly, D. P. (1994) A pleiotropic element in the medium-chain acyl coenzyme A dehydrogenase gene promoter mediates transcriptional regulation by multiple nuclear receptor transcription factors and defines novel receptor-DNA binding motifs. *Mol. Cell. Biol.* **14,** 4360–4372.

27

Identification and Cloning of RA-Regulated Genes by mRNA-Differential Display

Jay A. White and Martin Petkovich

1. Introduction

An important key to understanding the function of retinoids is the determination of genes whose expression they regulate. In the past, several techniques including differential screening *(1)*, subtractive hybridization *(2)*, reverse transcriptase-polymerase chain reaction (RT-PCR) analysis *(3)*, and RNase protection *(4)* have been utilized to study specific differences in gene expression following RA treatment. Although these techniques have proven useful, they are not without limitations; differential screening and subtractive hybridization are both technically demanding and lengthy procedures relying heavily on the quality of cDNA and genomic libraries used, RT-PCR and RNase protection analyses provide only quantitative information on previously characterized genes. Another drawback in the use of subtractive hybridization or differential screening is that these techniques only allow analyses of unidirectional, either upregulation or downregulation, changes in gene expression in one of the samples being compared. The recent development of a novel PCR-based procedure, mRNA-differential display (DD), provides a powerful alternative method that overcomes some of the shortfalls of other techniques *(5,6)*. DD is unique in that it permits the simultaneous comparisons of both positive and negative alterations in gene expression between multiple samples (*see* **Fig. 1**).

The DD procedure allows the characterization of changes in mRNA content facilitated by comparisons of PCR fragments, of a limited-size range usually between 200 and 500 bp in length. These PCR fragments of cDNAs correspond to carboxy-terminal ends of subpopulations of the total mRNA obtained from the samples of interest. The primers (5'-T_{12}VN-3') used in the initial step,

From: *Methods in Molecular Biology, Vol. 89: Retinoid Protocols*
Edited by: C. P. F. Redfern © Humana Press Inc., Totowa, NJ

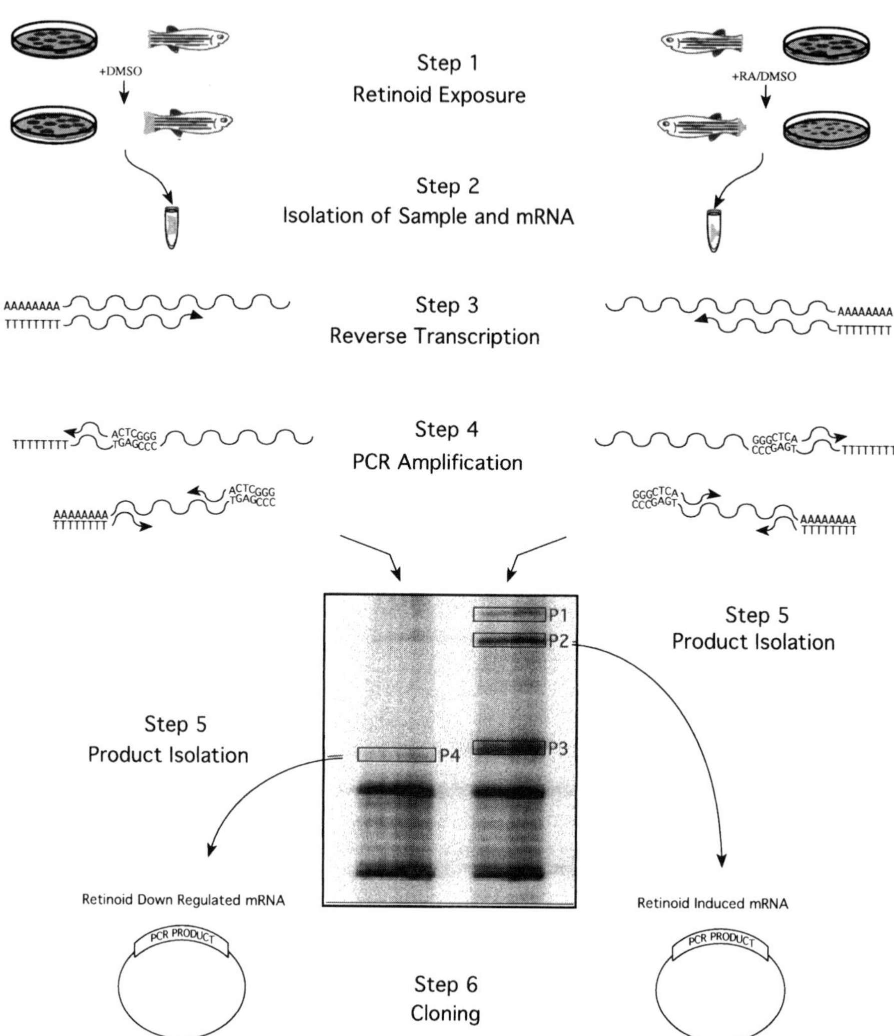

Fig. 1. Schematic representation of the steps involved in the isolation of retinoid-regulated genes using the differential display technique. The cloned products isolated in step 6 can then be used for sequencing, Northern blotting, or screening of cDNA libraries. P1, P2, and P3 correspond to fragments from RA induced mRNAs. P4 represents a PCR product from an mRNA that is downregulated.

the reverse transcription of mRNAs, are designed to anneal to the poly (A)$^+$ tail of a given mRNA subpopulation whereas the upstream primers are degenerate 10-mer oligonucleotides. The specific subpopulations of mRNAs targeted for reverse transcription are defined by the last two nucleotides V and N where V is either G, A, or C, and N is G, A, T, or C. PCR reactions including various combinations of these upstream and downstream primer sets can theoretically be utilized to represent the 15,000 or so different mRNA transcripts of a given cell or tissue sample *(7)*. After incorporating a radiolabeled deoxynucleotide during PCR amplification of the cDNA products, two or more different samples can be displayed for comparison on polyacrylamide gels and visualized by autoradiography. Following identification of potential differences in the mRNA content of the samples of interest the corresponding cDNAs can be subcloned and used in Northern blotting analyses to verify that the differences observed correspond to the actual differential representation of mRNA transcripts.

DD is well-suited for analyzing the complex time-dependent patterns of gene regulation resulting from retinoid treatment. Indeed alterations in RA-induced gene expression can be seen at various times following retinoid exposure. The analyses of expression of four genes, *RARβ, HNF3α, Stra*1, and *Mox*1 in P19 cells exemplifies this situation *(8)*. *HNF3α* is rapidly induced within 3–6 h following retinoid treatment after which time it begins to decline. *RARβ* shows a somewhat slower induction with maximal response observed at 12 h followed by a slight decline at 24 h. *Stra*1 RNA slowly increases until 6 h post-RA addition, at which point it increases more rapidly to a maximal level by 24 h. Finally *Mox*1 RNA shows the slowest response of the four genes with very little induction until 12 h when it is rapidly induced to a maximum at 24 h. In any DD experiments, care must be taken to ensure that the results obtained are interpreted properly with respect to both the time and duration of retinoid treatment. The most informative DD results can be obtained by comparing a variety of time points post-RA induction with the uninduced state ensuring that both rapid, transient and delayed changes in gene expression are detected.

Whereas this technique can be rather labor intensive owing to the many primer combinations and time-points, which must be used for significant representation of the mRNA populations of the samples being studied, it also has several advantages over other techniques. In addition, this type of PCR analysis is well-suited for use in situations where RNA is limiting such as in extracts from single cells. A newly described technique, serial analysis of gene expression (SAGE) *(9)*, provides a more quantitative analysis of the abundances of mRNAs in the cell populations being studied, but only provides short pieces (9 bp) of sequence information for further analyses.

This chapter details the basic steps including, mRNA isolation, reverse transcription, amplification, recovery, and cloning of PCR products, and subsequent verification, required to set up and utilize the mRNA DD technique to isolate RA-regulated genes. We have been successfully using DD to analyze changes in mRNA levels owing to exogenous-retinoid exposure during zebrafish caudal-fin regeneration in our laboratory (White, J. A., et al., manuscript in preparation). The techniques described can of course be applied to a variety of experimental situations where gene regulation in tissues or cells under different treatments are to be compared.

2. Materials

All reagents used prior to the PCR steps should be RNAse free. Solutions should be treated with 0.1% diethyl-pyrocarbonate (DEPC; VWR Scientific, Ontario, Canada) overnight and then autoclaved. Glassware to be used should be thoroughly washed and baked overnight in a 200°C oven. Gloves are to be worn at all times. Because this technique utilizes PCR, extreme care should be exercised to avoid DNA contamination of all supplies. We routinely use separate aliquots of enzymes, buffers, and so on, in order to be sure that all samples will be free of contaminating nucleic acids. In addition sterile, aerosol pipet tips which are DNase and RNase free (Diamed, Ontario, Canada) should be used for all manipulations up to and including the PCR.

2.1. Retinoids

1. Retinoid-stock solutions are stored at –20°C and kept in the dark at all times (*see* **Note 1**).

2.2. Isolation of mRNA from Tissue or Cells

1. High-quality mRNA can be isolated using a variety of protocols. Kits such as MicroFast Track (Invitrogen, CA) provide a good yield of pure poly (A)$^+$ RNA (*see* **Note 2**).

2.3. Reverse Transcription cDNA Synthesis

1. Poly (A)$^+$ RNA: 1-μL aliquot of 0.1 μg/μL mRNA dissolved in nuclease-free water. Store at –70°C.
2. Reverse Transcriptase: 200 U/μL Superscript II RNase H$^-$ Reverse Transcriptase (Gibco-BRL, Gaithersburg, MD). Store at –20°C (*see* **Note 3**).
3. 5X First Strand Buffer: 250 mM Tris-HCl, pH 8.3, 375 mM KCl, 15 mM MgCl$_2$. Store at –20°C.
4. dNTP mix: 200-μM stock solution containing 2′-deoxyadenosine-5′-triphosphate (dATP), 2′-deoxyguanosine-5′-triphosphate (dGTP), 2′-deoxythymidine-5′-triphosphate (dTTP), and 2′-deoxycytidine-5′-triphosphate (dCTP) prepared in nuclease-free water using lyophilized nucleotides (Pharmacia Biotech, Quebec, Canada). Store at –20°C.
5. DTT: 100 μM dithiothreitol stock solution (Gibco-BRL). Store at –20°C.
6. Oligonucleotide Primer: 50 pmol/μL 5'-T$_{12}$VN-3' primer dissolved in nuclease-free water. Store at –20°C (*see* **Note 4** and **Table 1**).
7. Nuclease-free water.

Table 1
**Sequences of Upstream Degenerate Oligonucleotides
and the Downstream Poly (T) Oligonucleotides That May
be Used in the Differential Display Procedure**

3' Poly(T) primers	5' Degenerate primers
5'-TTT TTT TTT TTT GG-3'	5'-AAG CGA CCG A-3'
5'-TTT TTT TTT TTT GA-3'	5'-TGT TCG CCA G-3'
5'-TTT TTT TTT TTT GT-3'	5'-TGC CAG TGG A-3'
5'-TTT TTT TTT TTT GC-3'	5'-GGC TGC AAA C-3'
	5'-CCT AGC GTT G-3'
5'-TTT TTT TTT TTT AG-3'	
5'-TTT TTT TTT TTT AA-3'	
5'-TTT TTT TTT TTT AT-3'	
5'-TTT TTT TTT TTT AC-3'	
5'-TTT TTT TTT TTT CG-3'	
5'-TTT TTT TTT TTT CA-3'	
5'-TTT TTT TTT TTT CT-3'	
5'-TTT TTT TTT TTT GC-3'	

2.4. PCR and Product Isolation

1. cDNA Template 1 µL of 20 µL cDNA synthesis-reaction mixture. Stored at –20°C.
2. *Taq* DNA Polymerase: 5 U/µL *Taq* DNA Polymerase (Gibco-BRL). Store at –20°C.
3. 10X PCR Buffer: 200 mM Tris-HCl, pH 8.4, 500 mM KCl. Store at –20°C.
4. dNTP mix: 20-µM stock solution containing dGTP, dTTP, and dCTP prepared in nuclease-free water using lyophilized nucleotides (Pharmacia). Store at –20°C.
5. Radioactive nucleotide: 10 µCi/µL (370 KBq/µL)α-[^{35}S]dATP (*Redi*vue, Amersham, Ontario, Canada). Store at 4°C.
6. MgCl$_2$: 50 mM MgCl$_2$ (Gibco-BRL). Store at –20°C.
7. Upstream Oligonucleotide Primer: 50 µM stock solution of the oligonucleotide dissolved in nuclease-free water (*see* **Note 5** and **Table 1**).
8. Poly-T oligonucleotide primer: 50 µM stock solution of the 5'-T$_{12}$VN-3' primer dissolved in nuclease-free water (*see* **Table 1**).
9. Paraffin Liquid: light grade (VWR).
10. 6X Loading Dye: 0.25% w/v bromophenol blue, 0.25% w/v xylene cyanol FF, 15% w/v Ficoll (Type 400; Pharmacia) in water. Store at room temperature.
11. 10X TBE Buffer: 0.045 M Tris-borate, 0.001 M ethylenediaminetetra-acetic acid (EDTA), pH 8.3.

12. 19:1 Acrylamide Solution: Liqui-Gel™ 19:1 (ICN Pharmaceuticals, Costa Mesa, CA), a premade 40% w/v acrylamide solution consisting of UltraPure Acrylamide (38%) and *bis*-acrylamide (2%) in deionized water.
13. *N,N,N',N*-Tetramethylethylenediamine, TEMED (ICN Pharmaceuticals).
14. Ammonium persulfate (ICN Pharmaceuticals): 10% w/v ammonium persulfate in deionized water.
15. Glass-distilled water.

2.5. Cloning

1. Isolated band solution: 100 µL supernatant of boiled product. Stored at –20°C.
2. dNTP mix: 200 µ*M* stock solution containing dGTP, dCTP, dTTP, and dATP in nuclease-free water. Stored at –20°C.
3. 10X PCR buffer: 200 m*M* Tris-HCl, ph 8.4, 500 m*M* KCl. Store at –20°C.
4. *Taq* DNA polymerase: 5 U/µL *Taq* DNA polymerase (Gibco-BRL). Store at –20°C.
5. $MgCl_2$: 50 m*M* $MgCl_2$ (Gibco-BRL). Store at –20°C.
6. Upstream-cloning primer: 5 pmol/µL stock oligonucleotide in nuclease-free water (*see* **Table 2**).
7. Downstream-cloning primer: 5 pmol/µL stock oligonucleotide in nuclease-free water (*see* **Table 2**).
8. Nuclease-free water.
9. Paraffin liquid: light grade (VWR).
10. Cloning vector: pBluescript SK+ (Stratagene, La Jolla, CA) or a suitable alternative. Store at 4°C.
11. Restriction endonuclease: 10 U/µL *Eag*I recombinant-restriction endonuclease (New England Biolabs, Ontario, Canada).
12. Restriction endonuclease buffer: 10X NEB 3 buffer; 1000 m*M* NaCl, 500 m*M* Tris-HCl, 100 m*M* $MgCl_2$, 10 m*M* DTT, pH 7.9.
13. Alkaline Phosphatase: 1 U/µL calf-intestinal alkaline phosphatase (Promega, Madison, WI).
14. Glycerol: sterile solution containing 60% glycerol in water. Store at room temperature.
15. Agarose: electrophoresis-grade agarose, ≥ 600 g/cm^2 (ICN Pharmaceuticals).
16. TAE buffer: 0.04 *M* Tris-acetate, 0.001 *M* EDTA, pH 8.5.
17. Phenol: 10 m*M* Tris, pH 7.6, 1 m*M* EDTA, pH 8.5 ($T_{10}E_1$), saturated phenol *(10)*.
18. Chloroform, omnisolve grade (VWR).
19. Sodium acetate: 3 *M* sodium acetate, pH 5.2.
20. Ethanol: absolute ethanol.
21. Gel purification kit (*see* **Note 6**).
22. Ligase: 0.5–2.0 U/µL T4-DNA ligase (Gibco-BRL).
23. Ligase buffer: 5X Ligase Buffer, 250 m*M* Tris-HCl, pH 7.6, 50 m*M* $MgCl_2$, 5 m*M* ATP, 5 m*M* DTT, 25% w/v polyethylene glycol-8000 (Gibco-BRL).

Table 2
Sequences of the *Eag*I Site-Containing Oligonucleotides Used to Facilitate Cloning into the pBluescript SK⁺ Plasmid Vector.

3' Poly(T) cloning primer	5' Degenerate cloning primers
5'-TTT TTT TTT TTT C<u>GC CGG C</u>GA TG-3'	5'-GTA G<u>CG GCC G</u>CA AGC GAC CGA-3'
	5'-GTA G<u>CG GCC G</u>CT GTT CGC CAG-3'
	5'-GTA G<u>CG GCC G</u>CT GCC AGT GGA-3'
	5'-GTA G<u>CG GCC G</u>CG GCT GCA AAC-3'
	5'-GTA G<u>CG GCC G</u>CC CTA GCG TTG-3'

The *Eag*I site is indicated by the underline.

2.6. Colony Hybridization

1. Competent *Escherichia coli:* JM109 electrocompetent bacteria (*see* **Note 7**).
2. Antibiotic Supplemented Agar Plates: 12% w/v bacteriological grade agar (ICN Pharmaceuticals) in LB + 100 µg/mL (final conc.) ampicillin (LBamp). Store at 4°C.
3. Nitrocellulose filters: 82-mm diameter nitrocellulose filters (Schleicher and Schuell, Keene, NH).
4. Blotting paper: VWR blotting paper 238 (VWR).
5. Template DNA: PCR product isolated in **Subheading 2.5.**
6. Probe-synthesis kit: random prime Prime-It II Kit (Stratagene) (*see* **Note 8**).
7. Size-exclusion column: Nap-5 columns (Pharmacia).
8. Prehybridization Solution: 50% formamide, 5X Denhardt's reagent, 5X SSPE, 0.1% w/v SDS, 100 µg/mL sonicated salmon sperm DNA (added fresh).
   a. 50X Denhardt's: 5 g Ficoll (Type 400; Pharmacia); 5 g Polyvinylpyrrolidine, water to 500 mL (store in 50-mL aliquots at –20°C). (*Note:* We do not add bovine serum albumin [BSA] to our 50X Denhardt's.)
   b. 20X SSPE: 175.3 g NaCl (ICN), 27.6 g NaH_2PO_4 (VWR), 7.4 g EDTA (ICN), pH to 7.4, water to 1L.
9. Hybridization Solution: Prehybridization solution containing denatured ³²P-labeled probe.
10. Wash Solutions: 2X SSC, 0.1% SDS. Stored at room temperature.
    a. 20X SSC: 175.3 g NaCl (ICN), 88.2 g Sodium Citrate (VWR), pH to 7.0, water to 1 L.
11. Autoradiography Film: Kodak X-Omat AR film.

2.7. Northern-Blot Analysis

1. Poly (A)⁺ RNA: isolated from samples corresponding to the conditions used for the initial differential-display reverse transcription.
2. Formaldehyde: 37% formaldehyde (VWR).
3. Agarose: electrophoresis grade agarose, ≥ 600 g/cm^2 (ICN Pharmaceuticals).

4. Ammonium acetate: 1 *M* ammonium acetate treated with DEPC and autoclaved prior to use to remove DEPC.
5. Charge modified-nylon membrane: Zeta Probe GT$^+$ nylon membrane (Bio-Rad, Hercules, CA).
6. Prehybridization solution: QuickHyb (Stratagene).
7. Template DNA: subcloned product isolated in **Subheading 2.6.**
8. Probe-synthesis kit: Random prime Prime-It II Kit (Stratagene) (*see* **Note 8**).
9. Size-exclusion column: Nap-5 columns (Pharmacia).
10. Hybridization solution: QuickHyb (Stratagene) + probe from **step 7**.
11. Wash solutions: 2X SSC, 0.1% SDS; 0.1X SSC, 0.1% SDS.
12. Autoradiography film: Kodak X-Omat AR film.

3. Methods

Since the technique of DD was first developed by Liang and Pardee *(5)*, a variety of modifications have been described. It is not our intention to review these modifications to the original DD technique; instead, we present a protocol which has been successfully used in our laboratory to isolate RA-regulated genes from the zebrafish (White, J. A., et al., manuscript in preparation).

3.1. Exposure of Samples to Retinoid

Duplicate tissue or cell samples should be prepared that can be treated with the retinoid of interest and the vehicle used to deliver the retinoid. The duration of treatment may vary depending on the specific conditions of interest, however; shorter exposure times may be helpful in limiting the number of genes which are not directly regulated by RA, but are involved further downstream in retinoid-signaling pathways. Samples should be kept in the dark during treatment to avoid breakdown of retinoids.

1. Add retinoid to the tissue or cell culture at an appropriate concentration and incubate for a suitable time (*see* **Note 9**).
2. Harvest samples, wash quickly in 1X PBS, and then flash freeze in liquid nitrogen. These samples can then be stored at –70°C or used immediately for mRNA isolation.

3.2. Isolation of mRNA from Tissue or Cells

Preparation of the purest possible mRNA is essential for the success of the DD protocol. Contaminants such as genomic DNA can provide priming sites for the oligonucleotides, thus giving rise to background smears or bands that may obscure real differences in the banding patterns observed with the PCR.

We have found that the use of commercially available kits, although rather expensive, has several advantages. The mRNA produced by these kits is generally very clean and free of contaminants. In addition, consistency in the preparation of RNA is essential not only between control and treatment samples, but from experiment to experiment. Kits such as MicroFast Track (Invitrogen) provide the required consistency and allow the user to isolate several µg of mRNA from very small sample sizes. In addition, kits usually guarantee their products to be RNase free and this eliminates a major concern for the user.

1. Prepare mRNA using the MicroFast Track Kit (Invitrogen), or equivalent, according to the manufacturers directions. Quantification of the yield of poly (A)$^+$ RNA is obtained by OD_{260} measurement.

3.3. Reverse Transcription

1. cDNA synthesis: 1.0 µL of poly (A)$^+$ RNA, 0.1 µg/µL, 4.0 µL of 5X First-strand buffer, 2.0 µL of 200 µ*M* dNTP mix, 2.0 µL of 100 m*M* DTT, 1.0 µL of 50 pmol/ µL 5'-T$_{12}$ VN-3' primer, 8.5 µL of nuclease-free water, 1.5 µL of Superscript II RNase H$^-$ Reverse Transcriptase, 300 U/µL.

The reaction should be incubated at 37°C for 1 h followed by heat denturation of proteins at 95°C for 5 min. The cDNA can then be stored at –20°C for at least 1 mo.

3.4. PCR and Product Isolation

Optimal-PCR amplification of a given target sequence is affected by a large number of parameters including the reagents and specific cycling conditions used. Presented in **steps 1** and **2** is a typical PCR protocol used successfully in our laboratory to amplify target sequences.

1. Mix the following constituents, on ice, in 0.5-µL microcentrifuge tubes: 1 µL of the cDNA-synthesis reaction from **Subheading 3.3.**, 2.0 µL of 10X PCR Buffer, 2.0 µL of 20 µ*M* dNTP mix, 1.2 µL of 50 m*M* mgCl$_2$, 2.0 µL of 50 µ*M* upstream 10-mer oligonucleotide primer, 1.0 µL of 50 µ*M* downstream 5'-T$_{12}$ VN-3' oligonucleotide primer, 9.3 µL of nuclease-free water, 1.0 µL of 10 µCi/µL α-[^{35}S]dATP, 0.5 µL of *Taq* DNA Polymerase, 5 U/µL.
2. Overlay the PCR mixes with 75 µL of paraffin oil and begin PCR cycling with the following conditions with the shortest ramping times possible: 1 cycle at 94°C for 5 min; 40 cycles: 94°C for 30 s, 42°C for 1 min, 72°C for 30 s; 1 cycle at 72°C for 5 min (to extend unfinished PCR products).
3. Recover the PCR sample from below the liquid paraffin using a pipet and remove any remaining paraffin oil by dispensing the sample onto a square of Parafilm (Greenwich, CT). The Parafilm will absorb the paraffin (rolling the drop around will aid in this separation). Samples can then be transferred to a clean microcentrifuge tube.
4. Analyze the samples by electrophoresis through a 0.4-mm-thick, 6% nondenaturing polyacrylamide gel (*see* **Note 10**). To the purified PCR sample add 2 µL of DNA loading dye. 4 µL of this sample can then be loaded on a

standard-length sequencing-type apparatus. We obtain good results using the Model-S2 Sequencing apparatus which supports gels 31 cm × 38.5 cm (Gibco-BRL). The gel should be electrophoresed at approx 60 W. Run the gel until the bromophenol blue dye has migrated 75% of the way toward the bottom of the gel.

5. Dry the gel for 1–2 h at 80°C using a gel drier such as the Model 583 (Bio-Rad) and expose to autoradiography film for 16–24 h by placing the dried gel directly in contact with the film. Several means can be used to mark the gel location on the film. We use small dots of luminescent (glow-in-the-dark) marker, placing three dots at the top, two on each side and three on the bottom of the gel. It is important to include enough marks so that precise alignment of the gel can be accomplished for later excision of the PCR products of interest. Failure to properly align the film with the gel may result in isolation of contaminating PCR products which will adversely affect subsequent steps in the procedure.

6. After aligning the film and the gel, mark the dried gel with a sharp pencil, and, using a clean razor blade, excise the portion of the dried gel and place it in a microcentrifuge tube. To the tube, add 100 µL nuclease-free water and let the sample sit for 10 min at room temperature and then boil for 15 min. It is important to remember that these samples will be used as a template for subsequent PCR, so it is necessary to ensure that sources of potential contaminating nucleic acids are eliminated.

3.5. Cloning

There are a variety of directions that can be taken at this point in the differential display procedure. Although we have chosen to clone the potential differential PCR products at this point, others move immediately to direct sequencing of the products. Our preference to proceed with the cloning of the PCR products is based on several criteria. First, direct sequencing of these products can be troublesome, owing to the fact that these products do not contain a homogenous population of transcripts at this point. Studies have shown that bands isolated using nondenaturing polyacrylamide gels may contain multiple DNA fragments *(7,11)*. In addition, any sequence generated at this time will likely correspond to 3' untranslated regions of the mRNAs owing to the design of the DD technique. Although these sequences are useful to determine if the PCR product corresponds to a previously isolated gene, sequence comparisons are limited to the species from which the products were generated, because interspecies conservation of 3' untranslated sequences is generally low. Cloning these potential differentially expressed products at this stage facilitates the further preparation of probes for Northern-blot analyses, *in situ* hybridization, and cDNA cloning of full-length products.

1. In order to clone the PCR products, reamplify an aliquot of the boiled, isolated band utilizing primers which include *Eag*I restriction-endonuclease sites (*see* **Table 2** and **Note 11**). PCR is performed as follows by mixing the following

constituents, on ice, in 0.5-µL microcentrifuge tubes: 4.0 µL of the boiled supernatant from **Subheading 3.4.**, 4.0 µL of 10X PCR Buffer, 4.0 µL of 200 µM dNTP mix, 1.0 µL of 50 mM MgCl$_2$, 2.0 µL of 50 µM upstream primer, 2.0 µL of 50 µM downstream 5'-GTAG<u>CGGCCG</u>CT$_{12}$-3' primer (*see* **Note 12**), 22.5 µL of nuclease-free water, 0.5 µL of *Taq* DNA polymerase, 5 U/µL.

2. Overlay the PCR mixes with 75 µL of paraffin oil and begin PCR cycling with the following conditions: 1 cycle at 94°C for 5 min; 40 cycles: 94°C for 30 s, 42°C for 1 min, 72°C for 30 s; 1 cycle at 72°C for 5 min.

3. Purify the PCR products as described in **Subheading 3.4., step 3**.

4. Analyze a 5-µL aliquot of the PCR reaction by electrophoresis through a 1.2% agarose gel. 2 µL of 60% glycerol and 2 µL of DNA loading dye are added to the sample prior to electrophoresis.

5. Precipitate PCR products by adding 0.1 vol of 3 M sodium acetate, pH 5.2, and 2 vol of ethanol overnight at –20°C. Collect the precipitated DNA by centrifugation at 12,000g for 15 min. Wash the pellet with 70% EtOH, air-dry, and resuspend in 5 µL nuclease-free water for subsequent restriction endonuclease digestion.

6. Restriction-enzyme digestion is performed in a 20-µL reaction volume consisting of: 5 µL of DNA; 1 µL of *Eag*I restriction endonuclease; 2 µL 10X NEB 3 Buffer; 12 µL of nuclease-free water; reaction should be mixed, and incubated 2 h at 37°C.

7. Inactivate the *Eag*I endonuclease by incubation at 65°C for 20 min. Store at –20°C. This preparation is then used directly in the ligation steps (*see* **Note 13**).

8. Prepare linearized pBluescript SK$^+$ cloning vector by digesting with *Eag*I and treating with alkaline phosphatase as follows: 5 µL of SK$^+$ DNA (approx 7.5 µg), 5 µL of *Eag*I, 10 µL of 10X NEB 3 buffer, 80 µL of nuclease-free water. Incubate this reaction for 2 h at 37°C then add 2 µL of calf intestinal-alkaline phosphatase (1U/µL) and allow the reaction to continue for a further 30 min.

9. Purify the digested cloning vector preparation using the GeneClean II Kit (Bio 101, Vista, CA) (*see* **Note 6**). This large-scale digest should generate enough vector for at least 100 ligations. Quantify the recovered DNA by OD$_{260}$ analysis and adjust the volume of the sample to an appropriate concentration (0.5 µg/µL). Store the purified product at –20°C.

10. Clone the amplified, *Eag*I digested, PCR products into the linearized vector using T4 DNA Ligase as follows: 1 µL of SK$^+$ linearized with *Eag*I, 0.1–0.5 µg; 2 µL of PCR product digested with *Eag*I, 0.1–0.5 µg; 4 µL of 5X Ligase Buffer; 2 µL of T4 DNA Ligase; 11 µL of nuclease-free water.
This reaction should be allowed to continue for 16–24 h at 16°C.

11. Bacterial transformation with the ligation products can be performed in a variety of ways. We use the Gene Pulser II (Bio-Rad) which we find to be efficient, reproducible, and fast. Transformations are performed using 1 µL of the ligation reactions and electro-competent JM109 *E. coli*. Using 0.2-cm electrode-gap electroporation cuvets (Bio-Rad), the Gene Pulser II should be set at 200 Ω resistance, 25 µFD capacitance and 2.5 kV. For more details on this method, *see* the manufacturer's directions.

12. Place transformed bacteria onto 100-mm LBamp plates (100 µg/mL) and allow to grow overnight at 37°C.

3.6. Colony Hybridization

The colony hybridization has been designed to select out those colonies which represent the most abundant species in the original amplified PCR band. In doing so, we hope to help eliminate false positives, which frequently compound the problems involved in mRNA-differential display.

1. Using sterile, autoclaved toothpicks, transfer the colonies to new LBamp plates in a grid pattern. By transferring the colonies to duplicate plates, one can be used to make filters for the hybridization, whereas the other provides a stock of the individual colonies for use after the results of the hybridization are completed.
2. After transferring the colonies to the new plates in a grid pattern, incubate the plates overnight at 37°C.
3. Colony lifts are taken by placing an 82-mm nitrocellulose filter on the plate for 5 min. Carefully remove the filter and place it colony side up on blotting paper.
4. Filters are denatured for 5 min by placing them colony side up on a piece of blotting paper soaked in a solution of 1.5 *M* NaCl, 0.5 *M* NaOH. Transfer the filters to blotting paper soaked in 1.5 *M* NaCl, 1 *M* Tris-HCl, pH 7.4, and incubate 5 min for neutralization. Filters are then washed twice by placing them on blotting paper soaked in 2X SSC and allowed to air-dry, before being baked under vacuum in an 80°C oven for 2 h.
5. Generate a radioactive probe using the Prime-It II kit (Stratagene) or equivalent. Probes should be generated using 2 µL of the *Eag*I-digested PCR product obtained in **Subheading 3.5., step 7**. We routinely purify our radioactive probes from unincorporated nucleotides using Nap-5 columns (Pharmacia) following the manufacturers directions.
6. Prehybridize filters in prehybridization buffer (3–4 mL/filter) for a minimum of 1 h and then hybridize at 42°C overnight by adding 200 µL of the purified-denatured probe to the prehybridization buffer.
7. Filters should be washed twice for 10 min each wash in 2X SSC, 0.1% SDS, followed by washing in 0.1X SSC, 0.1% SDS for 5–10 min or as required to reduce background signal which can be monitored by a hand-held radiation monitor (e.g., mini-monitor, Mini Instruments, Burnham on Crouch, UK). Filters are exposed to Kodak X-Omat AR autoradiography film overnight.
8. Positive colonies picked into 15-mL polypropylene tubes containing 3–4 mL LBamp can be grown up and plasmid DNA prepared using a variety of mini-preparation protocols. This DNA will be used in subsequent experiments as it represents a single, homogeneous, DNA fragment.

3.7. Northern Blot Analysis

Although somewhat time-consuming, the Northern blotting steps followed after the cloning of the PCR products are invaluable in screening out those products that are truly differentially expressed. It is useful to prepare RNA from additional samples treated identically to those used initially in the reverse-transcription cDNA synthesis (*see* **Subheading 3.2.**). The preparation of a second set of mRNA provides another level of control to ensure that differences in the levels of mRNA as shown in the differential display are genuine and not artifactual.

1. Prepare probes for Northern blotting using the Prime-It II kit (Stratagene) using RNase-free solutions.
2. Electrophorese 1 µg of poly(A)$^+$ RNA through a 0.66 M formaldehyde-agarose gel overnight at 12–15 V. Addition of 1 µL of 1 mg/mL ethidium bromide to the RNA samples prior to electrophoresis allows the gel to be photographed under ultraviolet light prior to blotting.
3. Soak the gel in 10X SSC (RNase free) twice for 20 min before blotting. We recommend using a charge-modified nylon membrane for the Northern blot (e.g., Zeta-Probe GT) as this allows repeated stripping and reprobing of the blots, which is necessary to screen large numbers of potential differentially regulated products.
4. Northern blots are both prehybridized and hybridized at 68°C using Quickhyb (Stratagene) according to the manufacturers directions.
5. Following hybridization, wash the blots twice for 10 min each in 2X SSC, 0.1% SDS at room temperature followed by washing at 60°C as required (*see* **Subheading 3.6., step 7**) in 0.1X SSC, 0.1% SDS. Expose blots to Kodak X-Omat AR autoradiography film overnight at –70°C.

4. Notes

1. Our stock solutions contain retinoids dissolved in either dimethy sulfoxide (DMSO) or absolute ethanol. Care must be taken to ensure that proper controls are performed in order to exclude any effects of the vehicle used to deliver the retinoid on the samples being examined.
2. It is imperative to start the mRNA DD protocol with the highest-quality poly(A)$^+$ RNA available. In our hands, we have found that the use of pure mRNA increases the reproducibility of the PCR-banding patterns observed later. In addition the use of a kit such as Microfast Track (Invitrogen) allows mRNA isolation to be accomplished using very small sample sizes.
3. We have tested several brands of reverse transcriptase in preparing cDNA for our DD protocol. Superscript II RNase H$^-$ Reverse Transcriptase available from Gibco-BRL appears to generate consistently good-quality cDNA as judged by the banding patterns observed in the differential display gels.
4. All the oligonucleotides used in these experiments were synthesized on a Beckman, Oligo 1000 oligonucleotide synthesizer. Oligos are cleaved and deprotected as described by the manufacturer. In our hands, it has not been necessary to further column-purify the oligos. Oligonucleotides used in the differential display are precipitated with 0.1 vol of 3 M sodium acetate, pH 5.2, 10 mM

MgCl$_2$, and 2 vol of absolute ethanol at –70°C for 30 min. Oligonucleotides are pelleted, air-dried, and resuspended in 300 µL nuclease-free water. After quantification by OD$_{260}$ readings, the oligos are diluted in the required volume of nuclease water to give the desired concentrations.

The following formula can be used to calculate the concentration of the oligonucleotide in pmol/uL:

$$[\text{oligo}] = (\# A_{260} \text{ U})(3.3 \times 10^4)/\text{mol wt oligo}$$

$$\text{mol wt oligo} = [(\#A)(312.2) + (\#G)(328.2) + (\#C)(288.2) + (\#T)(303.2)] - 61$$

5. The upstream primers used in this differential-display protocol (*see* **Table 1**) are as described originally by Liang et al. *(5)*.

6. There are currently numerous methods available for gel purification of DNA fragments from agarose gels. Any of these methods may be used for this purpose. We currently use the GeneClean II Kit (Bio 101) for isolating DNA fragments greater than 500 bp. Recently, modifications to the original GeneClean II protocol have been described (and verified in our laboratory), which enable this kit to be used to isolate smaller DNA fragments as well *(12)*.

7. The Gene Pulser (Bio-Rad) provides a fast, efficient way to perform the numerous bacterial transformations that must be undertaken with the differential-display technique. We prepare our own electrocompetent bacteria from frozen stocks as suggested by the manufacturer.

8. Any method may be used to generate the radioactive probes required at various stages of the protocol presented. The Prime-It II Kit available from Stratagene reliably generates high specific-activity probes useful for the hybridizations utilized in the colony hybridization and Northern blotting procedures.

9. We have chosen a retinoid concentration, 10^{-6} *M*, which we have previously shown *(13)* to cause morphological changes in relation to pattern formation during regeneration of the zebrafish caudal fin. Specific-retinoid concentrations may vary depending on the samples being studied.

10. Many published protocols for mRNA differential display entail the use of denaturing polyacrylamide gels for the separation of the PCR-amplified bands *(5,6,7)*. The use of nondenaturing gels avoids the problem of multiple banding patterns caused by the tendency of *Taq* and other DNA polymerases to add a single dTTP residue to the 3' end of amplified DNA sequences *(14,15)*.

11. We have utilized the *Eag*I restriction endonuclease in our cloning steps because it also cleaves within the polylinker of the SK$^+$ vector we routinely use for subcloning; however, any suitable restriction-endonuclease site can be incorporated into the primers.

12. Generation of the generic poly-T cloning primer 5'GTAGCGGCCGCT$_{12}$-3' allows products of all the combinations to be amplified using this primer in conjunction with the specific-upstream primer. This strategy avoids the need to synthesize multiple downstream primers.

13. In our hands, it has been adequate simply to heat-denature the *Eag*I endonuclease used to generate sticky ends of the PCR products before including these samples in the ligation. Purification of smaller (less than 300 bp) DNA fragments can be tedious and may result in a loss of product, making cloning steps more difficult and time consuming.

References

1. Uchida, T., Inagaki, N., Furuichi, Y., and Eliason, J. F. (1994) Down-regulation of mitochondrial gene expression by the anti-tumor arotinoid mofarotene (Ro 40-8757). *Int. J. Cancer* **58,** 891–897.
2. LaRosa, G. J. and Gudas, L. J. (1988) An early effect of retinoic acid: cloning of an mRNA (Era-1) exhibiting rapid and protein synthesis-independent induction during teratocarcinoma stem cell differentiation. *Proc. Nat. Acad. Sci. USA* **85,** 329–333.
3. Pan, J. B., Monteggia, L. M., and Giordano, T. (1993) Altered levels and splicing of the amyloid precursor protein in the adult rat hippocampus after treatment with DMSO or retinoic acid. *Mol. Brain Res.* **18,** 259–266.
4. Simon, H. G. and Tabin, C. J. (1993) Analysis of Hox-4.5 and Hox-3.6 expression during newt limb regeneration: differential regulation of paralogous Hox genes suggest different roles for members of different Hox clusters. *Development* **117,** 1397–1407.
5. Liang, P. and Pardee, A. B. (1992) Differential Display of Eukaryotic Messenger RNA by Means of the Polymerase Chain Reaction. *Science* **257,** 967–971.
6. Liang, P., Averboukh, L., and Pardee, A. B. (1993) Distribution and cloning of eukaryotic mRNAs by means of differential display: refinements and optimization. *Nucl. Acids Res.* **21,** 3269–3275.
7. Bauer, D., Muller, H., Reich, J., Reidel, H., Ahrenkiel, V., Warthoe, P., and Strauss, M. (1993) Identification of differentially expressed mRNA species by an improved display technique (DDRT-PCR). *Nucleic Acids Res.* **21,** 4272–4280.
8. Bouillet, P., Oulad-Abdelghani, M., Vicaire, S., Gamier, J., Schuhbaur, B., Dollé, P., and Chambon, P. (1995) Efficient Cloning of cDNAs of Retinoic Acid-Responsive Genes in P19 Embryonal Carcinoma Cells and Characterization of a Novel Mouse Gene, Stra1 (Mouse LERK-2/Eplg2). *Dev. Biol.* **170,** 420–433.
9. Velculescu, V. E., Zhang, L., Vogelstein, B., and Kinzler, K. W. (1995) Serial Analysis of Gene Expression. *Science* **270,** 484–487.
10. Sambrook, J., Fritsch, E. F., and Maniatis, T. (1989) Preparation of Reagents and Buffers Used in Molecular Cloning, in *Molecular Cloning: A Laboratory Manual,* vol. 2, Cold Spring Harbor Laboratory, Cold Spring Harbor, NY, pp. B.4.
11. Callard, D., Lescure, B., and Mazzolini, L. (1994) A method for the elimination of false positives generated by the mRNA differential display technique. *BioTechniques* **16,** 1096–1103.
12. Smith, L. S., Lewis, T. L., and Matsui, S. M. (1995) Increased yield of small DNA fragments purified by silica binding. *Biotechniques* **18,** 970–972.

13. White, J., Boffa, M., Jones, B., and Petkovich, M. (1994) A zebrafish retinoic acid receptor expressed in the regenerating caudal fin. *Development* **120,** 1861–1872.
14. Clark, J. M. (1988) Novel non-templated nucleotide addition reactions catalyzed by procaryotic and eucaryotic DNA polymerases. *Nucleic Acids Res.* **16,** 9677–9686.
15. Hu, G. (1993) DNA polymerase-catalyzed addition of nontemplated extra nucleotides to the 3' end of a DNA fragment. *DNA Cell Biol.* **12,** 763–770.

28

Gene Targeting of Retinoid Receptors

David Lohnes

1. Introduction

Gene targeting in embryonic stem (ES) cells is a powerful technique for the modification of the mouse genome *(1–3)*. Regarding the retinoid receptors, a number of laboratories have reported the phenotype of mice in which a given retinoic acid receptor (RAR) or retinoid X receptor (RXR) has been inactivated by this technique *(4–12)*. In addition, a number of cellular retinoid-binding protein-null mice have also been generated *(13–15)*.

Although analysis of these null mutants has generated a significant amount of information, our knowledge of the function of each component of the retinoid-signaling pathway is far from complete. With advances in gene targeting technology, it is now theoretically possible to introduce any mutation into any given gene in the mouse. Furthermore, tissue-specific knockouts and conditional mutations are now also feasible. With these new tools, problems inherent in straightforward gene knockout approaches, such as embryo or postnatal lethality, can now be circumvented. It is also possible to derive cell lines devoid of specific proteins, and RARα and RARγ have been successfully disrupted in F9 teratocarcinoma cells *(16,17)*. In theory, any diploid cell line should be amenable to gene-inactivation studies, thus permitting the derivation of numerous novel model systems for an extensive in vitro analysis of retinoid receptor function. Indeed, far from being complete, knockout studies of the retinoid signaling pathway are only in their initial phase.

In this chapter, some of the basic principles and techniques for gene targeting of RARs are described. Although some considerations for targeted inactivation of these particular receptors are discussed, the methods are generally applicable to any gene of interest. It should also be noted that a basic knowledge of the techniques and materials required for cell and molecular biology is

From: *Methods in Molecular Biology, Vol. 89: Retinoid Protocols*
Edited by: C. P. F. Redfern © Humana Press Inc., Totowa, NJ

presumed. Furthermore, the generation of chimeras from ES cells is largely technical and leaves little room for improvement at the moment, and is therefore discussed only briefly. Additional information regarding this aspect can be found in **refs.** *18* and *19*.

2. Materials

2.1. ES Cell Culture and Electroporation

The following reagents must be prepared to the highest possible standards.

1. Dulbecco's phosphate buffered saline, magnesium and calcium free (PBS; Gibco-BRL, Grand Island, NY). Prepare in distilled water. Sterilize by passage through a 0.22-µm filter. Store at 4°C or room temperature.
2. Trypsin: dissolve 0.5 g trypsin (1:250, Gibco-BRL) and 0.29 g EDTA (disodium salt, Gibco-BRL) in 1 L PBS. Sterilize by passage through a 0.22-µm filter. Freeze 50-mL aliquots; keep one working tube at 4°C. Prewarm briefly before use.
3. Gelatin: prepare a 10X stock solution by dissolving 10 g of gelatin (Sigma, St. Louis, MO) in 1 L of PBS, autoclave. Dilute 10-fold in PBS before use. Store at 4°C.
4. Fetal bovine serum (FBS or FCS), tested for ES culture. Store at –20 or –80°C.
5. Glutamine, 100X stock solution: 200 m*M* prepared solution from Gibco-BRL. Store at 4°C.
6. 2-mercaptoethanol, 100X stock solution: 10 m*M* 2-mercaptoethanol in PBS, filter sterilize. Keep at 4°C for 1 mo.
7. Freezing solution (1X): 70% DMEM, 20% FCS and 10% DMSO (all v/v). Store at 4°C for 1 wk.
8. Freezing solution (2X): 50% DMEM, 30% FCS and 20% DMSO. Store as per **item 7**.
9. Leukemia inhibitory factor (LIF): Gibco-BRL. Store at 4°C.
10. Mitomycin C, 50X stock: dissolve 2 mg mitomycin C (Sigma) in 4 mL PBS. Store at 4°C for 1 wk (make certain all crystals are dissolved prior to use).
11. Ganciclovir, 1000X stock: Dissolve 5 mg Ganciclovir (Syntex, Palo Alto, California) in 10 mL PBS. Heat to 65°C or add 0.1 *N* NaOH dropwise until dissolved. Filter sterilize. Store at 4°C; stable for at least 1 yr.
12. Geneticin (G418) 100X stock: Dissolve 150 mg of (active ingredient) G418 (Gibco-BRL) in 10 mL PBS. Filter sterilize and store at 4°C.
13. Complete medium for ES cells: DMEM high glucose (Gibco-BRL), prepared with 2.2 g/L tissue-culture grade sodium bicarbonate, supplemented with 15% lot-tested FCS, 2 mM glutamine, 0.1 m*M* 2-mercaptoethanol, 1000 U/mL LIF, and 50 ng/mL gentamycin sulfate (Gibco-BRL). Store at 4°C; use within 1 wk.
14. Complete medium for murine-embryo fibroblast (MEF) or STO cell culture: DMEM (Gibco-BRL) prepared with 2.2 g/L sodium bicarbonate, supplemented with 50 ng/mL of gentamycin and 10% FCS (it is not necessary to use lot-tested serum for these cultures).

15. CD1 or other outbred mice (male and female) for preparation of MEFs.
16. Electroporation apparatus and cuvets (Bio-Rad, Hercules, CA).

2.2. Genomic-DNA Extraction and Southern-Blot Analysis for Recombinants (standard reagents for molecular biology are not noted)

1. Proteinase-K digestion buffer: 250 mM NaCl, 50 mM Tris-HCl (pH 8.0), 5 mM EDTA, 1% SDS. Supplement with 0.5 mg/mL of fresh proteinase K (Fluka) immediately before use.
2. Hybond N+ (Amersham)
3. Prehybridization/Hybridization buffer: 40% deionized formamide, 0.9 M NaCl, 10 mM sodium phosphate, pH 6.5, 2 mM EDTA, 4X Denhardts solution, 5% dextran sulfate (optional), 1% SDS. Supplement with 0.1 mg/mL of sheared salmon-sperm DNA boiled for 5 min immediately before addition.
4. 20X SSC: 3 M NaCl, 0.3 M sodium citrate.
5. TE; 10 mM Tris-HCl, pH 8.0, 1 mM EDTA.
6. 50X TAE: for 1 L dissolve 242 g Tris base, 37.2 g EDTA (disodium salt) in 800 mL distilled water. Add 57 mL of glacial acetic acid and adjust to 1 L. pH should be approx 8.5.

2.3. Blastocyst Injection

1. Animals: vasectomized outbred male mice (e.g., CD1), C57BL/6 (males and females), and outbred CD1 females.
2. 2.5% Avertin: Prepare 100% avertin by mixing 10 mL t-amyl alcohol with 10 g of 2,2,2-tribromoethyl alcohol (both from Aldrich, Milwaukee, WI). Prepare 2.5% solution by dilution in 0.9% NaCl in water (vigorous agitation is necessary). Store at 4°C protected from light.
3. Standard dissection equipment.
4. Dissecting microscope.
5. M2 medium (Sigma): keep at 4°C for 1 mo.
6. Microscope equipped with two micromanipulators and two microinjectors (e.g., Nikon/Narishige or Leitz).
7. Needle puller (e.g., Sutter Instruments or Narishige) and microforge (e.g., De Fonbrune or Kramer Scientific).

3. Methods

3.1. Cell Culture and Electroporation

3.1.1. Overview of ES Cell Culture

ES cells are pluripotent cells derived from the inner cell mass of blastocysts, and are capable of colonizing all embryonic lineages upon reintroduction into mouse embryos. ES cells (if properly maintained) are thus capable of populating the germ line of chimeras, and hence can pass engineered genetic alterations to their progeny. A number of ES cell lines possessing this property have been described *(20–23)*. For laboratories about to embark on gene targeting studies, it is advisable to obtain an established line rather than attempt to derive ES cell cultures *de novo*. To confirm that conditions are adequate for maintenance of germ line potential, stock ES cells should be tested for germ line transmission by blastocyst injection (*see below*) prior to beginning any knockout experiments. Although this takes some time, there is no other reliable determinant for this property. Likewise, regular testing for karyotype and mycoplasma contamination should be performed.

All reagents for ES cell culture must be prepared to the highest possible standards. If there is any question as to the quality of reagents (notably water) to be used it is highly recommended that they be purchased as prepared solutions from a reliable source. The extra cost incurred is more than offset by the loss of time and materials that would result from suboptimal culture conditions. Likewise, it is important to optimize the FBS used for ES cell culture. This can be achieved by testing a number of lots from several suppliers, ideally comparing test lots to a proven batch of serum. Parameters that should be evaluated include plating efficiency, growth, and toxicity (culture in 30% FCS). Once a suitable lot has been identified, reserve a sufficient supply for your culture needs; we usually reserve enough serum to last at least 1 yr.

ES cells can be cultured on gelatinized tissue culture plates in the presence of LIF, a cytokine capable of maintaining ES cells in an undifferentiated state. More commonly, a layer of mitotically inactivated feeder cells is used, in conjunction with LIF supplementation, to suppress differentiation. Although germ line transmission has been achieved using ES lines manipulated for several passages in the absence of feeder cells, the continuous presence of a feeder layer greatly increases the chances of significant population of the germ line.

Feeder cells can be prepared from MEF or permanent fibroblastic lines such as STO or SNL76/7. The latter is an STO line producing LIF and expressing a neomycin-resistance gene (*neo*) for use in selection experiments employing G418 *(19)*. Alternatively, for the selection phase of culture, one can use feeders derived from a mouse knockout line. If using an established line for feeders, prepare sufficient cryogenic stocks, and do not allow cultures to overgrow; cells maintained at too high a density for prolonged periods will result in the outgrowth of deviant cells, which offer poor support for ES cell cultures.

ES cells must be maintained under optimum culture conditions to prevent the outgrowth of differentiated cells and/or loss of their normal euploid karyotype. Failure to follow strict culture regimes will likely result in the generation of poor chimeras which cannot pass ES-derived genetic alterations to their offspring. In general, cultures should be maintained and passaged at high density. Whenever passing cells, be certain that a single-cell suspension is obtained by repeated pipetting prior to replating; failure to do so will result in clumps of cells being deposited, which will cause differentiation into endoderm. Cells should be passaged at approx 80% confluence at a ratio of 1:3–1:6, usually every 2–3 d. Owing to the high cell density and metabolic demands of these cultures, medium should be changed daily. It is also important to keep track of the passage number as ES cells lose their pluripotency with time in culture. For this reason, early passage cultures should be expanded and sufficient frozen stock prepared for future experiments.

3.1.2. Derivation of Feeder Cells

Once prepared, feeders can be kept in culture for up to 2 wk prior to use, but must remain as an intact monolayer. Most researchers prepare feeders on gelatin-coated plates. Although this is not essential for the propagation of the fibroblasts themselves, it promotes tight adherence to the substratum, and is also essential for the adherent growth of ES cells to any regions devoid of fibroblasts.

1. To prepare gelatin-coated 100-mm cell culture plates, add enough 0.1% gelatin solution to cover the surface of the culture vessel and leave at ambient temperature for 2–3 h.
2. Aspirate the solution and air dry overnight in a culture hood (plates can be prepared in advance and stored at room temperature for subsequent use).
3. 13–15 d postcoitum (dpc) mouse fetuses (10–20 total) are aseptically obtained by cesarean section (the strain does not appear to be important; we use CD1 animals).
4. Remove the fetal liver and head, place in a small volume of PBS in a Petri dish, and use small scissors to mince the carcass into fine pieces.
5. Wash the tissue over a sieve with PBS and discard eluant.
6. Incubate the tissue at 37°C in 50 mL of trypsin solution with gentle agitation (using a rocker platform or magnetic stirrer; in the latter case, add a sterile stir bar). Replenish the trypsin solution twice at 30-min intervals.
7. Pass the cell suspension over a sieve and centrifuge the filtrate at $200g$ for 5 min.
8. Resuspend the pellet in 10 mL of medium and determine cell number using a hemocytometer or Coulter counter.
9. Plate gelatinized dishes with approx 5×10^5 cells each and culture at 37°C in a humidified incubator equilibrated with 5% CO_2 until confluent. If desired, cells can be passaged twice before use as feeders, and can be stored frozen at the first passage if not immediately required (*see below* for freezing cells).
10. Prepare mitomycin C in medium to a final concentration of 10 µg/mL.
11. Treat confluent cultures of MEF for 3 h with the mitomycin C solution in an incubator; STO or SNL76/7 feeders are prepared in an identical manner (*see* **Note 1**).
12. Wash cells 3–4 times with PBS and add fresh medium. Change to complete medium for ES cultures just prior to addition of these cells.

3.1.3. Culture of ES Cells

1. Rapidly thaw a stock vial of ES cells by immersion in a 37°C water bath.
2. Once thawed, transfer the cells to a centrifuge tube containing 4 mL of prewarmed complete medium.
3. Centrifuge 5 min at 200g.
4. Resuspend the pellet in medium and plate cells on a feeder layer; remember to change the feeder medium to complete medium for ES culture prior to plating. Culture in a 37°C, humidified incubator equilibrated to 5% CO_2.
5. Change medium daily until culture is approx 80% confluent (usually 2–3 d).
6. To passage cells, aspirate the medium and wash the layer with PBS.
7. Aspirate the PBS and add trypsin solution (usually 1–2 mL/100-mm plate). Place the plate in the incubator and monitor for cell detachment after 5 min.
8. When cells have detached, or can be easily dislodged by tapping the plate, neutralize the trypsin by addition of an equal volume of complete medium and create a single-cell suspension by pipetting using a micropipet.
9. Transfer the cell suspension to a 10-mL culture tube and centrifuge for 5 min at 200g.
10. Resuspend the cells in complete medium and pass the cells at a 1:3–1:6 ratio to new feeder plates containing complete medium. Make sure to note the passage number, previous and present passage date, and passage ratio on the culture plate.

3.1.4. Overview of Targeting Vector Design

Gene targeting is a labor-intensive and costly technique, and poorly designed targeting vectors can greatly exacerbate this process. Although an exhaustive description of strategies for gene targeting is beyond the scope of this chapter, some basic concepts in designing targeting vectors will be discussed.

The most frequent approach for straightforward gene inactivation employs replacement vectors using positive-negative selection strategies (**Fig. 1**; **refs. 24** and **25**). Typically this involves cloning a positive-selectable marker (usually a *neo* expression cassette) within critical coding regions, such that a probable null allele will be created following homologous recombination either by introduction of premature stop codons (within the *neo* sequences) or by deletion of critical regions in the targeting vector (*see* **Note 2**). A negative selectable marker, usually a herpes simplex virus thymidine kinase (HSV-*tk*) expression cassette, is cloned at one end abutting the genomic sequences (*see* **Note 3**). The HSV-*tk* sequences are often retained upon random integration, but are almost always deleted upon homologous recombination. Inclusion of ganciclovir (a nucleoside analog that is a specific substrate for the HSV-*tk* gene product) during selection results in metabolic conversion of ganciclovir to a toxic product, resulting in death of cells expressing HSV-*tk*. Using this strategy, enrichment of several orders of magnitude for homologous recombination have been reported (*see* **Note 4**).

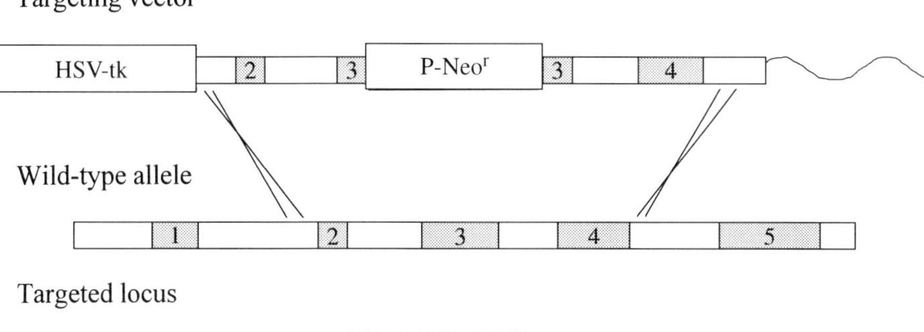

Fig. 1. Schematic representation of homologous recombination with a replacement-type vector employing positive-negative selection. Exons are shaded and intronic regions are represented by open boxes. Vector sequences are denoted by a wavy line. HSV-*tk*, herpes simplex virus thymidine kinase expression vector; P-Neo, promoter-neomyocin expression vector.

A number of alternative targeting strategies have been described that can be used not only for the creation of null alleles but also for the introduction of subtle mutations or conditional knockouts. One of these strategies, the "hit-and-run" approach, relies on the use of insertion vectors to insert subtle mutations within the genome *(21)*. Insertion vectors differ in that the targeting construct is linearized within a region of the genomic sequences and positive selection alone is used for initial targeting (*see* **Note 5**). Upon homologous recombination, this results in a partial duplication of the targeted allele. Once identified, targeted clones are replated and exposed to ganciclovir in the absence of positive selection. This results in resolution of the duplication, with a certain proportion (ideally 50%) of the surviving clones retaining the mutation (**Fig. 2**; *see* **Note 6**). Although this approach requires two sequential rounds of selection, mice null for *CRABP-1* or *Hoxb-4* have been generated using this strategy *(13,26)*. Clearly, this particular tactic is highly promising for the introduction of subtle mutations into the mouse genome, and could be used for fine mutational analysis of the retinoid receptors in vivo.

Perhaps the most efficient targeting method is the promoter-trap approach. This entails the use of a promoterless selectable marker, which is expressed only upon integration in an active transcription unit. Although this necessitates that the locus of interest be expressed in ES cells, a high level of basal transcription is not necessary, as exemplified by disruption of the *Hoxa-5* locus *(20)*.

One of the most promising recent approaches to manipulating the genome employs the Cre-*lox* system. Cre recombinase catalyses the recombination

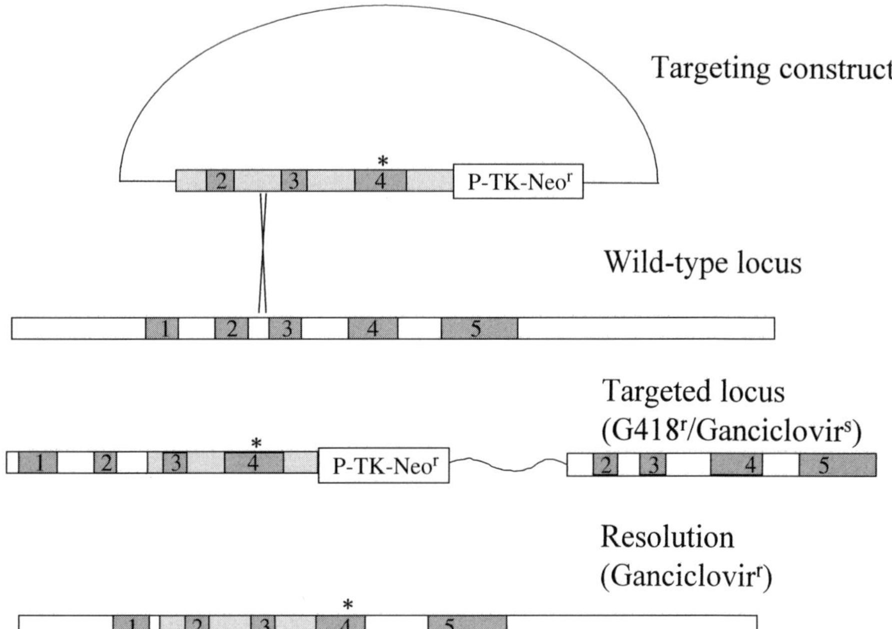

Fig. 2. Homologous recombination using an insertion-type targeting vector for the "hit-and-run" approach. Intron–exon designation is as per **Fig. 1**, with targeting vector sequences being stippled. The vector backbone is illustrated by a semicircle. Note that the positive/negative selection cassettes (P-TK-Neo) are cloned in a region abutting the genomic sequences and the vector is linearized within the genomic DNA (upper sequences). A preplanned mutation is indicated by an asterisk in exon 4. Following homologous recombination, the duplication can be resolved under selective pressure with ganciclovir, resulting in some surviving clones retaining only the preplanned mutation (lower allele).

between its substrate, the 34-bp *lox* site, resulting in several possible events depending on the relative orientation and location of the *lox* sequences. Of particular application to gene targeting is that sequences intervening two repeated *lox* sites are excised by Cre-mediated recombination (**Fig. 3**). This event functions with high efficiency in ES cells and deletions in the range of 100 kb have been reported *(27–29)*. This strategy can also be used to direct tissue-specific gene ablation by crossing *lox*-bearing animals (created by homologous recombination) to transgenic animals expressing Cre in a desired cell population *(30)*. This approach could be used to dissect retinoid-signaling in a tissue-specific manner, thus circumventing the lethality associated with certain RAR and RXR null mutants (*see* **Note 7**).

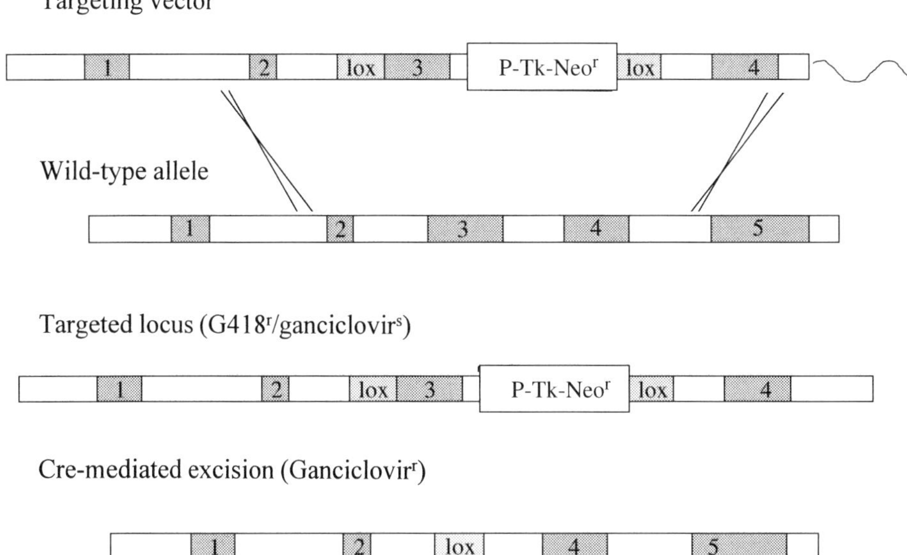

Fig. 3. Gene disruption using the Cre-*lox* approach. In the targeting vector, exon 3 is flanked by directly repeated *lox* sites with a positive–negative selection cassette within these sites. The ganciclovir-sensitive targeted cells can undergo Cre-mediated homologous recombination, resulting in excision of the selection vector and genomic sequences intervening the *lox* sites.

3.1.5. Factors affecting Gene Targeting

Although a number of loci, including several of the RARs, have been readily targeted at high frequency using constructs based on nonisogenic constructs, several reports clearly illustrate that the use of isogenic DNA can have a significant impact on targeting frequency *(32)*. Although the basis for this observation is unknown, a low frequency of polymorphism between mouse strains may underlie this finding. Because the majority of ES cell lines are derived from 129/Sv blastocysts, genomic clones to be used for constructing targeting vectors should be isolated from a library derived from this mouse strain (*see* **Note 8**). It has also been shown that an exponential relationship exists between targeting frequency and the amount of genomic DNA used in the vector *(32,33)*. Although no absolutes have been defined, we attempt to derive constructs with a minimum of 5–6 kb of genomic sequence.

Additional factors can also affect targeting efficiency. For the RARs, we have found that homologous recombination frequencies are greatly increased by using *neo* vectors devoid of polyadenylation signals (Lohnes, D., unpublished observation). This is presumably owing to utilization of the polyadenylation signal from the targeted transcription unit resulting in a stable *neo* message; such an event would only occur fortuitously in random integrants (*see* **Note 9**).

As a final note, a screening strategy for the identification of homologous recombinants is best contemplated while designing the targeting vector; the judicious deletion or addition of convenient restriction sites can greatly simplify analysis. Although PCR detection strategies may serve for initial screening, the only rigorous method to confirm predicted targeting events is by genomic Southern blot analysis. This entails using at least one genomic probe which lies outside the targeting sequences, and at least one internal probe to check for secondary integration events or rearrangement/duplication of inserts (*see* **Note 10**). A rapid and relatively simple method for genomic Southern blot analysis is described below.

3.1.6. Electroporation of ES cells

Although DNA can be introduced into ES cells by several means, electroporation has emerged as the predominant technique as it is efficient, reliable, and does not appear to affect germ line potential.

1. Prepare the targeting construct by a standard CsCl gradient protocol (two gradients).
2. Linearize the construct with an appropriate restriction enzyme. Monitor digestion by agarose gel electrophoresis. Once digestion is complete, phenol-chloroform extract the reaction and ethanol-precipitate the DNA. After washing with 70% ethanol, resuspend the linearized DNA to 1 µg/µL in sterile H_2O.
3. Actively dividing cells appear to integrate exogenous DNA more readily. To ensure that the cultures are in optimal growth conditions, pass the cells at a 1:2 dilution the day before and change the medium approx 4 h before electroporation.
4. Harvest the cells as described above and centrifuge for 5 min at 200*g*. Resuspend the pellet in complete medium and determine cell number using a hemocytometer.
5. Recentrifuge an appropriate aliquot and resuspend to a final concentration of 10^7 cells/800 µL in complete medium.
6. In an electroporation cuvet mix 25 µg of linearized-targeting vector with 10^7 cells (0.8 mL of cell suspension). Sit for 5 min and then electroporate at 960 µF, 250 V in a Bio-Rad gene pulser with a capacitance extender (*see* **Note 11**). Gently mix to disperse the pH gradient caused by the electroporation and sit 10–15 min at room temperature.
7. Aliquot the electroporated cells to 100-mm plates with feeder layers (note that the feeder cell medium must first be changed to complete ES medium). The number of plates required depends on the number of resultant colonies; crowding will

increase the chances of crosscontamination when picking clones. We routinely use 4–5 plates per experiment to maximize the dispersion of colonies (*see* **Note 12**).

8. Twenty-four hours after plating the cells, apply G418 at 150 µg/mL (active ingredient; *see* **Note 13**). Ganciclovir (1–2 µ*M*) is also added at this time if this negative selection is to be used. Refeed the cells daily for the first 6–7 d and every second day thereafter.
9. The bulk of the culture should exhibit extensive cell death 4–6 d after starting selection and macroscopic colonies should be apparent after 10–12 d. Once macroscopic colonies are observable, they can be picked (*see* **Note 14**).

3.1.7. Cloning and Freezing

1. Wash the plate to be picked twice with PBS, then cover with 10 mL of PBS.
2. Prepare a 96-well plate with 25 µL of trypsin in each well (U-shaped wells work best), and a second 96-well plate (flat wells) with feeders and 150 µL of complete medium.
3. Place a dissecting microscope in the culture hood. Place the plate containing the colonies on the microscope stage. Pick individual colonies with a micropipet and place in trypsin. Change tips between each clone and try to pick the entire colony in 10–20 µL.
4. Once 96 colonies have been picked, place the plate in an incubator for 5–10 min.
5. Neutralize the trypsin by addition of 25 µL of complete medium (a multichannel or repeater pipet is most efficient).
6. Using a multichannel pipet, disperse the ES cells by repeated pipetting (approximately five times) and transfer the cell suspension to the feeder layers; change tips between each transfer.
7. Change medium daily (selection is not required) until wells are approx 80% confluent (*see* **Note 15**).
8. When ready for passage, aspirate the medium, wash wells with PBS and add 25 µL trypsin. Incubate for 10 min at 37°C, then add 175 µL of complete medium. Using a multichannel pipet, disperse cells, and passage 100 µL to a fresh 96-well plate containing feeders; this plate will serve to prepare frozen stocks. Passage the second 100-µL aliquot to 24-well plates (with or without feeders) containing 500 µL complete medium; these cultures will serve for DNA analysis (*see* **Note 16**).
9. To freeze cells in 96-well plates, add medium to cultures approaching confluency. This provides the cells with a growth spurt which increases the recovery of viable cells.
10. Four to six hours after feeding, wash the wells twice with PBS, add 50 µL of trypsin solution and incubate 5–10 min at 37°C.
11. When cells have detached, add 50 µL of 2X freezing medium. Seal the plate with parafilm and place in a styrofoam box at –80°C overnight; the styrofoam box permits gradual cooling and increases viability upon thawing (*see* **Note 17**).
12. To recover cultures, place the plate in a 37°C incubator until all ice crystals have disappeared.

13. Remove the culture of interest with a micropipet and place cells in a tube containing 5 mL of prewarmed medium.

14. Centrifuge at 200*g* for 5 min. Aspirate the supernatant and transfer the cell pellet in 500 μL of complete medium to a 24-well plate with feeders. Monitor growth carefully and pass as necessary (*see* **Note 18**).

3.2. Genomic DNA and Extraction

3.2.1. DNA Extraction

1. When the cultures in the 24-well plates have reached confluence, aspirate the medium and wash once with PBS. Add 0.5 mL of digestion buffer and incubate in a humidified atmosphere at 60–65°C overnight. Humidity is best generated by placing the tissue-culture plates in an airtight container (e.g., Tupperware) containing saturated paper towels. The container is then placed in a suitable oven (hybridization ovens work very well for this purpose).

2. Prepare two 1.5-mL microtubes for each cultures. Transfer the cell digests to the appropriate tube and add 0.5 mL of phenol:chloroform (1:1).

3. Vortex and centrifuge 1–2 min at 12,000*g* in a microcentrifuge.

4. Transfer the supernatant to new microtubes and precipitate the DNA by addition of 1.0 mL ethanol (*see* **Note 19**).

5. Centrifuge 5–10 min at 12,000*g*, aspirate the supernatant, and add 1 mL of 70% ethanol.

6. Centrifuge for 1 min at 12,000*g*, aspirate the supernatant, and air-dry the pellet for about 30 min on the bench (do not overdry).

7. Add 50 μL of TE and heat at 65°C for 30 min to aid in redissolving the DNA (alternatively, allow the pellet to resuspend overnight at 4°C). After some practice, variances in yield can be compensated for by adding slightly more or less TE. This technique usually yields sufficient DNA for at least five digestions.

3.2.2. Restriction Digestion of Genomic DNA

1. Aliquot 10 μL of genomic DNA from the cell extracts into microtubes or 96-well plates (U-shaped).

2. Prepare a 2X digestion mix. Each 10-μL aliquot should contain 2 μL of 10X digestion buffer, 25 U of the appropriate restriction enzyme and H_2O to 10 μL (*see* **Note 20**).

3. Using a repeater pipet, aliquot 10 μL of the 2X restriction mix to the DNA samples. Centrifuge briefly (microtubes) or tap (96-well plates) to bring the mix to the bottom of the wells. Incubate overnight at 37°C (use a humidified atmosphere if digesting in 96-well plates).

4. Add 2 μL of 10X loading dye to each sample.

5. Prepare 0.8% agarose gels (without ethidium bromide) in 1X TAE; we use 15×20 cm gels with two 30-well combs per gel.

6. Load samples and electrophorese for 6 h at 100 V, or overnight at 20 V; we use 1-kb ladder markers (Gibco-BRL) for monitoring migration.

7. Stain gels with ethidium bromide (10 µg/mL) in 1X TAE for 20–30 min. Rinse once with distilled water and photograph the gel under UV light using a ruler to document migration of the markers. Alternatively, ladders can be end-labeled by kination or filling by Klenow in the presence of ^{32}P-dCTP; this will permit direct comparison of band size following blotting and autoradiography.

3.2.3. Southern Blotting

Although Southern blots are usually prepared using capillary or vacuum techniques, we have found that the following protocol gives comparable results and is much more convenient to set up.

1. Treat gels twice for 40–60 min with 0.5 *M* NaOH to denature the DNA.
2. Cut membranes to slightly larger than the gel (precision is not needed). We use Hybond N+ (Amersham) as a resilient support which is suitable for base fixation. Whatman 4 paper (2–3 sheets) is cut slightly larger than the membrane, and a bed of paper towels (3–4 cm) is prepared slightly larger than the Whatman paper. The order of placement is (from bottom to top) paper towels, Whatman paper, membrane. Finally, place the treated gel on top of the membrane (well-side up) and cover with Saran wrap.
3. Place a glass plate and a weight (approx 200–400 g; a partially full 500-mL buffer bottle works well) on top of the plate and leave overnight. The DNA will elute from the gel and the alkalinity ensures immediate and complete crosslinking of the DNA to the membrane if using Hybond N+ (or equivalent).
4. Rinse the membrane in 6X SSC 2–3 times for 5 min each to neutralize the blot.
5. Prehybridize for a minimum of 3 h at 42°C.
6. Prepare a genomic probe by random priming; specific activity should be $0.5–1.0 \times 10^9$ cpm/µg input DNA.
7. Mix probe and an appropriate aliquot of salmon-sperm DNA. Boil for 5 min and place on ice.
8. Remove prehybridization solution and prepare hybridization solution by mixing prewarmed prehybridization solution and the boiled probe/salmon-sperm DNA mix. Final probe activity should be approximately 1×10^6 cpm/mL buffer. Add to blot and hybridize overnight at 42°C.
9. Remove hybridization solution and wash blot either in a hybridization bottle or in a baking dish placed in an agitating water bath. The highest-recommended stringency is 68°C in $0.2 \times$ SSC/0.1% SDS, although most probes used in our lab do not require these rigorous conditions.
10. Expose blot overnight with an intensifying screen; low DNA yields will necessitate an increase in exposure time.
11. Using Hybond N+, blots can be readily stripped according to the manufacturers recommendations and reprobed a number of times. This is a particularly useful property for analysis of recombinants where several probes are required to document the authenticity of the recombination event.

3.3. Establishment of Mutant Mice by Blastocyst Injection

Only cultures that have the morphological characteristics of undifferentiated ES cells should be used for injection; try to test at least 3–4 lines if possible. If good chimerism is not obtained, new lines should be established and tested (*see* **Note 21**).

*3.3.1. Production of Blastocysts (see **Note 22**)*

C57BL/6 (B6) females are used to produce recipient blastocysts for injection. This strain does not superovulate well, and embryos obtained from superovulation are not well-synchronized. For these reasons, naturally mated females should be used for procuring blastocysts.

1. Examine females to determine which are in estrus (moist, slightly swollen vulva) and place these animals with B6 males. In order to optimize production, males should be mated with only one or two females at a time, a maximum of twice weekly.
2. Then next day, examine females for the presence of a vaginal plug. Plugs are usually readily visible as a white condensation of semen in the vagina; however, plugs are not always evident and a probe should be used to carefully examine the vagina. Noon of the day of plug is designated 0.5 dpc.
3. Three days after copulation, sacrifice the plugged females by cervical dislocation. Through a large ventral incision, locate the urogenital system and isolate the uterus, oviducts, and ovaries.
4. Excise the bladder from the cervix and the fat pads and mesenteric tissue from the uteri; this is best performed prior to removal of the uterine tract. Cut the uterus approx 1 cm from the ovaries and cut the cervix to remove the uterine tract from the abdominal cavity; place in a Petri dish containing M2 medium.
5. Insert a syringe containing M2 medium and equipped with a 25-gage needle into the cervix and toward one uterine horn and flush out the embryos.
6. Once collected, blastocysts can be pooled and kept in a 5% CO_2-equilibrated and humidified incubator at 37°C until ready to begin injections. Injections should be performed on fully expanded blastocysts (**Fig. 4**; *see* **Note 23**).

3.3.2. Preparation of Pseudopregnant Recipient Females

Following injection, blastocysts are reimplanted into 2.5-dpc pseudopregnant females. These animals must therefore be prepared on the day of plug for blastocyst collection. The strain of mouse is not important, as long as the females have a decent pseudopregnancy rate and are good foster mothers. F1 hybrids (B6 X CBA) or outbred CD1 females are often used for this purpose.

1. Prepare sterile stud males by vasectomization; F1 hybrids (B6 X CBA) or CD1 males are suitable. Vasectomized males should be test bred to ensure that the animal is sterile.
2. Introduce females in estrus and survey for vaginal plugs the next day. Place plugged females aside for reimplantation.

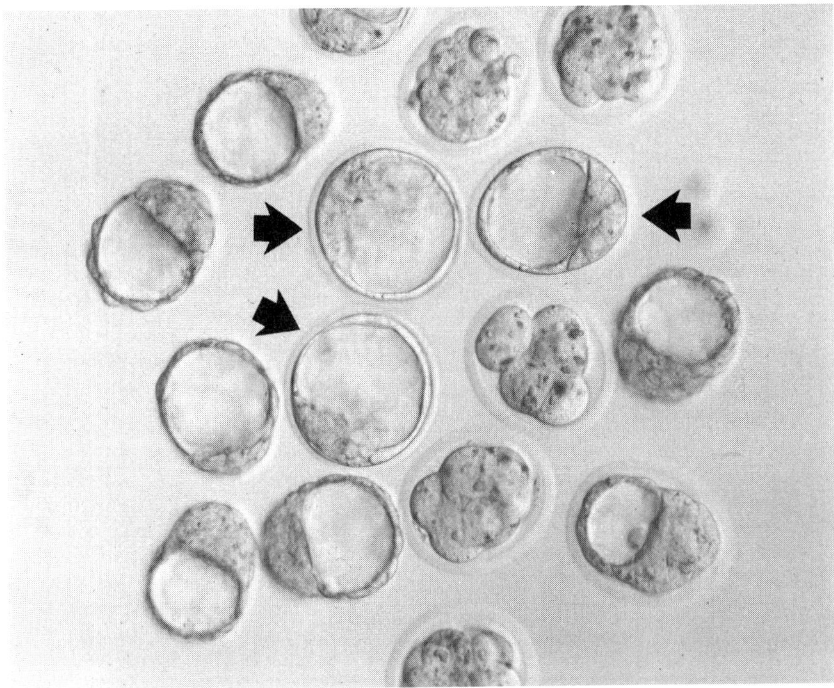

Fig. 4. Mouse embryos from superovulated C57Bl/6 females. Note the asynchronous development of the embryos from superovulated females. Fully expanded blastocysts, suitable for ES cell injection, are denoted by arrows.

3.3.3. Blastocyst Injection and Reimplantation

1. The key to successful injection is a properly prepared injection needle. This can be performed by using a microforge to generate the desired shape. Alternatively, needles can be pulled and placed on a strip of parafilm under a dissecting microscope. The desired tip can often be obtained by holding the needle at a slight angle and breaking the tip with a razor blade. The finished needle should have a beveled, smooth end, and be approx 20 μm in diameter. The holding pipet is prepared using a microforge, and should be 80–100 μm outer diameter and 20 μm inner diameter with a fire-polished end to prevent damaging the blastocyst (**Fig. 5**). We usually prepare several injection pipettes at a time in the event of breakage or clogging during use. A detailed description for the preparation of holding and injection pipets can be found in **ref. 34**.
2. Place a drop of M2 medium on a concave slide and cover with light paraffin oil.
3. Harvest ES cells for injection as described above; only a small culture is needed. Centrifuge the cells and resuspend the pellet in complete medium. Transfer an aliquot to the drop of medium on the slide.

Holding Pipette

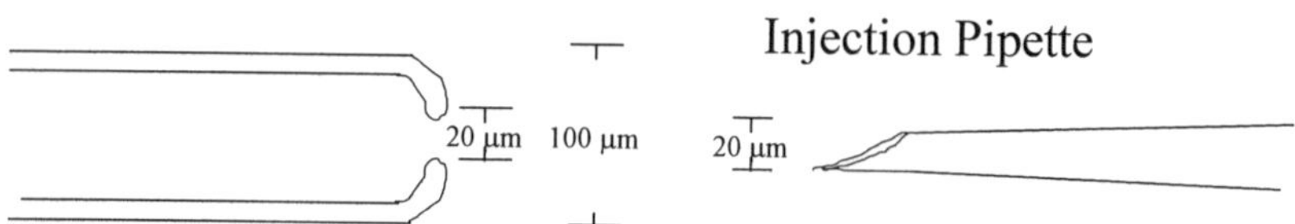

Fig. 5. Diagrammatic representation of holding and injection pipets, illustrating the correct dimensions and shape required for blastocyst injection.

4. Transfer blastocysts to the slide using a finely pulled and fire-polished Pasteur pipet equipped with a mouth suction device.
5. Using the holding pipet, pick up a blastocyst at the equatorial plane on the side of the inner cell mass.
6. Using the injection pipet, pick up some ES cells.
7. Focus on the equatorial plane of the immobilized blastocyst. Bring the injection needle into the focal plane and pierce the blastocyst until the tip of the needle is in the blastocoel (**Fig. 6**). Slowly inject 10–12 ES cells into the blastocoel.
8. Culture the injected blastocysts in a humidified incubator at 37°C. We usually perform injections in the late morning and allow the embryos to recover until late afternoon.
9. Anesthetize pseudopregnant females by intraperotineal injection with Avertin (approx 300 μL per animal). Place the anesthetized female face down and rinse the back with EtOH.
10. Make a small incision through the skin in the back of the animal approx 4 cm from the base of the tail and slightly lateral to the spine. Move the skin to a ventral position until the ovary (notable because of its pink coloration and associated fat tissue) can be seen through the body wall.
11. Make a small incision through the body wall and excise the ovary, oviduct, and a portion of uterus.
12. Use a 25-gage needle to make a small hole in the uterus about 2 cm from the oviduct. Transfer 6–8 injected blastocysts into the uterus using a fire-polished finely drawn Pasteur pipet. If a surplus of pseudopregnant females is available, implant one uterus only.
13. Close the incision in the body wall with a suture and seal the incision in the skin with a wound clamp.
14. Place the female under a heat lamp until recovered, then place in a cage, noting the appropriate information. Pups should be born in 17–18 d.

3.4. Screening Chimeric Offspring

Most ES cell lines have been derived from male 129Sv mice. This results in a phenotypic male sex bias of strong chimeras owing to "conversion" of female host blastocysts by the injected cells. Thus, one sign of a good ES cell line is a

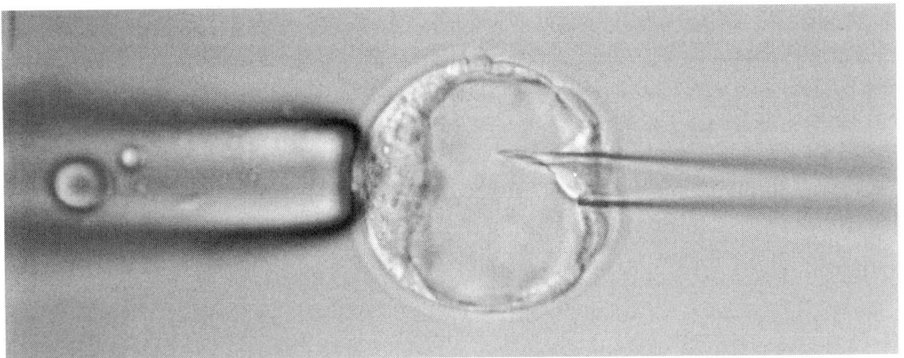

Fig. 6. Correct injection of blastocysts with ES cells. Note that the blastocyst is held by the side of the inner cell mass, allowing easy penetration of the blastocoel with the injection pipet.

predominance of male offspring among strong (50% or better by coat color) chimeras. These males should be test bred with B6 females; we do not test breed female chimeras because, in theory, these animals should not be germ line transmitters.

Because the 129Sv strain is an agouti, germ line transmission can be initially scored by coat-color assessment (e.g., when outbred with B6 females, offspring derived from the ES-cell genome will be agouti since this allele is dominant over black). However, only 50% of all F1 agouti offspring will carry the mutated allele (unless it is X linked), necessitating DNA analysis to determine which bear the mutated allele. This can be performed using the strategy employed to initially identify recombinant ES-cell clones. However, after having established the authenticity of germ line transmission by Southern blot analysis, it is usually more convenient to use a PCR-based assay for routine genotyping.

The F1 progeny from these initial crosses are on a mixed background. However, the phenotype of a number of null mutant mice, including *RARα (11)* can be affected (sometimes profoundly) by strain-dependent modifiers. For this reason, it is worthwhile to establish mutant lines on both outbred and inbred backgrounds. Although inbreeding usually represents a significant investment in time, lines can be established on a 129 background by simply mating *bone fide* germ line transmitting chimeras to 129 females. Although all offspring from this cross must be genotyped, strong chimeras often give rise to a high percentage of agouti offspring, and hence a good proportion of these animals should be heterozygous.

Once lines are established, heterozygous intercrosses can commence, and the fun truly begins.

4. Notes

1. γ-Irradiation can also be used for mitotic inactivation. Expose cells (either harvested or confluent cultures) to 3000–8000 rads of γ-irradiation. Replate or freeze cells for future use.

2. Deletions >10 kb have been reported using replacement vectors, with no apparent reduction in targeting efficiency *(35,36)*. This strategy allows the deletion of promoter or other critical regulatory or coding sequences, and in certain cases obviates concerns for "leaky" knockouts. However, deletions (or merely insertion of *neo* sequences) may have unforeseen impact on transcription of neighboring (or possibly even distant) transcription units. In addition to deleletions, one can also insert additional exogenous sequences (in addition to the selectable marker) by homologous recombination. This strategy can be used to place a reporter gene (e.g., β-galactosidase sequences) under the control of specific promoter elements *(37)*. A similar method can be used for mutagenesis by gene-trap approaches *(38)*.

3. Although it appears to make little difference as to which side the *tk* gene is cloned, a minimum of 1 kb of genomic sequence on the shorter arm of the targeting vector is recommended as shorter stretches of homology may significantly reduce targeting frequencies *(36,39)*. Note also that it is imperative that a unique restriction site be present in nonhomologous sequences abutting the targeting vector in order to linearize the construct prior to electroporation; supercoiled vectors are not suitable for homologous recombination.

4. Negative selection can be performed in a similar manner using a diphtheria-toxin A-chain cassette. This particular method does not require the addition of an exogenous substrate, and is reported to result in approx 10-fold enrichment for recombination events *(40)*. Furthermore, despite the success of negative selection using HSV-*tk* reported by other laboratories, we have found little, if any, enrichment using this method and have discontinued its use for routine targeting.

5. Several groups have found that targeting frequencies are slightly higher using insertion (as compared to replacement) vectors *(39,41)*. However, gene disruption using an insertion approach should be approached cautiously as the resulting duplication event usually allows for theoretical alternative-splicing events capable of generating a wild-type transcript.

6. For unknown reasons resolution of the duplication event in the second step of the hit-and-run strategy sometimes results in a bias toward either one of the two outcomes. Furthermore, a significant percentage of surviving clones may still contain the insertion vector, possibly owing to selection of inactive alleles of *tk*.

7. Despite great success in ES cells, Cre-mediated excision in transgenics may be incomplete *(30)*. This likely depends on the uniformity of Cre expression and the strength of the promoter used. Although in certain cases a mosaic knockout may not impede interpretation of the results, it is usually desirable to effect complete ablation in a targeted population. For this reason, tissue-specific in vivo knockout experiments using the Cre-*lox* system should be approached cautiously.

8. 129Sv genomic libraries are available commercially (e.g., Stratagene, Gibco-BRL). Although similar comparisons have not been reported, targeting efficiencies in other cell types would likely also increase through the use of isogenic constructs.

9. Because of the structure of the RAR loci, a positive selectable marker devoid of a polyadenylation signal must be employed for targeting downstream RAR isoforms in order to avoid premature-termination transcripts initiating from the upstream promoters. We have also found that the use of such *neo* vectors appears to increase the efficiency of obtaining double knockouts via culture in high concentrations of G418 (D., Lohnes, unpublished results; *see* **ref. *42***).

10. Before embarking on analysis of ES clones for recombination events, it is prudent to have devised an efficient strategy for Southern blot analysis of such clones. Although PCR strategies may suffice for initial identification of recombinants, confirmation of the correct targeting event can only be performed by Southern analysis. For initial screening, it is imperative that a genomic probe be used that lies outside the sequences used for the targeting construction; this strategy ensures that recombinants can be reliably detected. Because the *neo* cassette used for positive selection usually bears several rare restriction sites, homologous recombination often results in the fortuitous introduction of useful sites into the locus. It is advantageous to use these sites such that homologous recombination results in a decrease in the size of the restriction product. This is a useful property in that it eliminates false positives generated from partial (i.e., higher molecular weight) digests, an event that sometimes arises owing to "dirty" DNA. Once identified, additional diagnostic digests (and other probes) can be performed to confirm the correct targeting event. It is important to confirm that secondary random integration events have not occurred (using *neo* and *tk* sequences as probes). Note, however, that if secondary integration has occurred in a clone that would be very difficult to regenerate, mice can be derived and the random integration event segregated from the knockout allele by outbreeding, provided that it is not closely linked to the locus of interest.

    Derivation of genomic probes can sometimes be a cumbersome task. One approach to eliminate unsuitable sequences is to prepare a Southern blot of restricted genomic subclones (from regions 5', 3' or internal to the targeting sequences); choose a variety of enzymes giving a decent number of products in the 0.5- to 2.0-kb range. This blot can then be hybridized with a probe obtained from total genomic DNA. Bands that contain repeated elements will hybridize strongly and can be eliminated as potential probes. Bands that hybridize weakly can then be tested individually.

11. A number of conditions have been described for electroporation of ES cells; the protocol described here is one of the simpler approaches. Whatever method is to be employed, optimum conditions should be determined by empirical experimentation. We usually try two different conductivities with a range of voltages in 50-V increments. Because transient expression does not work well by electroporation in our hands (and is certainly not a direct measure of integration

of exogenous DNA), we use a resistance vector and assay resultant colonies as a measure of electroporation efficiency in these experiments.

12. The number of resultant colonies is dependent on the number of cells electroporated, the amount of DNA used and the type of targeting strategy employed. Most of the selection vectors we use are devoid of polyadenylation signals and yield 100–300 colonies per experiment. Identical constructs bearing polyadenylation signals typically yield 5–10 times more colonies. The promoter used to drive positive selectable markers can also affect the electroporation efficiency; we use the *PGK* promoter, which routinely yields 2–5 times more colonies than comparable vectors driven by SV40 enhancer sequences.

13. The concentration of G418 to be used depends on the targeting strategy; polyadenylation signal-less selectable markers require a lower concentration than markers bearing this sequence. Likewise, promoter-trap vectors can also often be selected in low concentrations (150 µg/mL) of G418. The use of higher concentrations may elicit faster death of the cultures, but may also result in some killing of recombinants. The use of low concentrations of G418 in vectors giving high levels of resistance may lead to slow killing and result in overgrowth and differentiation of the cultures. Empirical experiments with the selectable markers to be used for targeting vectors should be performed to determine optimum concentrations of drugs to be used for selection.

14. Often, macroscopic colonies are not visible after 10 d of selection. Rather than discarding these "negative" plates, discontinue selection and continue the cultures. Although the reason for this observation is unclear, it is possible that certain selection cassettes, in the context of certain genomic sequences, do not confer good growth properties on cells in the presence of G418, although cell survival does occur.

15. Although all 96 clones will not exhibit comparable growth, most will be in a reasonable window for passage. A certain percentage of these clones will be lost owing to passage at inappropriate densities; however, it is more desirable and efficient to maximize the number of lines to be analyzed by passaging the entire 96-well plate at one time. In order to normalize, somewhat, the growth characteristics of these clones, we attempt to pick large, medium, and small colonies on individual 96-well plates, if possible.

16. Some laboratories use 96-well cultures for DNA for Southern blot analysis. However, we have found that the quantity and/or quality of DNA obtained from these smaller cultures is often insufficient for unambiguous analysis. Culturing in 24-well plates, while slightly more expensive and time consuming, greatly increases the yield of DNA for analysis, and allows more reliable results in our hands. Note also that the quality of the 24-well cultures need not be high, as these cells will serve simply for DNA analysis and not for the generation of chimeric mice.

17. Cultures stored by this method rapidly lose viability after 1–2 mo at –80°C. For longer storage, plates should be placed at –135°C. In this case, the cultures should be overlaid with sterile mineral oil to prevent degassing during long-term storage.

18. Although the generation of chimeric mice can be performed with few cells, it is prudent to confirm that the recovered culture is indeed targeted and to prepare frozen stocks of lines of interest at as early a passage as possible. We routinely freeze a minimum of five aliquots of each important cell line from a 6-cm plate. This is performed by sequential passaging of the initial 96-well culture to a 24-well plate and subsequently to a 6-cm plate (always with feeders). This additional two passages should not greatly reduce the germ line colonizing capacity of the ES cells. Once the final plate is nearing confluence, the culture is fed with fresh medium, and 4–6 h later, the cells are harvested as described. Following centrifugation, the supernatant is aspirated and the cell pellet is resuspended in 5 mL 1X freezing solution. One milliliter aliquots are transferred to cryogenic vials, clearly labeled, and the cells are frozen at –80°C in a styrofoam box. After freezing overnight, transfer the frozen stocks to liquid nitrogen.

19. This protocol does not call for an intermediate extraction with chloroform prior to precipitation. The elimination of this step does not appear to greatly compromise the quality of the DNA. Furthermore, when extracting cells, do not be concerned with the viscosity of the samples as this also does not appear to affect restriction of the samples.

20. We use concentrated enzymes (50 U/µL; Gibco-BRL) for these analyses and use the buffer supplied by the manufacturer. In cases where the enzyme employs a buffer-containing salt, addition of spermidine-HCL (2 m*M* final) can greatly increase the efficacy of digestion.

21. Occasionally, germ line competent lines can be derived from poor ES cell lines by subcloning. This can be performed by plating at low density (200–500 cells/ 100-mm plate) and isolating lines as described above.

22. A number of investigators are now using an ES cell: embryo aggregation protocol to generate chimeric animals *(23,43)*. This approach does not require a micromanipulation apparatus.

23. Morulae, as well as blastocysts, can be injected to derive chimeras. However, in the former case, contribution by the ES cells is often extremely high. If the injected clone contains some variants, this can result in embryo-lethality.

References

1. Ramirez-Solis, R. and Bradley, A. (1994). Advances in the use of embryonic stem cell technology. *Cur. opin. biotechnol.* **5,** 528–533.
2. Joyner, A. L. (1991) Gene targeting and gene trap screens using embryonic stem cells: new approaches to mammalian development. *Bioessays* **13,** 649–656.
3. Bronson, S. K. and Smithies, O. (1994) Altering mice by homologous recombination using embryonic stem cells. *J. Biol. Chem.* **269,** 27,155–27,158.
4. Lufkin, T., Lohnes, D., Mark, M., Dierich, A., Gorry, P., Gaub, M. P., LeMeur, M., and Chambon, P. (1993) High postnatal lethality and testis degeneration in retinoic acid receptor alpha mutant mice. *Proc. Natl. Acad. Sci. USA.* **90,** 7225–7229.

5. Lohnes, D., Kastner, P., Dierich, A., Mark, M., LeMeur, M., and Chambon, P. (1993) Function of retinoic acid receptor gamma in the mouse. *Cell* **73,** 643–658.

6. Li, E., Sucov, H. M., Lee, K. F., Evans, R. M., and Jaenisch, R. (1993) Normal development and growth of mice carrying a targeted disruption of the alpha 1 retinoic acid receptor gene. *Proc. Natl. Acad. Sci. USA* **90,** 1590-1594.

7. Mendelsohn, C., Mark, M., Doll, P., Dierich, A., Gaub, M. P., Krust, A., Lampron, C., and Chambon, P. (1994). Retinoic acid receptor beta 2 (RAR beta 2) null mutant mice appear normal. *Dev. Biol.* **166,** 246–258

8. Kastner, P., Grondona, J. M., Mark, M., Gansmuller, A., LeMeur, M., Decimo, D., Vonesch, J. L., Dollé, P., and Chambon, P. (1994) Genetic analysis of RXR alpha developmental function: convergence of RXR and RAR signaling pathways in heart and eye morphogenesis. *Cell* **78,** 987–1003.

9. Kastner, P., Mark, M., Leid, M., Gansmuller, A., Chin, W., Grondona, J. M., Decimo, D., Krezel, W., Dierich, A., and Chambon, P. (1996) Abnormal spermatogenesis in RXR beta mutant mice. *Genes Dev.* **10,** 80–92.

10. Luo, J., Pasceri, P., Conlon, R. A., Rossant, J., and Giguère, V. (1995) Mice lacking all isoforms of retinoic acid receptor beta develop normally and are susceptible to the teratogenic effects of retinoic acid. *Mech. Dev.* **53,** 61–71.

11. Lohnes, D., Mark, M., Mendelsohn, C., Dollé, P., Dierich, A., Gorry, P., Gansmuller, A., and Chambon, P. (1994) Function of the retinoic acid receptors (RARs) during development. (I) Craniofacial and skeletal abnormalities in RAR double mutants. *Development* **120,** 2723–2748.

12. Mendelsohn, C., Lohnes, D., Décimo, D., Lufkin, T., LeMeur, M., Chambon, P., and Mark, M. (1994) Function of the retinoic acid receptors (RARs) during development. (II) Multiple abnormalities at various stages of organogenesis in RAR double mutants. *Development* **120,** 2749–2771.

13. Gorry, P., Lufkin, T., Dierich, A., Rochette-Egly, C., Décimo, D., Dollé, P., Mark, M., Durand, B., and Chambon, P. (1994) The cellular retinoic acid binding protein I is dispensable. *Proc. Natl. Acad. Sci. USA* **91,** 9032–9036

14. Fawcett, D., Pasceri, P., Fraser, R., Colbert, M., Rossant, J., and Giguère, V. (1995) Postaxial polydactyly in forelimbs of CRABP-II mutant mice. *Development* **121,** 671–679.

15. Lampron, C., Rochette-Egly, C., Gorry, P., Dollé, P., Mark, M., Lufkin, T., LeMeur, M., and Chambon, P. (1995) Mice deficient in cellular retinoic acid binding protein II (CRABPII) or in both CRABPI and CRABPII are essentially normal. *Development* **121,** 539–548.

16. Boylan, J. F., Lohnes, D., Taneja, R., Chambon, P., and Gudas, L. J. (1993) Loss of retinoic acid receptor gamma function in F9 cells by gene disruption results in aberrant Hoxa-1 expression and differentiation upon retinoic acid treatment. *Proc. Natl. Acad. Sci. USA* **90,** 9601–9605.

17. Boylan, J. F., Lufkin, T., Achkar, C. C., Taneja, R., Chambon, P., and Gudas, L. J. (1995) Targeted disruption of retinoic acid receptor alpha (RAR alpha) and RAR gamma results in receptor-specific alterations in retinoic acid-mediated differentiation and retinoic acid metabolism. *Mol. Cell. Biol.* **15,** 843–851.

18. Hogan, B., Beddington, R., Costantini, F., and Lacy, E., eds. (1994) *Manipulating the Mouse Embyro* (2nd ed.). Cold Spring Harbor Laboratory, Cold Spring Harbor, NY.

19. Ramirez-Solis, R., Davis, A. C., and Bradley, A. (1993) Gene targeting in embryonic stem cells. *Methods Enzymol.* **225,** 855–878.

20. Jeannotte, L., Ruiz, J. C., and Robertson, E. J. (1991) Low level of Hox1.3 gene expression does not preclude the use of promoterless vectors to generate a targeted gene disruption. *Mol. Cell. Biol.* **11,** 5578–5585.

21. Hasty, P., Ramirez-Solis, R., Krumlauf, R., and Bradley, A. (1991) Introduction of a subtle mutation into the Hox-2.6 locus in embryonic stem cells. *Nature* **350,** 243–246.

22. Gossler, A., Doetschman, T., Korn, R., Serfling, E., and Kemler, R. (1986) Transgenesis by means of blastocyst-derived embryonic stem cell lines. *Proc. Natl. Acad. USA* **86,** 9065–9069.

23. Nagy, A., Rossant, J., Nagy, R., Abramow-Newerly, W., and Roder, J. C. (1993) Derivation of completely cell culture-derived mice from early-passage embryonic stem cells. *Proc. Natl. Acad. Sci. USA* **90,** 8424–8428.

24. Thomas, K. R., Folger, K. R., and Capecchi, M. R. (1986) High frequency targeting of genes to specific sites in the mammalian genome. *Cell* **44,** 419–428.

25. Thomas, K. R., Deng, C., and Capecchi, M. R. (1992) High-fidelity gene targeting in embryonic stem cells by using sequence replacement vectors. *Mol. Cell. Biol.* **12,** 2919-2923.

26. Ramirez-Solis, R., Zheng, H., Whiting, J., Krumlauf, R., and Bradley, A. (1993) Hoxb-4 (Hox-2.6) mutant mice show homeotic transformation of a cervical vertebra and defects in the closure of the sternal rudiments. *Cell* **73,** 279–294.

27. Ramirez-Solis, R., Liu, P. T., and Bradley, A. (1995) Chromosome engineering in mice. *Nature* **378,** 720–724.

28. Gu, H., Zou, Y. R., and Rajewsky, K. (1993) Independent control of immunoglobulin switch recombination at individual switch regions evidenced through Cre-loxP-mediated gene targeting. *Cell* **73,** 1155–1164.

29. Sauer, B. (1996) Manipulation of transgenes by site-specific recombination: use of Cre recombinase. *Methods Enzymol.* **225,** 890–900.

30. Gu, H., Marth, J. D., Orban, P. C., Mossmann, H., and Rajewsky, K. (1994) Deletion of a DNA polymerase beta gene segment in T cells using cell type-specific gene targeting. *Science* **265,** 103–106.

31. te Riele, H., Maandag, E. R., and Berns, A. (1992) Highly efficient gene targeting in embryonic stem cells through homologous recombination with isogenic DNA constructs. *Proc. Natl. Acad. of Sci. USA* **89,** 5128–5132.

32. Hasty, P., Rivera-Perez, J., and Bradley, A. (1991) The length of homology required for gene targeting in embryonic stem cells. *Mol. Cell. Biol.* **11,** 5586–5591.

33. Deng, C. and Capecchi, M.R. (1992) Reexamination of gene targeting frequency as a function of the extent of homology between the targeting vector and the target locus. *Mol. Cell. Biol.* **12,** 3365–3371.

34. Miranda, M. and DePamphilis, M. L. (1993) Preparation of injection pipettes. *Methods Enzymol.* **225,** 407–412.

35. Mombaerts, P., Clarke, A. R., Hooper, M. L., and Tonegawa, S. (1991) Creation of a large genomic deletion at the T-cell antigen receptor beta-subunit locus in mouse embryonic stem cells by gene targeting. *Proc. Natl. Acad. USA* **88,** 3084–3087.
36. Zhang, H., Hasty, P., and Bradley, A. (1994) Targeting frequency for deletion vectors in embryonic stem cells. *Mol. Cell. Biol.* **14,** 2404–2410.
37. Mansour, S. L., Thomas, K. R., Deng, C. X., and Capecchi, M. R. (1990) Introduction of a lacZ reporter gene into the mouse int-2 locus by homologous recombination. *Proc. Natl. Acad. Sci. USA* **87,** 7688–7692.
38. Friedrich, G., and Soriano, P. (1991) Promoter traps in embryonic stem cells: a genetic screen to identify and mutate developmental genes in mice. *Genes Dev.* **5,** 1513–1523.
39. Deng, C., Thomas, K. R., and Capecchi, M. R. (1993) Location of crossovers during gene targeting with insertion and replacement vectors. *Mol. Cell. Biol.* **13,** 2134–2140.
40. Yagi, T., Ikawa, Y., Yoshida, K., Shigetani, Y., Takeda, N., Mabuchi, I., Yamamoto, T., and Aizawa, S. (1990) Homologous recombination at c-fyn locus of mouse embryonic stem cells with use of diphtheria toxin A-fragment gene in negative selection. *Proc. Natl. Acad. Sci. USA* **87,** 9918–9922.
41. Hasty, P., Crist, M., Grompe, M., and Bradley, A. (1994) Efficiency of insertion versus replacement vector targeting varies at different chromosomal loci. *Mol. Cell. Biol.* **14,** 8385–8390.
42. Mortensen, R. M., Zubiaur, M., Neer, E. J., and Seidman, J. G. (1991) Embryonic stem cells lacking a functional inhibitory G-protein subunit (alpha i2) produced by gene targeting of both alleles. *Proc. Natl. Acad. Sci. USA* **88,** 7036–7040.
43. Wood, S. A., Allen, N. D., Rossant, J., Auerbach, A., and Nagy, A. (1993) Non-injection methods for the production of embryonic stem cell-embryo chimaeras. *Nature* **365,** 87–89.

Index